NASA SP-4010

I0495753

ASTRONAUTICS AND AERONAUTICS, 1968

Chronology on Science, Technology, and Policy

Text by
Science and Technology Division
Library of Congress

Sponsored by
NASA Historical Division
Office of Policy

Scientific and Technical Information Division   1969
NATIONAL AERONAUTICS AND SPACE ADMINISTRATION
Washington, D.C.

Foreword

How the voyage of *Apollo 11* came to pass after seven years of concerted national effort will occupy the serious attention of competent historians for years to come. The moon walk of Astronauts Armstrong and Aldrin was an important achievement in the progress of man. This dramatic event was shared on live television by half a billion people around our small planet, involving the entire globe in a unique common historical event of all mankind. New questions concerning the place of man in the universe have been raised, which should stimulate inquiry into the philosophical aspects of space travel as well as the scientific, engineering, and organizational aspects. For such studies it is hoped that this chronology of the space events of the notable year of 1968 will be of high reference value.

The year preceding man's first landing and initial exploration of an extraterrestrial body was also the 11th year since Sputnik became the first manmade object lofted into earth orbit and the 10th anniversary of the creation of the U.S. National Aeronautics and Space Administration. For NASA it witnessed the conclusion of the great leadership of its second Administrator, James E. Webb, who organized and carried forward so vigorously America's effort to undertake a manned lunar landing in the decade of the 1960's. Two historic space flights in the final months of 1968 put the work for which he had been responsible to the final test: *Apollo 7*, the first manned Apollo flight, and *Apollo 8*, man's first trip from earth into orbit around another celestial body. After an 18-month hiatus in manned space flight, both flights were carried out flawlessly. *Apollo 8*'s 10-orbit voyage around the moon at Christmastime by Astronauts Borman, Lovell, and Anders and their safe return to "the good earth" was hailed as being in the historical and scientific tradition of Galileo and Kepler, Columbus and Magellan, Herschel and Newton, Verne and Tsiolkovsky, the Wright Brothers and Lindbergh, Goddard, and Gagarin. With this flight, extraterrestrial space was opened to man's exploration.

Other notable techno-scientific space triumphs were made in 1968, including the launching of the *Orbiting Astronomical Observatory II*, the successful conclusion of the X-15 rocket research airplane program, and the Soviet missions of the rendezvousing *Soyuz III* and unmanned circumlunar flights of *Zond V* and *Zond VI*. This chronicle for 1968 helps maintain historical perspective by including such events in their juxtaposition to other threads of history such as the budgetary and policy problems besetting NASA and its White House and Congressional overseers. The 1968 space program was carried out in a year of national unrest and turbulence in the United States. Major news events included tragic assassinations and urban riots in the United States, widespread campus disorders, the capture of the U.S.S. *Pueblo* by North Korea, continued combat in South Vietnam, war in the Middle East, the starving of thousands in Biafra, and a turbulent presidential campaign. To many Americans U.S. successes in space provided much-needed reassurance that this mighty nation could still accomplish great things.

If the engine of modern social change is technology, perhaps this first collation of space accomplishments in 1968 will promote greater in-depth understanding of our complex times. This, in turn, may help generate new solutions for current and future problems in those complex human areas which so badly need to progress as rapidly as our science and technology.

Thomas O. Paine
Administrator
National Aeronautics and Space Administration

August 20, 1969

Contents

	PAGE
Foreword	III
Administrator Thomas O. Paine	
Preface	VII
January	1
February	29
March	53
April	75
May	101
June	126
July	149
August	175
September	205
October	235
November	269
December	298
Appendix A: Satellites, Space Probes, and Manned Space Flights, 1968	339
Appendix B: Major NASA Launches, 1968	367
Appendix C: Chronology of Manned Space Flight, 1967–1968	371
Appendix D: Abbreviations of References	375
Index and List of Abbreviations and Acronyms	379

Preface

This chronicle covering aeronautical and space events in 1968 is, like its predecessor volumes, a tentative step in the process of documentation and writing of history. Admittedly a first skimming, it attempts to sort out the rising crescendo of events, decisions, and personalities into the sequence of their happening. It is largely a collection process on available sources rather than a product of research. Because of the virtual real-time basis on which it is compiled and written, it must rely heavily on the most immediate and available sources—newspapers, magazines, press releases, etc.—rather than on the in-depth documentation that full history requires. It is, in another sense, a holding action until the full history can be written, while at the same time providing early reference utility.

Within these limitations, we make a considerable effort to ensure accuracy and comprehensiveness. Our NASA Archives collects the current documentation. Under an exchange of funds agreement, the Science and Technology Division of the Library of Congress drafts the monthly segments in comment edition form. These are edited and augmented by the NASA Historical Division, published, and circulated for comment and use. At the end of the year the entire manuscript is reworked and augmented by the comments that have come in and by documentation that has become available since the comment edition was prepared. The Library also prepares the extensive index.

This annual volume is the result of a number of hands. The entire NASA Historical Division participated in source collection, review, and publication. At the Library Mr. Charles Thibault (through April 1968), Mrs. Patricia Davis (from May on), Mrs. Carmen Brock-Smith, and Miss Shirley Medley prepared the monthly texts which were circulated throughout NASA for comments as to completeness and accuracy on NASA items. Mrs. Davis, Mrs. Brock-Smith, and Mrs. Gay Arnelle then revised the monthly drafts for annual publication. Mr. Arthur Renstrom prepared the index. In the NASA Historical Division the general editor was Dr. Frank W. Anderson, Jr., Deputy NASA Historian. Technical editor was Mrs. Carrie Karegeannes. Appendix A, "Satellites, Space Probes, and Manned Space Flights, 1968," was prepared by Leonard C. Bruno of the Library. Appendix B, "Major NASA Launches, 1968," was prepared by Dr. Anderson. Appendix C, "Chronology of Manned Space Flight, 1967–1968," was prepared by William D. Putnam, Assistant NASA Historian for Manned Space Flight. Appendix D,

"Abbreviations of References," was prepared by Mrs. Brock-Smith. Creston Whiting of NASA's Information Services Branch, Scientific and Technical Information Division, kept the process abreast of Russian releases. At the NASA centers the historians and historical monitors submitted local material for the chronology. Validation was the work of many busy people throughout NASA and in other relevant branches of the Federal structure.

A chronology is but the first step toward history as an intellectual discipline and even it is never really completed. Comments, additions, and criticisms are always welcomed.

Eugene M. Emme
NASA Historian

January 1968

January 1: NASA announced it would conduct 29 major launches from ETR and WTR during 1968. Surveyor-G launch, last in Surveyor series, was scheduled for Jan. 7. Launch of OGO–E would include last scheduled flight of Atlas-Agena launch vehicle from ETR; future unmanned interplanetary probes and high-altitude earth-orbital launches would use hydrogen-fueled Centaur. Two unmanned tests of Apollo lunar module and first manned test of Apollo spacecraft would use new Saturn IB vehicle. Three Apollo tests employing Saturn V would be launched from new spaceport complex at KSC. In addition to OGO and Surveyor, 11 scientific satellites would be launched from ETR—four Intelsats, ATS–D, OAO, OSO, Biosatellite-D, HEOS (launched for ESRO), Pioneer-D and Skynet comsat (launched for U.K.). Another 10 scientific satellites would be launched from WTR—four TOS spacecraft (launched for ESSA), OGO–F, GEOS–B, Nimbus-B, IMP–G, Radio Astronomy Explorer-A, and International Satellite for Ionospheric Studies (joint American and Canadian program).

About 250 sounding rockets and scientific probes of upper atmosphere would be launched from NASA Wallops Station. (NASA Release 67–315; KSC Release 451–67)

- President Johnson announced stringent Government controls over private investments abroad and measures to reduce U.S. balance-of-payments deficit by $3 billion in 1968. At Johnson City, Tex., news conference, he said actions would reduce intolerable deficit that could "endanger the strength of the entire free world economy, and thereby threaten our unprecedented prosperity at home." (Harwood, *W Post*, 1/2/68, 1)

- Membership of NASA Historical Advisory Committee was designated for 1968–1969: Dr. Melvin Kranzberg, Case Western Reserve Univ., as Chairman; Dr. Eugene M. Emme, NASA Historian, as Executive Secretary; Dr. Raymond L. Bisplinghoff, MIT; Dr. James L. Cate, Univ. of Chicago; Dr. Earl H. DeLong, American Univ.; Dr. A. Hunter Dupree, Univ. of California at Berkeley; Dr. Joe B. Frantz, Univ. of Texas; Dr. Louis Morton, Dartmouth College; and Robert L. Perry, RAND Corp. (NASA NMI 1156.3A; NASA Special Release, 1/5/68)

January 2: FAA proposed rule requiring all jet aircraft to operate under instrument flight rules (IFR) when in controlled airspace within the contiguous 48 U.S. states. Aircraft would come under control of FAA air traffic control (ATC) facilities; pilots would file flight plan and observe ATC instructions which provided for separation between IFR aircraft. Under visual flight rules (VFR), pilots maintained separation on a "see and avoid" basis. FAA noted that rapid growth of total flight operations was "creating demands for substantial expansion and improvement in the nation's air traffic control system . . . [so that] proposed action

should be taken to maintain the desired level of safety pending long range solutions." (FAA Release 68–1)
- F–111A aircraft crashed near Edwards AFB, Calif., on test flight. Crewmen, Col. Henry W. Brown (USAF) and L/Col. Joe B. Jordon (USAF), parachuted to safety in escape compartment. Crash was fourth for F–111; one F–111B and two F–111As had crashed previously. (AP, *NYT*, 1/4/68, 74)
- AFSC awarded United Aircraft Corp. $3,300,000 initial increment to $33,476,000 contract for advance development program to demonstrate high-performance reusable oxygen-hydrogen rocket engine. (DOD Release 4–68)
- Commercial irradiation of foods by gamma rays for protection against spoilage had begun at Nuclear Materials Equipment Corp. (NUMEC), Apollo, Pa. Bacon would be processed initially, although FDA had also approved irradiation for potatoes, wheat, and wheat flour. Bacon would be treated by exposure to 4.5 million rads of gamma rays to destroy all bacteria; product could be stored indefinitely without refrigeration. Neither texture nor wholesomeness would be affected, according to AEC and Army tests conducted over past 15 yr. (*NYT*, 1/2/67, 58)

January 3: Development of blood-pressure sensors small enough to pass through dog's artery into heart—invention of ARC's Grant W. Coon—was announced by NASA. Sensors showed promise as diagnostic and monitoring instrument for human beings, particularly infants. (NASA Special Release; NASA Tech Brief 67–10669; AP, *NYT*, 1/4/68; UPI, P *EB*, 1/4/68)
- NASA announced presentation of awards of $25 each to 68 NASA and contractor employees for technical innovations that might be useful to non-aerospace industry. Awards were expected to stimulate reporting of "useful new materials, processes, products, tools, computer programs, and other space technology developed in the course of NASA work." (NASA Release 68–3)
- *Cosmos CLXXXIV*, launched Oct. 24, 1967, was identified by U.S.S.R. as operational meteorological satellite which also surveyed Arctic shipping conditions. Satellite had completed its first 1,000 orbits. Information obtained from mission would be shared with U.S. *Cosmos CLXXXIV* was fourth known Soviet meteorological satellite. (UPI, *W Post*, 1/4/68, A14; *SBD*, 1/4/68, 13)

January 4: NASA Administrator James E. Webb announced personnel changes for new Apollo Lunar Exploration Office [establishment announced Dec. 19, 1967]: Capt. Lee R. Scherer (USN, Ret.), Assistant Director for Lunar Programs and Lunar Orbiter Program Manager, OSSA, would transfer to OMSF to direct new office; Capt. William T. O'Bryant (USN, Ret.), Program Manager for Apollo Lunar Surface Experiment Package, OSSA, would head Flight Systems Development Div.; and Dr. Richard J. Allenby, Jr., Deputy Director of Manned Flight Experiments, OSSA, would head Lunar Science Div. (NASA Release 68–5)
- Robert F. Packard, Director of State Dept.'s Office of Space and Environmental Science Affairs, described prospects for international cooperative programs in space before meeting of National Capital Section of AIAA: ". . . our foreign policy objectives clearly call for an ongoing and successful space program [and] to a surprising degree the American position in the world today may be affected by the manner in

which we conduct our space program. . . . There may be opportunities to use space technology for arms verification and arms control, and to create special opportunities for cooperation between the major space powers. There will clearly be opportunities to extend the economic and social benefits which this technology offers and to use these applications, as well as scientific research projects, as a means for increased international cooperation." (Text)

January 5: Resignation of Dr. Robert C. Seamans, Jr., as Deputy Administrator of NASA, announced Oct. 2, 1967, became effective. (NASA Release 67-257)

- USAF announced launch of 200th Minuteman ICBM from Vandenberg AFB. Since September 1962, Vandenberg had been launch site for missiles fired over Pacific Ocean for crew training, reliability tests, and developmental work. (Boeing Release)
- Computer services for 1968 would be provided to major contractors operating at Michoud Assembly Facility by Ling-Temco-Vought's Range Systems Div. under $2.7-million MSFC contract, MSFC announced. Second of three one-year renewed options, contract now had total value of $7.6 million. Services would include operation and maintenance of a large complex of digital and analog computers, data transmission systems, data reduction systems, and related electronic equipment. (MSFC Release 68-1)

January 7: NASA's *Surveyor VII* (Surveyor G), last spacecraft in NASA's Surveyor Program to analyze lunar surface, was successfully launched from ETR by Atlas-Centaur (AC-15) booster on 67-hr lunar intercept trajectory. Primary mission for 2,293-lb (at launch) spacecraft was to softland on the moon and obtain postlanding TV pictures of lunar surface. As secondary mission spacecraft would determine relative abundance of chemical elements in lunar soil with alpha-scattering instrument; manipulate lunar material with surface sampler in view of TV camera; and obtain touchdown dynamics, thermal, and radar reflectivity data [see Jan. 9-22]. *Surveyor VII*—which also carried magnets attached to landing pads, mirrors for viewing beneath spacecraft and for stereo photography, and seven special dust-detection mirrors—was first in Surveyor series to carry both alpha-scattering instrument and surface sampler.

Launch sequence for *Surveyor VII* was flawless, and spacecraft performed nominally. First of two scheduled midcourse maneuvers was so precise that second correction was not necessary; maneuver directed spacecraft toward a point 1.6 mi from center of target area.

Surveyor VII was seventh in series of seven spacecraft designed to prove out design, develop technology of lunar softlanding, obtain postlanding TV pictures of lunar surface, and provide basic scientific and engineering data in support of Project Apollo. Surveyor program was directed by OSSA Lunar and Planetary Programs Div.; project management was assigned to JPL; Atlas-Centaur launch vehicle was managed by LeRC; and prime contractor for spacecraft development and design was Hughes Aircraft Co. (NASA Proj Off; NASA Release 67-316K; AP, NYT, 1/8/68, 14; UPI, W *Star*, 1/8/68, 1; AP, B *Sun*, 1/8/68, A3)

- USAF and United Technology Center announced that hybrid rocket engine using solid propellant and liquid oxidizer had successfully fired Sandpiper missile from F-4 aircraft in secret test Dec. 12, 1967. Ac-

January 7: Surveyor VII, *launched Jan. 7 for softlanding on moon Jan. 9, photographed crescent earth from lunar surface Jan. 20 and recorded two laser beams (arrow) aimed toward landing site from earth. Spacecraft was last in* NASA's *Surveyor program.*

cording to UTC manager Douglas D. Ordahl, engine had performed well during five-minute flight over Eglin AFB, Fla., aerial test range and "proved for the first time that hybrids are practical and can fly." (*W Post*, 1/7/68, A7; Wilford, *NYT*, 1/8/68, 3)

January 8: Two-volume Review Board report, *Status of Actions Taken on the Apollo 204,* was submitted to Senate Committee on Aeronautical and Space Sciences and House Committee on Science and Astronautics' NASA Oversight Subcommittee by NASA Associate Administrator for Manned Space Flight, Dr. George E. Mueller. According to report, major changes made in Apollo project included: installing quick-opening hatch; eliminating most combustible materials in cabin; remaking spacesuit, primarily of glass fiber instead of nylon; equipping and training launch pad workers for fire fighting; placing metal shielding over exposed electrical wiring; and replacing aluminum oxygen pipes with more-fire-resistant stainless steel pipes. NASA reported spacecraft had been modified by North American Rockwell Corp. so that it could use ordinary air or oxygen-enriched air, as well as 100% oxygen, while on launch pad. (Text; Wilford, *NYT*, 1/20/68, 1)

- Spain had submitted "declaration of intent" to withdraw from ESRO, *Space Business Daily* reported. Under March 20, 1964, eight-year organization plan, Spain would have contributed 2.66%, $7.98 million, of ESRO's total budget of $300 million. Nine countries would remain in ESRO: Belgium, Denmark, West Germany, France, Italy, The Netherlands, Sweden, Switzerland, and U.K. (*SBD*, 1/8/68, 27)
- Five or six meteorites landed on earth every day, usually undetected, National Geographic Society reported. Some 8 billion meteors actually entered earth's atmosphere daily, but intense heat from air friction vaporized all but largest. (*NYT*, 1/8/68, 47)

January 9: NASA announced creation of Research and Technology Advisory Council to assist in planning and evaluation of research and technology for aeronautics and space. Council would assess relative importance of ongoing research, suggest additional work, and advise on methods for developing national resources.

Headed by Dr. Raymond L. Bisplinghoff, Head of MIT Department of Aeronautics and Astronautics, and supported by committees representing seven major technology areas, Council included: Dr. David Altman, United Technology Center; Dr. Allen V. Astin, Director of National Bureau of Standards; Dr. Loren D. Carlson, Univ. of California School of Medicine; Dr. Howard W. Emmons, Harvard Univ.; Gen. James Ferguson, Commander of AFSC; Dr. Nicholas J. Hoff, Head of Stanford Univ. Dept. of Aeronautics and Astronautics; Dr. Hans W. Liepmann, Cal Tech; Dr. John G. Linvill, Head of Stanford Univ. Electrical Engineering Dept.; Matthias E. Lukens, Deputy Executive Director of Port of New York Authority; Dr. Ronald Smelt, Vice President and Chief Scientist of Lockheed Aircraft Corp.; Dr. Chauncey V. Starr, Dean of Univ. of California Engineering School; and Edward C. Wells, Vice President, Product and Development, Boeing Co. Dr. Allan E. Puckett of Hughes Aircraft Corp. accepted appointment later. (NASA Release 68–4; NASA OART)

- NASA said one mile of coaxial cable had been laid in two-foot-deep trench across Ascension Island in South Atlantic Ocean to link two antennas—30-ft-dia parabolic reflector for contact with Apollo spacecraft and 42-ft-dia dish to link with *Intelsat II–C*—as part of chain of Manned Space Flight Network (MSFN) tracking, telemetry, and voice communications facilities. Also integrated in NASA communications circuits were Satellite Tracking and Data Acquisition Network (STADAN) facilities for scientific satellites and Deep Space Network (DSN) facilities to support lunar and planetary probes. Ascension Island cable was "final link" in two million miles of NASA Communications (NASCOM). (NASA Release 68–1)
- Dr. Robert C. Seamans, Jr., former NASA Deputy Administrator, was sworn in as consultant by NASA Administrator James E. Webb. Webb

said Dr. Seamans would "lend his talents and experience to the complex problems we face as the space program moves into its second decade." (NASA Release 68-7)
- AEC scientists had produced man-made atoms of transuranium elements —artificial elements of californium, einsteinium, and fermium that were heavier than uranium, heaviest of the 92 elements that occurred naturally. Man-made atoms had numerous potential uses in medicine and space, such as providing power for satellites and for electrical heart stimulator. (Spivak, *WSJ*, 1/9/68, 1)
- Aerospace industry sales in 1967 totaled $27.3 billion, 13% increase over 1966 sales, Aerospace Industries Assn. reported. (*W News*, 1/9/68, 38)

January 9-22: NASA's *Surveyor VII*, launched Jan. 7, became fifth U.S. spacecraft to softland on moon when it touched down in Tycho crater ejecta after 66-hr 35-min flight and began transmitting first of 21,274 detailed television pictures to JPL Deep Space Facilities, Goldstone, Calif.

Landing sequence began when *Surveyor VII* shifted its normal cruising attitude to position main retrorocket. Triggered by radar, main retromotor slowed spacecraft to $3\frac{1}{2}$ mph; retromotor then ejected. Vernier engines cut off at 13 ft above lunar surface and spacecraft landed.

First photos transmitted showed *Surveyor VII* was resting in rock-strewn area. Later photos consisted of wide and narrow angle surveys, panoramas, and special area surveys of spacecraft and landing area, including closeups of rocky debris on rim of Tycho crater and rugged landscape with prominences and boulders estimated to be three stories high. Stereo pictures were obtained with special mirrors; polarization pictures were taken of lunar surface, geologically interesting rocks, and earth; and photos of Jupiter and earth were obtained. Camera also photographed laser beams directed at spacecraft from Table Mountain, Calif., and Kitt Peak, Ariz., observatories, two of six U.S. stations which directed laser beams. Test, first use of light to communicate over such great distance, was considered highly significant for future use of lasers in communications and measurements in space and on earth and was expected to permit distance between points on moon and on earth to be measured with accuracy of six inches. Special surveys of dust-detection mirrors showed no accumulation of dust. Footpad magnets did not show a significant adherence to particles, but material did accumulate on surface sampler scoop magnet.

Only flaw in mission occurred Jan. 10 when alpha-scattering instrument failed to contact lunar surface after command to deploy had been issued. Photos revealed that although release squib had fired, instrument had remained in background count position. Efforts to dislodge it by rocking solar panel were unsuccessful. After satisfactory bearing-strength data had been obtained, engineers positioned surface sampler on top of alpha-scattering instrument, applied downward force, and successfully released instrument without damage. Alpha-scattering instrument, moved by surface sampler to its initial position over soil, to second position over rock, and to third position over one-half-inch-deep trench, obtained data for 63 hr 2 min. Surface sampler conducted six bearing-strength tests; excavated several trenches, including one 16 in long and over 6 in deep; turned over large rock for examination by TV camera; and attracted gravel-size rock with its magnet. On Jan. 22, communications with spacecraft were halted to preserve battery power during cold of lunar night.

Performance of *Surveyor VII*, final mission in NASA's Surveyor program to analyze lunar surface, was excellent. *Surveyor I* (launched May 30, 1966), *Surveyor III* (launched April 17, 1967), *Surveyor V* (launched Sept. 8, 1967), and *Surveyor VI* (launched Nov. 7, 1967) had all softlanded successfully and transmitted photos to earth. *Surveyor II* (launched Sept. 20, 1966) had failed to softland because of an ignition failure. Communications with *Surveyor IV* (launched July 14, 1967) were lost seconds before spacecraft was scheduled to softland, and its condition could not be determined. (NASA Proj Off; NASA Release 67–316K; AP, *W Post*, 1/10/68, A1; UPI, *W Star*, 1/10/68, A3; AP, *B Sun*, 1/11/68, A1; *SBD*, 1/11/68, 56; Hill, *NYT*, 1/12/68, 4; *W Post*, 1/13/68, A5; Clark, *NYT*, 1/22/68, 17).

January 10: Vice President Hubert H. Humphrey, as Chairman of National Aeronautics and Space Council, had asked BOB to place $20 million in FY 1969 NASA authorization for Earth Resources Observation Satellite (EROS). White House budget-cutters had eliminated EROS from new budget, George C. Wilson said in *Washington Post*. Secretary of the Interior Stewart L. Udall and NAS scientists also were expected to press for restoration of funds. NAS report to NASA would urge EROS funding as "one of the potentially big pay-offs for the national space program." (Wilson, *W Post*, 1/10/68, B7)

- Dr. Frederick Seitz, National Academy of Sciences President, was awarded DOD's Distinguished Civilian Service Medal for outstanding contribution to DOD as Chairman of Defense Science Board for past four years. (NAS–NRC–NAE *News Report*, 3/68, 6; NAS Info Off)

January 11: NASA successfully launched *Explorer XXXVI*—also called *Geos II* (GEOS–B) Geodetic Earth Orbiting Satellite—from WTR by Thrust-Augmented Delta booster into orbit with 978.4-mi (1,574.5-km) apogee; 670.8-mi (1,079.5-km) perigee; 112.18-min period; and 105.8° inclination.

The 460-lb, gravity-gradient satellite, launched to contribute to completion of NASA-managed U.S. National Geodetic Satellite Program (NGSP), carried six geodetic systems for development of a more precise model of earth's gravitational field and improvement of knowledge of earth's size and shape: (1) flashing-light beacons with 6,620-candle-sec-per-flash combined candle emittance to be photographed against background of stars; (2) corner cube quartz reflectors to pinpoint satellite's position by reflecting a laser beam; (3) three radio transmitters to help determine satellite's line-of-sight velocity using Doppler shift principle and USN TRANET equipment; (4) radio transponder to provide distance between satellite and interrogating USA SECOR network ground station; (5) range and range-rate transponders for simultaneous determination of satellite-to-ground-station range and line-of-sight velocity, using NASA's STADAN system; and (6) C-band transponder systems at STADAN stations for calibration and experimentation to determine applicability of C-band radar tracking system to satellite geodesy. Primary mission objectives were to obtain 90 days of precision spacecraft position data, support geodetic positioning of 40 reference control points and 64 densification observation sites, and evaluate accuracy of ground-based C-band radars.

Explorer XXXVI was fifth satellite launched in NGSP series and second in GEOS series. Mission of first GEOS satellite, *Explorer XXIX* (launched Nov. 6, 1965), had ended in December 1966, but satellite was still transmitting weak signals on Doppler shift. *Geos II* was designed, fabricated, and tested by Johns Hopkins Univ.'s Applied Physics Laboratory. Overall mission responsibility was assigned to OSSA. (NASA Release 68–2K; NASA Proj Off; UPI, *NYT*, 1/11/68, 30)

- SR–71, USAF's fastest and highest flying aircraft, crashed in northern California. L/C Robert G. Sowers (USAF), instructor, and Capt. David E. Fruehaf (USAF), trainee, ejected safely. (UPI, *W Post*, 1/13/68, 4)
- Malfunction in test equipment apparently caused failure of two modified Apollo drogue parachutes during flight test at USN's aerospace research facility, El Centro, Calif. Test objective to demonstrate structural integrity of parachutes to slow and stabilize spacecraft before main parachute deployment was not realized; MSC announced repeat of flight test would be made Feb. 5. (*Aero Tech*, 1/29/68, 10)
- NASA announced award of $4,521,671 contract to Honeywell, Inc., for management and engineering services during 1968 for Centaur launch vehicle guidance system. Atlas-Centaur booster had successfully launched seven Surveyor spacecraft toward the moon and was scheduled to launch OAO, ATS, and two Mariners. (LeRC Release 68–3)

January 12: XB–70 research aircraft, flown by Fitzhugh L. Fulton, Jr., and Donald L. Mallick, reached mach 2.55 and 67,000-ft altitude and accomplished 70% of primary test objectives during 1-hr 54-min flight from Edwards AFB. Purpose of flight was to take wing tuft photos and check stability, control, and handling qualities; inlet performance; boundary layer noise; friction; fuselage bending. (XB–70 Proj Off)

- Most realistic way to reduce sonic boom from supersonic aircraft was to make successive small reductions, according to *Generation and Propagation of Sonic Boom*, report of Subcommittee on Research, NAS Committee on SST-Sonic Boom. Reductions could be achieved by "refinements in conventional aircraft design, a better understanding of theory, and improvements in propulsive efficiency and operating procedures." Future aircraft designs might yield significant reductions. Studies should be undertaken by both Government and industry on "less conventional configurations." (NASA Release; NAS–NRC–NAE *News Report*, 1/68, 1–2)
- NASA awarded $7,178,937 contract to General Dynamics Corp. Convair Div. to continue for 15 mo contract now in force for launch support services for Atlas-Centaur launch vehicle. LeRC had management responsibility for Centaur. (NASA Release 68–8)
- DOD announced termination of $175-million Mark 17 reentry vehicle program—for which $45 million had already been expended—"before additional R&D or heavy production costs are incurred." Funds originally designated for Mark 17 would be used for design changes on Mark 11. (DOD Release 43–68; AP, *NYT*, 1/14/68, 50)
- Postdoctoral research opportunities under one-year research associateships to work in AFSC's laboratories and research centers would be offered by AFSC in 1968 and 1969. Awards would be made by NRC's selection board around April 1, 1968. (AFSC Release 4.68)

January 15: NASA announced personnel changes: Samuel H. Hubbard, OMSF Special Assistant, Gemini Program, moved to Apollo Applications Program Office as Special Assistant for Programs, responsible for ensuring adequate communications and documentation were provided by AAP in support of organizations with program interfaces.

Col. Maynard E. White (USAF, Ret.), Director, MSF Program Control, was named Director of MSF Management Operations. Jerald K. Kubat of Apollo Program Control would succeed him. (NASA Ann)

- USAF Air Defense Command was redesignated Aerospace Defense Command. (*SBD*, 1/3/68, 1; *AFHF Newsletter*, 2/68)
- Terms for development, production, and launch of two comsats for U.K. under DOD contract announced March 8, 1967, were made final in $7,535,000 fixed-price-incentive-fee USAF contract with Philco-Ford Corp.'s Space and Re-entry Systems Div. First satellite would be launched into synchronous, equatorial orbit with Thor-Delta booster from ETR in 1968; second satellite would be backup. Philco-Ford would design satellites "to satisfy certain defense communications requirements of the United Kingdom." Satellites would have station-keeping capability to maintain specific positions over earth. Space and Missile Systems Organization would be contracting agency. (AFSC Release 195.67)

January 16: U.S.S.R. successfully launched *Cosmos CXCIX*. Orbital parameters: apogee, 221 km (137 mi); perigee, 159 km (99 mi); period, 88.2 min; inclination, 65.5°. Satellite reentered Feb. 1. (*Krasnaya Zoez*, 1/18/68; GSFC *SSR*, 1/31/68; 2/15/68)

- Two NASA Nike-Cajun sounding rockets launched from NASA Wallops Station carried payloads to 85-mi (137-km) and 88-mi (141-km) altitudes to investigate D region electron density during solar x-ray flare. Rocket and instrument performance was satisfactory. Preliminary data indicated launches were accomplished during period of decreasing solar activity. (NASA Rpt SRL)
- France had successfully test-fired LEX, improved, single-stage experimental hybrid sounding rocket which used liquid oxidizer and inert solid propellant. Developed by Office National d'Etudes et de Recherches Aérospatiales (ONERA), French space agency, LEX could carry 14-lb payload with meteorological experiment to 115-km (71-mi) altitude. (*SBD*, 1/16/68, 74)
- NASA awarded NAR Rocketdyne Div. $14,796,400 cost-plus-incentive-fee contract for engineering support services for H–1 engines to power Saturn IB 1st stage. Contract covered period of July 1967 through June 1971. (MSFC Release 68–9)
- British Prime Minister Harold Wilson, speaking to House of Commons on long-range spending cuts to restore financial confidence, announced withdrawal of U.K.'s military forces east of Suez by end of 1971 and cancellation of agreement to purchase 50 F–111 aircraft from U.S. for $875 million. F–111 aircraft order cancellation was expected to have adverse effect on cooperation between U.K. and U.S. in defense production. To obtain U.K. purchase agreement of $2.9 billion for F–111, F–4, and C–130 aircraft plus Polaris spare parts, U.S. had agreed to purchase $425 million worth of defense items from U.K. by FY 1977 and to help U.K. obtain additional $400 million in cooperative arms sales to third countries. (Lewis, *NYT*, 1/17/68, 1; Sheehan, *NYT*, 1/17/68, 15)

January 17: USAF launched unidentified satellite from Vandenberg AFB by Thrust-Augmented Thor-Agena D booster into orbit with 335-mi

(539.1-km) apogee, 285-mi (458.6-km) perigee, 94.5-min period, and 75.1° inclination. (*Aero Tech*, 1/29/68, 11; *Pres Rpt 68*)

- Primary objectives of *Mariner V* mission (launched June 14, 1967) had been fulfilled and mission had been adjudged a success by NASA. Communications with spacecraft had been terminated Dec. 1, 1967, when signal level had been lowered because of antenna pointing angle; spacecraft systems had operated normally during telemetry reception, Nov. 21, 1967. Spacecraft was expected to survive solar heat and return within range of DSN antennas in August or September 1968. Total lifetime would be three to six years. (NASA Proj Off)

- A 21.5-ft-dia, 40-ft-long mockup of Saturn V 3rd (S-IVB) stage was flown to MSFC for crew station design review; it would be launched as 2nd stage of Saturn IB and as flight orbital workshop, with aluminum-grid partitions rearranged to create separate rooms. Modification provided two-story arrangement with separate rooms for sleeping, food preparation, and control center and large open laboratory area for experiments. (MSFC Release 68–11)

- In Moscow, U.S.S.R. and France agreed to place French instruments on board Soviet lunar orbiter satellite at undetermined future date, in third joint Soviet-French space project, Evert Clark reported in *New York Times*. Projects, most significant cooperative effort U.S.S.R. had undertaken with any country, included launch of French satellite on interplanetary physics mission in 1972 and use of Soviet Molniya comsats to exchange experimental color TV broadcasts. (*NYT*, 1/18,/68, 52)

- AFCRL geodesists hit *Explorer XXII* satellite with multipulse ruby laser beam in daylight for second time. First success in using satellite as target for reflection in daylight was on Dec. 19, 1967, by AFCRL. *Explorer XXII*, launched by NASA Oct. 9, 1964, carried special reflectors for laser beams. Experiment was designed to improve measurements of distance between widely separated points on earth's surface and improve knowledge of earth's size and shape. (OAR *Research Review*, 5/68, 13; *Instruments and Spacecraft*, NASA SP-3028)

January 17–19: Rep. Emilio Q. Daddario (D-Conn.), Chairman of House Committee on Science and Astronautics' Science, Research, and Development Subcommittee, at hearings on environmental pollution, stressed crucial role for science and technology in pollution abatement. He said hearings would focus on research programs of Federal agencies which "must produce an expanded basis of facts and technical options to make our pollution laws work." Testimony would show "that it is shortsighted indeed to consider goals for restoring and maintaining the quality of the environment apart from the supporting research and development." (Transcript)

January 18: USAF launched unidentified satellite from Vandenberg AFB by Titan III-B-Agena D booster. Satellite entered orbit with 254-mi (408.8-km) apogee, 77-mi (132.9-km) perigee, 89.8-min period, and 111.4° inclination and reentered Feb. 4. (*SBD*, 1/19/68, 98; GSFC SSR, 1/31/68; *Pres Rpt 68*)

- Study of Northern Lights, auroras, and polar cap airglow by coordinated use of aircraft flights from Churchill Research Range, sounding rocket launches, satellite overpasses, and ground observations was begun by NASA. During two extended periods (Jan. 18–Feb. 8 and Feb.

21–March 12) NASA airborne laboratory, Convair 990 jet aircraft, would fly over Churchill area and beyond, from Alaska to Greenland. Three flights would be coordinated with sounding rocket launches from Churchill; many flights would be coordinated with passes of *Ogo IV*, containing 12 experiments for studying auroral and polar cap phenomena. Aircraft would carry spectrometers, photometers, wide-angle cameras, radio frequency receivers, and magnetometer to study magnetic field activity and time and space variations of auroras and polar gap airglow. NASA's 1968 Airborne Auroral Expedition would be directed by ARC's Airborne Science Office and managed by ARC's Louis C. Haughney; 14 universities and research organizations in Canada and U.S. would participate. (NASA Release 68-9)

- Complete draft of treaty to ban spread of nuclear weapons was submitted by U.S. and U.S.S.R. to 17-nation U.N. Disarmament Conference, which would report to U.N. General Assembly by March 15. Agreement had been reached on international inspection and controls to detect any violations of treaty's provisions. West Germany and other nations contended proposed controls would interfere with peaceful atomic development. European Atomic Energy Community (EURATOM) had refused to take part in negotiations and said it would not sign treaty. (*NYT*, 1/19/68, 1; 1/20/68, 8)

- Proposal for four additional basic Block II Apollo spacecraft command and service modules would be made by NAR Space Div. in accordance with NASA request. Action would bring total purchase to 19 for flights on Saturn IB and Saturn V launch vehicles. Delivery would start in 1970. (NASA Release 68-12)

- Northrop Corp. scientists G. M. Andrew and M. S. Cahn told news conference discharge of electricity into air ahead of supersonic aircraft might not only lessen sonic boom intensity but also reduce aerodynamic drag, saving fuel. Scientists said NASA and Boeing Co. were interested in new line of SST research. Electrical discharge—up to 30,000 volts in tests to date—would repel molecules of air and remove them from aircraft's path; molecules would then flow smoothly around aircraft instead of bunching ahead to cause sonic boom. Scientists admitted electrical discharge could interfere with radio and TV broadcasts and with communications between aircraft and ground. (AP, *W Post*, 1/19/68, A5)

- Lockheed Missiles & Space Co. received $5,700,000 USAF contract for Agena launch services at Vandenberg AFB from Oct 1, 1967, through Sept. 30, 1968. (DOD Release 64-68)

January 19: U.S.S.R. successfully launched *Cosmos CC*. Orbital parameters: apogee, 538 km (334 mi); perigee, 518 km (322 mi); period, 95.1 min; inclination, 74°. (*Pravda*, 1/21/68; GSFC *SSR*, 1/31/68)

- Fiftieth anniversary of USAF School of Aerospace Medicine, dedicated to ensuring that man could perform efficiently and safely in air and in space. Established in 1918 at Hazelhurst Field near Mineola, N.Y., school was currently headquartered at Brooks AFB, Tex., under command of Col. George E. Schafer. (*AFHF Newsletter*, 2/68)

- Holder of world manned aircraft speed record, X-15 pilot Maj. William J. Knight (USAF), received senior astronaut wings and Distinguished Flying Cross from M/G Hugh B. Manson, Commander of AFFTC. Maj.

Knight set 4,534-mph speed record Oct. 3, 1967, in special ablative-coated X–15; two weeks later he reached 277,000-ft altitude to qualify as an astronaut. On June 29, 1967, after experiencing complete engine and power failure, he had piloted X–15 to safe landing to earn DFC. (AFFTC Release 68–1–9)

- President Johnson named Clark M. Clifford, lawyer and intimate adviser to three Democratic Presidents, to succeed Robert S. McNamara as Secretary of Defense. President praised Clifford as "a counselor on most of the important decisions made in many of the international fields from defense to strength to weapons to actions." (Frankel, *NYT*, 1/20/68, 1; *PD*, 1/22/68, 84–5)
- NASA announced award of $8.5-million contract to Bendix Field Engineering Corp. to provide logistics support services for NASA's consolidated worldwide spacecraft tracking networks, which included 50 worldwide sites, ships, instrumental aircraft, communications switching centers, spacecraft control centers, mobile units, and a training center. Bendix would establish consolidated logistics support system by combining certain functions previously performed separately for each network. Both networks were operational responsibility of GSFC, under NASA's Office of Tracking and Data Acquisition. (NASA Release 68–10)

January 20: NASA General Counsel Paul G. Dembling told Rutgers Univ. audience: "... we are moving toward the end that the rule of law will prevail in space." He explained space law treaty, which entered into force Oct. 10, 1967, and praised effectiveness of U.N. "as a vehicle for the development of international law." Significance of space law and space rescue treaties was that they "indicate that international law can develop ... even among nations having widely divergent ideologies and national interests ... [and] can evolve by gradually codifying the ground rules which are considered by States to be in their common interests." (Text)

January 21: FRC announced cuts in FY 1969 budget dictated closing out X–15 aircraft program in fall 1968. Major setback was Nov. 15, 1967, crash which resulted in death of pilot Maj. Michael J. Adams (USAF). Main task in 191 flights to date had been study of problems of manned controls in high-speed, high-altitude aircraft. Remaining tasks concerned testing of atmospheric reentry conditions. As successor to X–15 —holder of speed record of 4,534 mph and altitude record of 354,200 ft —officials would consider aircraft capable of taking off from ground under own power, achieving earth orbit, reentering, and landing at virtually any point on earth. (AP, *C Trib*, 1/22/68)

- Telescope at Univ. of Arizona's Planetary Laboratory photographed *Lunar Orbiter V* as it appeared beyond left limb of moon's face. Scientists believed experiment to be first success in sending observable light signals from moon's vicinity to earth. Some 80 photos were made through 61-in telescope by team of three astronomers under Laboratory's Director, Dr. Gerard Kuiper, in experiment conducted for NASA and spacecraft's designer and builder, Boeing Co. Satellite was visible on 52 photos. Boeing engineers oriented *Lunar Orbiter V* so its shiny panels reflected sunlight toward earth. Astronomers reduced stray light near very bright moon to prevent multiple reflections within telescope from reaching photographic plates. Dr. Kuiper explained that spacecraft resembled 12th-magnitude star in brilliance. These reference data, used with radar tracking data, would enable scientists to locate more precisely center of mass of moon with respect to its visible limb. *Lunar Orbiter V* was crashed on the moon Jan. 31. (NASA Release 68-34; *NYT*, 2/20/68, 19)

January 22: NASA's *Apollo 5* (AS-204) unmanned earth orbital mission, delayed for nearly four hours because of ground equipment malfunctions, was successfully launched from KSC Complex 37 at 5:48 pm EST in fourth flight of Saturn IB and first flight of 31,700-lb lunar module (LM-1), designed to land two astronauts on the moon. Primary mission objectives were to verify operation of LM ascent propulsion system (APS), descent propulsion system (DPS)—including restart—and spacecraft structures; evaluate LM staging; and evaluate 2nd-stage (S-IVB) and instrument-unit (IU) orbital performance.

Launch phase occurred as planned; S-IVB ignited to insert spacecraft into orbit with 138-mi (222-km) apogee, 101-mi (163-km) perigee, 88.3-min period, and 31.63° inclination; nose cone jettisoned; spacecraft coasted for 43 min 52 sec; and LM separated from spacecraft LM adapter. LM entered orbit with 138-mi (222-km) apogee, 104-mi (167-km) perigee, 88.4-min period, and 31.63° inclination. Initial 39-sec DPS burn, designed to simulate deceleration for descent to lunar surface, was automatically shut down after only 4 sec because of overly conservative computer programming. Ground controllers switched to minimum requirement sequence, an alternate flight plan with shorter total DPS firing time and no provision for lunar landing simulation. Second DPS engine firing occurred successfully, with 26-sec burn at 10% thrust level and 7 sec at maximum thrust. Third DPS engine firing 32 sec later consisted of 26-sec burn at 10% thrust, 2 sec at maximum thrust, and ascent stage fire-in-the-hole (FITH) burn during which the two stages separated and APS engine was ignited simultaneously while DPS was being shut down, simulating abort during landing phase. Duration of initial APS burn during abort staging was 60 sec. APS engine fired second time for 6 min 23 sec until fuel was depleted. At end of 11-hr 10-min test period, both stages of the LM were left in orbit eventually to reenter and disintegrate.

Apollo 5 mission, adjudged successful by NASA in spite of premature DPS engine shutdown, proved out structural integrity of LM; verified in-space operation of DPS and APS; and proved value of contingency planning. LM was last major piece of Apollo hardware to have its first test in space. Command module (CM) had been tested during AS-203 mission (July 5, 1966) and with service module (SM) on AS-202 (Aug. 25, 1966) and *Apollo 4* (AS-501) (Nov. 9, 1967). Final decision on whether *Apollo 5* results justified omitting second unmanned test and scheduling next mission to be manned would be deferred until March, pending final mission review. (NASA Proj Off; NASA Releases 68-6K, 68-19; MSC Release 68-4; AP, Strothman, *W Post*, 1/23/68, 1; Fahnestock, *W Star*, 1/23/68, 1; Wilford, *NYT*, 1/24/68, 20)

- NASA Nike-Tomahawk sounding rocket launched from NASA Wallops Station carried GSFC payload to 184-mi (296-km) altitude to measure intensity and polarization distribution of hydrogen Lyman alpha at night. Although rocket performance was satisfactory, instruments failed to function. (NASA Rpt SRL)

- Flammability characteristics of mixed gas atmosphere (60% oxygen and

January 22

40% nitrogen at 16 psi) in Apollo command module had been evaluated by MSC in 24 "enriched air" tests, with "all but two ignition points" showing self-extinguishing characteristics. Two ignition points on circuit breaker panels exhibited what MSC termed "moderate fire propagation" because fire spread "beyond the point of self extinguishment." Dr. Robert R. Gilruth, MSC Director, retained three major options for "mixed gas" for Apollo launch phase: regular air, enriched air using oxygen and nitrogen, or pure oxygen. (*Aero Tech*, 1/29/68, 10)

- MSC had awarded $2-million contract extension to MIT's Div. of Sponsored Research for design and developmental support of Apollo guidance and navigation systems, including flight test and operational support for Apollo command and service modules. Contract extension brought total estimated value of MIT contract since April 1965 to $48.7 million. (NASA Release 68-15)

- In statement to press, Secretary of the Navy Paul R. Ignatius and Chief of Naval Operations Thomas H. Moorer refuted rumors USN wanted to drop controversial F-111B aircraft in favor of another type: "In early October 1967, representatives of the Grumman Aircraft Engineering Company submitted to the Navy a design layout of a new fighter type aircraft. The design proposal sought to meet future fleet air defense and fighter interceptor requirements [and] would build on the F-111B technology and incorporate most of its basic components, including the engines, and the PHOENIX missile [and] variable sweep wing." They stated that three other proposals had been received from aircraft industry but added that "evaluation . . . of proposals from industry . . . is a continuing process. . . . The Navy continues to support the on-going aircraft programs and the funds in the FY 69 budget request for production of both F-4 and F-111B aircraft, as well as the VFAX concept." (DOD Release 70-68)

January 23: NASA Nike-Tomahawk sounding rocket launched from Churchill Research Range carried Univ. of Alaska payload to examine spatial distribution of ionospheric currents near visual auroral forms with flashing light on rocket used to locate position of rocket on TV film of aurora. Rocket and instrument performance was satisfactory. Radar did not track vehicle to sufficient altitude to obtain peak altitude data. (NASA Rpt SRL)

- At AIAA's 6th Aerospace Sciences Meeting in New York City, NASA Associate Administrator, Dr. Homer E. Newell, compared current space activities with those of past: "Today, as we note the anniversaries of Sputnik, Explorer, and Vanguard, we do well to take stock of where we are and where we are going in Space, for after a decade of unparalleled success we find ourselves having to resell our fellow citizens this greatest adventure of mankind."

  Ten years ago "we could only assert from intuition and prophecy the value of space techniques to science. Today, we can point to a profound influence that space has had on the geosciences, is having on astronomy, and is beginning to have in life sciences." Ten years ago, planning for future required organizational and management, as well as disciplinary and technical, capability to achieve objectives. Now it could be based on established capability in science, engineering, and administration; proved reliability of wide variety of launch vehicles; demonstrated ability to use automatic techniques in space and to construct and fly new large systems successfully; and demonstrated capability of man to operate in space. Major problem of mastering space technology was solved, failures were few, and success, once "rare and precious commodity," became routine.

Budget reductions and postponement of new space missions would continue until resolution of Vietnam war, but long-term outlook for space explorations, research, and applications was good: (1) "the intrinsic worth of the program merits the interest and support of the ... people"; (2) post World War II babies raised in midst of space age "will in the 70's become a powerful force of intelligent, highly educated people ... [who will] want to continue" space exploration; (3) U.S.S.R. would continue vigorous, expanding program, presenting challenge U.S. would have to meet; (4) as U.S. economic growth continued and military expenditures decreased, resources would be available and need for vigorous U.S. space program recognized; and (5) gap between professionals and laymen would be bridged by space-oriented younger generation.

NASA's key policies were directed toward preserving U.S. space capability and laying groundwork for vigorous program in 1970s. NASA would seek approval from Congress and Executive Branch to initiate programs, increase emphasis on aeronautics and space applications and practical applications on earth, and continue to support development of key advanced technologies. (Text)

- AIAA's highest award, Goddard Award, was presented jointly to General Electric Co. project managers Donald C. Berkey and James E. Warsham and Ernest C. Simpson, Chief of Turbine Engine Div. at USAF's Aero Propulsion Laboratory, for "an outstanding contribution to aircraft propulsion in relation to development of the high bypass ratio turbofan engine" [TF39 engine which would power USAF's C-5A jet transport aircraft]. Other awards: Sylvanus Albert Reed Award to William H. Cook, Boeing Co., for his influence on development of Boeing 727 and SST; Space Science Award to Prof. Kinsey A. Anderson, Univ. of California at Berkeley, for "a consistent series of major contributions to the development of space science and the improvement of our understanding of solar cosmic radiation, the aurora and the magnetosphere"; and 1968 Dryden Research Lecture Award to Hans W. Liepmann for research on laminar instability and transition.

    LaRC Director Dr. Floyd L. Thompson was elected AIAA President, succeeding Harold T. Luskin. (AIAA *News*, 1/5/68, 1/15/68, 1/23/68; AIAA *Facts*)

- Honored at Chicago's National Conference on Industrial Research as Man of the Year in research by *Industrial Research* magazine, JPL Director, Dr. William H. Pickering, told press at Illinois Institute of Technology, "The United States has superiority in space exploration but the Congressional cutbacks may enable the Russians to get well ahead of us." (Kotulak, *C Trib*, 1/24/68)

- Rep. George P. Miller (D-Calif.), Chairman of House Committee on Science and Astronautics, said on floor of the House, ". . . these are times of stringent budgetary considerations [and] times when fiscal allocations for our space program must be evaluated." He referred to Jan. 2 *Oakland Tribune* editorial which stated: "After the frustrations and fears brought on by the Soviet Union's early space successes, this nation resolved to bear the burdens and financial strain of winning the space race. We were not going to continue to be second in space. That commitment is as worthy today as it was in the early years of this decade. At stake is not only national prestige but critical scientific knowledge and military advantages. Fickle second thoughts at this late date are inexcusable." (*CR*, 1/23/68, E139)

January 23–24: House Committee on Science and Astronautics opened Ninth Annual Seminar with Panel discussion of applied science and its relationship to world economy. President of International Bank for Reconstruction and Development George D. Woods said that "technologies do . . . provide the developing countries of the world with the basis for a satisfactory rate of economic growth." We should concentrate efforts on education, agriculture, population control, development efforts on VTOL and STOL aircraft, communications and geodetic satellites, and sea water desalination.

House Speaker John W. McCormack said, "If scientific knowledge is applied indiscriminately without regard to social and world costs, the results over a period of time can be disastrous [and] Congress must have a means for judging the consequences of science."

Dr. Anthonie T. Knoppers, Senior Vice President of Merck & Company, Inc., speaking of transfer of marketable technology, said that "everybody recognizes . . . problems of the future the large corporation cannot solve" and called for "new firms in which the corporation, the university, and the government have to work to better solve some of our sociological problems." (Transcript)

January 24: USAF launched two unidentified satellites from Vandenberg AFB by Long-Tank Thrust-Augmented Thor (LTTAT)-Agena D booster. One satellite entered orbit with 269-mi (432.9-km) apogee, 112-mi (180.2-km) perigee, 90.6-min period, and 81.5° inclination and reentered Feb 27. Other satellite entered orbit with 338-mi (543.9-km) apogee, 294-mi (473.1-km) perigee, 94.9-min period, and 81.7° inclination. Flights were fifth and sixth since LTTAT vehicle was activated on May 9, 1967. (*SBD*, 1/26/68, 137; *Pres Rpt 68*)

- NASA's *Explorer XXXVI* (*Geos II*) satellite, launched Jan. 11, had completed initial checkout tests and was ready for operational use. Supporting ground systems were expected to be ready to operate with satellite in late February. All six geodetic systems had responded to ground commands. Satellite had been stabilized by gravity gradient boom so that it faced earth constantly with ±5° stability. (NASA Release 68–16)
- ARC scientists Dr. Cyril A. Ponnamperuma and Fritz H. Woeller said gigantic red spot on Jupiter's atmosphere might be millions of square miles of red organic dye. Scientists had made large number of electrical energy discharges (lightning) in simulated Jupiter atmosphere of ammonia and methane, atmosphere mix agreed to by most scientists, and had produced quantities of amino acids and other organic materials. Most common product had been organic dyestuff with ruby red translucency. From experiments and spectroscopic and theoretical studies, scientists concluded red spot might result from giant meteor craters in solid hydrogen surface of Jupiter creating vortex in Jupiter's atmosphere. Since atmosphere might be largely red dye, upwelling caused by vortex would create red spot in top of dense surrounding white clouds. Jupiter was like huge dynamo, rotating once every 10 hr; rapid

alternation of day and night, hence warm and cold, was believed to produce great atmospheric turbulence and electrical energy transfer. (ARC Release 68–3)

- *Pravda* released description of preparations for launching of Maj. Yuri A. Gagarin in history's first manned space flight April 12, 1961. Songs by 110-member folk music choir—instead of one man's voice—had been selected for radio tests to avoid arousing rumors that U.S.S.R. had put a man into orbit. *Pravda* reports were excerpts from *The First Four Stages*, book recognizing leading role in Soviet manned space flight of Sergei P. Korolev. U.S.S.R. acknowledged malfunction had occurred during first unmanned test of Vostok spacecraft launched May 15, 1960, as *Sputnik IV*. Because of improper alignment, spacecraft was hurled deeper into space when retrorockets fired instead of braking for reentry. (Anderson, *NYT*, 1/25/68; *A&A*, 15–60, 123, 147)
- JPL announced appointment of R/A John E. Clark (USN, Ret.) as JPL Deputy Director, effective Feb. 19. Adm. Clark had retired as Commandant, Twelfth Naval District, in September 1967. (JPL Release)

January 26: NASA Aerobee 150 sounding rocket launched from WSMR carried GSFC payload to 99-mi (160-km) altitude to check instrumentation. Random failure occurred in control system, but pointing system acquired two of three target stars. Rocket performance was satisfactory. (NASA Rpt SRL)

- NASA Administrator James E. Webb announced resignation of Edmond C. Buckley as Associate Administrator for Tracking and Data Acquisition, and Buckley's appointment as his Special Assistant and Vice Chairman of NASA Post Apollo Advisory Group (headed by LaRC Director Dr. Floyd L. Thompson). Gerald M. Truszynski, Deputy to Buckley for seven years, would succeed him. Webb said Buckley had been "one of the architects of this nation's great competence in tracking and data acquisition." (NASA Release 68–17)
- UCLA physicist Dr. Willard F. Libby believed icecaps, similar to those on earth, covered Venus' poles and might extend over as much as half of planet. U.S.S.R.'s *Venus 4* probe had found surface temperatures reaching 540° F during landing of instrument package, Oct. 18, 1967. Dr. Libby, Nobel Laureate, explained that temperatures between hot equatorial belt and icy polar regions must grade off into area of moderate warmth where plant life might exist; gaseous carbon dioxide cloud covering Venus did not permit animal life as known on earth. (UPI, *NYT*, 1/26/68)

January 27: First flight test of Apollo lunar module (*Apollo 5*) Jan. 22–23 was adjudged success. NASA officials were gratified with maturity of spacecraft's hardware; studies indicated LM showed more maturity in its first flight than many previous spacecraft, including some designed to be manned. Although overly conservative programming of guidance computer caused early shutdown of first descent propulsion system burn, data indicated no problems with computer or system's engine. (NASA Release 68–19)

January 28: Airborne laboratory, Convair 990 jet aircraft *Galileo*, had completed more than one week of flights above Alaskan-Canadian area in NASA's 1968 Airborne Auroral Expedition, based at Churchill Research Range, NASA reported [see Jan. 18]. Scientists had obtained hundreds of unique color photos of auroras and had reported unusually clear night views of towering auroral displays. Aircraft had provided superior data at altitudes previously reached only by balloons. Expedition scientists intended to establish more precisely width and extent of Auroral Oval region—belt usually located below 80° north latitude and often extending as far south as 55° north latitude. (NASA Release 68-18)

- Pakistan Space and Upper Atmosphere Research Committee had successfully installed and tested Pakistan's first satellite tracking station at Dacca, East Pakistan. Automatic Picture Transmission (APT) station capable of receiving cloud-cover photographs via U.S. Nimbus and ESSA satellites would enable meteorologists to forecast cyclones, which frequently struck Pakistan. (AP, *NYT*, 1/29/68; AP, *NYT*, 2/20/68, 61)

January 29: President Johnson sent message to Congress on FY 1969 budget.

Praising effectiveness of NASA's cost reduction efforts, he noted that NASA by utilizing idle, excess, and surplus Government property had avoided expenditures of over $22 million for new equipment and facilities and had saved over $16 million by improving procurement practices. However, to meet urgent national needs in other areas, further reductions still had to be made. "New obligational authority requested for [NASA] . . . is about $220 million below the 1968 amount. Expenditures will be $230 million below 1968, $850 million below 1967, and over $1.3 billion less than in 1966. This reduction reflects our progress beyond the costly research and development phases of the manned lunar mission, as well as the immediate need to postpone spending for new projects wherever possible.

"Based on a careful examination of priorities, the 1969 budget provides increases in some areas to prepare for important advances in future years, while deferring other less urgent, new projects. The production of our large Saturn-class space boosters is continued but at a reduced rate. The development of a nuclear rocket engine to increase the capability of our Saturn V launch vehicle is also continued, but at a smaller size and thrust than originally planned, to reduce development cost."

Planetary exploration would be continued with development of "a new spacecraft for launch in 1973 to orbit and land on Mars. This new Mars mission will cost much less than half the Voyager program included in last year's budget. Although the scientific result of this new mission will be less than that of the Voyager, it will still provide extremely valuable data and serve as a building block for planetary exploration systems of the future."

Request for DOD, increased "to assure that our defense capabilities remain equal to any challenge or threat," included funds to: (1) maintain strategic deterrent by converting from Minuteman II to Poseidon missiles with multiple warheads and modernizing manned bomber force with additional F-111B aircraft and improved short-range attack missiles; (2) proceed with procurement of Sentinel missile defense system for defense against possible Communist Chinese threat and revamp air defense; (3) augment firepower, mobility, and readiness of general-purpose forces by improving air defenses with new fixed-wing aircraft, helicopters, and weapon systems; (4) improve airlift-sealift capability by purchasing additional C-5A aircraft and procuring fast deployment logistics ship; and (5) continue vigorous R&D effort. (*The Budget of the United States Government, FY 1969,* 26–9)

- President Johnson submitted $186.1-billion FY 1969 budget request to Congress, including $6.76-billion total space budget. Of this sum, NASA would receive $4.37 billion, smallest amount since 1963; DOD space efforts, $2.216 billion; AEC space applications, $143 million; and ESSA satellite system, $30.6 million.

    NASA FY 1969 budget, $218 million less than for FY 1968, provided for $3.677 billion to be spent for R&D (down $233 million from FY 1968 and $557 million from FY 1967); $45 million for construction of facilities; and $648 million for administrative operations. Project Apollo would be kept on schedule, development of NERVA I would be continued, and launch plans would be made for two pioneer flights toward Jupiter and one Sunblazer probe. Research on Earth Resources Observation Satellite would continue, but development would be postponed. OGO program would be phased out after OGO–F. NASA's $2.5-billion Voyager program would be replaced with a $500-million, four-mission Mars orbiter project. Although NASA's $2.039-billion Apollo budget request accounted for 47% of total NASA budget, amount for Apollo was $517 million less than for FY 1968, reflecting declining expenditures as program neared its completion. Requested $439.6 million for Apollo Applications (AA) program was less than half of amount originally sought and necessitated cancellation of planned 14-day AAP–1A flight. Some $76.9 million—a $10.1-million increase over FY 1968—was allotted for NASA's aeronautics program, with most of increase attributed to additional supporting research in subsonic aircraft technology. XB–70 and X–15 research programs would be phased out by December 1968. Space science and applications were allotted $538.2 million, with 18 major NASA launches and 13 non-NASA launches scheduled for 1968.

    Major portions of DOD space budget would be spent on MOL—$600 million, compared with $431 million requested for FY 1968. Some $60.4 million was allocated for defense and tactical satellite communications programs, and $10.5 million for Vela nuclear test detection satellite program.

    AEC space budget—$11 million higher than for FY 1968—included $72 million for Project Rover, with $39 million for NERVA system. Most of ESSA's funds would be used for development of sensors and satellite system in support of World Weather Watch and improved techniques for warning services. FAA's request included $351 million for SST development, $251 million increase over FY 1968 and one of largest increases given to any domestic program. It was principal item in DOT's $449-million request for R&D. (Text; DOD Budget Summary; *W Post,* 1/16/68, A5; AP, *B Sun,* 1/16/68, A3; McNamara Statement; DOD Background Briefing; NASA Release, 1/29/68; *W Star,* 1/29/68, A7; Clark, *NYT,* 1/30/68, 16; *Aero Tech,* 2/12/68, 17–35)

- NASA released transcript of Jan. 27 background briefing on NASA FY 1969 budget, in which Administrator James E. Webb explained how Apollo program had been kept close to schedule in spite of severe budget cuts. Although budget had been reduced $600-million in 1964, NASA had been given great deal of flexibility to reprogram funds. "This in effect permitted the driving forward of the program even though there were substantial reductions in the plan. I think if you examine all of the very large systems development in the country, you will not find a one that has kept more on schedule, given more for the money, and more nearly realized its goals on time than the Apollo, in spite of the fire, in spite of the reductions. The kind of rolling readjustment that we have had to make . . . has involved very large problems. And the projections we made as to our capability to meet problems of that kind have pretty well been borne out. . . . The fact is within this short period of time the Saturn V has flown. The heat shield on the Apollo has been improved. Service module has been tested. LM has been tested. And we have I think moved about as rapidly as any program—certainly more rapidly than any program of a comparable complexity."

Webb said NASA and DOD were studying closer cooperation "in the area of orbiting laboratories," such as MOL and Saturn V Workshop. He envisioned production of "a basic capability of a rather primitive sort, something like an Antarctic base flown on the Saturn V which could be used for any national purpose." (Transcript)

- In statement on DOD FY 1969 budget released by Senate Armed Services Committee, Secretary of Defense Robert S. McNamara said DOD was requesting $79 million more for R&D in FY 1969 than in FY 1968 to "support more vigorously many scientific fields that show great promise and clear relevance to our future security." DOD would continue working closely with NASA to ensure maximum interchange of personnel, ideas, technology, and hardware and to avoid wasteful duplication of effort in national space program. Under new budget work on MOL development would be increased substantially. "FY 1969 is expected to be a peak year of activity in the MOL program, including the completion of a major portion of the structural test programs on flight hardware, continued fabrication of hardware for the first three flights, developmental test firings of the seven-segment solid motors for the Titan III–M, and installation of the ground equipment in the launch complex." (Text)
- Lee B. James, Deputy Director of Apollo Program in OMSF, returned to MSFC's Industrial Operations as Deputy Director for Special Activities, MSFC Director of Industrial Operations B/G Edmund F. O'Connor (USAF) announced. Before assignment with OMSF, James had been manager of Saturn I and Saturn IB programs at MSFC. (MSFC Release 68–16; *Marshall Star*, 2/7/68, 4)

January 30: President Johnson transmitted to Congress his annual report, *United States Aeronautics and Space Activities, 1967.* "The fruits of . . . technology have not been limited to space exploration alone. The knowledge built through our space program has benefited our earthbound lives. It has: revolutionized communications throughout the world; given us better weather information and more accurate navigational and geographic data; brought improved medical instruments and techniques, advanced education, and added to our store of scientific knowledge; spurred the development of more sophisticated aircraft and improved flight safety; [and] strengthened both the security of this nation and our leadership in the search for a peaceful and secure world."

The President was heartened to see that the space program had "moved forward in a spirit of international cooperation, giving new

January 30: *Engineering mockup of Saturn I orbital workshop readied for crew station review at Marshall Space Flight Center. Space-suited astronaut leaves full-size model of Saturn S–IVB rocket stage which will contain individual sleeping rooms, kitchen-dining area, laboratory work spaces, and waste-management area.*

    hope that the conquest of space can contribute to the establishment of peace." (Text)
- MSFC reported "all mission objectives were met" by Saturn IB launch vehicle used in first unmanned flight test of Apollo lunar module, Jan. 22. Flight was fourth consecutive success for Saturn IB and 15th out of 15 for MSFC-developed Saturn launch vehicles. Flight events had

January 30

- been "exactly on time or within a few hundredths of a second until second stage but still within tolerance limits." Trajectories of both stages had been within expected tolerances, and orbit insertion conditions had been met satisfactorily. (MSFC Release 68–17)
- ARC scientists Dr. Gordon W. Hodgson and Dr. Cyril A. Ponnamperuma had demonstrated by synthesizing molecules in porphyrin group, of which chlorophyll was a member, means by which chlorophyll might have first appeared on earth. By subjecting ammonia, methane, and water in a simulated primitive atmosphere to continuous lightning charges, scientists created new molecular combinations, including some tentatively identified as porphyrins. Such a process of chemical evolution to form of self-sustaining life probably occurred during billion years of earth's 4.5 billion years. Scientists confirmed synthesis with tests, but warned that materials might not be true porphyrins but structurally related compounds. (ARC Release 68–4)
- Shipment of LM-2, second Apollo lunar module, and associated Saturn IB launch vehicle to KSC had been postponed pending further evaluation of *Apollo 5* mission results [see Jan. 27 and 30]. Initial evaluations of first LM flight, Jan. 22–23, had indicated that second unmanned flight might not be required to qualify spacecraft for manned flight. NASA announced that further "detailed review of *Apollo 5* flight data and deliberations by a [NASA] design certification review board in March will determine the final decision." (NASA Release 68–21).
- Orbital workshop engineering mockup was readied for five-day crew station review Feb. 12–16 at MSFC. It was second of three design reviews. First had been conducted Dec. 11–14, 1967. In mockup, recently returned to MSFC from McDonnell Douglas Corp., astronauts would carry out tasks on ground that would later be done in orbit under zero gravity conditions, including experiment installation and operation. MSFC announced that third and final critical review in July would precede production of flight model of orbital workshop. (MSFC Release 68–18)
- NASA announced award of $7.4-million contract to Link Group, General Precision Systems, Inc., for maintenance and modification support of MSC's simulator complex, continuing services provided by Link Group since March 1964. Apollo mission simulators, at MSC and at KSC, had provided flight training for astronaut crews in which nearly every detail of flight except weightlessness could be simulated. (NASA Release 68–20).

January 31: Tenth anniversary of first U.S. satellite, *Explorer I,* launched by ABMA–JPL team with Jupiter C booster. In defiance of original prediction of three-year lifespan, 30.8-lb stovepipe-shaped spacecraft had traveled 1½ billion miles and was expected to continue orbiting through 1968, though silent. *Explorer I,* currently in orbit with 771-mi (1,241-km) apogee, 215-mi (346-km) perigee, and 100-min period, had stopped transmitting data May 23, 1958, after supplying data on cosmic rays, micrometeoroids, and temperature and discovering one of earth's radiation belts later named after Dr. James A. Van Allen, designer of satellite's radiation-detection experiment.

At National Press Club Ceremony in Washington, D.C., commemorating anniversary, Dr. William H. Pickering, JPL Director, reviewed

January 31: *Tenth anniversary of first U.S. satellite, Explorer I, shown in 1958 photo with JPL scientists before ABMA–JPL launch. Still in orbit, 30.8-lb 80-inch assembly of instrumented upper section and final-stage solid-propellant rocket had traveled 1½ billion miles. Left to right are John Small; the late Dr. J. E. Froelich, JPL project director; Dr. Albert R. Hibbs; Karl W. Linnes; and Robert Victor. (U.S. Army photo)*

American and Soviet space statistics: "The U.S. has placed about 500 vehicles in Earth orbit to 250 for the USSR. We have had 13 successful missions to the Moon, the USSR 8. Our reconnaissance spacecraft have obtained about 100,000 high-resolution photographs of the Moon; USSR craft have returned about 100. The U.S. has had three successful missions to the near planets in five attempts. Russia has successfully landed an instrumented package on Venus after an estimated 19 planetary attempts. We have accumulated nearly 2,000 hours of manned space flight time, compared with about 530 for the USSR. Our astronauts have performed multiple rendezvous and docking operations; the Soviets have demonstrated one automatic docking. We have logged over 12 hours of extravehicular activity; the USSR shows 20 minutes."

Dr. Wernher von Braun, MSFC Director, critical of cuts in NASA budget, noted: "NASA has not been able to plan for post-Apollo projects with any degree of certainty. NASA has attempted production of the Saturn IB and Saturn V at the rate of two each per year. Frankly, this rate is too low to maintain the progress and momentum so painstakingly acquired in the space program over the past decade.

"The dismantling of the high competence built up over the years at JPL and [MSFC] has already begun. Both organizations are already losing valuable, highly trained people because of the insecurity and the lack of challenging work for them to do. And we face the grim reality of even further reductions and cutbacks. . . . The exploration of space has proven its worth, and has become deeply ingrained into the everyday fabric of our society. . . ." Space program would be examined closely by Congress, Dr. von Braun said, but "NASA's record of accomplishments and the ability of its management will bear up well under the scrutiny. In determining what our space program should be during the coming years . . . members of Congress . . . [should] give careful consideration to where we are today in space, at the end of our first decade, and where we could be ten years from now, by building on the accomplishments of the past." (EH; Texts; MSFC Release 68–14; Clark, *NYT*, 2/1/68, 15)

- NASA announced end of Lunar Orbiter project. *Lunar Orbiter V*, launched Aug. 1, 1967, had obeyed spacecraft controllers and crashed on lunar surface after 1,200 orbits. Final operation of spacecraft had consisted of 18.9-sec burn of velocity control engine at apolune to decrease speed by 64 mph for orbit decay. Impact occurred at 2:58.5 am, EST, near equator on moon's western limb. Destruction of *Lunar Orbiter V* had become necessary because its supply of attitude control gas was nearly depleted. In concentrating on 36 areas of moon's face, spacecraft had completed coverage necessary for full photographic survey of moon's hidden side, photographing and transmitting 212 telephoto and 212 wide-angle pictures of lunar surface. Spacecraft had also been used to provide tracking target to MSFN stations and for crew training and computer program verification.

Lunar Orbiters I, II, and *III* had been crashed on Oct. 29, 1966, Oct. 11, 1967, and Oct. 9, 1967, as their control gas became exhausted. Communications were lost with *Lunar Orbiter IV*, and it had apparently crashed on lunar surface. During Lunar Orbiter project, since first launch Aug. 10, 1966, 6,034 orbits of moon had been completed and more than 99% of lunar surface had been photographed by orbiting cameras to provide data for selection of eight candidate landing sites for Apollo project. Lunar Orbiter program was directed by OSSA and managed by LaRC. Boeing Co. was prime spacecraft contractor. (NASA Proj Off; NASA Release 68–23)

- Apollo 204 accident was reviewed in Senate Report No. 956, issued by Senate Committee on Aeronautical and Space Sciences: "The thorough investigation by the Apollo 204 Review Board . . . determined that the test conditions at the time of the accident were extremely hazardous. However, the test was not recognized as being hazardous by either NASA or the contractor prior to the accident. The committee can only conclude that NASA's long history of successes in testing and launching space vehicles . . . led to overconfidence and complacency." Committee recommended that "safety . . . be considered of paramount importance in the manned space flight program even at the expense of target dates" and urged NASA to keep "appropriate Congressional committees informed on significant problems arising in its programs."

    Individually, three Committee members—Sen. Walter F. Mondale (D-Minn.), Sen. Edward W. Brooke (R-Mass.), and Sen. Charles H. Percy (R-Ill.)—filed more critical separate reports. Sen. Brooke and Sen. Percy stated: ". . . NASA's curious reticence to supply these facts and figures [Phillips Report] relevant to a thorough evaluation of Apollo program management brought the credibility of NASA and its top management into sharp question [and] this initial lack of candor as to the existence and then the status of the Phillips report threatened one of the essential assets of the space program—the confidence of the American public and their elected representatives." Sen. Mondale said NASA "has an unfortunate habit of swamping Congress with engineering details and starving it for policy and management information. . . . Congress should be able to count on frank answers to pertinent, responsible, and legitimate inquiries."

    Criticizing Committee's report, Rep. William F. Ryan (D-N.Y.) said in news release later, "Although the Senate Committee does charge NASA with such misdemeanors as overconfidence and complacency, the report remains a whitewash and an evasion of the real issues involved." (Texts)

- Senate Report No. 957, *Aeronautical Research and Development Policy*, sponsored by Senate Committee on Aeronautical and Space Sciences, concluded NASA should increase its aeronautical effort, particularly in the development phase, and carry it on to "proof-of-concept testing as a means of providing a larger variety of options to aircraft designers and systems engineers." Report called for NASC to act as focal point for development of more comprehensive and better coordinated aeronautical R&D policy and recommended that NASA's aeronautical activity be upgraded to major office level and directed by an Associate Administrator. (Text)

- In *Annual Report to Congress of the Atomic Energy Commission for 1967*, AEC announced fifth launch of twin Vela nuclear detection satellites was planned, with increased detection capabilities. Earth-oriented spacecraft would be launched by Titan III–C booster.

    Among other programs, large heat source for power system for space applications in mid-1970s was being studied at NASA's request, with scheduled July 1971 delivery of 25-kw heat-source subsystem consisting of high-temperature fuel capsules integrated into safe flight package. Second Phoebus test reactor, Phoebus-2A, was being assembled at Nuclear Rocket Development Station in Nevada; testing was scheduled to begin in second quarter of year. Work on XE cold-flow engine and testing of first "hot" (uranium-fueled) ground experimental engine were scheduled for second quarter. Power testing of second-generation SNAP-8 reactor was planned and more advanced reactor systems for space use were being examined, including thermionic fuel elements.

For treatment of "heart block," AEC was developing plutonium-fueled cardiac pacemaker. Studies were also under contract on isotopic "engine" for pumps to assist or replace functions of damaged heart. (Text; *InteraviaAirLetter,* 2/1/68, 11)

- Fifth anniversary of ComSatCorp was noted by Sen. Warren G. Magnuson (D-Wash.), Chairman of Senate Committee on Commerce, on floor of Senate: "Intelsat has grown from its original membership of 11 nations to a total of 61, including 40 which qualify as among the less developed countries.... The commercial utilization of space for communications purposes—a dream for the future when the Congress passed the Satellite Act—is today a reality." (NASA *LAR* VII/7; *CR,* 1/31/68, S676)

- President Johnson announced his intention to nominate Dr. Thomas O. Paine, manager of General Electric Co.'s Technical Military Planning Operation (TEMPO), as NASA Deputy Administrator to succeed Dr. Robert C. Seamans, Jr., who resigned Jan. 5. (*PD,* 2/5/68, 176)

- NASA announced award of $1,769,200 contract to Lockheed Missiles & Space Co. for "adapting the Agena second stage rocket for use on the SERT II mission to test ion engines in earth orbit." SERT II (Space Electric Rocket Test) for at least six months of flight would evaluate "in-flight performance of electron bombardment ion engines ... and analyze the possible effects of the electric thrustors and their associated electric fields on other spacecraft components." Entire Agena 2nd stage would be used in 1969 launch as bed of spacecraft SERT II; Agena would be equipped with 1½-kw solar cell array to provide power for ion engines and other systems. LeRC had management responsibility for launch vehicle and spacecraft in SERT II mission. (LeRC Release 68–8)

- NASA personnel changes: Thomas B. Shillito of LeRC's Office of Development Evaluation and Management Review was appointed Supersonic Transport Program Coordinator for NASA in Washington, D.C. He would have offices at FAA and would "stay abreast of technical progress in the SST program [and] inform NASA ... of any impending requirements." Shillito would report to NASA Deputy Associate Administrator for Advanced Research and Technology (Aeronautics) Charles W. Harper.

Harvey Sherman and James M. Beggs were named consultants to NASA Administrator James E. Webb. Sherman, Port of New York Authority's Director of Organizations and Procedures Dept., would be member of senior group of advisors on management matters. Beggs, Westinghouse Electric Corp.'s Assistant Director of Purchases and Traffic, would advise on management concepts and policies. (LeRC Release 68–7; NASA Release 68–22)

During January: New knowledge of movement and state of electrons in Van Allen radiation belts that explained movement of electrons into and out of belts and trapped electron activity was reported from three new discoveries. Bell Telephone Laboratories used energetic particle collectors aboard NASA's *Ats I*, launched Dec. 6, 1966, to show that high-energy outer Van Allen belt electrons exhibited bunching phenomenon as they drifted around earth in its magnetic field. Bell scientist Charles S. Roberts presented theory that "radio noise in the charged particle gas" surrounding earth was "responsible for the loss of electrons from the Van Allen belts, rather than the mere presence of lightning-bolt-initiated whistlers alone, as previously believed." At Nov. 16, 1967, meeting, Dr. James A. Van Allen had presented satellite-collected data suggesting that "both electrons and protons enter the outer Van Allen radiation belt" drawn by high voltage generated across earth's magnetic field, caused by earth's spin between dawn and evening sides of earth-surrounding magnetic envelope. (Strasser, *Aero Tech*, 1/29/68, 18–20)

- U.S. Arms Control and Disarmament Agency released *Technological Innovation in Civilian Public Areas*, study by Analytic Services Inc. which urged Government to continue support of advanced R&D on domestic problem-solving to prevent economic decline if disarmament agreement were reached. Unless such R&D was funded, study said, it was questionable whether large demonstration projects to alleviate domestic problems could be undertaken quickly enough to offset substantial reductions in defense spending. Defense industry could help solve civilian programs, study concluded, but it questioned whether industry noted for technological innovation would not encounter serious obstacles in orienting its activities to areas which often had resisted innovation in any form. (Text; *Aero Tech*, 1/15/68, 43)

- India could be first developing country to orbit her own artificial satellite, according to recommendations of five-man mission of UNESCO. Satellite would be first of "distribution" communications design, more powerful than current point-to-point satellites such as *Early Bird I* and requiring less complex ground facilities. Operating under long-term plan to end in 1981, India would build 56 ground stations which would bring TV to 25% of population. Network of 160 stations could reach 80% of India's population. Mission recommended India produce 50,000 home TV receivers by 1970–1971 and enter into bilateral agreement with major space power to purchase satellite and launching services. (*NYT*, 1/28/68)

- Payoff from space investment (in satellite applications in extended weather forecasting, reduced cost, and extended range for telecommunications) was discussed in *Space/Aeronautics*. Government policy on future satellite applications would depend on four "key study efforts": (1) study by Presidential Task Force on Communications Policy on status of INTELSAT and its role in global and foreign domestic comsat systems; (2) study by NAS on potential of applications satellites and "an effective cost/benefit strategy"; (3) recommendation by President's Commission and Council on Marine Sciences for consolidation of Federal efforts in marine engineering and oceanography, and creation of policy framework hospitable to applications satellites; and (4) European organization of firms' formulation of arguments advocating four regional systems rather than one global network. (*S/A*, 1/68, 80)

- History of lunar nomenclature and tradition in naming lunar features was reviewed by Director of Fels Planetarium, Dr. I. M. Levitt, in *Air Force and Space Digest*. Knowledge from space research had created complications for specialists who had to name lunar features. In past, craters, mountains, rills, and seas on moon had been named for scientists, mathematicians, and philosophers. In 1647, Johannes Hevelius had published *Selenographica* in Danzig, setting stage for naming lunar features. Ground rules were set down in 1932 by International Astronomical Union (IAU), which formed committee of astronomers to select names.

U.S.S.R. had asked 1967 General Assembly of IAU to pass on 153 more names for features on moon's hidden side. Levitt recommended that names of dead astronauts—both American and Russian—be accepted for lunar nomenclature; later, living astronauts could be similarly honored. Also pioneers in space research—"such as Goddard, von Braun, Gilruth, Newell, Ley, Lovelace, Flickinger, Stehling, and so on" —and astronomers and benefactors of mankind could be considered. (*AF/SD*, 1/68, 66-9)

- William Leavitt in *Air Force and Space Digest* asked: "Are antiballistic missile systems necessarily provocative and 'destabilizing' in the sense that they will inevitably set off a new spiral in the arms race? Many— but by no means all—scientists answer 'yes' to these questions, despite impressions to the contrary." Director of Oak Ridge National Laboratory Dr. Alvin Weinberg, who would answer "no" to ABM question, was quoted: "The deployment of ABMs on both sides has been deplored as the first step in the unending arms spiral that eventually will consume everything, including our vision of abundance. But suppose ABMs and other defensive measures turn out to be effective, and at the same time there is no escalation of offense in unending spiral. The knife-edge of delicately balanced terror would then be blunted." (Leavitt, *AF/SD*, 1/68, 72-4)

February 1968

February 1: NASA Nike-Cajun sounding rocket launched from NASA Wallops Station carried payload to 65-mi (104.4-km) altitude to obtain vertical profile of atmosphere parameters between 22- and 59-mi (35- and 95-km) altitudes with exploding-grenade and falling-sphere techniques. Rocket and instrumentation performed satisfactorily except for one of the two spheres; other sphere was tracked for 23 min. Globe Exploration Corp. and Superior Engineering Co. experiments were correlated with firings from Kiruna, Sweden; Fort Churchill, Canada; and Point Barrow, Alaska. (NASA Rpt SRL)

- *Ats III*, launched by NASA Nov. 5, 1967, was again returning color and black-and-white pictures to ground stations; color camera had been turned off Nov. 29, and black-and-white camera, Dec. 9. Latter had been returned to active photographing of cloud cover Dec. 30. Color camera had operated since first week of January, but ground equipment had lacked precise adjustment. GSFC believed camera problems, possibly caused by gas seepage, had been corrected. (NASA Release 68-24)

- Secretary of Defense Robert S. McNamara, in annual report to Congress, stated before joint session of Senate Armed Services Committee and Defense Appropriations Subcommittee that U.S.S.R. had nearly closed nuclear missile gap in 1967 by more than doubling its IBM force. According to intelligence estimates, Soviet buildup could give U.S.S.R. capability of delivering about 1,000 nuclear warheads and bombs; U.S. could deliver up to $4\frac{1}{2}$ times that number. Both U.S. and U.S.S.R. possessed strategic forces capable of withstanding surprise attack and retaliating overwhelmingly against the other. "It is precisely this mutual capability to destroy one another, and conversely, our respective inability to prevent such destruction, that provides us both with the strongest possible motive to avoid a strategic nuclear war." Communist China's medium-range missiles, and ICBMs could be operational as a "modest" force in mid-1970s. (Beecher, *NYT*, 2/2/68, 2; Marder, *W Post*, 2/2/68, A1; *NYT*, 2/6/68, 42)

- V/A Charles E. Weakley (USN, Ret.), former Commander of U.S. Atlantic Fleet Anti-Submarine Warfare Force and recipient of Navy Distinguished Service Medal, became NASA Assistant Administrator for Management Development and took over staff of Organization and Management Planning Div., formerly under Associate Administrator for Organization and Management. (NASA Release 68-11)

February 1-2: Tenth anniversary of *Explorer I* [see Jan. 31] was commemorated by JPL and Cal Tech with two-day program including symposium on First and Second Decades of Space Research and JPL open house. JPL Director, Dr. William H. Pickering, leader of first Explorer scientific task force, was chairman. Speakers included Dr. James A.

Van Allen of State Univ. of Iowa, who had verified radiation belts bearing his name; L/G Austin W. Betts (USA), Army Chief of Research and Development; Dr. John W. Findlay, NASA Lunar and Planetary Missions Advisory Board Chairman; and Dr. Joseph Kaplan of UCLA, U.S. Chairman of IGY when *Explorer I* was launched Jan. 31, 1958.

In next 10 yr, "as an increasingly important area of scientific and technical research a balanced program in space must certainly be part of the activities of our Nation," Dr. Kaplan said. "No more noble and useful service can be conceived than one which gives strength to . . . cooperation among the nations of the earth. Space certainly qualifies for such a role."

Dr. Findlay outlined possible missions "of strong scientific interest" to Mercury, Venus, Mars, and Jupiter in next decade. Gen. Betts voiced concern about attitudes toward exploratory efforts; he was "not convinced that we remember, the lesson" of *Sputnik I*. "Only some two weeks" after its launch, numerous programs were proposed for U.S. satellite—possible only because "far-sighted technical people had been planning for various applications of space for some time, and even though firm military requirements were *not* available." (JPL Releases, 1/22/68, 1/31/68; Texts; Wilford, *NYT*, 2/2/68, 4; *SBD*, 1/26/68, 38)

February 2: NASA launched two Aerobee 150 sounding rockets from WSMR. First carried American Science and Engineering, Inc., experiment to 86.5-mi (139-km) altitude to collect data on location and flux levels of celestial x-ray sources in 1- to 20-kev range using collimator, proportional counters, aspect camera, and attitude control system (ACS). ACS performance was poor because of timer malfunction; performance of rocket and instrumentation was satisfactory.

Second rocket carried GSFC payload to 92.8-mi (149.3-km) altitude to measure spectral irradiance of four early-type stars in 1,100- to 4,000-Å interval, using UV stellar spectrometer and STRAP system for attitude control. Rocket, instrumentation, and STRAP performed satisfactorily. Good data were obtained. (NASA Rpt SRL)

- Sen. Clinton P. Anderson (D-N.Mex.), Chairman of Senate Committee on Aeronautical and Space Sciences, on behalf of himself and Sen. Margaret C. Smith (R-Me.), ranking member, introduced S. 2918, FY 1969 NASA authorization bill. Measure was referred to Committee. Total authorization of $4.37 billion would provide R&D, $3.677 billion; construction of facilities, $45 million; and administrative operations, $648.2 million. (NASA *LAR* VII/8)

- S–II 2nd stage for fifth Apollo Saturn V mission left Seal Beach, Calif., onboard USNS *Point Barrow* en route to Mississippi Test Facility (MTF), where stage would undergo static testing before shipment to KSC. Also onboard ship, to save $6,000 in transportation charges, was F–1 rocket engine. Engine would be unloaded for inspection at Michoud Assembly Facility in New Orleans, then transferred to barge for remainder of trip to MTF. (MSFC Release 68–23)

- India's Thumba Equatorial Rocket Launching Station (TERLS) was dedicated as international facility by Prime Minister, Mrs. Indira Gandhi. NASA was represented by Assistant Administrator for International Affairs Arnold W. Frutkin, Director of Space Applications Programs Leonard Jaffe, and NASA Wallops Station Assistant Chief of Flight Test

Div. Robert T. Duffy. Duffy had served as advisor, and Wallops had assisted by helping to develop range, training scientists and engineers, and lending launching and tracking equipment.

TERLS had been established Nov. 21, 1963, by Indian National Commission for Space Research (INCOSPAR) and to date had launched 52 rockets: 17 American Nike-Apaches, 29 American Judi-Darts, and 6 French Centaures. Rockets launched as part of dedication included two Nike-Apaches, one Judi-Dart, and one Centaure. Payloads contained experiments involving French, Soviet, and American collaboration with India. India's Dept. of Atomic Energy planned Centaure manufacturing unit at TERLS, and INCOSPAR would build space science and technology center at Veli (near TERLS) to design and develop space research systems, including vehicle, payload, instrumentation, and ground support.

Mrs. Gandhi said: "This center represents only one facet of the tremendous industrial and technological revolution in which we are involved. . . . Technology is a key. It is a key to knowledge that opens the door to plenty as well as power. In a sense it is a key to independence, for it was the failure to advance technologically which made Asia and Africa dependent and poverty-stricken." (NASA Release 68–14; WS Release 68–1; *India News*, 2/2/68, 4; *LA Times*, 2/3/68; B *Sun*, 6/30/68)

- Cornell Univ.'s association with Cornell Aeronautical Laboratory had been ended by Univ.'s Board of Trustees on recommendation of special committee headed by Univ.'s law professor W. David Curtiss. Reasons cited included overlap and potential conflict between Laboratory's overseas research projects and Univ.'s expanding program of international studies. (*Science*, 2/2/68, 515)

February 3: Clifford C. Furnas, President of Western New York Nuclear Research Center and member of NAE, was appointed to newly created position of Vice Chairman of National Research Council (NRC), in which he would "assist the Chairman, NAS President Frederick Seitz, in the general administration of the NRC and help coordinate activities" between NAE, NAS, and NRC. (NAS–NRC–NAE *News Report*, 3/68, 5: NAS PIO)

February 5: NASA Tomahawk sounding rocket launched from NASA Wallops Station carried GSFC payload to 73.3-mi (118.5-km) altitude to evaluate vehicle performance and measure payload environmental instrumentation, including three-axis acceleration, lateral and longitudinal vibration, three-axis gyro, 10 temperatures, and motor pressure. Rocket performed as predicted and was recommended for acceptance by NASA. All instruments operated to impact. (NASA Rpt SRL)

- Nike-Tomahawk sounding rocket launched by NASA from Churchill Research Range carried Univ. of New Hampshire-Univ. of California at San Diego payload. Objectives were to detect and identify energetic charged particles that caused aurora, measure local electron density and correlate it with ground measurements of ionization, study aurora light emissions at several wavelengths and detect related electric fields, and investigate auroral ionospheric currents and magnetic fields driving them in ranges 1–15 kev and 2–30 kev and protons from 10–40 kev with electrostatic curved plate analyzers and channel multipliers. Launch was first in series of five to be launched from Churchill. Exact performance was not calculated because of early radar loss; time correlated performance indicated rocket functioned as predicted. Telemetry

February 5

signals were strong throughout flight. All experimental objectives were achieved. (NASA Rpt SRL)

- Dr. Edward C. Welsh, NASC Executive Secretary, testified before House Committee on Appropriations' Subcommittee on Independent Offices and HUD that there would be no duplication of experiments on DOD's Manned Orbital Laboratory (MOL) and NASA's Orbiting Workshop. They would operate in different orbits. NASC, Welsh said, was stressing "those types of space activities from which we can see relatively direct benefits to the general public." (Transcript; *Aero Daily*, 4/23/68, 11–F)
- Rep. George P. Miller (D-Calif.), Chairman of House Committee on Science and Astronautics, introduced H.R. 15086, FY 1969 NASA authorization bill. Measure was referred to Committee. (NASA *LAR* VII/9)

February 6: U.S.S.R. launched *Cosmos CCI* into orbit with 327-km (203-mi) apogee, 202-km (126-mi) perigee, 89.7-min period, and 64.9° inclination. Satellite performed satisfactorily and reentered Feb. 14. (*SBD*, 2/7/68, 210; GSFC *SSR*, 2/15/68)

- Senate confirmed nomination of Dr. Thomas O. Paine as NASA Deputy Administrator [see Jan. 31]. (Transcript; NASA *LAR* VII/10; UPI, *NYT*, 2/8/68, 4)
- NASA announced appointment of five members to its Aerospace Safety Advisory Panel: M/G Carroll H. Dunn, Director of Military Construction in Office of Chief of Army Engineers; Dr. J. A. Hornbeck, President of Sandia Corp.; Dr. Henry Reining, Jr., Dean of von KleinSmid Center of International Affairs, USC; Dr. Eberhard F. M. Rees, Special Assistant to Apollo Spacecraft Manager (detached from position as Deputy Director, MSFC); and Bruce T. Lundin, Associate Director for Development, LeRC. Appointments implemented Sec. 6, P.L. 90–67, NASA Authorization Act for FY 1968. (NASA Release 68–26)
- Dramatic reversal in balance of U.K. aerospace trade with U.S. had occurred in 1967, *InteraviaAirLetter* reported. Whereas in 1966, U.K. exports to U.S. of $187 million (predevaluation) exceeded imports by a favorable $137.2 million (predevaluation), U.K. aerospace industry's 1967 dollar earnings of $118.3 million (new rate) were swamped by imports from U.S. totaling $149.3 million (new rate). This was highest figure ever recorded and included $127.2 million for aircraft and $22.1 million for engines. (*InteraviaAirLetter*, 2/6/68, 1)

February 6–8: Five nations—Brazil, India, Italy, Romania, and Sweden— had misgivings about draft treaty to prohibit further spread of nuclear weapons, submitted jointly at Geneva disarmament conference by U.S. and U.S.S.R. Jan. 18.

On Feb. 8, India, through authoritative source, announced she feared neither U.S. nor U.S.S.R. would supply "an airtight guarantee of its territorial integrity and inviolability once Communist China achieves a second-strike intercontinental nuclear capability," according to *Los Angeles Times*. Sweden appealed to U.S.S.R. to match offers by U.S. and U.K. to permit international inspection of their peaceful nuclear activities. (Hamilton, *NYT*, 2/7/68, 8; *LA Times* in *W Post*, 2/9/68; *NYT*, 2/9/68, 16)

February 7: NASA Nike-Cajun sounding rocket launched from WSMR carried special parachute and payload to 41.7-mi (67-km) altitude to test parachute deployment at high altitude and to determine ozone distribution in 12- to 40-mi (20- to 65-km) region. Rocket and instruments performed satisfactorily; parachute deployed at approximately 216,000 ft. Recovery attempts, initially unsuccessful because of 65-mi off-range drift, would be continued in attempt to recover onboard camera. (NASA Rpt SRL)

- President's authorization request for NASA's FY 1969 budget was presented by NASA Administrator James E. Webb to House Committee on Science and Astronautics. At $4.37 billion, authorization request was $700 million below FY 1968 request, almost $500 million less than FY 1968 authorization, and $200 million below FY 1968 appropriation. Webb stated: "NASA's expenditures for FY 1969 will be down $230 million from this year, $850 million below last year, and $1.3 billion less than in FY 1966. The NASA program has been cut. I hope you will decide it has been cut enough and will approve the full amount recommended by the President.

"The FY 1969 request does not meet at all our Nation's needs in aeronautics and space. It is a compromise—one which I fully support—between needed work toward advances in aeronautics and space which we can and should make, and other overriding requirements. . . . During this period when we are reducing our effort by one-third the U.S.S.R. is still increasing its effort. . . . In terms of scientific advance and in applications of immediate economic use, such as meteorological and communications systems, our program has contributed more than theirs. But in terms of the use of large launch vehicles and in the rate of which future greater capabilities are being developed they are and will remain ahead, at our 1969 budget level. The hard fact we now face is that just as we have begun to catch up in large-scale booster operations . . . we are cutting back our program while they continue to advance."

He listed guidelines within "necessary fiscal constraints," which determined decisions on FY 1969 program and budget estimates. NASA would continue 1968 operating plan for Apollo program while using all-up test concept for Saturn V and Apollo. Complete success in all remaining eight Saturn V flights would be needed for manned lunar landing by "end of 1969." NASA would reduce numbers of Saturn IB and V and work out with DOD future requirements for large launch vehicles under "sharply limited" launch program. Launch of Saturn I Workshop in 1970 would be interim step toward Saturn V Workshop and would be coordinated with later launch of Apollo Telescope Mount. "Practically all" programs had been stretched out, and a number of projects would be phased out.

Scope of new starts would be reduced, with NERVA I replacing larger NERVA II and "less costly Mariner class spacecraft" replacing Voyager mission. Urgent activities would be expanded in aeronautics and in space applications; for example, noise reduction, VTOL and STOL, aircraft for remote measurement of earth resources, and Nimbus D for charting atmosphere parameters.

NASA Centers would be reorganized "to retain . . . after reductions [in work force] are made, a limited but strong and well-balanced team of scientists, engineers, program and project managers."

Webb responded to questions: "There are no signs that the Soviets are cutting back as we are. New test and launch facilities are steadily added . . . and a number of spaceflight systems more advanced than any heretofore used are nearing completion." Webb forecast Soviet booster with thrust greater than that of Saturn V, resumption of manned space flights, and landing of Soviet instrument packages on Mars in 1969 and 1971. Budget reductions in FY 1968 had compelled revisions, reductions, delays and cancellations in NASA programs. "We can conduct a viable and useful program at these lower levels, but it will be a sharply reduced one."

Despite reductions allowing almost no flexibility to work around problems, "we still have the possibility of making the lunar landing before 1970." (*CR*, 2/7/68, H926–7; 2/12/68, E708–9)

- Testimony on NASA's FY 1969 Budget request was presented to House Committee on Science and Astronautics by three NASA Associate Administrators.

  NASA Associate Administrator for Space Science and Applications, Dr. John E. Naugle, presented OSSA's budget for "an austere program for the immediate future," yet one that would make "effective use in the years ahead of the capabilities we have developed in the past decade" and that would advance scientific knowledge and useful applications. "New starts are at a minimum," he said, "but if this budget request is approved, we can continue the most useful of our on-going programs, we can avoid abandoning the field of planetary exploration, and we can hold together teams and capabilities to meet future national needs. . . . " Authorization request for OSSA for 1969 was for $538.2 million, $14.7 million below FY 1968 operating plan, with "the decrease . . . primarily in the Lunar and Planetary Programs budget, reflecting the completion of the initial automated phase of lunar exploration and cancellation of Voyager." Space applications budget request had increased for FY 1969 (to $112.2 million) because of "demonstrated value of communications and meteorological programs and the potential value of applying space technology on a global basis."

  NASA Associate Administrator for Advanced Research and Technology, Dr. Mac C. Adams, described OART program as "the key to tomorrow's leadership in aeronautics and space." He explained: ". . . our FY 69 budget request for the R&D line items . . . is $336,800,000, up 5.7 percent over FY 68. The increase is primarily due to increases in effort on aircraft technology [up to $10.1 million] and nuclear rockets [up $6 million]. . . . Major changes also included more V/STOL research, noise research for subsonic jets, and development of NERVA I nuclear rocket engine at thrust level of 75,000 lb.

  NASA Associate Administrator for Tracking and Data Acquisition Gerald M. Truszynski presented FY 1969 budget request of $304.8 million for R&D and $21.8 million for construction of facilities and reported that "50 satellites by the end of 1968" would be supported by network while "some 16 new missions" would be launched in FY 1969. For FY 1969, "the great majority of funds requested will be directed toward operating and maintaining present network capability. Much of the existing equipment has been operated almost continuously for at least eight years and must be refurbished or replaced."

  Construction of facilities budget would fund: two 210-ft-dia antennas

for Spain and Australia; power generating plant for Fairbanks, Alaska, tracking station; and Sunblazer antenna. (Testimony)
- NASA awarded Boeing Co. $3,226,374 supplement to its Saturn V systems engineering and integration effort, bringing this portion of Boeing's three-part Saturn V contract to $194.8-million total. Boeing would be responsible for propulsion systems' preflight and postflight performance analysis for first 10 launch vehicles. (MSFC Release 68–28)

February 8: NASA management officials continued to testify on their programs, within NASA's FY 1969 budget request, to House Committee on Science and Astronautics.

NASA Associate Administrator for Manned Space Flight, Dr. George E. Mueller, said OMSF's $2.815-billion FY 1969 request was lowest request since FY 1963, $332.8 million below FY 1968, and "almost $700 million below the peak Manned Space Flight budget year, Fiscal Year 1966." Dr. Mueller included $2.483 billion for R&D, $18.7 million for construction of facilities, and $313 million for administrative operations. Over 80% of total R&D report was committed to Apollo. "The program of Apollo and Apollo Applications Flights, with the development of the Saturn V Workshop and the Advanced Missions studies leading to a space station of longer duration and greater earth orbital capabilities, constitute a minimum effort to continue manned space flight into the early 1970s." Dr. Mueller warned against failing "to capitalize on the investment in the Apollo Program" and said it was "economically prudent to proceed now with the Apollo Program and the definition study of the . . . long duration space station . . . keystone in the future of manned space flight."

NASA Associate Administrator for Organization and Management Harold B. Finger, citing 27% decrease in procurement for first three months of FY 1968 below same period in 1967, predicted 94% of procurement dollars going to industry would decrease to less than 90% by end of FY 1969 as industry-developed systems came into NASA for testing and launch. Presenting $10-million Sustaining University program budget—equal to 1968 budget but one-third 1967's $30 million—Finger stressed need for university contributions and combining universities' "demonstrated competence in physical sciences with their social science strengths to deal with the total impact of the nation's aerospace effort." Finger submitted statement of Dr. Richard L. Lesher, Assistant Administrator for Technology Utilization, describing programs "to experimentally test and develop methods for bringing about the multiple use of the new knowledge gained as a result of NASA activities."

NASA Assistant Administrator for Administration William E. Lilly charged $7.8-million increase in FY 1969 administrative operations budget request (from $640.4-million budget for FY 1968) to cost of 1967 Pay Act and conversion to civil service of certain GSFC support services. He cited personnel reduction of 1,704 positions since freeze on hiring new employees on Aug. 24, 1967. (Testimony)
- Five ellipse-shaped, three- by five-mile lunar landing areas were selected for astronaut safety considerations by NASA Apollo Site Selection Board. First two sites were in Sea of Tranquility, third was in Central Bay, and fourth and fifth were in Ocean of Storms. Sites contained features of scientific interest, including small craters, raw material ejected

from craters, ridges, and faults. Criteria for selection were area smoothness, approach path, propellant conservation, countdown recycling time, free-return trajectory, optimum lighting, and slope. Board had studied Lunar Orbiter high-resolution photos and Surveyor close-up photos and surface data. (NASA Release 68-25; AP, *NYT*, 2/5/68, 13)

- NAS released *Space Applications Summer Study, 1967 Interim Report*, Volume I, Central Review Committee Conclusions and Recommendations and Summaries of Panel Reports. Chaired by Dr. W. Deming Lewis, study treated aspects of space technology "likely to produce practical benefits to large segments of the American and world economies." Report urged extensive and coherent program and said NASA should double $100 million it currently spent to develop technology of space applications. (Text; Clark, *NYT*, 2/9/68, 6)
- NSF announced scientists and engineers from USA's Cold Regions Research and Engineering Laboratory in Hanover, N.H., had for first time successfully drilled through Antarctic icecap, obtaining continuous cores to depth of 7,100 ft at Byrd Station. Later studies of cores and samples of underlying rock were expected to provide insight into climatic and atmospheric history. Engineers had reached their immediate goal Jan. 29; drilling would be resumed in October. (NSF Release 68-108; Reinhold, *NYT*, 2/8/68, C3)
- Soviet Prof. Georgy Petrovich, writing in *Trud*, praised U.S. space achievements but pointed to 20 Soviet "firsts" since *Sputnik I*. Past neutral appraisals had generally agreed with Petrovich that U.S.S.R.'s payloads had been heavier and U.S. hardware more sophisticated. Recent contradiction had been intricate softlanding for Soviet *Venus IV*, Oct. 18, 1967, and successful U.S. test of Saturn V, world's largest booster, Nov. 9, 1967. Petrovich praised NASA's broad application of space equipment in practical commercial use of satellite technology for communication, meteorology, and navigation. (Winters, B *Sun*, 2/9/68, A3)

February 9: Aerobee 150 sounding rocket launched by NASA from Churchill Research Range carried Univ. of Pittsburgh-Johns Hopkins Univ. payload to 96.3-mi (155-km) altitude to perform coordinated auroral experiments with spectrometers and photometers. Rocket and instruments performed satisfactorily. (NASA Rpt SRL)

February 10: Second-stage (S-II) engine for fourth Apollo/Saturn V mission was successfully static-fired for its full duration of six minutes at MTF, with engine developing equivalent thrust of over 1 million lb. (*Marshall Star*, 2/14/68, 1)

- Sixteen scientist-astronauts, from two groups selected in June 1965 and August 1967, had complained of dimming prospects for scientific space work because of slowdown in spaceflight schedule. UPI said astronauts also feared falling behind their colleagues outside space program because flight training left little time for scientific research. MSC Public Affairs Officer Paul Haney had announced designation of MSC's Science Director Dr. Wilmot N. Hess to find means for relieving complaints. (UPI, *NYT*, 2/11/68, 55)

February 11: Electric signals out of phase with jet engine noise could be introduced into exhaust flame and converted into sound waves to cancel out engine noise, according to United Technology Center scientist, Dr.

A. G. Cattaneo. Idea, in principle, had been demonstrated by recent laboratory experiments in which acetylene torch flame converted electric signals into sound with high fidelity. Evaluation of combustion efficiency during rocket engine tests could be additional application. (Wilford, *NYT*, 2/11/68, 30)

February 12: GSFC technicians had moved NASA's *Ats III* satellite to new position over Pacific off Ecuador's coast to aid in U.S. Coast and Geodetic Survey (C&GS) mapping project. After movement from position on Equator at 95° west longitude to location at 85° west longitude, satellite could provide precise timing signals to C&GS team on South Atlantic islands of South Georgia and Tristan de Cunha. *Ats III*, carrying communications, meteorological, and navigational experiments, had been launched Nov 5, 1967, into synchronous equatorial orbit at 22,300-mi altitude. (NASA Release 68–30)

- "Space Activities and the National Well-Being" were discussed in speech before Detroit's Economic Club by Dr. George E. Mueller, NASA Associate Administrator for Manned Space Flight. "It is a measure of how far we have come in space in 10 years," he said, "to consider that, with the flight of the Apollo-Saturn V last November, and the flight of the lunar module on January 22, we have now successfully flight-tested and proven every piece of the Apollo-Saturn space vehicle, the vehicle the United States will use . . . for the exploration of space for years to come."

  He warned that while U.S. was decreasing its rate of investment in space activities, "Soviets are spending 50 per cent more in terms of real purchasing power than we are. In percentage of gross national product, they are spending 2 or 3 times as much as we are."

  He noted that "when we launched our first satellite, Explorer I, the cost was several millions of dollars per pound of payload to get it up in orbit. With the Saturn V, we are now able to put payloads into orbit for only about $500 per pound. Considering that reduction from millions of dollars per pound to hundreds of dollars per pound in orbit over a period of only 10 years, one can foresee reducing the cost to $50 per pound or even $5 per pound for getting into and out of orbit."

  Saturn V development should move U.S. ahead during next year, but U.S.S.R. would soon surpass U.S. "as the full impact of the resources they are putting into their space program at this time becomes evident. And I would expect that by 1971 or 1972 we will be significantly behind them in terms of all of the aspects that we now know will be characteristic of space flight." (Text)

- Radar map of planet Venus had been produced by Cornell Univ. scientists from computer-assembled data. Radar-astronomy data had been gathered by world's largest radiotelescope at Arecibo Ionospheric Observatory in Puerto Rico. Principal scientist Raymond F. Jurgens said map was equivalent to what largest optical telescope on earth could obtain were it not for thick cloud cover obscuring Venusian surface. Measurements had been made in 1964 and 1967 when Venus was nearest earth, 26 million miles away. Cornell's Center for Radiophysics and Space Research Director Thomas Gold concluded from measurements that Venus' surface material was denser than moon's. Scientists confirmed that Venus rotated with same face toward earth each time it passed nearest earth. Radar observations revealed rough areas near planet's equator and probable steep mountain range in southern hemisphere. One-third of Venus had already been mapped; data needed for map of almost entire planet were available. (AP, *NYT*, 2/13/68, 30)

- RFPs for two series of advanced comsats, Intelsats III½ and IV, and for design and feasibility study of rotary joint for use in Intelsat III½ satellites had been issued by ComSatCorp. Two 725-lb Intelsat III½ satellites planned for mid-1969 launch would use directive antenna system to increase radiated power directed toward eastern North America and Western Europe and would have 1,900 two-way voice circuits. Rotary joint would transmit four RF signal channels across interface between spinning spacecraft and mechanically despun antenna. Four 2,430-lb Intelsat IV satellites planned for launch beginning in mid-1970 would be larger, more sophisticated comsats with more than 5,000 voice circuits and greater operational flexibility and versatility. Both satellite models would have onboard apogee motors and both would be launched into synchronous circular equatorial orbits. (ComSatCorp Release 68–4)
- XB–70 research aircraft flown by NASA test pilot Fitzhugh L. Fulton, Jr., and L/C Emil Sturmthal (USAF) reached 41,000-ft altitude and mach 1.18 during 2-hr 43-min flight from Edwards AFB. About 80% of primary objectives were accomplished, testing stability, control, handling, canard loads, engine performance, airspeed system calibration, fuselage bending photos, and runway noise. (XB–70 Proj Off)

February 13: President Johnson, presenting National Medal of Science to 12 scientists, said: "In a democratic society, the public attitude toward science must always be a real concern of the scientific community. If that attitude is to be favorable, science must be prepared to play its part in correcting the flaws in our environment."

Award winners were: Jesse W. Beams, professor of physics, Univ. of Virginia; Francis Birch, professor of geological sciences, Harvard Univ.; Gregory Breit, professor of physics, Yale Univ.; Paul J. Cohen, professor of mathematics, Stanford Univ.; Kenneth S. Cole, senior research biophysicist, National Institutes of Health, and visiting professor of biophysics, Univ. of California at Berkeley; Harry F. Harlow, professor of psychology, Univ. of Wisconsin; Louis P. Hammett, retired professor of chemistry, Columbia Univ.; Michael Heidelberger, professor of immunochemistry, New York Univ.; George B. Kistiakowsky, professor of chemistry, Harvard Univ.; Edwin H. Land, president, Polaroid Corp.; Igor I. Sikorsky, retired engineering manager, Sikorsky Aircraft Div. of United Aircraft Corp.; and Alfred H. Sturtevant, professor of biology, emeritus, Cal Tech. National Medal of Science was Government's highest award for distinguished achievement in science, mathematics, and engineering. (*PD*, 2/19/68, 285–6)

- NASA Deputy Associate Administrator for Aeronautics, OART, Charles W. Harper presented 31st Wright Brothers Lecture, "Prospects in Aeronautics Research and Development," at AIAA meeting in Washington, D.C. He said potential of air transportation had not been realized, partly because aeronautics R&D had not fully met its challenges, and suggested: (1) aeronautics researchers and socioeconomists cooperatively analyze possible air transportation approaches to provide basis for most effective R&D program; (2) emphasis on theoretical analysis in all sciences of concern to aeronautics be greatly increased; and

(3) increased experimental effort be directed toward solving problem of system integration. (Program; AIAA Paper 68-217; NASA *LAR VII/14*)

- Air Force Chief of Staf Gen. John P. McConnell announced renaming of Bunker Hill AFB, Ind., to Grissom AFB in honor of Astronaut Virgil I. Grissom (L/C, USAF) who was born in Mitchell, Ind., graduated from Purdue Univ., was first astronaut or cosmonaut to make a second space flight, and died in Apollo fire Jan. 27, 1967. (DOD Release 151-68)

- NASA issued 20 RFPs for final design of two ATS satellites. Of two contracttors selected to develop final design, one would be designated to build spacecraft. Planned for launch in early 1970s, two satellites carrying communications, navigational, and meteorological experiments would be placed in synchronous equatorial orbit at 22,300-mi altitude by Atlas-Centaur booster. Current program of five ATS launches had two successes in three launches; remaining two (ATS-D and -E) would be launched in June 1968 and in mid-1969. (NASA Release 68-31)

- NASA awarded Boeing Co. Space Div. $3,064,946 cost-plus-incentive-fee contract, effective through September, for continued prelaunch systems analysis and integration for first manned Apollo/Saturn V launch vehicle. Contract brought total for Saturn V systems integration contract to $200 million. (MSFC Release 68-29)

- ERC scientists Dr. Lothar Frenkel and Thomas E. Sullivan and Bell Telephone Laboratories scientists M. A. Pollack and T. J. Bridges had measured frequency of laser light with error margin of about 20 parts in 1 billion. They viewed experiments as important step toward new measurement of speed of light, a fundamental physical constant, and of distances of space. (ERC Release 68-3)

February 15: Some 25 Mexican and Brazilian scientists began four-month study course at MSC to learn how aircraft-borne sensors could enable them to map natural resources. Program, initial step toward development of earth resources satellites which would carry sensors for international exchange of data, included 12-wk technical course taught by Univ. of Michigan instructors under $92,000 contract; 6-wk field trip to Government and university sensor development centers; and additional 2-wk study at MSC. After completing program, scientists would initiate test programs for remote sensing techniques in their countries, first with NASA aircraft and then with native-owned aircraft. (*H. Chron*, 2/15/68; AP, *NYT*, 2/16/68, 1)

- John D. Hodge, MSC, and Dr. George F. Pezdirtz, LARC, received Arthur S. Flemming Award, presented annually to 10 outstanding young men in the Federal Government. Pezdirtz had developed pyrones, new family of polymer plastics, and Hodge had been flight director of manned flight missions beginning with Project Mercury. (NN; *PD*, 2/19/68, 289-90)

- INTELSAT comsats had carried heaviest load of traffic across Atlantic in their history during interruption of service on two transatlantic cables, ComSatCorp announced. With cooperation of earth stations overseas, ComSatCorp used 177 additional circuits on *Intelsat I* and *II-C*. Satellites and stations functioned precisely. (ComSatCorp Release 68-6)

- Current U.S. aerospace test facilities were becoming outdated, AFSC Commander, Gen. James Ferguson, told Los Angeles Chapter of

American Ordnance Assn. Warning that U.S. was "stretching present-day facilities to a thin point and reaching a limit in . . . ability to 'make do' on a year to year basis," Gen. Ferguson urged that an "imaginative, comprehensive, long-range plan" be instituted for design, development, and acquisition of new facilities. Major facility needs, he said, included: wind tunnel for testing VTOL aircraft over entire performance range, 200-ft vacuum chamber capable of simulating space environment, shock interaction facility to simulate aerodynamic flow and strong shock generated over warhead models and true-temperature wind tunnel (TRIPLTEE) to duplicate flight conditions at hypersonic speeds and temperatures. (Text)

- Dr. Stephen J. Lukasik, Director of Nuclear Test Detection in DOD's Advanced Research Projects Agency (ARPA) and Acting ARPA Deputy Director, was appointed ARPA Deputy Director. (DOD Release 157–68)
- USAF Space and Missile Systems Organization awarded General Electric Co. $2,600,000 increment to a previously awarded contract for reentry vehicle flight testing. (DOD Release 162–68)

February 16: President Johnson at news conference announced that resignations of A. B. Trowbridge as Secretary of Commerce and Charles Murphy as member of Civil Aeronautics Board had been accepted. With Senate approval, Trowbridge would be succeeded by C. R. Smith, Chairman of Board of American Airlines, and Murphy, who would become consultant to President, by John H. Crooker of law firm Fullbright, Crooker, Freeman & Bates. (PD, 2/19/68, 295–300)

- National Sporting Aviation Council, representing over 80,000 sporting aviation enthusiasts, was formed within NAA to promote progress and development of sport aviation in U.S. and to encourage worldwide participation through FAI. (NAA *News*)
- Flight of Soviet heavy turboprop Bear aircraft had made most southern penetration over North American waters, coming close to Newfoundland coast in early February, Charles W. Corddry reported in Baltimore *Sun*. According to DOD officials, flight was closest to date but was not first; several Soviet aircraft had been detected 75–100 mi from Newfoundland coast since January. (Corddry, B *Sun*, 2/17/68, 1)

February 17: President Johnson announced appointment of Dr. Philip Handler, Chairman of Dept. of Biochemistry at Duke Univ. Medical Center, and Dr. Herbert F. York, Jr., professor of physics at Univ. of California (San Diego), to four-year terms on President's Science Advisory Committee. (PD, 2/26/68, 312–3)

February 18: U.S.S.R. would increase pace of its space program in preparation for manned circumlunar flight, John N. Wilford reported in *New York Times*. Although no official Soviet flight schedules had been announced, informed sources speculated that U.S.S.R. would conduct manned earth-orbital flight, possibly within three months, during which cosmonauts in redesigned Soyuz spacecraft would practice rendezvous with another vehicle; unmanned circumlunar flight during summer 1968; and manned flight in fall 1968, in which cosmonauts or animals would circle moon in Soyuz spacecraft and return to earth. (Wilford, *NYT*, 2/18/68, 18)

February 19: Goddard Space Flight Center fired its 1,000th sounding rocket. Aerobee 150 launched from WSMR carried Lockheed Missiles & Space Co. payload to 97.9-mi (160.5-km) altitude to obtain quantitative measurements of spectrum and intensity of solar x-ray flux in 2- to 30-kev interval with eight proportional counter x-ray sensors, solar aspect sensors, and yo-yo despin unit. Rocket and instrumentation performed satisfactorily. (NASA Rpt SRL; GSFC Historian)

- "The successful completion of the current phase of the automated exploration of the Moon and our planetary successes to date provide the capability, experience, and framework for the next step forward in the exploration of the planets," NASA Associate Administrator for Space Science and Applications, Dr. John E. Naugle, told House Committee on Science and Astronautics' Space Science and Applications Subcommittee. Advanced Planetary Mission Technology (APMT) effort, begun in Fiscal Year 1968 at the conclusion of the Mariner V, Lunar Orbiter, and Surveyor Programs, and deferral of Voyager was "directed at planning and technology for potential planetary missions in the early 1970's." Emphasis had shifted from automated to manned exploration and return of lunar samples. "Regardless of missions planned by the U.S.S.R. during the opportunities in 1969, 1971, and 1973," Dr. Naugle said, NASA's "systematic approach . . . will be meaningful and more likely to be complemented by the U.S.S.R. missions than duplicative." (Testimony)

- NASA Associate Administrator for Manned Space Flight, Dr. George E. Mueller, in statement before House Committee on Science and Astronautics' Manned Space Flight Subcommittee, stressed urgency of funding Apollo Applications and advanced missions programs in continuing "this country's position as a space pioneer." These programs, he said, "are an investment in our national posture and the future of manned space flight . . . [and provide] the opportunity to stabilize the manned space flight program so that it can effectively and efficiently respond to the challenges of the next decade. We have the resources, the facilities, the people, and the knowledge of Apollo upon which to build, and the crucial question posed by this minimum request for Apollo Applications is whether we are going to pursue the manned exploration of space. . . . Man will prevail in space. On that there can be no serious question. The only question is whether they will be Americans." (Testimony)

- Dr. Abe Silverstein, LeRC Director, described progressive decline of LeRC funding from peak $389.2 million in FY 1965 to $242.4 million in FY 1967, to House Committee on Science and Astronautics' Advanced Research and Technology Subcommittee. Dollar value of procurement had decreased, from $324.5 million in FY 1964 to $211.6 million in FY 1967, while workload was maintained and lead time reduced. Power usage had increased "but costs had declined significantly." However in last several years, "funds available for maintenance had been less than those believed necessary for proper upkeep and repair." In long run, Dr. Silverstein said, delays "may prove more costly to the Government." (Testimony)

- USN Aquanauts Fernando Lugo and Don C. Risk, wearing standard neoprene wet suits and Mark VIII breathing apparatus, dived to record 1,025-ft simulated ocean depth and, with three other aquanauts, accumulated record 48 hr each at 825-ft depth during tests at Washington, D.C., Navy Yard. Experiment was in preparation for USN's 60-day Sea-

February 19

lab III experiment in underwater living scheduled to begin in fall 1968. (DOD Release 180-68)

- R/A John E. Clark (USN, Ret.), former Commandant of Twelfth Naval District in San Francisco, became JPL Deputy Director. He had been Commander of Naval Air Missile Test Center (1954), ARPA Deputy Director (1958-61), and Commander of PMR (1961-65). (*JPL Laboratory*, 2/68, 3)
- Jodrell Bank Observatory Director, Sir Bernard Lovell, claimed explosion of Soviet spacecraft in orbit during 1962 Cuban missile crisis led U.S. to believe U.S.S.R. was launching massive ICBM attack and warned that World War III could be triggered by misidentified space debris falling to earth. (*W News*, 2/20/68, 3; *SBD*, 2/20/68, 281)
- U.S.S.R. was closing its missile gap and could equal U.S. 1,054-ICBM force by mid-1969, according to DOD sources. Between October 1966 and October 1967 Soviet ICBM force had increased from 340 to 720 missiles; currently, U.S.S.R. had more than 720 ICBM sites in operation and about 280 under construction. In addition, U.S.S.R was reportedly developing new 16-tube nuclear submarines, missiles that could travel long distances underwater, mach 3 interceptor aircraft, and mobile, solid-fueled ICBMs. (Beecher, *NYT*, 2/19/68, 1; 2/20/68, 18; Wilson, *W Post*, 2/20/68, 1)
- Preliminary to joint NASA-USAF flight testing, X-24 manned lifting-body vehicle built by Martin Marietta Co. had been sent to ARC for full-scale wind-tunnel tests, to begin Feb. 26, FRC announced. Prime purpose was to verify aerodynamic predictions obtained from small-scale model tests. (FRC Release, 5-68)

February 20: U.S.S.R. launched two Cosmos satellites. *Cosmos CCII* entered orbit with 456-km (283.3-mi) apogee, 211-km (131.1-mi) perigee, 91.2-min period, and 48.4° inclination and reentered March 24. *Cosmos CCIII* entered orbit with 1,203-km (747.5-mi) apogee, 1,186-km (736.9-mi) perigee, 109.2-min period, and 74° inclination. Both satellites performed satisfactorily. (UPI, *P Inq*, 2/21/68; *SBD*, 2/21/68, 294; 2/23/68, 300; GSFC *SSR*, 2/29/68; 3/31/68)

- NASA announced Apollo 6 would be launched no earlier than March 21 in second unmanned test flight of command and service modules on Saturn V. First Saturn V had been launched successfully Nov. 9, 1967.

  Ten-hour, earth-orbital mission would include (1) 5½-min second burn of 3rd-stage engine to provide 279,000-mi flight into space on 16-day elliptical earth orbit; (2) separation and flight of Apollo spacecraft to 13,824-mi altitude, using 4-min retro-burn of spacecraft's main propulsion system; and (3) high-speed spacecraft reentry into earth's atmosphere simulating lunar mission return. (NASA Release 68-37)
- *Surveyor VII*, on the moon, stopped operating 22 hr before nominal sunset time. Spacecraft, launched Jan. 7, had responded to turn-on commands Feb. 12, transmitted 45 200-line TV pictures, and obtained 22 hr of useful data from alpha-scattering instrument, but performance before signal loss indicated that appreciable functional degradation had occurred during preceding lunar night. (NASA Proj Off; AP, *B Sun*, 2/22/68, A5)
- *Explorer XXXVI* (*Geos II*), launched Jan. 11, became fully operational, having achieved orbit well within specifications. With launch of this spacecraft, Thrust-Augmented Delta had accomplished record 23rd con-

secutive, successful launch. Spacecraft would support 128 global observation stations which used both electronic and optical geodetic instrumentation. (NASA Proj Off)

- Dr. Bruce C. Murray, Cal Tech associate prefessor of planetary science, told Space Science and Applications Subcommittee of House Committee on Science and Astronautics it would be tragic if U.S. were to "ignore the challenge and the opportunity" of planetary exploration. He urged sustained exploration program competitive with U.S.S.R. "We need not always be first, but we must not always be second." He saw "real possibility of Soviet Mars lander attempts as early as 1969" and other attempts at planetary firsts. (NASA Auth Hearings; Randal, W Star, 3/7/68, A12)

- Lockheed Aircraft Corp. engineer John W. Jones was granted patents 3,369,455 and 3,369,485 for new rocket-launching technique with potential military and space applications. Rockets encased in liquid-filled plastic sheaths were fired from a gun, after which sheath fell away, solid propellant ignited, and rocket continued on its own power. Jones said high-altitude research probes could be conducted using seven- or eight-inch gun at one-half to one-fourth cost of present methods. Protective sheath and liquid permitted use of new thin-walled military rockets which had better flight performance and traveled twice as fast as standard artillery shell. (Jones, NYT, 2/24/68, 37; Patent Off PIO)

- National Academy of Engineering announced award of its third Founders Medal to Dr. Vladimir K. Zworykin, Honorary Vice President of Radio Corporation of America and technical consultant to RCA Laboratories, in recognition of his many contributions to engineering and to betterment of human society. Dr. Zworykin, known as "Father of Television" for invention of iconoscope, first practical picture transmission tube, was cited also for "his role in developing the first commercial electron microscope in the western hemisphere . . . promoting the cause of traffic safety through the imaginative concept of an automated 'electronic highway,' and . . . working . . . to bring about a union of electronics and medicine." Medal would be presented April 24 during NAE's 4th Annual Meeting. (NAE Release, 2/20/68)

- USA announced plans to establish Advanced Ballistic Missile Defense Agency which would combine some elements of DOD's Advanced Research Projects Agency (ARPA) Office of Ballistic Missile Defense and Nike-X advanced development. Dr. Patrick J. Friel, Director of ARPA office, would be appointed USA Deputy Assistant Secretary and Director of new agency. He would be replaced by Dr. David E. Mann. (DOD Release 176–68)

- Administration would soon announce plans to slow pace of SST development, Evert Clark reported in New York Times. He said industry sources believed that technical reasons would be cited as cause, but "the real reason was chiefly political—an attempt to reduce the request for funds . . . for [FY] 1969 to a level acceptable to Congressional critics of the program." (Clark, NYT, 2/21/68)

February 21: NASA Nike-Tomahawk sounding rocket was launched from Churchill Research Range carrying GSFC experiment to study auroral activity as it developed and dissipated for comparison with data from similar Nike-Apache launch. Flight was unsuccessful because of unde-

February 21

termined failure related to pitch/roll couple. Radar and loss of telemetry indicated payload breakup. (NASA Rpt SRL)

- Sixty-day simulated earth-orbited mission, using Douglas Missile & Space Systems Div.'s 12-ft-dia, 40-ft-long space cabin simulator, had begun for four California college students—Jack G. Angaran, Dennis Giroux, Guy H. King, and Robert B. Zeuschner. NASA OART had awarded $200,000 contract to McDonnell Douglas Corp. to conduct what was believed first test in which crew would subsist in space cabin on "closed water and oxygen loop" for lengthy period. Students would use reclaimed water and oxygen from human waste. Additional missions to evaluation of integrated life support system were evaluation of waste management; crew's ability to maintain, service and repair life support system; and crew's physiological and psychological reactions. Life support system resembled that needed for three-year spacecraft mission. Crew's contact with outside world would be by radio. (DAC Release 67–172; *St. Louis G–D*, 2/21/68)

- Explaining decision to develop flight-rated NERVA engine at 75,000-lb rather than 200,000-lb thrust level, NASA Manager of Space Nuclear Propulsion Milton Klein told House Committee on Science and Astronautics' Advanced Research and Technology Subcommittee that lower level would be suitable for most missions in which nuclear rocket was significantly advantageous. It was "substantially off-optimum for manned planetary landing missions," but with manned exploration delayed, "less emphasis is now properly placed upon that mission, and the 75,000-pound-thrust NERVA class is appropriate. . . . The 75,000-pound-thrust NERVA gives somewhat better payload performance for orbital operations missions which start from low Earth orbit and involve a single Saturn V launch," but its use for manned planetary missions "entails a significant performance penalty." Further development might be required "to extend operating time and to accommodate more extensive clustering and staging if such a mission were to be undertaken." (Testimony)

- Within limited budget, NASA was "attempting increased emphasis on Astronomy" using manned spaceflight capabilities, NASA Associate Administrator for Space Science and Applications, Dr. John E. Naugle, told House Committee on Science and Astronautics' Space Science and Applications Subcommittee. Advanced studies and supporting R&T programs would be expanded in FY 1969 to lay "ground work for a major program in space astronomy." Flight program in space physics was being "supported at reduced levels," primarily because of OGO phaseout, but resources would be needed "to capitalize on analysis of data obtained." Decline in data and experiments, despite output from Explorer and Pioneer programs, might be stemmed, but NASA "considered it much more important to use our requested resources to develop a program which will enable us to fly new experiments, to make new investigations, and to explore new regions of interplanetary space." Development of two new Pioneer spacecraft would be initiated in FY 1969, to reach past Mars and toward Jupiter, with first launch proposed for 1973. (Testimony)

- NSF announced award of first nine grants under National Sea Grant College Program Act of 1966 to encourage development of national marine resources. Grants, which totaled nearly $2 million, included three

to help higher education institutions develop broadly based major programs for increasing utilization of marine resources and six to support individual sea grant projects. (NSF Release 68–112)
- USAF awarded Big Three Industrial Gas and Equipment Co. $1,156,845 fixed-price contract for production of propellant nitrogen to support Project Apollo at KSC. (DOD Release 183–68)

February 21–22: NASA successfully launched six Nike-Apache sounding rockets carrying chemical cloud experiments from NASA Wallops Station between 6:17 pm and 6:02 am EST. Seven launches had been scheduled, but second launch was postponed because of payload problems [see Feb. 26]. Rockets ejected vapor trails during descent between 50- and 90-mi (81- and 145-km) altitudes to measure wind velocities and directions at various altitudes in upper atmosphere. Nike-Apache launched at dawn carried sodium experiment which created reddish-orange trail. Other five payloads consisted of one triethylborane (TEB) and four trimethyl-aluminum (TMA) experiments which formed pale blue and green clouds. Rocket carrying TEB was equipped with photometer to observe airglow in sunlight for vertical profile of atomic oxygen. Other five payloads carried Langmuir probes for measuring electron energy distribution. Data were obtained by photographing continuously motions of trails from five ground-based camera sites. Launches were conducted for GCA Corp. under GSFC contract. (WS Release 68–3; NASA Release 68–32; NASA Rpt SRL)

February 22: SST prototype construction would be delayed one year because of design changes "which would result in significant improvements in the production airplane," Boeing Co. President William M. Allen announced. Detailed plans of recommended changes were being prepared for FAA approval. Although present design could result in airworthy flight-test aircraft, FAA and airline reviewers had concluded that prototype should have more substantial range and payload "to assure it is a sound foundation for an economically successful commercial aircraft." First flights now would be in 1972. (Boeing Release; *W Post*, 2/23/68, A3; *WSJ*, 2/23/68, 4)

February 23: NASA Nike-Tomahawk sounding rocket launched from Churchill Research Range carried Univ. of Alaska experiment to obtain data on horizontal and vertical spatial variation of auroral light emissions and relationship between their intensities and volume emission rates. Rocket and instrumentation performance was satisfactory and experiment was successful. Peak altitude was not determined because radar lost track. (NASA Rpt SRL)
- President Johnson announced appointment of Gen. Maxwell D. Taylor (USA, Ret.) as Chairman of President's Foreign Intelligence Advisory Board to succeed Clark M. Clifford, who would become Secretary of Defense. (*PD*, 2/26/68, 345–6)
- President Johnson nominated B/G Edmund F. O'Connor (USAF), MSFC Director of Industrial Operations, for promotion to Major General. Assigned to NASA through special NASA–DOD personnel exchange agreement in 1964, Gen. O'Connor had been responsible for technical and administrative management of Saturn IB and Saturn V launch vehicles and Apollo Applications program. (MSFC Release 68–33)
- Soviet Defense Minister, Marshal Andrey A. Grechko, speaking at Moscow ceremony marking 50th anniversary of Soviet armed forces, con-

firmed reports that U.S.S.R. had greatly expanded its ICBM force [see Feb. 19] and emphasized that Soviet armed forces had been modernized to emerge victorious in nuclear or non-nuclear war. (Anderson, *NYT*, 2/24/68, C8)

February 23–24: Results of 13 experiments carried on board *Biosatellite II* (launched Sept. 7 and recovered Sept. 9, 1967) were discussed by scientists at NASA–NAS symposium in Washington, D.C. Experimenters reported that radiation in weightless state caused greater damage to plant and animal organisms than radiation on earth. Affected most severely by weightlessness and radiation were young and actively growing cells and tissues and cells with high metabolic activity. Animal cells were least affected by weightlessness. Generally, plants had difficulty maintaining proper orientation; some plant structures, mechanisms, and biochemical activities were affected.

Analyses of effects of 45-hr flight on individual experiments, compared with control group on earth, revealed that pepper plant leaves twisted and curled downward and wheat seedling roots grew upward and sideways. Radiation experiments showed: wasp nurse cells and primitive egg cells slowed activities, allowing time for repair of radiation damage; two strains of bacteria grew substantially faster and tolerated radiation better, and viruses hosted by these bacteria appeared to be less effective in harming bacteria than on earth; twice as many flour beetle offspring died, and beetles suffered 50% more of characteristic wing defect; tradescantia plant had greater cell death and pollen abortion; and both adult and larval stages of vinegar gnats suffered greater chromosome damage. (NASA Release 68–35)

February 24: House Government Operations Committee said in report that despite acute need to eliminate U.S. balance of payments deficit, Government continued to make grants for foreign research projects of about $20 million yearly—more than $15 million being spent in developed countries. Committee recommended rigorous scrutiny of dollar-financed projects in Japan, Canada, U.K., Sweden, Italy, Australia, France, and West Germany. (W *Star*, 2/24/68, B9)

• U.S.S.R. inaugurated fifth permanent Antarctic research station, Bellingshausen station on King George Island, according to Moscow newspapers. U.S.S.R. thus joined U.K., Argentina, and Chile in operating scientific bases in northernmost and warmest part of Antarctica. (*NYT*, 3/6/68, 49)

February 25: U.S.S.R. would probably launch several spacecraft, possibly containing dogs, on circumlunar missions and return them to earth before launching manned lunar spacecraft, Cosmonaut Valery Bykovsky said in Hungarian Army newspaper interview. Quoted in *Nephadsreg*, he said that first Soviet passengers around moon might be offspring of space dog Laika. Manned spacecraft would not be launched until necessary experience had been gained. (AP, B *Sun*, 2/26/68, A3)

February 26: NASA Nike-Apache sounding rocket launched from NASA Wallops Station ejected triethylborane (TEB) vapor trails during descent from 95.2-mi (153.5-km) altitude to study nighttime airglow and variations in wind structure, determine vertical profile of atomic oxygen with photometer, and measure electron energy distribution in normal nighttime ionosphere with Langmuir probe. Rocket and instrumentation performance was satisfactory. Launch, postponed Feb. 21 to check instrumentation for high voltage breakdown, was one of seven launches in series conducted for GCA Corp. under GSFC contract [see Feb. 21–22]. (NASA Rpt SRL)

- NASA Deputy Associate Administrator for Aeronautics, OART, Charles W. Harper, in statement before House Committee on Science and Astronautics' Subcommittee on Advanced Research and Technology, explained NASA's increased emphasis on aeronautics. Air transportation "has grown to dominate the long range transportation systems" and "has created an industry of great importance to the economic strength of the nation," showing $850 million in export value in first 10 mo of 1967, more than any other product, according to Dept. of Commerce. NASA planned to direct major effort at providing noise suppression, increasing aerodynamic and propulsion efficiency and advanced avionics, while avoiding "major new airport complexes" by increasing development efforts for V/STOL aircraft.

  Harper presented FY 1969 aircraft technology budget: $91.3 million for R&D, $20.6 million for "other NASA support applicable to aeronautics," and $54.5 million for administrative operations (AO) (for both OART and non-OART). He said, "A large part of the most fundamental aeronautics research is carried out by NASA scientists using NASA facilities. . . . Without this continued support the foundation of the whole program is in jeopardy and the research supported by R&D funds is much less effective." (Testimony)

- Research on concepts for high lift capability for STOL aircraft and special hover and speed requirements for VTOL aircraft would continue in FY 1969, NASA Director of Aeronautical Vehicles, OART, Albert J. Evans reported to House Committee on Science and Astronautics' Advanced Research and Technology Subcommittee. Aircraft noise program would pursue efforts to suppress noise by acoustic treatment of engine nacelles (with results expected from McDonnell Douglas Corp. and Boeing Aircraft Co. investigations by October 1969 and January 1970), by engine redesign (with construction of engine components in Quiet Research Engine Program initiated in FY 1969), and by operational techniques. Major NASA program in flight dynamics and propulsion in support of SST program would be continued. Phase III of Hypersonic Research Engine (HRE) program, test of HRE on X–15 in flight as well as on ground, would not be conducted because of X–15 program termination. Major emphasis in general-aviation technology would be on flight safety, with flight tests of "typical twin-engine vehicle" at FRC in FY 1969. (Testimony)

- NASA Associate Administrator for Space Science and Applications, Dr. John E. Naugle, presented NASA's bioscience programs to House Committee on Science and Astronautics' Space Science and Applications Subcommittee. Biosatellite flights D and F, 30-day earth-orbiting missions to study effects of space environment on living organisms, were scheduled for second and fourth quarters of 1969 and 21-day flights C and E for 1970 and 1971. Flights would test reactions of central nervous system, cardiovascular system, metabolism, performance, and behavior of pigtailed monkey. Gas management system for two-gas atmosphere would be flown for first time in U.S. satellite. (Testimony)

- Discovery by MSFC officials of tiny welding flaws in Saturn V 2nd stage had prompted scheduling of new test series before flight could be manned, AP reported. Although none of flaws was considered serious enough to cause failure, cryogenic prooftesting would be conducted March 18–25 with Saturn V fourth flight version. (AP, B *Sun*, 2/27/68, A3; AP, W *Star*, 2/27/68, A5)

- Harold T. Luskin, chief advanced design engineer at Lockheed-California Co. and former AIAA President, was named NASA Deputy Associate Administrator for Manned Space Flight (Technical), effective March 18. Luskin would be responsible for ensuring technical excellence and would share responsibility for overall planning and direction of Manned Space Flight Program with Associate Administrator for Manned Space Flight and his other deputies. (NASA Release 68–39)

February 27: NASA Administrator James E. Webb began testimony on President's authorization request for NASA's FY 1969 budget before Senate Committee on Aeronautical and Space Sciences. He outlined significant NASA achievements and said "remarkable series of successes shows how far we have come since the beginning of the Space Age ten years ago. . . . Today . . . success is treated almost routinely, no matter how difficult the task or how significant the achievement."

Describing budget as "a compromise," he said President "was forced, in spite of his conviction as to the importance of a larger effort in aeronautics and space, to accept reductions. . . . This means that for NASA 1968 and 1969 are . . . years in which we will be completing programs started in previous years and endeavoring to make limited further advances. Under these conditions we will devote a major effort to stabilizing our organization and the resource base we have built." (Testimony)

- NASA's university program, operating at lower level because of reduced funding, was also changing to meet changing requirements, NASA Associate Administrator for Organization and Management Harold B. Finger told Senate Committee on Aeronautical and Space Sciences. Emphasis in predoctoral trainee grants had shifted from earlier need to replenish "national reservoir of engineers and scientists who had advanced training in space sciences" to "current and future needs . . . for engineers trained in design and development of complex engineering systems . . . and for people trained in management . . . of large-scale research and development programs." Finger declared "importance of the university participation in the nation's space program is equal to or possibly greater now than in the early 1960's," but cited decreases in predoctoral trainee grants to 75 in FY 1968 from about 800 in 1967 and over 1,300 in 1966. No new funding awards for new university facilities were anticipated for FY 1968 or FY 1969. Multidisciplinary research had been reduced to about one-half 1967 level.

In Technology Utilization program, Finger believed some of greatest benefits would come from "the storehouse of information that we are building that permits easy public access to the large masses of data and information" in many disciplines. (Testimony)

- NASA Assistant Administrator for Administration William E. Lilly presented NASA's FY 1969 administrative operations (AO) budget to Senate Committee on Aeronautical and Space Sciences. He said $648.2 million was requested and "the same stringent measures that were re-

- quired in FY 1968 to operate at a level of $628 million have been continued into FY 1969." (Testimony)
- NASA Assistant Administrator for Space Science and Applications, Dr. John E. Naugle, testifying before House Committee on Science and Astronautics' Space Science and Applications Subcommittee, said Nimbus B would be launched in meteorological flight program in spring 1968. Nimbus D continued on schedule for 1970 launch as major step in charting earth atmosphere with new techniques. Nimbus E and F were planned for 1971 and 1973. First of next-generation meteorological satellites (Tiros M) would be available for launch for ESSA in 1969, providing in single spacecraft both stored picture data for global use and local readout of cloud photos, day and night. Dr. Naugle foresaw "in more distant future the possibility that several economic applications of satellite technology can be combined on single, multiple purpose satellites, thus achieving economy through the sharing of many basic spacecraft systems." (Testimony)
- British Minister of Technology Anthony W. Benn announced in House of Commons that U.K. would underwrite production of Anglo-French Concorde supersonic aircraft with $180- to $240-million loan for working capital, bringing total U.K. commitment to more than $1 billion. He later revealed first test flight would be delayed until summer 1968. First flight of U.S. SST had been postponed until 1972 [see Feb. 22]. (Lee, *NYT*, 2/28/68, 5)
- MIT physicist Dr. Irwin I. Shapiro, speaking at American Physical Society meeting in Boston, said he and associates had successfully tested refined radar technique that might prove validity of Einstein's general theory of relativity. Using 20-ft-dia Haystack, Mass., dish antenna, scientists observed impulses which they bounced off of Venus and Mercury as planets passed behind sun. Results, which confirmed theory's prediction that signals would be slowed down slightly by gravitational pull, were more precise than those from previous tests but were not clear enough to resolve completely debate on theory's validity. Dr. Shapiro believed greater precision could be achieved. (Sullivan, *NYT*, 2/28/68, 22; O'Toole, *W Post*, 2/28/68, 1)
- NAA Executive Director, M/G B. E. Allen (USAF, Ret.), in Washington, D.C., ceremony presented awards to Allen Bourdon, William Diehl, and Ernest Hall, last three living civilian flight instructors of World War I air service, for "their patriotism, devotion to duty, and capability as pilots [who performed] an outstanding service toward our achievement of victory in World War I." (NAA *News*)

February 28: NASA Nike-Tomahawk sounding rocket launched from Churchill Research Range carried Univ. of Alaska experiment to 183-mi (295-km) altitude to obtain data on horizontal and vertical spatial variation of auroral light emission and relationship between their intensities and volume emission rates. Rocket and instruments performed satisfactorily. Experiment was successful. (NASA Rpt SRL)
- NASA Aerobee 150 MI sounding rocket launched from WSMR carried Cornell Univ. experiment to 106-mi (170.6-km) altitude to study far infrared in spectral range from five to several hundred microns, using mercury-, copper- and gallium-doped germanium and antimonide detec-

tors. Nikon F camera monitored instantaneous rocket aspect. Rocket and instrumentation performance was satisfactory. (NASA Rpt SRL)

- XB-70 research aircraft flown by NASA test pilot Fitzhugh L. Fulton, Jr., and L/C Emil Sturmthal (USAF), reached mach 0.50 and 18,500-ft altitude. Primary objectives were not accomplished because main landing gear valve malfunctioned; low speed stability data were obtained during 1-hr 51-min flight. (NASA Proj Off)
- Second anniversary of Essa II, first spacecraft in Tiros Operational Satellite (TOS) system and first to carry Automatic Picture Transmission (APT) equipment for cloud-cover photos. Satellite was still operating satisfactorily. (GSFC Historian)
- Press conference on preliminary scientific results from NASA's *Surveyor VII*, which landed on moon Jan. 9, revealed spacecraft could make safe landing in highland area with "major ejecta blanket" (rim of debris ejected from center of crater). Data indicated Tycho region of moon's surface contained larger rocks, fewer craters, and thinner debris layer than did maria. Fine particles and rocks near Tycho crater had higher albedo, or lighter color. Iron content was lower, accounting for lower density. In Tycho region, where iron group of elements was less concentrated, reflection of light from moon's surface was greater than from area where iron group elements existed in greater concentration. Low iron content was "probably strongest direct evidence" that moon had undergone chemical fractionation, suggesting it "has been hot, has been melted at least partially, and has been differentiated into different types of rock." (Transcript; Sehlstedt, B *Sun*, 2/29/68, A8; *W News*, 3/21/68)

February 29: NASA Associate Administrator for Space Science and Applications, Dr. John E. Naugle, described OSSA programs before Senate Committee on Aeronautical and Space Sciences. Three-year budget history of Space Applications program was one of very few in OSSA "showing a steady rise" and increase for FY 1969 was primarily due to increase in Earth Resources Survey Program.

"Together with user agencies—the Departments of Agriculture, Commerce, Interior, and Navy—we are studying the feasibility of applying space technology and techniques to such Earth resources disciplines as agriculture and forestry resources; hydrology and water resources; geography, cartography, and cultural resources; and oceanography and marine resources. Data are now being obtained by flying with electronic and electro-optical sensors over geographical areas . . . to permit correlation of remote sensor data with actual conditions. . . . To complement the acquisition of sample data in preparing for future systems, we intend to initiate a definition and economic benefit study in Fiscal Year 1969 of an automated spacecraft system called Earth Resources Technology Satellite." (Testimony)

- NASA Associate Administrator for Tracking and Data Acquisition Gerald M. Truszynski, testifying on FY 1969 budget before House Committee on Science and Astronautics' Advanced Research and Technology Subcommittee, cautioned: "The many actions we have taken to reduce operating costs have, however, lowered the level of support we provide to the flight projects. In our judgment, further reductions are not possible without reducing network reliability to the point of jeopardizing the success of flight missions." (Testimony)

- NASA Associate Administrator for Manned Space Flight, Dr. George E. Mueller, in statement before Senate Committee on Aeronautical and Space Sciences, pointed to "establishment of a Manned Space Flight Safety Office . . . to focus all our safety activity at the highest level of management." NASA, he said, was ensuring "that quality assurance and reliability officials have direct access to contractor management and NASA management." (Testimony)
- *Apollo and Apollo Applications,* staff study released by House Committee on Science and Astronautics' Subcommittee on NASA Oversight, included statement by Rep. Olin E. Teague (D-Tex.), Subcommittee Chairman: ". . . the general posture of the programs is good [but] the Apollo program, with its inherent complexity, has had and can be expected to have a variety of problems. Yet this study identifies the fact that confidence in the governmental-industrial team accomplishing this work is well placed."

  Report found "that the NASA industry team is continuing to effectively employ its resources in the solution of those problems which currently face the program. NASA has indicated that, given the stretchout in the Apollo program caused by the Apollo 204 accident, NASA is not constrained by total funding for the Apollo program during [FY] 1968. The continued reduction in total effort in the Apollo program as the program passed its peak effort in fiscal year 1967 is beginning to cause dislocation in the contractor and NASA center effort to the extent that personnel that will be needed during the flight portion of the program may not be available in the event a major flight problem occurs. Timely support for the Apollo Applications program and other post-Apollo effort could have a major effect on minimizing this problem." (Text)
- President Johnson presented Presidential Medal of Freedom, highest civilian award, to Secretary of Defense Robert S. McNamara at White House ceremony. McNamara, who would head World Bank, was cited for administering DOD for seven years and "unifying our strength so that we might respond effectively wherever the security of our free world was challenged." (AP, *NYT*, 2/29/68, 3; *PD*, 3/4/68, 387–8)
- Iowa State Univ. scientist Dr. James A. Van Allen said that although he was not outright opponent of manned space flight, so far man had done nothing in space but survive—although "that is not unimportant." He advocated Jupiter exploration using unmanned flyby, which he said could be accomplished in 1972. He felt there was little chance Jupiter could have life forms of any sort, but he wanted this large planet (10 times diameter of earth) "put on an equal footing with Mars and Venus in U.S. exploration." His opposition to man in space was based on dominance of manned flight in national funding to detriment of unmanned missions that could have returned major scientific benefits. (Miles, *LA Times*, 2/29/68)
- LaRC Director Dr. Floyd L. Thompson was appointed Special Assistant to NASA Administrator James E. Webb and head of NASA Hq. Interim Working Group to evaluate future manned spaceflight projects. LaRC Deputy Director Charles J. Donlan would serve as Acting Director in Dr. Thompson's absence. (NASA Release 68–41; *W Post*, 3/1/68, A7)

During February: NASA published *1967 Summer Study of Lunar Science and Exploration* (NASA SP-157), results of July 31–Aug. 13, 1967, conference at Univ. of California in Santa Cruz. Conference recommended NASA: (1) develop lunar flying unit (LFU), Saturn V dual-launch capability, and dual-mode local scientific survey module (LSSM) for lunar surface mobility; (2) make available in 1970–75 Block II Surveyor or another system capable of deploying numerous experiments; (3) increase total returned payload to 400 lb so that at least 250 lb of lunar samples could be returned on Apollo Applications (AA) missions; (4) increase flexibility of ALSEP stations' design, possibly by adopting modular concept; (5) provide for continuous telemetry coverage of all payload elements and operations; (6) develop subsatellite system for deploying instruments in close lunar orbit; (7) implement any extension of Apollo science program by open solicitation of experiments from scientific community; (8) undertake strong programs in scientific instrument definition and development and in lunar supporting research and technology; (9) establish position of Project Scientist within manned spaceflight program; (10) include ability in field geology as a major requirement for astronauts who would land on moon; (11) conduct immediate and intensive program of detailed mission analyses for prime lunar landing sites and traverses suggested by the conference; and (12) include in lunar surface studies observations of micrometeoroid environment, radio-frequency noise levels, surface impedance and conductivity, lunar ionosphere, gamma rays and x-rays, soil mechanics, thermal effects, and contaminants. (NASA SP-157; AP, *NYT*, 2/19/68, 17)

- Discovery of gamma radiation from celestial sources by AFOSR-supported group at Rice Univ., using "gamma-ray telescope," was reported by Dr. R. C. Haymes in OAR *Research Review*. Telescope tracked celestial sources in diurnal motions across sky with pointing system also developed by Rice group. The two gamma-ray sources, discovered in 1967 flight from Scientific Balloon Flight Station of National Center for Atmospheric Research, Texas, were found in Crab Nebula and Cygnus. (OAR *Research Review*, 2/68, 1–2)

- U.K. was urged to join "the mainstream of space development" by Council of the British Interplanetary Society. Council specifically criticized 13th report from Parliamentary Estimates Committee, "Space Research and Development," for its statement that U.K. "should not take part in the CETS [Conférence Européene sur les Télécommunications par Satellites] programme for a television distribution satellite."

  Council said U.K. should assess fully future opportunities for collaboration with U.S. and U.S.S.R. and recommended multiaction course for U.K. to conduct international and national space activities. Concluding that "lack of purpose at the political level, and a stubborn refusal to establish a viable European space programme" had caused present confused, unplanned, and ineffectual situation, Council urged that U.K. seize opportunity to create family of satellite launchers of maximum utility to extend "many branches of space competence with other types of applications satellites." (*S/F*, 2/68, 56–7)

March 1968

March 1: USAF launched navigational satellite from Vandenberg AFB by Scout booster into orbit with 711-mi (1,444.2-km) apogee, 640-mi (1,030.0-km) perigee, 106.9-min period, and 89.9° inclination. (GSFC SSR, 3/15/68; *Pres Rpt 68*)

- NASA test pilot William H. Dana flew X-15 No. 1 rocket research aircraft to 104,500-ft altitude and 2,898 mph (mach 4.36) from Edwards AFB to check out aircraft's electrical systems and test newly developed spray foam insulation planned for use on 2nd stages of Saturn V rockets beginning with vehicle eight. Foam was much lighter than insulation previously used to maintain low temperatures required for cryogenic propellants used in Saturn V. (X-15 Proj Off; MSFC Release 68-69)

- House Committee on Science and Astronautics decided on $153-million cut in NASA's authorization request of $4.37 billion for FY 1969. Reductions included cut from $48.3 million to $11.7 million for nuclear rocket program [Advanced Research and Technology Subcommittee had recommended cut of entire $48.3 million], $36.5 million from unmanned space program, and $60.3 million from request of $2.8 billion for manned space flight program. (AP, W *Star*, 3/2/68, A3; *W Post*, 3/2/68, A10; HR 1181, 3/19/68)

- President Johnson, accompanied by NASA Administrator James E. Webb, visited MSC and announced plans for new Lunar Science Institute to be constructed under $580,000 grant to NAS and operated by NAS and Rice Univ. in Houston. "We have invested billions of dollars during the past 10 years in our efforts in space . . . [and] I am certain that as future generations look back on our incredible decade, they will be unanimous in their belief that the treasure that we have dedicated . . . was the most significant and important investment ever made by any people. We are truly reaching for the stars." (*PD*, 3/11/68, 410-2; Kilpatrick, *W Post*, 3/2/68, 1; AP, *NYT*, 3/2/68, 21; MSC *Roundup*, 3/15/68, 1)

- Former NASA Deputy Administrator Dr. Robert C. Seamans, Jr., became visiting professor in MIT's Dept. of Aeronautics and Astronautics and at Sloan School of Management. He would continue as part-time consultant to NASA. (*Boston Globe*, 3/6/68; *SBD*, 3/11/68, 51)

- Clark M. Clifford was sworn in by U.S. Chief Justice Earl Warren as Secretary of Defense, replacing Robert S. McNamara, new head of World Bank. (Sheehan, Finney, *NYT*, 3/3/68, E3; *PD*, 3/4/68, 395-6)

March 2: Nike-Tomahawk sounding rocket launched by NASA from Churchill Research Range carried Univ. of California payload to 161.9-mi (260.6-km) altitude to study flux-energy-spectrum pitch-angle distribution and time and space variation of mirroring and precipitating charged particles in auroral zone. Launch was one of series of four. Apogee was 11% below predicted; telemetry was satisfactory. Electro-

March 2

- static analyzers performed satisfactorily and good data were obtained. Plasma experiment on boom was 100% successful. Boom deployed late, providing satisfactory data from last 30% of flight. Electric field did not produce necessary data. (NASA Rpt SRL)
- NASA Javelin sounding rocket launched from Churchill Research Range carried Rice Univ. payload to 500-mi (805-km) altitude to investigate auroral fluxes and backscattered particles. Rocket and instruments performed satisfactorily. (NASA Rpt SRL)
- U.S.S.R. successfully launched *Zond IV* automatic research station, Tass announced. Station was placed in planned flight from parking orbit of artificial satellite and was flying on trajectory close to calculated one to study "outlying regions of near-earth space." Heinz Kaminski, Director of Institute for Satellite and Space Research at Bochum, West Germany, said mission "must be regarded as a preliminary phase for a planned flight to the moon." Sir Bernard Lovell, Director of Jodrell Bank Experimental Station, said he had "hunch" *Zond IV* was not intended to fly around moon. Apparently station made no attempt to orbit moon. One source said that after journey of about 3½ days, station reached "apogee . . . comparable to lunar altitude."

    Zond I, launched April 2, 1964, failed in attempt to reach Venus, and *Zond II*, launched Nov. 30, 1964, suffered communications failure on route to Mars. *Zond III*, launched July 18, 1965, obtained photos of far side of moon. (*W Post*, 3/3/68, A17; 3/4/68, A14; 3/8/68, A12; AP, *W Star*, 3/4/68, A3)
- President Johnson—watching world's largest aircraft, USAF C–5A Galaxy jet, roll off production line—called event "a long leap forward in the effective military might of America." President praised USAF, Lockheed-Georgia Co., and General Electric Co. for carrying out "a great new adventure" in aviation and completing the job "on time." (AP, *NYT*, 3/3/68, 1; AFSC *Newsreview*, 3/68, 1)

March 3–9: Radio signals which scientists speculated possibly could have been from extraterrestrial civilization had been received by Cornell Univ.'s antenna at Arecibo Ionospheric Observatory in Puerto Rico. U.K. astronomers during February had proposed such signals might originate with pulsating neutron stars. Arecibo's Director, Dr. Frank D. Drake, would investigate all possibilities of natural origin. He believed four- to five-month study might determine whether signals came from planet in orbit around another star. Dr. Drake said Arecibo observations had confirmed all signal properties reported by U.K., including regularity and variable intensity of signals. (Sullivan, *NYT*, 3/10/68, 1)

March 4–16: NASA's 1,347-lb *Ogo V* (OGO–E) Orbiting Geophysical Observatory, carrying 24 experiments in most comprehensive study of particle-wave processes in space during period of maximum solar activity, was successfully launched at 8:06 am EST from ETR by Atlas (SLV–3A)-Agena D booster, being used for first time. Spacecraft entered highly elliptical earth orbit with 92,078-mi (148,186-km) apogee, 168-mi (271-km) perigee, 63.3-hr period, and 31.3° inclination. Primary mission objective was to acquire correlative scientific data at high information rates in magnetic fields, energetic particles, and plasma disciplines through leading quadrant of magnetosphere into geomagnetic tail, both in interplanetary medium and within large sections of magnetosphere. Secondary objectives were to demonstrate technological capability of OGO's three-axis stabilization system in highly elliptical orbit for extended operations of 135 days or more and to conduct electric-field, gamma-ray, and radioastronomy observations.

On March 16, second of two 30-ft-long extendable antennas was deployed to complete post-launch checkout and NASA reported *Ogo V* was operating satisfactorily. All experiments except one, low-energy electrons and protons experiment, which had stopped operating March 11 when input power apparently failed, were acquiring data effectively. Data provided first observations of hydrogen cloud surrounding earth (geocorona), first detailed measurements of electric fields at shock and magnetospheric boundaries, and first spark chamber observations of gamma rays.

Ogo V was fifth spacecraft in NASA's six-mission OGO series, last OGO planned for highly eccentric orbit, last spacecraft to be launched by Agena launch vehicle, and last to be launched from Complex 13. Agena would be replaced by Atlas-Centaur, and Complex 13 would revert to USAF. Scientific instrumentation for *Ogo V*'s 24 experiments had been provided by six U.S. universities; four groups from U.K., France, and the Netherlands; four Government centers; and two private companies. *Ogo V* joined three other operational OGOs—*Ogo I* (launched Sept. 4, 1964), *Ogo III* (launched June 6, 1966), and *Ogo IV* (launched July 28, 1967)—in providing data for studies of earth's environment and solar-terrestrial interactions during period of maximum solar activity. *Ogo II*, launched Oct. 14, 1965, had been turned off in November 1967. Of 85 experiments assigned to four missions still operating, 68 were still obtaining useful data, including 12 for which instrumentation was working after 42 mo in space.

OGO program was managed by GSFC under OSSA direction. LeRC was responsible for Atlas-Agena launch vehicle, and KSC for launch operations. (NASA Proj Off; NASA Releases 68-38, 68-53; LeRC Release 68-14; UPI, *M Her*, 2/25/68; *W Star*, 2/27/68, A9; *NYT*, 3/5/68, 17; *W Post*, 3/6/68, A7)

March 4: NASA Assistant Administrator for Policy, Dr. Alfred J. Eggers, Jr., told New Orleans AIAA conference on technology for manned planetary missions, "the manned planetary mission of first priority is the mission of man on planet Earth." He explained that only "minute" part of NASA's funding went toward research on manned planetary flight and described overall NASA program as "lean, but . . . by no means lifeless. It is planned to emphasize the space applications and the supporting research and technology programs." He called for best in talent for achieving manned planetary flight "in the course of man's exploration of space." (Text; AP, *NYT*, 3/30/68, 31M)

- Gen. Bernard A. Schriever (USAF, Ret.), former AFSC Commander, delivered Dr. Robert H. Goddard Lecture to National Space Club in Washington, D.C. He warned the U.S.S.R. had moved forward steadily with new and more diversified strategic threat highlighted by development of orbital missile and sophisticated space systems. U.S. comparative missile strength, he noted, was declining steadily, with emphasis still on updating existing force. "What is missing so far in our military space program is the necessary high priority research and development aimed at protecting us in the high ground of space. In space, as elsewhere, the

Soviets say one thing—even sign agreements—but consistently do whatever is in their own self-interest. The orbital missile is merely the latest example of their policy. In light of their development of an orbital missile, it would be totally irresponsible . . . to assume that we do not need to initiate any R&D to prepare us to meet a Soviet space threat."

Attributing loss of endorsement of national space program to change in national value scale because of Vietnam and domestic problems, Gen. Schriever said: "The old criteria, national security and national prestige, are no longer the only justifications for our space efforts, important though they are." If public understood "the practical, dollars and cents worth of the program to this country and to the world," they would support it. U.S. was on verge of receiving annual return larger than its annual investment in space, and within one decade, "those returns should be several times our annual investment." (Text; *Aero Tech*, 3/11/68, 16)

- Current knowledge of effects of sonic boom on building materials and structures was insufficient to assess accurately kinds of damage SST might produce, NAS Physical Effects Subcommittee of NAS Committee on the SST-Sonic Boom report concluded. Assuming SSTs would fly over land and sonic booms along flight paths would be low enough in intensity to be publicly acceptable, probability of structural damage under normal operating conditions would be "very small." Subcommittee stressed, however, that more meaningful physical response was necessary, including simulator studies, laboratory tests of glass, and acceleration and expansion studies of atmosphere's bending and magnifying effects on sonic boom shock waves. (*NYT*, 3/5/68, 3; *Science*, 3/8/68, 1081)

- Sen. Claiborne Pell (D-R.I.), member of Senate Committee on Foreign Relations, released proposed draft of international treaty to govern exploration and exploitation of "ocean space." He said there was "probable danger of anarchy" beneath seas because "man's sea technology has brought him to the verge of total undersea capability." Treaty would apply to undersea areas same freedom-of-the-seas principles now governing ocean surfaces, create licensing body and sea guard, seek to prevent development of weaponry on ocean floor, and provide for disposal of radioactivity in seas. (Clark, *NYT*, 3/5/68, 19; SR 263, 3/5/68)

March 5: NASA launched 198-lb *Explorer XXXVII* (Solar Explorer B), second joint project of NRL and NASA to measure solar emissions, from NASA Wallops Station by four-stage Scout booster. Orbital parameters: apogee, 545 mi (877 km); perigee, 324 mi (521 km); period, 98.8 min and inclination 59.4°. Primary NASA objective was to place satellite in planned orbit and provide tracking and telemetry support. NRL objectives were to continue and augment overall solar x-ray monitoring program into ascending portion of 11-yr solar cycle; perform temporal measurements of x-ray emission intensity and spectral quality of solar flare emission; correlate measurements with optical and radio ground-based observatories; and provide real-time solar monitoring information. International scientific community had been invited to acquire solar radiation data from satellite.

Although *Explorer XXXVII* did not enter planned circular orbit,

orbit achieved would satisfy scientific objectives. Expected lifetime was one year. First NRL–NASA solar monitoring project, *Explorer XXX* (IQSY Solar Explorer), had been launched Nov. 18, 1965, and had ceased operating Nov. 20, 1967. Explorer program was directed by OSSA. Wallops Station was responsible for vehicle integration and launch operations. NRL was responsible for satellite development, construction, and testing. (NASA Proj Off; NASA Release 68–43; WS Release 68–5; AP, *NYT*, 3/7/68, 10; *W Post*, 3/7/68, D21)

- U.S.S.R. successfully launched two Cosmos satellites. *Cosmos CCIV* entered orbit with 843-km (524-mi) apogee, 271-km (168-mi) perigee, 95.7-min period, and 70.9° inclination. *Cosmos CCV* entered orbit with 292-km (181-mi) apogee, 197-km (122-mi) perigee, 89.3-min period, and 65.6° inclination. *Cosmos CCV* reentered March 13. *Cosmos CCIV* reentered March 2, 1969. (AP, *NYT*, 3/7/68, 10; GSFC *SSR*, 3/15/68; 3/15/69; *SBD*, 4/1/68, 174)

- NASA Associate Administrator for Advanced Research and Technology, Dr. Mac C. Adams, told Senate Committee on Aeronautical and Space Sciences that House Committee on Science and Astronautics cut of $48.3 million from request for FY 1969 NERVA program "would essentially wipe out the entire program. With the $11.7 million authorized by the House Committee [see March 1]—it is not clear that would even cover termination of the contracts. . . . We would immediately have to start termination of not only the engine development, but also termination of the technology work. I feel that we would waste a very valuable investment."

  Dr. Adams described "substantial progress" in OART programs in 1967: initial acoustic absorber tests showed reduction in jet transport noise; parawing (steerable parachute) showed promise for spacecraft and precise military landings; static firing of 260-in-dia solid rocket motor had reached 5.9-million-lb thrust in 80-sec test; NERVA had been successfully tested at full power 60 min; all major components of SNAP–8 nuclear turbo-generator system had completed 1,800 hr of endurance tests, one exceeding 13,000 hr; and new Beta cloth for astronaut clothing had passed wear and flammability tests. In "beneficial crossflow between engineering and life sciences," he named development of garment to give physicians three-dimensional data on heart activity, new blood velocity meter, and sensor to measure blood pressure in heart—small enough to be threaded through hypodermic needle and derived from instruments developed for small flight models for wind-tunnel tests. (Transcript)

- President Johnson presented Goddard Memorial Trophy for 1968 to Dr. Robert C. Seamans, Jr., former NASA Deputy Administrator and now NASA consultant and MIT visiting professor, in White House ceremony. National Space Club award named for the late Dr. Robert H. Goddard, "father of American rocketry," was given Dr. Seamans for helping "develop the policies, plans and programs that have led to the outstanding achievements and United States leadership in the field of rocketry and astronautics." President Johnson had received trophy in 1966. (*PD*, 3/1/68, 475; Program, Goddard Memorial Dinner, 3/5/68; *W Post*, 3/6/68)

- MSFC Director, Dr. Wernher von Braun, discussed "The Outlook for Space Exploration" addressing National Space Club's Goddard Memo-

March 5: *Rep. George P. Miller (D-Calif.), Chairman of House Committee on Science and Astronautics, reads citation on Goddard Trophy which President Johnson (right) presented to Dr. Robert C. Seamans, Jr. (center), former NASA Deputy Administrator.*

rial Dinner in Washington, D.C. "The public's knowledge and understanding of space exploration have been broadened by the remarkable achievements of the first decade of the space age, but a great many people in the space program are deeply concerned about an apparent decline in popular support of space activities." Space achievements had been well publicized but "the capability that made these feats possible, the underlying science and technology, the inherent value of delving deeper into the mysteries of the atmosphere and space, and the mechanism by which increased scientific knowledge enhances economic and social progress are apparently little understood or appreciated on a wide basis.

"Up to now we have had to devote our full energies to working on the means by which we reach into space. Now that the pipelines are filling with space hardware, more and more of our preoccupation is turning toward the question of what can be done to assure that the people who have supported our program . . . receive the full range of benefits which can be derived from space exploitation." (Text; *CR*, 3/12/68, E1769)

- Rep. Joseph E. Karth (D-Minn.), Chairman of House Committee on Science and Astronautics' Manned Space Flight Subcommittee, told Sixth Goddard Memorial Symposium in Washington, D.C., he saw "grave danger to our economic and social progress, as well as to our position in the world, if 'let technology wait' attitude becomes too prevalent." He advocated building sound foundation for economic growth by long-term investment in science, technology, and education and by maintaining technological leadership "as a necessary basis for our national security and economic strength." He called for "a space program better balanced between scientific, exploratory and economic payoff missions." (Text; Clark, *NYT*, 3/6/68, C33; *SBD*, 3/6/68, 25)

- NASA awarded $26,116,200 fixed-price-incentive-fee contract to McDonnell Douglas Corp. to provide launching and launch support services, including inspection and checkout, for improved Delta boosters. Contract would cover 20 launches from ETR and WTR over 21-mo period. (NASA Release 68-42; *WSJ*, 3/6/68, 20)

March 6: NASA awarded $30.1-million contract extension to Bendix Corp. for operation and maintenance services for launch facilities at KSC, bringing total contract value to $76.3 million. NASA also awarded $11.6-million extension to Aerojet-General for work on nuclear power plant, making total of $76.1 million awarded Aerojet-General through Aug. 30, 1969. (*WSJ*, 3/6/68, 20)

- LeRC awarded $11,600,440 cost-plus-award-fee contract to Aerojet-General Corp. to continue through Aug. 31, 1969, development of SNAP-8 nuclear space power system. (LeRC Release 68-16)
- USAF Space and Missile Systems Organization awarded North American Rockwell Corp. $1-million initial increment to $4-million fixed-price contract for production of Thor propulsion systems. (DOD Release 215-68)

March 7: In draft U.N. Security Council resolution at Geneva Disarmament Conference U.S., U.S.S.R., and U.K. formally committed themselves to take "immediate action" against nuclear attack or threatened attack on any country that renounced nuclear weapons. Object of big-power cooperation was to reassure governments asked to ban spread of nuclear weapons by treaty signature. (Hamilton, *NYT*, 3/8/68, 1; Middleton, *NYT*, 3/10/68, 8; *NYT*, 3/8/68, 40; 3/11/68, 14)

- Astronauts James A. Lovell, Jr., Charles M. Duke, Jr., and Stuart A. Roosa were chosen prime crew for 48-hr at-sea checkout of Apollo spacecraft, scheduled to begin March 18 in Gulf of Mexico. Apollo's at-sea post-landing systems for first manned mission would be checked from deck of NASA's motor vessel *Retriever*. (*H Chron*, 3/7/68)
- MSFC awarded $11,096,282 contract extension to Feb. 1969 to Sperry Rand Corp. for engineering support in applied research, testing, and design at MSFC's Astrionics Laboratory. (MSFC Release 68-37)
- New salt fog chamber at Naval Missile Center's environmental laboratory in Point Mugu, Calif., had improved and speeded tests of missiles and rockets. Chamber provided 70–120° F environment for weapons up to 14 ft long and operated automatically for round-the-clock tests. (PMR Release 280-68)
- U.K. Defence Equipment Minister Roy Mason advised Commons that military aircraft programs in 1968–69 would be worth more than £100 million ($240 million) for R&D alone. Employment in the industry had fallen from 268,000 in December 1963 to 264,000 in December 1967, but output was up by £133 million ($349 million) and exports by £90 million ($216 million), and Government assistance for

civil transport aircraft had quadrupled. Mason added that current offset agreements with U.S. would not now be changed to affect existing contracts which had provided high-level sales for aircraft and engine industries. (*InteraviaAirLetter*, 3/8/68, 3)

March 8: NASA Nike-Tomahawk sounding rocket launched from NASA Wallops Station carried GSFC experiment to 180-mi (290-km) altitude to verify performance of quadrupole ion-mass spectrometer and measure electron density by CU propagation technique. Rocket and instruments performed satisfactorily. (NASA Rpt SRL)

- NASA launched first two boosted Arcas I sounding rockets from TERLS carrying GSFC experiments to obtain data on equatorial electron density in ionosphere D region by means of radio propagation experiments. Rockets reached 52.6-mi (84.5-km) and 57.6-mi (93.7-km) altitudes and performed satisfactorily. Good data were obtained. (NASA Rpts SRL)
- ELDO announced unanimous election of Gen. R. Aubiniere of France as new president to succeed A. Paternotte de la Vailee of Belgium, named Belgium's ambassador to Lebanon. (*SBD*, 3/8/68, 43)
- U.K. estimated space expenditures for FY 1967–68, ending March 31, 1968, would total $73.2 million. Detailed amounts: ESRO, $9.9 million; ELDO, $23.3 million; and INTELSAT, $1.4 million. Largest other amount was $17.9 million for U.K.'s Skynet military comsat system. (*SBD*, 3/8/68, A3)

March 10: NASA's *Echo I* (launched Aug. 12, 1960), world's first passive reflector comsat, was being driven closer to denser regions of earth's atmosphere and would reenter before summer 1968. The 100-ft globe of aluminum-coated mylar plastic had been battered by space dust and meteoroids, its skin wrinkled, and its benzoic acid and anthraquinone inflating gas lost. *Echo I*'s nearly circular orbit 800 mi above earth had become egg-shaped in more than 7½ yr of operation. It had probably been seen by more people than any other man-made object in space. (NASA Release 68–44; P *EB*, 2/13/68; UPI, *NYT*, 2/14/68, 8)

March 11: The *International Biological Program—Its Meaning and Needs*, released by House Committee on Science and Astronautics' Science, Research, and Development Subcommittee, considered "one of the most crucial situations to face this or any other civilization—the immediate or near potential of man to damage, perhaps beyond repair, the ecological system of the planet on which all life depends." International Biological Program asked for $200 million over five years to attack problem; Subcommittee recommended $3–5 million for FY 1969. (Text; *W Post*, 4/5/68, A24)

- President Johnson sent to Congress *Marine Science Affairs—A Year of Plans and Progress; The Second Report of the President to the Congress on Marine Resources and Engineering Development* and reported on FY 1969 budget, which included $516 million for marine science and technology programs. Increased funding was requested for program on application of spacecraft technology in oceanography and improved observation and prediction of ocean environment. (*PD*, 3/18/68, 489–90)

March 12: NASA awarded $13,748,200 contract to General Electric Co.'s Missile and Space Div. for integration and test of 1,400-lb Nimbus D meteorological satellite. Nimbus D, fourth in series, would be launched

in 1970 carrying 10 experiments to improve long-range weather forecasting and further study of earth's atmosphere. (NASA Release 68-46)
- MSFC awarded $5,779,884 cost-plus-fixed-fee contract to Chrysler Corp., bringing total to $14.7 million, for ground support and engineering equipment for Saturn IB program. (MSFC Release 68-41)
- DOD awarded $456-million cost-plus-incentive-fee contract to Lockheed Missiles & Space Co. for development and production of Poseidon (C-3) missile system, making definite March 1966 letter contract for $26.6 million. Addition of $25 million had been awarded in October 1967 modification and total funding allocated to date was $507.7 million. About $1.2 billion would be spent through 1971. (DOD Release 230-68; AP, *NYT*, 3/13/68, 8)
- AFSC Aeronautical Systems Div. awarded Bendix Corp. $677,500 initial increment to $1.4-million contract for reentry vehicle systems reliability testing. (DOD Release 231-68)
- Analysis of *Luna IX* (launched Jan. 31, 1966) TV photos showed lunar mass was decreasing because of micrometeoroid bombardment, Soviet scientist A. A. Gurshteyn reported. He explained that speed of particles dislodged by falling micrometeoroids often exceeded second escape velocity for moon of only 2.4 km per sec. (*SBD*, 3/12/68, 61)

March 13: USAF launched unidentified satellite from Vandenberg AFB with Titan III-B-Agena D booster. Satellite entered orbit with 260-mi (418.5-km) apogee, 82-mi (132-km) perigee, 89.9-min period, and 99.9° inclination and reentered March 24. (*Pres Rpt 68; SBD*, 3/14/68, 75)
- GSFC selected Philco Ford Corp. for negotiations on one-year, $1,200,000 contract with two one-year options for engineering and related duties for NASA's 2-million-circuit-mi communications network (NASCOM). (GSFC Release G-8-68)

March 14: USAF launched two unidentified satellites from Vandenberg AFB using Long-Tank Thrust-Augmented Thor (LTTAT)-Agena D booster. First entered orbit with 242-mi (481.0-km) apogee, 114-mi (183.5-km) perigee, 90.2-min period, and 83.1° inclination and reentered April 10. Second entered orbit with 326-mi (524.6-km) apogee, 299-mi (448.2-km) perigee, 94.6-min period, and 31.6° inclination. (*Pres Rpt 68; SBD*, 3/18/68, 91)
- U.S.S.R. successfully launched *Cosmos CCVI*, carrying television and infrared cameras for meteorological observations. Orbital parameters: apogee, 640 km (398 mi); perigee, 598 km (372 mi); period, 97.1 min; and inclination, 81.2°. (GSFC *SSR*, 3/15/68; AP, *NYT*, 3/20/68, 10)
- Spacecraft for first manned Apollo mission would use 60% oxygen and 40% nitrogen cabin atmosphere while on launch pad and pure oxygen in orbit, NASA announced. Astronauts would continue to breathe pure oxygen in their spacesuits before and during launch phase, at higher pressure than that of cabin to avoid leakage into suits. Spacecraft environmental control system in orbit would gradually replace mixed cabin atmosphere with pure oxygen.

Since October 1967, more than 140 flammability tests on full-scale simulated spacecraft at MSC had shown that spacecraft modifications and installation of fire extinguisher and new quick-opening hatch for crew egress had drastically reduced fire hazards.

Detailed physiological review of new mixed gas atmosphere had included considerations of operational characteristics of spacecraft and life support equipment. Astronauts would be adequately protected physiologically during all phases of atmosphere change. Crew procedures during period when diluted oxygen atmosphere would be in cabin were only slightly affected. (NASA Release 68–47; O'Toole, *W Post*, 3/6/68, 1; UPI, *NYT*, 3/15/68, 41)

- NASA's *Ats III* (launched Nov. 5, 1967), in synchronous orbit above equator over mouth of Amazon River, was photographing cloud patterns with multicolor spin-scan camera to identify possible tornado-breeding storm clouds. Scientists would prepare movies from photos in effort to determine whether tornado breeding situations could be identified from characteristic cloud motions before tornadoes developed. *Ats* III was capable of photographing cloud patterns over northern hemisphere every 15 min. (ESSA Release 68–20; *NYT*, 3/14/68, 88)

- Sen. Frank Carlson (R-Kan.), ranking member of Senate Committee on Post Office and Civil Service, addressed Space Age Law Conference in Cape Kennedy, Fla. "The continuing progress of our nation in trying to solve the age old problems of poverty, ignorance and disease simply cannot be achieved if we neglect the basic strength of our national economy—the advancing frontier of technology. The basic point to remember is not whether this nation can have its space program—meet the crisis in the cities and fight a controversial war in Vietnam—but rather how we can possibly solve these problems unless we push ahead into the frontiers of technology." Sen. Carlson praised Earth Resources Observation Satellite program which would provide "new ability to improve harvests—prevent crop diseases—attack air and water pollution—inventory our agricultural resources—and perhaps even control floods. Frankly, I think this program alone would justify all that the space effort is costing!" (*CR*, 3/21/68, S3130–1)

- MSFC awarded $49,985 nine-month study contract to Lockheed Aircraft Corp. to assess possible radiation damage to film used on ATM to record solar activity and to consider radiation hazards to astronauts in Saturn I orbital workshop. (MSFC Release 68–43)

- Geneva disarmament conference completed draft of nuclear nonproliferation treaty and referred it to U.N. General Assembly special session. (Hamilton, *NYT*, 3/15/68, 18; Egli, *W Post*, 3/15/68, A19)

March 15: NASA launched two Aerobee sounding rockets from WSMR. One carried American Science & Engineering, Inc., payload to 96.2-mi (154.8-km) altitude to obtain high-resolution x-ray pictures of active regions on sun and general x-ray emission of solar corona. Ten percent loss in rocket performance suggested sustainer ruptured; instruments performed satisfactorily. Data would be degraded by pointing control failure.

Second rocket carried GSFC payload to 95-mi (152.8-km) altitude to collect data on x-rays from Crab Nebula and its proximity, using six collimated x-ray detectors. Rocket and instrumentation performance was satisfactory. Entire payload was recovered in excellent condition. (NASA Rpt SRL)

- NASA Nike-Tomahawk sounding rocket launched from NASA Wallops Station carried GSFC payload to 180-mi (290-km) altitude to verify performance of quadrupole ion mass spectrometer in D and E regions of ionosphere and measure electron density using CW propagation technique. Rocket and telemetry performed satisfactorily and good results were obtained from spectrometer, but propagation experiment was lost at 43.4-mi (70-km) altitude. (NASA Rpt SRL)

- NASA's HL-10 lifting-body vehicle, flown for first time Dec. 22, 1966, made its second flight, following vertical tail fin modifications to direct more airflow over its control surfaces. Piloted by Maj. Jerauld R. Gentry (USAF), HL-10 was air-launched from mother ship flying at 45,000-ft altitude and 400 mph. Rocket engine for craft was not used during 4½-min flight and 6,000-lb HL-10 was piloted through "U" pattern to make a 220-mph glide landing. Built by Northrop Corp.'s Norair Div., HL-10 was flight-tested in joint NASA–USAF study of potential as spacecraft capable of maneuvering in flight to a ground landing under pilot control. (FRC Release 8–68)
- At background briefing on Apollo 6 mission, NASA Apollo Program Mission Director, William C. Schneider, described flight planning activities for Apollo missions up to landing on the moon. Apollo 6 mission, originally scheduled for launch March 21, would be postponed until April because of "erratic behavior" in service module's helium check valves. Since *Apollo 5* had been so successful, Apollo 6 objectives had been "downgraded to secondary" and primary objective would be to evaluate launch vehicle. New optimum injection method in which S–IVB would be launched toward moon would be used but rest of mission would be basically same as *Apollo 5*, he said. NASA would continue mission planning, "so that we have the option to select a lunar orbit mission if as a result of previous flights we felt it was technically the best thing to do." (Transcript; Sehlstedt, B *Sun*, 3/16/68, 1)
- Univ. of Illinois sociologist Alexander Vucinich in *Science* described problem of relationship of science to morality which prompted leading Soviet scientists to search for broader cultural autonomy of science. "Moral law has become a by-product of science; science, in official Soviet ideology, is a structural component of Soviet society, while the moral code is only a superstructural derivation." He traced growth of critical reassessment made by individual members of Soviet Academy of Sciences who looked to "the broader cultural effects of modern science and the ongoing technological revolution" under stimulation of considerable relaxation of political and ideological controls which resulted from Stalinist policies. (Vucinich, *Science*, 3/15/68, 1208–12)

March 16: U.S.S.R. successfully launched *Cosmos CCVII* into orbit with 342-km (213-mi) apogee, 210-km (130-mi) perigee, 89.8-min period, and 65.6° inclination. Satellite reentered March 24. (*Krasnaya Zvezda*, 3/19/68, 1; *SBD*, 3/19/68, 100; GSFC *SSR*, 3/31/68)

- Demonstrated success of lunar module's (LM) initial unmanned test flight in earth orbit Jan. 22 and of subsequent ground test analysis by NASA Manned Space Flight Safety Office had eliminated need for second unmanned LM flight, NASA announced. First manned LM flight using Saturn V launch vehicle was planned for late 1968. (NASA Release 68–50; UPI, W *Star*, 3/17/68, A4)

March 17: Tenth anniversary of second U.S. satellite, *Vanguard I*, 3¼-lb, 6½-in-dia spacecraft that had proved earth was slightly pear shaped and examined composition of upper atmosphere. Satellite had stopped

March 17

March 17: *Ten years ago Vanguard I, second U.S. satellite, was launched from Cape Canaveral (now Cape Kennedy) into orbit expected to last another 200 years. Satellite, 6½ inches in diameter, proved earth was pear shaped and continued transmitting until May 1964. (Official U.S. Navy photo)*

transmitting in May 1964, but was expected to remain in orbit at least 200 yr longer. (KSC Release 63–68; AP, *M Her*, 3/17/68)

March 17–18: NASA Wallops Station engineers launched three Nike-Apache and three Nike-Tomahawk sounding rockets from Vega Baja launch site in Puerto Rico to detect diurnal changes in atmosphere and ionosphere. Seventh planned firing was postponed because of rocket and radar tracking difficulties. Measurements taken by rocket-borne experiments, furnished by GSFC and Univ. of Michigan, would be compared with those taken by radiotelescope at Arecibo Ionospheric Observatory in Puerto Rico, and by Canadian *Alouette II* satellite during overpass for three-way comparison of rocket, satellite, and observatory data. (WS Release 68-7)

March 18: NASA launched Nike-Tomahawk sounding rocket from Churchill Research Range carrying Univ. of California experiments to study flux-energy-spectrum, pitch-angle distribution and time and space variation of mirroring and precipitated charged particles in auroral zone with variety of detectors. Good data were received from all experiments except electric field measurement experiment which had two damaged booms. (NASA Rpt SRL)

- NASA announced personnel changes effective May 1: Deputy Associate Administrator for Manned Space Flight Edgar M. Cortright would become LaRC Director, replacing Dr. Floyd L. Thompson, who would serve as Special Assistant to the Administrator until his retirement Nov. 25 at age 70 and would then serve as part-time consultant. Cortright would be replaced by Charles W. Mathews, Director of Apollo Applications Program. Harold T. Luskin, who would serve as Deputy Associate Administrator for Manned Space Flight (Technical) until May 1, when he would succeed Mathews, would be replaced by LaRC Deputy Director Charles J. Donlan. (NASA Release 68-51)

March 19: NASA Aerobee 150 sounding rocket launched from WSMR carried ARC payload to 98.7-mi (159-km) altitude to check Solar Pointing Aerobee Rocket Control System (SPARCS) and map flight-path magnetic field. Rocket and instrumentation performed satisfactorily. (NASA Rpt SRL)

- FRC engineers had used small, inexpensive, radio-controlled model spacecraft to evaluate concepts for possible advanced spacecraft recovery systems in over 100 successful flights, FRC's planning engineer Robert D. Reed revealed. In status report to AIAA's 2nd Flight Test, Simulation and Support Conference in Los Angeles, Reed said flight tests of models, including heavy volume and slender lifting-body vehicles, with various advanced flexible-wing and gliding-parachute recovery systems, were being tested to determine their suitability for ground landings. (FRC Release 9-68)

- NASA Associate Administrator for Tracking and Data Acquisition Gerald M. Truszynski, in statement before Senate Committee on Aeronautical and Space Sciences, summarized 1967 activities and technical problems and presented FY 1969 funding requirements for tracking and data acquisition program. He explained that FY 1968 funding limitations had made it necessary to defer important equipment procurement affecting network reliability and recounted efforts to realize savings and reduce operating costs. (Testimony)

- NASA Manager for Space Nuclear Propulsion Milton Klein explained to Senate Committee on Aeronautical and Space Sciences that AEC's nuclear space power work would "establish technology in advance of missions, with specific system development undertaken as mission requirements dictate." He presented a sampling of AEC's activities as it worked closely with NASA and DOD "to keep in focus the types and likely timing of future mission needs." (Testimony)

- ESSA had received its first meteorological photos from U.S.S.R.'s *Cosmos CCVI*, launched March 14, *Space Business Daily* reported. All seven pictures were "good quality." (*SBD*, 3/20/68, 109)
- D. P. Nerry warned in *Data* that space research, candidate to help alleviate many national problems, was "placed on a back burner of the nation's stove of priorities. . . . For the sad fact is that after landing men on the moon this nation will only have a cursory space program." Nerry called for assessment of new outlook on space for 1970s. "If, for example, Congress should decide that there should be a permanent large-scale reduction in the resources allocated to space, NASA then should be firmly advised to plan accordingly." But if slowdown would be temporary, NASA should plan to pursue space exploration "vigorously" at end of present delays. Nerry urged that press employ "strong debate and factual reporting" to provide "a reappraisal of U.S. emptiness in space in this Guns-Butter Society." (Nerry, *Data*, 3/19/68, 7)

March 19–20: More than 100 scientists and engineers attended Saturn I Workshop design review board meeting at MSFC to discuss results of previous reviews. Workshop, scheduled for launch in 1970, would consist of Saturn IV–B stage modified for living and working in space for better understanding of permanent, man-made, island-in-space requirements. (MSFC Release 68–45)

March 20: MSFC Director Dr. Wernher von Braun, addressing National Capital Area Chapter of American Society for Public Administration, praised U.S. space program for "unequaled competence and unlimited opportunity." After 10 yr in space, he noted, U.S. had launched 514 spacecraft into earth orbit and 28 to moon and other planets. "The moon has been completely mapped, the composition and texture of its soil analyzed, and sites . . . selected for manned lunar landings. Our scientific knowledge of Mars and Venus . . . has been advanced tremendously by . . . unmanned, instrumented probes. . . . American astronauts, who have a total of almost 2,000 hours in space to their credit in the Mercury and Gemini Projects, will begin the third phase of manned space flight this year with the first manned launches in Project Apollo. These achievements in space, fruits of the advanced science and technology of our times, mark the first decade of the Space Age as one of the finest periods in American history. And they are forerunners of even greater discoveries and widespread applications to come during our second decade in space."

Dr. von Braun credited NASA Administrator James Webb's direction with adaptation of systems approach to space research without which "we would never be able to go to the moon." Approach, Dr. von Braun believed, could also be tested on river pollution and satellite communications. (Text; *W Post*, 3/21/68, H14)

- AIAA named Maj. William J. Knight (USAF) winner of Octave Chanute Award for notable contribution to hypersonic and reentry flight. Maj. Knight flew X–15 research aircraft at mach 6.72 (4,534 mph) Oct. 3, 1967. Annual award of $500 and travel stipend of $1,000 would be presented at AIAA Second Flight Test, Simulation and Support Conference in Los Angeles March 26. (AFNS Release 3–8–68–165; AIAA Release, 3/20/68; AFSC *Newsreview*, 5/68, 9)
- M/G J. C. Maxwell, director of SST development for FAA, told Wings Club in New York SST would have "tremendous significance" on U.S.

balance of payments in international trade. One SST would "pay for 20,000 Volkswagens." He explained Boeing Co.'s Feb. 22 announcement of one-year delay in prototype construction: "Our prototype design simply wasn't good enough, not yet. . . . We are convinced that we can get what we need but we want to stop now to make certain that we do get it."

Very reason for existence of SST program was belief "that this is next logical step in development of civil aeronautics." Supersonic travel was first step toward hypersonic flight, he said. "Unless we take it now it's going to be many years before civil aviation advances beyond subsonic flight—if ever." Biggest problem facing SST program was sonic boom. "We have made all our program decisions . . . on assumption that supersonic flights over land may not be permitted. We are reasonably certain . . . we can operate over the oceans on an inter-continental basis," and have sufficient market to assure profitable program." (Text; *Seattle Times*, 3/21/68; Hudson, *NYT*, 3/21/68, 3/24/68, 23)

- PMR Aero-mechanical Branch announced development of new cold-gas rocket-launching system consistently more reliable, less expensive, and more efficient than old cartridge system. New system, being used for launches of Arcas meteorological rockets, consisted of modified launcher door incorporating 550-cu-in tank of compressed air. Ignition forces opened valve which released air into launch chamber, increasing liftoff power and peak altitude. (PMR Release 350–68)

March 21: U.S.S.R. successfully launched *Cosmos CCVIII*. Orbital parameters: apogee, 278 km (173 mi); perigee, 202 km (126 mi); period, 89.2 min; and inclination, 64.9°. Satellite reentered April 2. (UPI, *NYT*, 3/22/68, 5; *SBD*, 3/22/68, 125; GSFC *SSR*, 3/31/68; 4/15/68)

- NASA Nike-Tomahawk sounding rocket launched from Churchill Research Range carried Univ. of Maryland experiment to 150-mi (241-km) altitude to study dissipation processes and electron acceleration mechanisms in aurora. Four experiments measured energy spectrum of electrons from 1 ev to 200 kev. Photometers, UV emissions, and plasma probes were included. Three of four electron experiments returned data on energy spectra; no data were acquired from optical and plasma experiments because clamshell nose cone failed to deploy. (NASA Rpt SRL)

- XB–70 research aircraft was flown to 15,500-mi altitude by Col. Joseph F. Cotton (USAF) and NASA test pilot Fitzhugh L. Fulton, Jr. Primary goals were not achieved because main landing gear valve malfunctioned. Low-speed, gear-down handling qualities, propulsion system, and performance tests were accomplished during 2-hr 30-min flight from Edwards AFB. (XB–70 Proj Off)

March 22: *Cosmos CCIX* was successfully launched by U.S.S.R. into orbit with 945-km (587-mi) apogee, 871-km (541-mi) perigee, 103.1-min period, and 65.3° inclination. (*SBD*, 3/25/68, 136; GSFC *SSR*, 3/31/68)

- NASA Aerobee 150 sounding rocket launched from WSMR carried GSFC payload to 129.4-mi (208.2-km) altitude to measure spectral irradiance of Gamma star in constellation Vela and of Zeta star in constellation Puppis using UV stellar spectrograph. Rocket performance was satisfactory. Instrumentation performance was marred by loss of fine pointing 17 sec before end of second exposure. (NASA Rpt SRL)

- NASA authorized partial restoration of primary structure of M2–F2 lifting-body vehicle so it could be removed from inspection jig used to determine damage sustained in landing accident May 10, 1967. It would be returned by builder, Northrop Corp.'s Norair Div., to FRC by late summer. OART would determine future research work on M2-F2 from "flight results and other experience obtained from both the M2 and the HL-10." (FRC Release 10-68; ARC *Astrogram*, 3/28/68, 2)

- Sen. Gaylord Nelson (D-Wis.), member of Senate Committee on Interior and Insular Affairs, on Senate floor praised construction of Wisconsin Regional Space Center as "a place where the layman can acquire information about the developments and our accomplishments in the space age." He inserted in *Congressional Record* article by Barbara E. Kocjan, stenographic coordinator of projects for Center, describing Center as "a highly effective teaching laboratory that can readily be tied into the on going programs of tens of thousands of elementary and secondary schools and hundreds of colleges and universities." (*CR*, 3/22/68, E2151)

March 24-25: NASA launched series of three Nike-Cajun sounding rockets from Natal, Brazil, carrying GSFC payloads to obtain data on diurnal temperature variations as a function of latitude using grenade technique. Rockets performed satisfactorily and grenades exploded at planned intervals. (NASA Rpt SRL)

March 25: Solar Wind Composition Experiment, developed by Swiss scientific team headed by Univ. of Berne's Dr. Johannes Geiss, was first foreign experiment accepted for NASA's Apollo Lunar Surface Experiments Package (ALSEP). Swiss experiment would require astronaut to deploy on lunar surface aluminum foil sheet oriented toward sun to entrap solar wind ions; foil sheet would be packaged for return to earth by astronauts, providing in this one-pound experiment first opportunity to capture ions for earth analysis.

Swiss National Committee for Space Research had already built prototype. NASA would provide qualification tests for prototype and construction and acceptance tests for flight hardware. (NASA Release 68-55; *Marshall Star*, 3/27/68, 10)

- Dr. Thomas O. Paine was sworn in as NASA Deputy Administrator in NASA Hq. ceremony. Nomination had been confirmed by Senate Feb. 7. (NASA Ann)

- MSFC awarded Univ. of California $929,000 contract to develop and test prototype balloon flight system for NASA's High Altitude Particle Experiment (HAPPE) program. Primary objective of HAPPE was to use naturally occurring radiation in investigating interactions of elementary particles at 90,000-ft altitude. Balloon payload for prototype flight, scheduled for late summer, would be 50 ft tall and 8 ft in dia and would weigh 10,000 lb. (MSC Release 68-26; *Aero Tech*, 4/8/68, 3)

- Use of 260-in solid rocket motor as 1st stage for advanced manned launch vehicle was advocated by MSFC Director, Dr. Wernher von Braun, *Aerospace Technology* reported. Dr. von Braun, whose support for motor was revealed in Oct. 11, 1967, testimony released by House Committee on Science and Astronautics' Subcommittee on NASA Oversight, said concept was advantageous for resupply of flights to space stations and for space rescue systems where 80% of emergencies in manned spacecraft in earth orbit were "time critical." NASA and DOD, in joint study, were considering use of 260-in motor and its development

March 25: NASA *Administrator James E. Webb shakes hands with Dr. Thomas O. Paine (left) after swearing him in as Deputy Administrator of* NASA.

but questioned whether there would be enough missions for booster to justify high development costs. (Text; *Aero Tech*, 3/25/68, 12)

- Four F–111A jet aircraft flew first mission from Ta Khli AFB in Thailand under radar control to target areas northwest of Dong Hoi, North Vietnam, according to unofficial sources, AP reported. Aircraft conducted successful nighttime strike and encountered no enemy aircraft or surface-to-air missiles. (AP, B *Sun*, 3/26/68, A2)

March 26: ComSatCorp in fourth annual report announced 1967 net income of $4.6 million. Highlights in 1967 included successful launch of three satellites (in four attempts) into synchronous orbit to extend full-time service to Pacific and expand capability in Atlantic, beginning of construction on three U.S. ground stations, completion of new ground stations by Italy and Spain, and increase in INTELSAT membership to 61 with addition of Uganda in January. (Text; ComSatCorp Release 68–11)

- Director of Defense Research and Engineering, Dr. John S. Foster, Jr., in statement before Senate Committee on Aeronautical and Space Sciences, described DOD activities in space and aeronautics. "Military launches," he stated, "were intended to implement in part the functions of navigation, communication, nuclear detection, space defense, and meteorology, or to furnish scientific and/or development support." NASA's work, he said, had contributed "heavily" to science and technology base for DOD effort. In MOL activities "considerable progress" had been made; feasibility of the MOL system was affirmed, technical risks were assessed in some detail, and full scale development was initiated."

He described coordinated DOD–NASA range and network efforts and other cooperative programs in which DOD made its resources available to NASA on reimbursable basis. (Testimony; *Aero Tech*, 4/8/68, 10–11)

March 27: Cosmonaut Yuri A. Gagarin—who became first man in space when he orbited earth once in U.S.S.R.'s *Vostok I* April 12, 1961—and engineer Col. Vladimir S. Seryogin were killed when their MiG–15 jet aircraft crashed northwest of Moscow during training flight. Gagarin, who had been commander of Soviet Corps of Cosmonauts and officer in charge of cosmonaut training, was second cosmonaut to die in an accident. Cosmonaut Vladimir M. Komarov had died when *Soyuz I* crash-landed after reentry April 24, 1967. Bodies of Cosmonaut Gagarin and Col. Seryogin would be cremated and buried in Kremlin wall. (UPI, *W Star*, 3/28/68, 1; Reston, *W Post*, 3/29/68, 1)

- First launch of British Aircraft Corp.-built Skylark sounding rocket from ESRANGE near Kiruna, Sweden, carried London Imperial College experiment to 105-mi (170-km) altitude to investigate relationship between auroral events and polar "electrojet." Rocket and instrumentation performance was satisfactory. (*SBD*, 4/4/68, 199)

- USAF was testing thermal preconditioner with promethium-147 radioisotope heat source to reduce aircraft guidance-system errors caused by temperature changes, AFSC announced. Unit, which eliminated warmup time required by orthodox heaters, had been developed by AEC and USAF. (AFSC Release 7.68)

- Rep. Henry S. Reuss (D-Wis.), Chairman of House Committee on Government Operations' Research and Technical Programs Subcommittee, released *Scientific Brain Drain from the Developing Countries* (dated March 28). Immigration to U.S. of scientific manpower from developing countries had more than quadrupled in past 12 yr. Report, based on Subcommittee hearings held Jan. 23, said, "the long-sustained U.S. foreign aid program has devoted substantial sums and given high priority" to educating and training professional manpower. When these countries "suffer an emigration drain of the very skills and talents they are attempting to increase, an important part of the foreign aid program is undermined." (Text; *NYT*, 3/28/68, 33)

March 28: NASA Nike-Apache sounding rocket launched from NASA Wallops Station carried GSFC experiment to 100.2-mi (161.2-km) altitude to test flight models of impedance measuring instruments to be flown on RAE–A satellite. Rocket and instrumentation performance was satisfactory. (NASA Rpt SRL)

- Apollo 6 mission, second flight of Apollo/Saturn V space vehicle, would be launched from ETR on or after April 3, NASA announced. Primary mission objective would be qualification of launch vehicle for future manned flights; spacecraft objectives, including recovery, would be secondary. (NASA Release 68–54)

- ARC Director H. Julian Allen had been elected Fellow of U.K.'s Royal

Aeronautical Society for outstanding contributions to aeronautics, ARC *Astrogram* announced. Allen, who had served NACA and its successor NASA since 1936, had received NACA Distinguished Service Medal for originating concept of bluntness as aerodynamic technique for avoiding severe reentry heating, AIAA Sylvanus A. Reed Award, Air Force Assn.'s Air Power Trophy, and NASA Exceptional Scientific Achievement Medal. (ARC *Astrogram*, 3/28/68, 1)

- DOD announced loss of F-111A aircraft with two crew members, presumably on mission to North Vietnam. According to Hanoi radio broadcast, F-111A, one of six based at Thailand, had been shot down near Laotian border. (AP, B *Sun*, 3/29/68, 1; AP, W *Star*, 3/29/68, A4; Kumpa, B *Sun*, 3/30/68, A2; Wilson, W *Post*, 3/30/68, A14)

- USN announced compromise plan to purchase 8 F-111B aircraft instead of original 30 planned and to consider building alternate aircraft lighter and more maneuverable than F-111B. However, Senate Armed Services Committee voted 11 to 2 against additional funds for F-111B, dimming chances of continuing program at all. (Finney, *NYT*, 3/29/68, 1; Maffre, W *Post*, 3/29/68, 1; *WSJ*, 3/29/68, 6)

March 27-28: GSFC team headed by Dr. Henry H. Plotkin hit orbiting satellite *Explorer XXXVI* (*Geos II*) with laser beam but failed for two successive nights to relay Morse code on laser beam to satellite for replay to earth. First test to carry modulated 13-kc signal into space for detection by spacecraft was unsuccessful because satellite was so low on horizon it passed out of range before scientists had time to send coded message. Second attempt at message delivery failed because of overcast. (NASA Release 68-56; W *Star*, 3/28/68, 1; W *Post*, 3/29/68, A18)

March 29: FAA released 12-yr forecast of aviation activity. Number of airline passengers was expected to triple, from 126 million in FY 1967 to 444 million in FY 1979; revenue passenger miles would quadruple, from 86 billion to 342 billion; and airline fleet would increase from 2,272 aircraft to 3,860. General aviation fleet would double, from 104,706 aircraft on Jan. 1, 1967, to 203,000 by 1979, and its flying hours would increase from 21.9 million to 40.5 million hr. Total flights under instrumented flight rules (IFR) would increase from 5.8 million in 1967 to 18.7 million in 1979. (FAA Release 68-21)

- NAA announced award of 1967 Frank G. Brewer Trophy to New York Univ. Professor Emeritus Dr. Roland H. Spaulding for "continuous, outstanding, and pioneering contributions in aerospace education to the youth of the nation. . . ." (NAA *News*, 3/29/68)

- Sen. Karl E. Mundt (R-S.D.), ranking member of Senate Committee on Government Operations, praised March 28 decision of Senate Committee on Armed Services to disapprove additional funds for F-111B aircraft and urged that F-111As in combat zone "be brought back immediately . . . and not returned to combat until they are fully tested and ready." He said it was "imperative that the production line be held up until the design of the airplane number 160 configuration is completed and tested and can be incorporated into the very next plane the Air Force buys. If this drastic step is not taken, then we will truly be committing another billion dollar blunder in this TFX program which already has cost the American taxpayers many, many fruitless billions of dollars." (*CR*, 3/29/68, S3657-60)

March 29
- USA and Western Electric Co., prime contractor, signed six-month, $85,480,628 initial production contract for Sentinel ABM system. Western Electric would receive $28 million of total; balance would be shared by eight subcontractors. Contract would be managed by Sentinel System Command, Redstone Arsenal, Ala. (DOD Release 284–68)
- Lockheed Aircraft Corp. announced orders for 144 triple-jet L–1011 airbuses totaling $2.16 billion. Eastern Airlines, Trans World Airlines, and U.K.'s Air Holdings, Ltd., ordered 50, 44, and 50 aircraft. Order, largest in history for aircraft, was dramatic reentry for Lockheed into commercial aircraft industry, which it left in 1962 at conclusion of turboprop Electra production. U.K.'s Rolls-Royce, Ltd., was selected to build engines for 250- to 275-passenger aircraft, which would be delivered starting in fall 1971. (Hudson, *NYT*, 3/30/68, 1)
- U.N. Secretary General U Thant presented to U.N. Economic and Social Council *Resources of the Sea,* urging world action to regulate use of sea's largely unexplored riches. (Brewer, *NYT*, 3/30/68, 16)
- Nearly 200,000 mourners filed past biers of Cosmonaut Yuri A. Gagarin and Col. Vladimir Seryogin in Soviet Army House in Moscow. Premier Alexey Kosygin, President Nikolay V. Podgorny, and Communist Party leader Leonid I. Brezhnev briefly joined honor guard of soldiers near biers. (Winters, B *Sun*, 3/30/68; *NYT*, 3/31/68, 21)

March 30: USA successfully conducted first test-firing of Spartan long-range interceptor missile at Kwajalein Test Site in Marshall Islands. (DOD Release 289–68; UPI, *W Post*, 3/2/68, A7)

- Second combat loss of F–111A aircraft in Vietnam conflict occurred when aircraft crashed in Thailand after an "in-flight emergency." Two crewmen were rescued. First F–111A was reported missing March 28. (*NYT*, 3/31/68, 1; 4/1/68, 1; *W Post*, 3/31/68, 1; AP, W *Star*, 3/31/68, 1; Wilson, *W Post*, 4/1/68)

March 31: NASA's six-week 1968 Airborne Aurora Expedition had accomplished most intensive studies ever made of northern lights and had proved dramatically value of high-altitude observatory jet transport aircraft to "stop time" near poles [see Jan. 18]. Expedition, based at Fort Churchill, Canada, with cooperation of National Research Council of Canada and managed and directed by ARC's Louis C. Haughney, had ended in mid-March. Convair 990 jet transport, carrying 13 experiments, followed underneath auroras at 550 mph, making frequent trips to north magnetic pole, north of Greenland; took 40,000 auroral photos; and recorded instrument readings on 180,000 ft of magnetic tape. Jet aircraft canceled out speed of earth's rotation by flying against it and holding constant position on night side of earth opposite sun at latitudes above 60° north.

Aircraft on three occasions crossed same spot on arctic north pole end of earth's magnetic field as did NASA's *Ogo IV* satellite at south pole end, in 400-mi-altitude orbit over Antarctica. Measuring instruments on both satellite and aircraft were nearly identical. In another aircraft-satellite combination, on six passes aircraft measured group of auroras from below at 40,000-ft altitude while *Ogo IV* measured same group from above. (NASA Release 68–45; ARC Release 68–8)

- Cost of spaceflight was discussed by space writer Arthur C. Clarke. As with aviation, cost would decrease as techniques improved. Reusable spacecraft, orbital refueling, and nuclear propulsion would make travel to moon, at least, "comparable in cost to that of global jet transport today."

If conquest of space served no other purpose, it would provide "new mental and emotional horizons which our age needs more desperately than most people yet realize." Yet there were signs U.S. space effort was "grinding to a halt, as its initial crisis-induced momentum is exhausted." U.S. might well be "Sputniked" again in early 1970s. (Clarke, *H Chron*, 3/31/68; *LA Times*, 3/31/68)

During March: Special ESSA task force recommended environmental science and service agencies take early, joint steps toward national effort for development and use of earth-oriented space technology. In *Man's Geophysical Environment: Its Study from Space*, task force predicted future space platforms would be able to acquire global geophysical data on unprecedented scale for environmental disciplines. It recommended combination of manned and unmanned space vehicles—rather than either alone—and warned that orbiting environmental observatories might provide data too rapidly for effective use unless data handling and display improvements were begun in immediate future. It rated highly spacecraft capability to service and repair manned spacecraft in orbit, to provide semiautomatic mode to operate manned spacecraft after flight crew left, and to launch subsatellites, special probes, and recoverable capsules from orbit.

Among proposals for missions were continuous monitoring of space disturbances to predict spaceflight hazards, global ionospheric mapping, global noise and interference survey, global measurements of absolute ground and sea surface temperatures and surface roughness, and surveys of snow areas, river and lake ice distribution, rain and river gauging, shoals, and sea states. (Text)

- NRC Committee on Polar Research, established in 1958 to continue research begun by NAS during International Geophysical Year, began review of significant results of past research efforts, to pinpoint scientific questions that should be studied in either of polar regions during next few years, and to make recommendations on national research goals. (NAS–NRC–NAE *News Report*, 3/68, 2)

- Strong arguments in favor of Europe's making comsats "focal point" of space activity were presented in *Spaceflight* by spacewriter Arthur C. Clarke, former chairman of British Interplanetary Society. "Reliable domestic radio services are not available over most of the world. Long distance services are of poor quality . . . [and] by 1970, there will be 130 million VHF sets in the world, many of which could pick up direct radio broadcasts from satellites." U.K., he said, "certainly cannot do *everything* in space. But what we should not tolerate is the apparently invincible ignorance of those who think that nothing in space is worth doing. . . . Our space achievements will be our greatest legacy to the future. Indeed they will create that future. They will make it possible to have a future." (Clarke, *SF*, 3/68, 78–84)

- Effects of manned space flight program on communities surrounding NASA space flight centers were discussed in *Monthly Labor Review*. South had gained most from 1967 decision to proceed with Apollo Program. Total civil service and contractor employment increase of 66,000 in five states bordering Gulf of Mexico was about 5% of area's increase in total nonagricultural employment. However, economic significance of program had not been uniform. Employment at Mississippi Test Facility exceeded half of total employment in Hancock County in 1966, while space employment in Houston, Texas, was less than 2% of total employment. Space employment accounted for slightly more than 7% of total employment growth in Alabama, Louisiana, and Mississippi since 1961.

Employment growth brought population growth, caused expansion of school facilities and faculties, raised per capita income, and increased retail sales. Beginning decline of employment in program in 1966 had moderated economic growth, and if funding continued to decline after 1968 communities would have to adjust to sharply contracting employment. (Holman, Konkel, *Monthly Labor Review*, 3/68, 30–6)

April 1968

April 1: Dr. Robert R. Gilruth, Director of NASA's Manned Spacecraft Center, was elected member of National Academy of Engineering "in recognition and in honor of his important contributions to engineering and of his leadership in the field." He was cited for development and operation of manned spacecraft. (NAE PIO; MSC *Roundup*, 4/12/68, 1)

- USAF F-111A aircraft had been grounded by USAF pending results of investigation of March 28 and 30 crashes, Associated Press reported. (W *Star*, 4/1/68, A3)
- *New York Times* editorial on F-111A aircraft: "The difficulties that have beset the controversial F-111 swing-wing plane recently provide new evidence of the folly of allowing political factors to veto or dilute technical judgment.

    "These additional blows to a plane that still has great potential promise emphasize what most experts have been saying for seven years —the F-111 has been built the wrong way from the beginning. It reflected former Secretary of Defense McNamara's insistence, in the name of 'commonality' and savings, that the Navy and the Air Force buy one plane for two entirely different missions.

    "But, though the Navy's version may never—as Congress believes—meet the Navy's needs, the Air Force model has great potential capabilities as a supersonic high- and low-level all-weather fighter-bomber . . . [and] must be developed, tested and utilized to its fullest capability." (*NYT*, 4/1/68)
- AEC–NASA Space Nuclear Propulsion Office awarded Aerojet-General Corp. extension of cost-plus-fixed-fee contract for nuclear propulsion work. Extension covered period through Sept. 30, but NASA funding would be restricted to effort through July 31 pending Congressional action on NASA's FY 1969 budget request. Extension brought total estimated cost to $59,413,790 for Oct. 1, 1967, through Sept. 30, 1968, including $25,845,000 NASA share. (NASA Release 68–57; *WSJ*, 4/2/68, 12; *SBD*, 4/2/68, 179)
- MSFC contract awards: $2,056,360 contract modification to RCA for continued support of RCA 110A computers for use in checkout and launch of Saturn IB and Saturn V launch vehicles, bringing total contract value to $12.7 million; and $11,750,000 follow-on contract to Sanders Associates, Inc., to provide logistics and engineering support to Saturn V operational display systems at MSFC, bringing total value of contracts to $3,899,548. (MSFC Releases 68–57, 68–58)
- USAF awarded RCA $100,000 initial increment to $1.5-million fixed-price contract for study, evaluation, and testing of advanced electro-optical techniques for surveillance of high-altitude space vehicles. (DOD Release 288–68)

April 2: NASA and German Federal Ministry for Scientific Research (BMWF) were conducting series of four sounding rocket launches from

Thumba Equatorial Rocket Launching Station (TERLS) to study upper atmosphere near Equator. NASA Nike-Apache sounding rockets ejected barium clouds between 90- and 120-mi altitudes to investigate electric fields in upper atmosphere region of intense electric current. Results would be available to world scientific community. BMWF was responsible for chemical payloads, photographic equipment, and cloud observation; Indian National Committee for Space Research provided launch services; and NASA supplied sounding rockets and rocket launcher. (NASA Release 68–58)

- President Johnson, in letter transmitting to Congress *Fifth Annual Report on Communications Satellite Act of 1962*, said: "[Report] reflects . . . steady progress toward the ultimate goal of providing mankind with new capabilities for worldwide communication. In the brief span of five years, satellite technology has grown dynamically. The possibilities envisioned in 1962 have been greatly exceeded." Communications, he said, must provide " 'network for knowledge' so that all peoples can share the scientific, educational, and cultural advances of this planet. . . .

  "Failure to reach these goals can only contribute to apathy, ignorance, poverty and despair in a very large part of the world. Success in our telecommunications policies can be a critical link in our search for the understanding and tolerance from which peace springs. Communication by satellite is a tool—one of the most promising which mankind has had thus far—to attain this end." (Text; AP, W *Star*, 4/3/68, A7; *PD*, 4/8/68, 637)

- Senate Armed Services Committee, after hearing DOD witnesses in closed session, voted to appropriate $297 million—including $170 million for contract definition of Navy VFX–1 as possible replacement for F–111B—for continuation of USN's fighter-bomber program and procurement of F–4J Phantom jet aircraft. Committee had voted March 28 to deny the funds. Contract definition phase was expected to take 8–12 mo. (W *Star*, 4/3/68, A8)

- Dr. John C. Houbolt, Executive Vice President of Aeronautical Research Associates of Princeton, Inc., received AIAA Structures and Materials Award for his "original, definitive, and continuous research leading to the use of random processes in aircraft gust loads design." (*SBD*, 4/4/68, 199)

April 3: U.S.S.R. successfully launched *Cosmos CCX* into orbit with 374-km (232.4-mi) apogee, 198-km (123-mi) perigee, 90.2-min period, and 81.3° inclination. Satellite reentered April 11. (*SBD*, 4/4/68, 197; GSFC *SSR*, 4/15/68)

- Nike-Apache sounding rocket launched by NASA from Churchill Research Range carried GSFC payload to 81-mi (129-km) altitude to gather data on charged particle flux associated with aurora and to investigate distribution of electric fields in ionosphere, occurrence of radio noise, and ionospheric electron densities during auroral displays. Experimental results were 70% successful. (NASA Rpt SRL)

- National Academy of Sciences president, Dr. Frederick Seitz, was elected president of Rockefeller Univ. to succeed Dr. Detlev W. Bronk, who would retire July 1. Dr. Seitz would divide his time between NAS and University until early 1969, when he would assume his full-time educational duties. Member of President's Science Advisory Committee and of DOD's Defense Science Board, which he chaired four years ending in March, Dr. Seitz had succeeded Dr. Bronk as NAS president in 1962. (Farber, *NYT*, 4/4/68; NAS–NRC–NAE *News Report*, 4/68)

- Dr. Harold A. Rosen, Assistant Manager of Hughes Aircraft Co. Space Systems Div. and Manager of Hughes Satellite Systems Laboratories, was named recipient of AIAA's first Aerospace Communications Award for his "leadership in making synchronous satellite communications a global reality, thereby opening a new challenge for the progress of mankind." Award also honored late Don Williams, former Chief Scientist for Communications Satellite Systems at Hughes, for "his early recognition, technical judgment, inventiveness, and singular dedication in pioneering the development and design of synchronous communications satellites." (AIAA *News;* AIAA PIO)
- USAF was flight-testing tactical photographic image transmission (TAPIT) subsystem which would enable tactical fighter aircraft to perform as reconnaissance vehicles. TAPIT, self-contained in pod mounted under aircraft wing with small control box in cockpit, took panoramic pictures from low altitudes; developed film in seven seconds; electronically scanned photos; and transmitted signals to ground stations within 100-mi radius. (AFSC Release 24.68)
- Marshall Space Flight Center awarded IBM's Space Guidance Center $1,303,758 contract for spare parts and logistics support of instrument units that guided Saturn IB and Saturn V launch vehicles. (MSFC Release 68–63)

April 4: NASA's *Apollo 6* (AS–502) was successfully launched from KSC Complex 39A at 7:00 am EST on mission to qualify Saturn V launch vehicle for future manned space flights. Primary objectives were to demonstrate structural and thermal integrity and compatibility of launch vehicle and spacecraft; confirm launch loads and dynamic characteristics; demonstrate S–II/S–IC and S–IVB/S–II stage separations; verify operation of propulsion (including S–IVB restart), guidance and control (optimum injection), and electrical systems; evaluate performance of emergency detection system (EDS) in closed-loop configuration; and demonstrate mission support facilities and operations required for launch, mission conduct, and command module (CM) recovery.

Launch vehicle 1st-stage performance was near nominal, but two of five 2nd-stage J2 engines shut down prematurely, causing remaining 2nd-stage engines and 3rd-stage engine to burn longer than planned. As result, spacecraft and 3rd stage entered elliptical parking orbit with 223.1-mi (395.1-km) apogee, 107-mi (172.1-km) perigee, and 89.8-min period instead of planned circular orbit of 115-mi (175-km) altitude. When 3rd stage failed to reignite on command after two orbits as planned, NASA switched to alternate mission, firing service propulsion system (SPS) to place spacecraft into trajectory with 13,823-mi (22,225.4-km) apogee. Since insufficient propellant remained after extended burn, second SPS burn was not attempted and CM reentered at 22,376 mph, just under planned 25,000-mph rate. Spacecraft splashed down 50 mi off target in Pacific 9 hr 50 min after launch and was recovered in good condition by U.S.S. *Okinawa*. Preliminary assessment indicated four of five primary objectives were attained, even though

April 4

launch vehicle performance and S–IVB restart and guidance control (optimum injection) were not fully successful [see April 11 and 24].

Apollo 6 was second flight for Saturn V launch vehicle and boilerplate lunar module (LM) and fourth for operational Block I command/service module (CSM). Spacecraft had been modified to include Block II heatshield and instrumentation for unmanned configuration, delete crew provisions, incorporate new unified quick-operating hatch and movie camera to record launch escape system (LES) jettison and reentry conditions, and relocate sequence camera for earth landmark photography. *Apollo 4* (launched Nov. 9, 1967) and *Apollo 5* (launched Jan. 22, 1968) had both been highly successful, completing inflight tests of all major pieces of Apollo hardware. Apollo program was directed by NASA Office of Manned Space Flight; MSC was responsible for Apollo spacecraft development, MSFC for Saturn launch vehicle development, and KSC for launch operations. Tracking and data acquisition was managed by GSFC under overall direction of NASA Office of Tracking and Data Acquisition. (NASA Proj Off; NASA Release 68–54K; *W Post*, 4/5/68, A18; UPI, *W Star*, 4/5/68, A3)

- NASA test pilot William H. Dana flew X–15 No. 1 to 187,500-ft altitude and 3,546 mph (mach 5.27) to test spray-on foam insulation, much lighter than previously used insulation, for use on Saturn V 2nd stage. Test, from Edwards AFB, was satisfactory, with X–15 performing in maximum-heating design trajectory close to that of Saturn V and sustaining temperatures of up to 1,500°F. (X–15 Proj Off; MSFC Release 68–69; AP, *P Inq*, 4/5/68)

- Nike-Tomahawk sounding rocket was launched by NASA from Churchill Research Range carrying GSFC payload to gather data on charged particle fluxes associated with aurora and to investigate distribution of electric fields in ionosphere, occurrence of radio noise, and ionospheric electron densities during auroral displays. Rocket and instruments performed satisfactorily; good data were acquired. (NASA Rpt SRL)

- ARC scientists Dr. William L. Quaide and Verne R. Oberbeck had developed method of calculating lunar soil depths using measurements based on Lunar Orbiter photos and Surveyor photos and surface analyses, NASA announced. Studies indicated that many of moon's smaller craters and much of soil and fragmental material on lunar surface were result of meteoroid impacts. By simulating impacts in laboratory and comparing results with photos of lunar craters scientists identified four crater types: (1) craters with up to 12-ft dia, round bottoms, and depths 25% of their diameter; (2) craters with 12- to 22-ft dia, flat bottoms, and central mound; (3) craters with 22- to 30-ft dia, flat bottoms, and no mound; and (4) craters with diameter greater than 30 ft with second crater gouged in flat bottom. Thick layer of fragmented material, calculated by new method to be up to 20 yd deep, coincided with densely cratered areas to support impact theory. (NASA Release 68–59; *SBD*, 4/5/68, 202)

- Model of wheel-shaped planetary landing craft, sterilized by heat and dropped from 250-ft altitude by Jet Propulsion Laboratory, operated successfully after impacting dry lake in Mojave Desert at 80 mph—major step in demonstrating feasibility of sending lightweight scientific landing capsule to Mars. Craft's radio transmitter turned on 30 sec after craft struck surface and operated 20 min. Anemometer deployed automatically 3 min after impact, to measure wind velocity. Following mission profile identical to projected Mars surface operations, radio turned on again 22 hr after initial transmission (when earth would again be in view). Signals from three-watt transmitter were received for another 40 min to conclude test. Craft was powered by 12-cell, silver-zinc battery, first known to survive both heat sterilization and high-velocity impact. (NASA Release 68–69; JPL Release 473; JPL PIO)

- USAF's *Lincoln Experimental Satellite* (*Les V*) (launched July 1, 1967), first all solid-state UHF band comsat, had been used in first network of tactical terminals to include a comsat, first air-to-air link via satellite relay, and first communications link from high latitudes via satellite as part of USAF program to improve communications between aircraft. *Les V* was testing UHF teletype system which relayed 60-wpm messages over ground distances of up to 8,000 mi. Satellite's 20,000-mi-altitude orbit allowed line of sight stretching nearly halfway around the world. USAF proposed using system for communications between low-altitude attack aircraft and rear area controllers, for USAF worldwide logistic control and status reporting system, and for strike and reconnaissance reporting. (AFSC Release 23.68)

- Dr. William H. Pickering, Director of Cal Tech's Jet Propulsion Laboratory, spoke at Space Forum sponsored by American Institute of Aeronautics and Astronautics, American Astronautical Society, and Institute of Environmental Sciences in Washington, D.C. Describing first decade in space as "most productive . . . in history of technology," he forecast manned lunar operations including lunar laboratories before end of second decade; tour of Jupiter, Saturn, Uranus, and Neptune by single spacecraft in 1977; and dramatic yield from growing applications of near-earth satellites.

  He urged initiation of "orderly planning cycle" to replace major programs being phased out. Emphasis of next phase was likely to be "gleaning more benefits" from space dollar expenditure. National Space Council estimated annual return from space would markedly exceed expenditures in 10 yr. (Text)

- NASA would negotiate $3.5-million, one-year, cost-plus-fixed-fee contract with General Electric Co.'s Apollo Systems Div. for Apollo Applications engineering support. GE, under direction of NASA Hq. Apollo Applications Program Office, would provide engineering support in areas of quality and reliability, configuration and data management, test, and checkout. (NASA Release 68–61)

- MSFC contract activity: RCA was awarded $1,293,640 contract to modify RCA 110 computer module boards, by systematically incorporating improved solder design.

  IBM was issued $1,523,282 supplemental agreement for adjustment and implementation of configuration management for fabrication, assembly, checkout, and delivery of 27 Apollo/Saturn instrument unit stages and other support equipment.

  Air Products and Chemicals, Inc., received $2,249,364 contract extension to supply 12 million lb liquid hydrogen by March 31, 1969, to MSFC, purchasing agent for Government agencies and their supporting contractors in eastern U.S.

  Three one-year contract renewals, effective through March 31, 1969, were awarded for MSFC support services: $10.5 million to Brown Engineering Co. for services in Propulsion and Vehicle Engineering Laboratory, $4,504,000 to SPACO Inc. for services in Quality and Reliability Assurance Laboratory, and $2,273,000 to Hayes International for services in Manufacturing Engineering Laboratory. (MSFC Releases 68-64, 68-65, 68-66, 68-67)

April 5: ComSatCorp, on behalf of International Telecommunications Satellite Consortium (INTELSAT), leased antenna and related facilities at Fucino, Italy, earth station from Telespazio, Italian space communications company. Fucino facilities, approved by INTELSAT's Interim Communications Satellite Committee (ICSC), would be used for tracking, telemetry, and command duties for INTELSAT comsats. (ComSatCorp Release 68-15)

- MSFC announced award to Harvard Univ. of $1,942,300 supplementary contract for development of UV scanning spectrometer to be flown as solar experiment on first launch of Apollo Telescope Mount. Award increased total value to $6,458,544 for experiments for use with manned solar observatories. (MSFC Release 68-68; *SBD*, 4/8/68, 210)

- Crash of F-111A in Thailand March 30 had been caused by failure in terrain radar guidance system, newspapers said reliable sources reported. Aircraft had reportedly bucked and gyrated severely, forcing two crew members to eject. USAF team was conducting on-site investigation and was expected to report findings in two weeks. (Horton, AP, *W Star*, 4/5/68, A6; Beecher, *NYT*, 4/6/68, 10)

April 6: USAF launched two Orbiting Vehicle research satellites pickaback from Vandenberg AFB by Atlas-F booster. *OV I-13* entered orbit with 5,789-mi (9,316.2-km) apogee, 346-mi (556.7-km) perigee, 199.5-min period, and 100.5° inclination to determine flexibility of cadmium sulfide solar cells in space and to measure proton/electron energy spectra and angular distribution of electrons. *OV I-14*, launched to obtain high-resolution measurements of proton/electron flux, spectra, decay, and time variations, entered orbit with 6,173-mi (9,934.2-km) apogee, 348-mi (560.0-km) perigee, 207.8-min period, and 100.0° inclination. (UPI, *C Trib*, 4/8/68; *SBD*, 4/9/68, 220; GSFC *SSR*, 4/15/69; *Pres Rpt 68*)

- Third anniversary of launch of 85-lb (*Early Bird*), world's first commercial comsat, owned by INTELSAT and managed by ComSatCorp. Originally designed as experimental-operational satellite with 18-mo life expectancy, comsat launched by NASA into 22,300-mi-altitude synchronous orbit over Atlantic, was still providing service between North America and Europe with 100% reliability. *Intelsat I* had received and transmitted more than 200 hr of TV and thousands of telephone calls, data and record messages, and other general communications without satellite service outage. TV use of *Intelsat I* increased from 31 programs consuming 31 hr leased time in 1965 to 160 programs and 125 hr in 1967. Highlights of TV broadcasts included live coverage of Atlantic splashdowns of Gemini spacecraft, sports events, public affairs, and news programs. (ComSatCorp Release 68-16)

- *New York Times* editorial on *Apollo 6* mission: "What was illustrated . . . was the extraordinary difficulty of assuring that every one of the literally millions of components in such an extremely complicated system as the Saturn 5 works perfectly. But the complexity of the total Apollo mechanism for the planned manned voyage to the moon . . . is even greater. . . . This fact argues for a slow but sure approach to future Apollo tests, rather than an adventuresome policy aimed primarily at completing the job by the end of 1969.

"Regrettable as were Saturn 5's deficiencies as demonstrated in this week's test, they provide a useful warning against renewed overconfidence and the costs it could again impose." (*NYT*, 4/6/68, 36)

- In *New Republic*, Louis J. Halle wrote "Why I'm for Space Exploration." It was less than 12 yr since life on earth had emerged "from our planet's atmospheric envelope into outer space." He was puzzled to find "marked lack of enthusiasm . . . at the prospect of man's liberation from this earthly prison." Many, he felt, were moved by "spiritual horror" at notion of looking beyond earth. "Now . . . that we are at last beginning to escape from our native confines, there is no telling what light we may find in the larger universe to dissipate the darkness of our minds." There was also possibility "that we may begin to populate new planets as, after 1492, we began to populate a new continent. Suddenly man's fate seems boundless." (*AF/SD*, 7/68, 51–4)

April 7: U.S.S.R. successfully launched *Luna XIV* unmanned spacecraft toward moon "to conduct further studies of near-lunar space," Tass announced. All systems were functioning normally and spacecraft was traveling close to planned trajectory. [See April 10 and 11.] (Anderson, *NYT*, 4/8/68, 1; AP, W *Star*, 4/8/68, A3; GSFC *SSR*, 4/15/68)

- Long-nosed USAF C-131 research aircraft was being developed for Air Force Systems Command by Cornell Aeronautical Laboratory, Inc., as unique flying simulator to test controls, instruments, and aircraft configurations of advanced aircraft such as Advanced Manned Strategic Aircraft (AMSA), military C-5A cargo and passenger aircraft, and SST. Total In-Flight Simulator (TIFS) configuration—with nose length varying to simulate advanced aircraft and with second cockpit below and ahead of main cockpit and six independent controls—would realistically reproduce handling conditions of modeled aircraft and enable USAF to determine inexpensively in advance correct design and instrumentation for advanced aircraft. (AFSC Release 45.68)

- In *Perception of Space and Time in the Cosmos*, published in U.S.S.R., Cosmonaut Aleksey A. Leonov, who made first space walk from *Voskhod II* spacecraft March 18, 1965, and Soviet space medicine specialist Vladimir Lebedev claimed pilots' distorted perceptions of dimensions and time were frequent factors in aircraft accidents. In space flights such miscalculation could cause incineration during reentry or, conversely, divert spacecraft into an orbit of no return. Authors divided people into three categories: those who suffered no ill effects from weightlessness, those who required time for adjustment, and those who were unable to adjust and should be permanently grounded. They suggested spacecraft of future protect against monotony of unbroken routine by carrying well-stocked libraries, cinema, and discotheque. (Stevens, W *Star*, 4/7/68, F6)

- Washington *Sunday Star* editorial on *Apollo 6* mission: ". . . Saturn 5's latest performance . . . suggests that our astronauts may not be able to carry out their lunar mission until considerably later than optimists have suggested—possibly not until 1971, if then. . . . However . . . it is better to be safe than sorry. Saturn 5's deficiencies must be eliminated, no matter how long the job takes, before it is used to lift a manned Apollo spacecraft to the moon. Despite loose talk about a Soviet-American 'space race,' there should be no all-out drive, no senseless rush, to score a first in this field." (W *Star*, 4/7/68, F1)

April 8: NASA announced publication of *Teleoperators and Human Augmentation* (NASA SP-5047), NASA–AEC technology survey issued by NASA Office of Technology Utilization and written by Edwin G. Johnsen and William R. Corliss. It covered work of both agencies in development of teleoperators, including history of the robot-like machines, assessment of their impact on technology, and explanation of principal subsystems of those in use. Since 1948 "some 3,000 manipulator arms" had been built in U.S.— "the advance guard of an army of man-machine systems now assembling to serve man in a variety of ways." (NASA Release 68-62)

- NASA selected Teledyne Systems Co. for negotiation of $950,000 15-mo contract to design and construct prototype airborne digital computer unit for Centaur launch vehicle's guidance and control system. Contract, which would include option for five additional units with required ground support equipment and spare parts, would be managed by LeRC. (NASA Release 68-64)
- Harold D. Babcock, 40-yr member of Mt. Wilson and Mt. Palomar Observatory staffs, died. He was specialist in study of spectra of sunspots and discoverer of fact that magnetic field of sun reversed polarity periodically. (NAS–NRC–NAE *News Report*, 5/68, 10)

April 9: U.S.S.R. successfully launched *Cosmos CCXI*. Orbital parameters: apogee, 1,545 km (960 mi); perigee, 199 km (123.6 mi); period, 102.1 min; and inclination, 81°. Satellite reentered Nov. 10. (*SBD*, 4/10/68, 266; GSFC *SSR*, 4/15/68; 11/15/68)

- NASA launched two Javelin sounding rockets from NASA Wallops Station. One carried GSFC payload to 497-mi (800-km) altitude to observe helium ionization levels in exosphere with vacuum-ion chamber and to observe helium and oxygen-ion resonance dayglow with filtered photometer. Rocket and instrumentation performance was satisfactory. Telemetry signal was received for 16 min 40 sec. Second rocket carried Syracuse University Research Corp. vacuum-ion chamber to observe helium ionization levels in exosphere and Univ. of Southern California filtered photometer to observe helium- and oxygen-ion resonance dayglow to 497-mi (800-km) altitude. Rocket and instruments performed satisfactorily. (NASA Rpt SRL)
- FAA awarded $3.8-million contract to IBM's Federal Systems Div. to modernize air traffic control at 100 U.S. facilities by installing printers and keyboards for faster coordination and reduction of controllers' oral and manual workload. Delivery of equipment, to begin April 15, would be coordinated with delivery of other automation components for National Airspace System. (FAA Release 68-24)

April 9–10: Electronic signals on medical condition of USMC volunteer patient in Tokyo were transmitted between Tokyo, Houston, and Washington, D.C., via satellite and ground equipment to show how worldwide diagnosis of complex medical problems could be achieved by advanced communications. *Intelsat-II F-2 (Pacific I)* comsat and AT&T landline facilities were used in demonstration for 1968 National Telemetry Conference of IEEE in Houston. Signals were relayed from Brewster Flat, Wash., earth station to Conference and to computer centers at

U.S. Public Health Service in Washington, D.C., and Univ. of Texas. Demonstration was directed by ITT World Communications, Inc., with cooperation of ComSatCorp and Kokusai Denshin Denwa Co., Ltd., Japan. (ComSatCorp Release 68-17; AP, W *Star*, 4/10/68, A18; UPI, *W Post*, 4/12/68, A15)

April 10: U.K.'s Jodrell Bank Experimental Station reported U.S.S.R.'s *Luna XIV* spacecraft had apparently entered lunar orbit and was transmitting telemetry but no photographic signals. U.S.S.R. had made no official statement since April 7 launch. (UPI, *NYT*, 4/11/68, 4; Cohn, *W Post*, 4/11/68, A27)

- Ernest W. Brackett, Special Assistant to NASA Assistant Administrator for Industry Affairs, was appointed Chairman of NASA Board of Contract Appeals, succeeding E. M. Shafer, who became Chairman of NASA Contract Adjustment Board. GSFC's Matthew J. McCartin was appointed Vice Chairman. (NASA Release 68-65)

April 11: U.S.S.R.'s *Luna XIV* had entered orbit around moon "close to the calculated one" to study correlation between earth and moon and collect data necessary for landing cosmonauts on moon, Tass announced in first official statement since April 7 launch. Satellite had entered lunar orbit April 10 with 870-km (540.6-mi) apolune, 160-km (99.4-mi) perilune, and 2-hr 40-min period. (UPI, W *Star*, 4/11/68, A3; *SBD*, 4/12/68, 239-40; Reuters, *NYT*, 4/14/68, 8)

- MSFC issued report on preliminary results of April 4 *Apollo 6* flight. Although "basic source of the difficulties" had not yet been determined, scientists and engineers speculated that wires carrying cutoff commands to the malfunctioning engines were interchanged. First stage had performed as planned and stage thrust was near predicted during first portion of flight. Second stage had performed satisfactorily through 1st-stage boost, 2nd-stage ignition, and early portion of 2nd-stage powered flight. First indications of anomaly were decreasing temperatures on main oxidizer valve and its control line on fifth engine and steady decrease in second engine's yaw actuator pressure. Third stage performed satisfactorily through first burn and orbital coast. Although engine and stage prestart conditions appeared normal, engine received start signal, and valves opened properly, engine did not restart. Initial data suggested that leak in one of two propellant lines to engine's augmented spark igniter might have caused insufficient or inadequately mixed propellant for proper start condition. Investigations were continuing on longitudinal oscillation of vehicle. Guidance and other instrumentation functions, telemetry performance, and onboard TV camera operation were satisfactory. (MSFC Release 68-74; AP, *NYT*, 4/12/68, 20)

- USAF and NASA had agreed to consolidate their photographic operations at ETR under one contractor to save estimated $1 million first year. Single contractor, selected by competitive bid, would report to ETR contract manager. USAF and NASA each would provide one technical manager to monitor performance. New operation would be effective Jan. 1, 1969. (KSC Release 68-151)

- V/A Hyman G. Rickover (USN), testifying before House Committee on Banking and Currency hearings on H.R. 15683 to renew Defense Production Act of 1950 as amended, warned against emergence of "fourth branch of government," partnership of Federal bureaucrats and giant corporations "with men exerting power without political responsibility." DOD's industry-oriented philosophy, lack of inhouse capability, and absence of standardized accounting procedures permitted Government subsidization of civilian business of defense contractors and cost U.S. taxpayers billions of dollars, he said. (Transcript: Porter, *W Post*, 5/2/68, G2)

April 12: NASA would negotiate $900,000, one-year, cost-plus-fixed-fee contract with Chrysler Corp.'s Space Div. to study needs and configuration alternatives for an intermediate payload launch vehicle in post-1973 space operations. Payloads under consideration included long-duration manned operations in low earth orbit, unmanned satellites in synchronous orbit, logistic support for manned lunar exploration, and unmanned planetary and deep space probes. Contract would be managed by OMSF. (NASA Release 68–67; Sehlstedt, B *Sun*, 4/13/68)

April 14: *Cosmos CCXII* was successfully launched by U.S.S.R. into orbit with 200-km (124.3-mi) apogee, 180-km (111.8-mi) perigee, 88.3-min period, and 51.6° inclination. Satellite docked with *Cosmos CCXIII* April 15 and reentered April 19. (AP, B *Sun*, 4/15/68; UPI, *NYT*, 4/15/68, 86; GSFC *SSR*, 4/15/68; 4/30/68)

April 15: U.S.S.R. successfully launched *Cosmos CCXIII* into orbit with 254-km (157.8-mi) apogee, 186-km (115.6-mi) perigee, 89.1-min period, and 51.6° inclination. At 1:21 pm Moscow time (3:21 Baikonur time) satellite was automatically docked with *Cosmos CCXII* (launched April 14). Tass later announced that satellites used "special closing-in systems, radio, technical and computing devices, to carry out an automatic mutual search, closing-in, docking, and rigid coupling to each other." Maneuver was second automatic docking in space and was filmed by TV cameras on board both satellites. U.S.S.R. had successfully accomplished first automatic docking Oct. 30, 1967. First successful manned docking had been conducted by U.S. March 16, 1966. *Cosmos CCXII* and *Cosmos CCXIII* remained docked in near-circular orbit 3 hr 50 min and were then separated automatically by ground command and placed into different orbits. *Cosmos CCXII* reentered April 19 and *Cosmos CCXIII*, April 20. (*W Post*, 4/16/68; *NYT*, 4/16/68; B *Sun*, 4/16/68; *SBD*, 4/16/68, 255–6; GSFC *SSR*, 4/15/68; 4/30/68)

- Flight reenactment had revealed that USAF F–111A aircraft crash March 30 had been caused by malfunction of flight control system, press said informed sources reported. Second F–111A, which North Vietnam claimed to have shot down, was still missing; U.S. officials speculated that aircraft had crashed in Thailand jungle area. (UPI, *NYT*, 4/16/68, 22; *W News*, 4/16/68, 7)

- Defense Projects Support Office (DPSO) was established in Special Programs Office at NASA Hq. to manage specialized tasks where NASA's unique capabilities could provide needed support to a limited number of DOD projects. M. J. Raffensperger, Director of Advanced Manned Missions Planning and Operations, Office of Manned Space Flight, was appointed Deputy Director of Special Programs Office and Acting Director of DPSO. (NASA Release 68–66; NASA Ann, 4/17/68)

- Lawrence A. Hyland, Vice President and General Manager of Hughes Aircraft Co., was selected by NAA to receive Robert J. Collier Trophy for significant achievement in aeronautics and astronautics in 1967 as Hughes Surveyor program director. Trophy would be presented May 7. (NAA Release; AP, *W Post*, 4/16/68, A3)

- Tass announced issuance of three stamps commemorating Soviet space achievements: March 18, 1965, space walk by Aleksey A. Leonov; Oct. 30, 1967, automatic docking of two Cosmos satellites; and Oct. 18, 1967, softlanding of *Venus 4* on Venus. (*W Post*, 4/15/68, C20)

April 16: NASA Associate Administrator, Dr. Homer E. Newell, summarized Earth Resources Survey program at Fifth Symposium on Remote Sensing of Environment at Univ. of Michigan's Institute of Science and Technology in Ann Arbor. Prospects in field were promising. Greatest use of satellites for earth survey to date was for meteorological data, including global cloud-cover photos, cloud motion, and ocean temperatures, but U.S. still lacked "much of the data essential for worldwide long-range weather forecasting," such as data on three-dimensional fields of density, wind velocity, temperature, and water vapor within the atmosphere.

Major contributions expected from research in other fields included: completion of geodetic programs which would permit determination of relative positions of any two points on earth with improved accuracy; monitoring of sea surface state, evaluation of marine biological resources, and surface observations of conditions of interest to oceanographers; and improved identification of spectral signature of various species for agriculture, forestry, and geology. (Text; *SBD*, 4/19/68, 280)

- John N. Wilford described in *New York Times* front page article decline in U.S. space expenditures since 1966: "Under pressure from the war in Vietnam, civilian space spending has dropped from $5.9-billion in the peak year of 1966 to $4.8-billion this year, and it is expected to drop much lower in the fiscal year starting in July. Employment in space work at private companies, universities and Government centers has declined from 420,000 in 1966 to fewer than 300,000 today, and it is still dropping at the rate of 4,000 a month." Signs of decline were clearly visible, in "ghost towns" that were once test sites, and in removal of numerous projects from NASA's post-Apollo plans. Fortunately, impact of cutback was softened because NASA had not replaced many personnel who ordinarily left agency each year and because personnel dismissed were absorbed by growing aircraft industry and expanding military space program. But there was a growing feeling "that once astronauts have landed on the moon, they will have no other place of significance to go for several years because of sharp budget cuts. These cuts have trimmed to the bone all preparations for future missions. It is as if the astronauts are heading for a dead-end on the moon." (*NYT*, 4/16/68, 1)

- U.K. Minister of Technology Anthony W. Benn announced that U.K. would withdraw from European Conference on Satellite Communications and would make no new commitments to ELDO, though it would increase its contribution to ESRO by up to 6%. U.K. officials reportedly said decision not to participate in proposed project for experimental European TV relay satellite was made in effort to avoid unrealistic projects and concentrate on nonspace aircraft and computer industries. (Shuster, *NYT*, 4/17/68, 79; Mott, *W Post*, 4/17/68, A11)

- MSFC awarded Ball Brothers Research Corp. $134,500 contract for six solar sensor systems, including one prototype and five flight units, for

April 16

Apollo Telescope Mount pointing control system. (MSFC Release 68–76)

April 17: USAF launched unidentified satellite from Vandenberg AFB using III–B-Agena D booster. Satellite entered orbit with 262-mi (421.6-km) apogee, 79-mi (127.1-km) perigee, 89.9-min period, and 111.4° inclination and reentered April 29. (*Pres Rpt 68*; UPI *W Post*, 4/18/68; *SBD*, 4/18/68, 275; GSFC *SSR*, 4/30/68)

- NASA Marc 42A2 Arcas booster launched from NASA Wallops Station carried GSFC payload to 4.8-mi (7.8-km) altitude in ballistic performance evaluation test. Booster and instruments performed satisfactorily; ballistic parameters agreed closely with predictions. (NASA Rpt SRL)

- Charles W. Mathews, Director of NASA Apollo Applications Program, told National Space Club in Washington, D.C., NASA's manned space plan, beyond first Apollo landing, "contemplates a balanced activity of lunar exploration and extension of man's capabilities in earth orbit." Program had been designed for flexibility so activities could be conducted in harmony with available resources. "We are also prepared to move forward at an increased pace when it is desirable and possible to do so." Both civil benefits and national security implications of space program warranted continued strong support. Contingency planning would leave more room for budgetary or goal changes, thus placating critics in Congress who claimed NASA had not provided them with sufficient flexibility. (Text: Lannan, W *Star*, 4/18/68, A5; AP, B *Sun*, 4/18/68, A11)

- Sen. Margaret C. Smith (R-Me.), ranking member of Senate Committee on Aeronautical and Space Sciences, on Senate floor presented General Accounting Office review of source selection and award of major subcontract by NASA and its prime contractor, Grumman Aircraft Engineering Corp., for development of landing and rendezvous radar equipment for Apollo lunar module, compiled at her request. Sen. Smith explained that although another electronics firm had expressed interest in performing under fixed-price contract, RCA had received contract because of agreement between Grumman and RCA "before the requirements and specifications for the radar components had been defined." Noting that RCA estimated cost of $23.4 million had now increased to about $112 million, she suggested that if Grumman radar subcontract was illustrative of how NASA "maintains surveillance over its appropriated funds, it would appear that substantial savings could be realized merely by strengthening the agency's contracting practices." (NASA *LAR* VII/36; *CR*, 4/17/68, S4138–46; AP, B *Sun*, 4/23/68, A5)

- NASA had awarded Aerojet-General Corp.'s Space Div. $316,776 contract to perform preliminary design of spacecraft for basic research on frog's balance mechanism (otolith) under weightlessness and repeated acceleration. Project, initial step in NASA's Human Factor Systems program to investigate primary balance mechanism within inner ear, would be managed by NASA Wallops Station under direction of Office of Advanced Research and Technology. ARC would be responsible for otolith experiment package designed by Johns Hopkins Applied Physics Laboratory. (WS Release 68–8; NASA Release 68–71)

- MSFC awarded nine-month, $99,000 contract to Raymond Loewy/William Snaith, Inc., to conduct habitability studies of planned earth orbital space stations. Basic goal would be to ensure that workshop configurations were comfortable and functional structures in which to live and work. (MSFC Release 68-79)

- Naval Research Laboratory scientist, Dr. Richard C. Henry, speaking at dedication of NRL's new E. O. Hulburt Center for Space Research in Washington, D.C., presented data strongly supporting closed universe concept. Aerobee sounding rocket launched from White Sands Missile Range Sept. 7, 1967, carrying soft x-ray detector, had detected radiation from large, unexpected amount of thinly spread intergalactic hydrogen gas, evidence of existence of intergalactic matter previously supposed but undetected. Amount detected indicated presence of 100 times as much matter between galaxies as in all stars in universe—enough to fill up all space and satisfy all theoretical requirements for a closed universe. (Text; Cohn, *W Post*, 4/18/68, 1)

- Editorial comment on Soviet space achievements, including successful orbiting of *Luna XIV* and docking of *Cosmos CCXII* and *CCXIII*, urged U.S. to acknowledge challenge:

    "The Soviet Union's sense of purpose in space is apparently as steady and unwavering as it ever was. The U.S. . . . after coming up fast from behind in a wave of feverish anxiety and enthusiasm, now seems to have lost interest. That could be a dangerous—even fatal—tendency in an age where space is of key importance to the security of the Nation. It should be reversed, before great harm is done." (*P Inq*, 4/17/68)

    "If we muff what now looks like a good chance to beat the Russians to the moon with manned space ships, the prestige loss to this nation will probably be immeasurable. And if our space people aren't paying at least as much attention as the Russians to the military possibilities of space, then we are in grave danger and growing more so. All of which adds up . . . to a whole string of dire warnings to Congress not to be stingy about space projects of any description." (NY *News*, 4/17/68)

- USAF F-111A aircraft crash March 30 in Thailand had been caused by "a mechanic's mistake, not by a flaw in design," U.S. military command announced. Recovered aircraft revealed that pilots lost control of aircraft because tube of sealant normally used to seal fuel tanks was left loose in aircraft, hardened during low-temperature flight, and jammed flight-control system. Loss of another F-111A March 28 remained mystery and search in Thailand area where it presumably crashed had ended unsuccessfully. (UPI, *W Post*, 4/16/68, A8; Carroll, B *Sun*, 4/18/68, A5)

April 18: U.S.S.R. launched *Cosmos CCXIV* into orbit with 370-km (229.9-mi) apogee, 199-km (123.6-mi) perigee, 90.1-min period, and 81.3° inclination. Satellite reentered April 26. (GSFC *SSR*, 4/30/68)

- LaRC selected Northrop Corp.'s Norair Div. for negotiations on $2-million contract to design and construct differential maneuvering simulator. System, two identical piloted flight simulators linked electronically through central computing equipment, would be used to study future aerospace vehicle concepts. (NASA Release 68-74)

- NASA announced appointment of two new members to Aerospace Safety Advisory Panel: Dr. C. D. Harrington, President of Douglas United Nuclear, Inc., for six-year term; and S. T. Harris, Officer of the Board, Texas Instruments, Inc., for four years. (NASA Release 68-72)

April 19: U.S.S.R. launched *Cosmos CCXV* into orbit with 403-km (250.4-

mi) apogee, 255-km (158.4-mi) perigee, 91.1-min period, and 48.4° inclination. Satellite reentered June 30. (SBD, 4/22/68, 292; GSFC SSR, 4/30/68; 7/15/68)

- Sixty-day simulated earth-orbital mission for four UCLA students ended when they left McDonnell Douglas Corp. Missile & Space Systems Div. space cabin simulator [see Feb. 21]. Although students had tired of food and missed female companionship, attending doctor said they remained in good health. Leaving simulator they first noticed extreme humidity and "myriad smells and odors in normal air." Experiment had included cycles of rest and work, testing air-water samples, and manning scientific equipment. (AP, B Sun, 4/20/68, A3)
- Page Communications Engineers, Inc., and government of South Vietnam were negotiating agreement to permit Page to finance and construct $7-million earth station in Vungtau. Station, which would be used with ComSatCorp satellite to be launched in November, would have 60-channel capacity initially and would be able to expand to 120. South Vietnamese government would receive 20% of gross revenues—expected to total $4–5 million annually—for first five years and 50% for second five years, after which operation would be turned over to a South Vietnamese corporation. (Page PIO; Wilson, W Post, 4/19/68)
- NATO's Nuclear Planning Group had concluded that construction of European ABM defense system was not justified under present circumstances, Robert C. Doty reported in *New York Times*. Recommendation, he said, "which appears certain to be endorsed by the alliance as a whole, ends for the foreseeable future European interest in any multibillion-dollar project to match the antimissile screen now under construction by the Soviet Union." U.S. decision in 1967 to build Sentinel ABM system to protect against possible Chinese Communist attack had promoted NATO review of Europe's nuclear defense. (NYT, 4/20/68, 2)

April 20: U.S.S.R. successfully launched *Cosmos CCXVI*. Orbital parameters: apogee, 265 km (164.7 mi); perigee, 195 km (121.2 mi); period, 89.1 min; and inclination, 51.8°. Satellite reentered April 28. (Anderson, NYT, 4/21/68, 28; W Star, 4/21/68, A5; GSFC SSR, 4/30/68)

- NASA Astrobee 1500 sounding rocket launched from NASA Wallops Station carried Univ. of Minnesota experiment to 776-mi (1,250-km) altitude to study levels of electric and magnetic field variations in magnetosphere, check operation of antenna systems for use on satellite, and verify vehicle design changes. Rocket performed satisfactorily. Instrumentation suffered partial failure, but cause had not been determined. (NASA Rpt SRL)

April 20–22: Technical review of Saturn launch vehicles, attended by about 140 scientists, engineers, and administrators, and held at MSFC. Participants investigated status and flight schedule of Saturn launch vehicles. (MSFC Release 68–77; UPI, W Star, 4/21/68, A5)

April 21: U.S.S.R. successfully launched eighth Molniya I comsat, *Molniya I–8*. Orbital parameters: apogee, 39,719 km (24,680.2 mi); perigee, 414 km (257.2 mi); period, 11 hr 53 min; and inclination 65°. (UPI, NYT, 4/23/68, 34; GSFC SSR, 4/30/68)

- Soviet scientists reported that automatic docking of *Cosmos CCXII* and *CCXIII* April 15 had occurred only 47 min after pursuit vessel was launched. *Cosmos CCXII* (launched April 14) was orbiting earth at 225.3-km (140-mi) altitude and 17,500 mph when *Cosmos CCXIII* entered orbit three miles from it. According to Tass, satellites repeatedly changed their orbits, reoriented, maneuvered in space, and conducted various scientific experiments during four days in orbit. Soviet scientists had reportedly developed three launch systems for spacecraft intended for automatic docking in orbit: (1) simultaneous side-by-side launch with docking maneuvers beginning immediately after spacecraft separated from launch vehicles; (2) separate launches from same or different sites with second spacecraft launched as close as possible to first spacecraft in orbit as it passed over launch site [method used for April 15 docking]; and (3) separate launches of spacecraft into same plane, but with distances between them great enough to require several orbital corrections and maneuvers to close gap. (Anderson, *NYT*, 4/22/68, 9; *SBD*, 4/23/68, 296-7)

- NASA announced appointment of Dr. Henry J. Smith, Deputy Director of Physics and Astronomy Programs, OSSA, as Deputy Associate Administrator for Space Science and Applications (Science), replacing Dr. John E. Naugle who was named Associate Administrator for Space Science and Applications Oct. 1, 1967. Dr. Smith would be Chief Scientist for OSSA, responsible for obtaining and implementing scientific advice for the national space program. (NASA Release 69-70)

- JPL soil sciences group, headed by Dr. Roy E. Cameron, reported tests and cultures of Antarctic soil samples in JPL's walk-in freezer laboratory to determine what micro-organisms lived in extreme cold and to help determine whether life existed on Mars. Samples had come from high, dry valleys in Victorialand near U.S. base at McMurdo, Antarctica. Scientists discovered bacteria, yeasts, molds, and algae, which began to grow within two weeks when Antarctic soil kept laboratory-frozen for over one year was subjected to temperature 68° F or above. Studies were sponsored by NASA and NSF. (NASA Release 68-73)

- Dept. of Commerce announced that "the exploration of Antarctica has now been virtually completed," with conclusion of two-month, 815-mi scientific journey led by C&GS geophysicist Norman W. Peddie. "There are now no major areas [of Antarctica] which have not been explored," Peddie said.

  Expedition, which started Dec. 5 and ended Jan. 30, was made by nine-man party of Belgian, Norwegian, and American scientists. Trip covered region in Queen Maud Land between South Pole and Princess Ragnhild Coast in direction of Africa (*W Post*, 4/21/68, A12)

- Dr. Kurt H. Debus, KSC Director, and JPL Director William H. Pickering had been named cowinners of American Astronautical Society's 1967 Space Flight Award for outstanding achievement in advancement of space flight and space science. (AP, *W Star*, 4/21/68, A5)

April 22: Representatives of 43 nations signed space rescue treaty at separate ceremonies in Washington, D.C., London, and Moscow. At State Dept. ceremony in Washington, D.C., President Johnson said he hoped treaty would end wasteful competitive spacemanship between U.S. and U.S.S.R. and that next decade in space would increasingly become a partnership. Treaty, which provided for assistance to astronauts in emergency and safe return of astronauts and space hardware, had been unanimously approved by U.N. General Assembly Dec. 19, 1967. Secretary of State Dean Rusk signed for U.S. It would become effective when ratified by U.S., U.S.S.R., U.K., and two other countries. (Ward, B *Sun*, 4/23/68, 1; *PD*, 4/29/68, 1108)

- *New York Times* editorial on cuts in NASA FY 1969 budget: "Now that the desired space research capability has been created, it is merely good sense to shift some of the resources thus employed to other and more urgent national needs . . . [such as] cleaning up the nation's polluted air and water, providing high-speed land transportation, or working out faster and cheaper ways to build new housing to replace the noxious and overcrowded slums. . . .

  "None of this means . . . that the United States will or should abandon the effort to explore space and exploit space technology. . . . But for the moment the new relatively svelte—though still very adequate—space program meets the nation's needs quite generously." (*NYT*, 4/22/68; *CR*, 5/1/68, E3646)

April 23: NASA launched two sounding rockets from Churchill Research Range. Nike-Tomahawk carried Univ. of New Hampshire-Univ. of California at San Diego payload to 166.5-mi (268-km) altitude. Objectives were to measure electric field, ionospheric currents, auroral light intensity and location, and proton and electron fluxes in 1- to 10-kev region while payload was passing through or close to visible auroral display. Rocket and instruments performed satisfactorily; all scientific objectives were achieved. Launch was second in series of four [see Feb. 5].

Nike-Apache carried Univ. of Michigan payload to 6-mi (9.6-km) altitude to study atmospheric parameters of temperature, pressure, and density using Pitot-static-probe technique and measure extreme UV emission to determine energy disposition with altitude of incoming particles. No scientific data were obtained because of vehicle malfunction. (NASA Rpt SRL)

- Aerobee 150 MI sounding rocket launched from White Sands Missile Range carried Univ. of Colorado experiment to 111.9-mi (185.4-km) altitude to measure intensity of spectral lines in 3,400–1,100Å band. Rocket and instrumentation performed satisfactorily. (NASA Rpt SRL)

- Senate Committee on Aeronautical and Space Sciences heard testimony in support of NASA FY 1969 budget from U.S. scientists. Dr. Harry H. Hess of Princeton Univ. presented statement for record by National Academy of Sciences President Dr. Frederick Seitz. Space program, Dr. Seitz said, was "the latest and one of the greatest human exploratory adventures in a long sequence that has enriched mankind. It offers us the promise of extending the range of our domain . . . to the entire solar system. We can expect many benefits along the way, some of conceptual and some of direct material value . . . but those which will prove to be the most rewarding are probably, in the main, still hidden from us over the horizon."

Dr. John A. Simpson of Enrico Fermi Institute and Univ. of Chicago noted: "Researches in space have made, through the bold programs which NASA established with universities in the early 1960s, major contributions to the critical problems of generating, developing and retaining first-class scientific manpower. . . . The great fear at present . . . is that the momentum established will be dissipated by the preferentially deep budgetary cuts made by NASA in those areas which most affect the universities." He stressed "deeply felt conviction" that U.S. was "in danger of unwittingly destroying what we wish to save and need . . . pre-eminence in science and technology which is crucial for each major problem of the nation, from poverty to war." Actions taken in Congress in 1968 might "largely determine whether the U.S. will retain its leadership in the space sciences."

Dr. Simpson urged early start of Pioneer concept since it was "absolutely clear that discoveries important for the progress of science and technology may be made by . . . experiments and observations on spacecraft moving outward from the orbit of the Earth . . . the program is not a gamble and a hope, but an objective of high importance and certain to produce fruitful results." (Testimony; *NYT*, 4/24/68, 24; *W Star*, 4/24/68, A21)

- Eleventh Saturn IB booster was successfully static fired at Marshall Space Flight Center at 1.6-million-lb thrust for 145 sec by Chrysler Corp. personnel. It would be returned to Michoud Assembly Facility for post-static checkout. (MSFC Release 68–85)

- At American Physical Society Meeting in Washington, D.C., Stanford Univ. physicist, Dr. William M. Fairbank, described experiments on superconducting accelerators that would enable scientists to accelerate electrons faster and for longer periods and, possibly, to produce 10 times as much energy as world's most powerful existing accelerator, 2-mi-long, 20-bev Stanford Linear Accelerator (SLAC). By immersing accelerator in liquid helium cooled to absolute zero, energy loss could be reduced so much that electrons could be fired continuously and accelerator kept at constant temperature. SLAC currently could be fired for only 0.001 sec because of excess heat generated by pulse. Experiments, preliminary to construction of $5-million, 500-ft-long prototype accelerator, had been conducted on 5-ft model. (Sullivan, *NYT*, 4/24/68, 26; O'Toole, *W Post*, 4/24/68, A17)

  R. F. Taschek of AEC's Los Alamos (N. Mex.) Scientific Laboratory presented "Space-Based Detection of Radiations from Nuclear Detonations and Other Sources." Eight test-detection satellites—launched two at a time in joint DOD–AEC Vela Hotel program, Oct. 17, 1963; July 17, 1964; July 20, 1965; and April 28, 1967—were still orbiting earth at 69,000-mi altitude. They had not spotted any detonations to date but provided "undoubtedly the best information" about solar winds, Taschek said. Information would "eventually allow us to understand and perhaps control our environment more readily." (Test; EH; *P Inq*, 6/5/68, 5)

- Capt. J. Laurence Pritchard (RAF), pioneer in British aircraft industry, died in U.K. at age 83. First textbook on airplane structural analysis, *Aeroplane Structures*, which he wrote jointly with A. J. Sutton Pippard, was used all over the world and served as model for many later books on subject. He was editor of *Journal of the Royal Aeronautical Society* 26 yr. (A&A, 7/68, 69)

- MSFC announced award of $1,400 to MSFC Test Laboratory Engineer John A. Hauser for invention of five-module system for purifying and filtering gas to purity necessary in development of Saturn rockets. (MSFC Release 68–82)

- NASA announced swearing in of Dr. Waino W. Suojanen, Chairman of Univ. of Miami's Dept. of Management, as a consultant to Administrator James E. Webb. Dr. Suojanen would serve as member of NASA Man-

April 23

agement Advisory Panel which reviewed NASA's pattern of administration and advised on specific aspects of organization and management. (NASA Release 68-78)

- MSFC announced appointment of M. Keith Wible, chief of MSFC's Manpower Utilization and Administration Office, as head of new manpower utilization system for NASA Hq. Operations Management Office, OMSF. He would be succeeded by Paul L. Styles, head of MSFC's Labor Relations Office. (MSFC Release 68-83; *Marshall Star*, 4/24/68, 1)

- NASA announced that Astronaut Brian T. O'Leary had withdrawn from astronaut training program because he disliked piloting aircraft. Dr. O'Leary, who had completed 15 hr flying time in training program at Williams AFB, Ariz., hoped to remain with space program as researcher in planetary astronomy. Astronaut F. Curtis Michel had received special permission to spend 80% of his time teaching and studying at Rice Univ. and 20% in astronaut training for one year. (MSC Release 68-32; AP, W *Star*, 4/24/68, A2; W *Post*, 4/24/68, 4)

- U.S. and U.S.S.R. had included "little-publicized sanctions" in proposed nuclear nonproliferation treaty, John W. Finney reported in *New York Times*. "Unless they sign the treaty or accept its requirements for international inspection over all their atomic activities, nations may find themselves cut off from assistance in developing the peaceful uses of atomic energy." Such nations would not be able to purchase atomic power plants or to obtain nuclear fuel from U.S. or U.S.S.R. Further, European Atomic Energy Community would not receive fuel unless it entered into inspection agreement within two years with International Atomic Energy Agency. (*NYT*, 4/24/68, 1)

April 24: U.S.S.R. successfully launched *Cosmos CCXVII* into orbit with 182-km (113.1-mi) apogee, 150-km (93.2-mi) perigee, 87.6-min period, and 62.2° inclination. Satellite reentered April 26. (GSFC *SSR*, 4/30/68)

- NASA Apollo Program Director M/G Samuel C. Phillips (USAF) told press at NASA Hq. briefing that *Apollo 6* mission, in spite of anomalies, was "a safe mission from a crew safety standpoint" as demonstrated by spacecraft's recovery in excellent condition after performing an alternate mission. He cited three substantial technical problems—J-2 engine failure because of fuel leak, amplitude of oscillations during 1st-stage burn (pogo effect), and apparent separation of large piece of paint or skin from lunar module adapter during ascent—and one procedural problem—premature shutdown of second of two 2nd-stage engines because of wiring error made by North American Rockwell Corp. which was not discovered by NASA in prelaunch tests. He said all could be corrected.

From demonstrations of *Apollo 4* (launched Nov. 9, 1967) and information gained from *Apollo 6* Gen. Phillips said he had determined "the course of action . . . necessary to correct and demonstrate the correction of the problems and . . . recommended to the Administrator of NASA that we proceed with preparations for the manned flight of 205 with the 101 spacecraft which is planned to be the first manned flight in Apollo, and . . . a Saturn IB." He also recommended that NASA prepare third Saturn V (No. 503) for manned flight in late 1968 with option to revert to unmanned mission if necessary corrections did not meet requirements to ensure crew safety on manned mission. NASA

Administrator James E. Webb's decision on Gen. Phillips' recommendation was expected shortly. (Transcript; *W Post*, 4/25/68, A9)

- NASA Administrator James E. Webb urged Senate Committee on Aeronautical and Space Sciences to restore $48.3 million cut by House from NASA FY 1969 budget request for nuclear rocket program [see March 1]. Webb stressed importance of proceeding with U.S. development of nuclear rocket propulsion as part of total capability in aeronautics and space to (1) meet potential civil or military requirements for space vehicles and missions, (2) avoid short-sighted cutoffs or constraint on promising new technological developments because they had no specific justification in advance, (3) prove that U.S. "does not intend to limit its development of large launch vehicles and payload capabilities" to Saturn V class, and (4) serve as "central focus for continuing advance in nuclear and other technologies involved."

  Responding to questions, Webb cited recent Soviet development of fractional orbital bombardment system, automatic docking flights, and maneuvering of heavy payloads in orbit as evidence U.S.S.R. was "not neglecting any important capabilities. . . . Everything I know . . . indicates they are still probing for those areas that will put them ahead the fastest and give them the lead over us that we cannot overcome in a short time." (Testimony; *SBD*, 4/25/68, 309; *NYT* 4/25/68, 16)

- Univ. of Wisconsin professor Dr. William Kraushaar, speaking at dedication of new $4.3-million Center for Space Research at MIT, reported discovery by NASA's *Oso III* of high intensity of gamma rays flowing from center of Milky Way. Dr. Kraushaar said finding was first observation to support theory that galaxy centers were rich reservoirs of cosmic rays. (Wilford, *NYT*, 4/27/68, 40)

- U.S. leadership in physics "very likely" would be overtaken soon by U.S.S.R. and Western Europe, Dr. Marvin L. Goldberger, professor of physics at Princeton Univ., said at 105th Annual Meeting of National Academy of Sciences in Washington, D.C. Dr. Goldberger, chairman of symposium on current advances in high-energy physics, and other physicists attributed threatened loss of leadership to budget cutbacks and U.S. failure to develop apparatus for producing colliding beams of high-energy particles which would permit exploration of realms of physics inaccessible by other experiments. Plans for accelerators at Stanford Univ. and at Weston, Ill., provided for storage rings for experiments, but there seemed to be no early prospect for their construction. (Text; Sullivan, *NYT*, 4/25/68, 17)

- ComSatCorp reported $1.8-million net income (18 cents per share) for first quarter of 1968—$569,000 (6 cents per share) more than for first quarter of 1967—and operating revenues of record $6.9 million. As of March 31, ComSatCorp was leasing, full-time, equivalent of 754 half circuits, 453 more than on March 31, 1967. Of number leased in 1968, 421 were through two Atlantic satellites and 333 were through two Pacific satellites. One year ago only two satellites were in service, one over Atlantic and one over Pacific. (ComSatCorp Release 68–19)

April 25: U.S.S.R. launched *Cosmos CCXVIII* into orbit with 209.2-km (130-mi) apogee, 143.2-km (89-mi) perigee, and 50° inclination. Period was not disclosed. Satellite reentered same day. Simultaneously, U.S.S.R. disclosed April 24 launch of *Cosmos CCXVII*.

There was widespread speculation that U.S.S.R. would soon attempt

April 25

new space spectacular. Evert Clark had suggested in *New York Times* that U.S.S.R. was secretly testing "a maneuverable rocket stage that could be used to guide bombs down from orbit or to send instruments to the moon." AP said Soviet failure to reveal period of *Cosmos CCXVIII* suggested spacecraft might have reentered before completing one orbit to test fractional orbital bomb system (FOBS) described by Secretary of Defense Robert S. McNamara Nov. 3, 1967. (Clark, *NYT*, 4/3/68, 1; AP, B *Sun*, 4/26/68, 2; GSFC *SSR*, 4/30/68)

- NASA Nike-Tomahawk sounding rocket launched from Churchill Research Range carried TRW, Inc., experiment to measure: total flux energy, including spectrum of precipitated energetic H atoms; spectrum of precipitated energetic protons and electrons; fluctuating and DC electron fields; H light-intensity altitude profile; and location and intensity of ionospheric current systems. Rocket and instrumentation performance was satisfactory. (NASA Rpt SRL)
- In statement to Senate Committee on Aeronautical and Space Sciences, Milton Klein, Manager of NASA–AEC Space Nuclear Propulsion Office, summarized progress of nuclear rocket program. Major milestone has been achieved with "operation in late 1967 of a single reactor for 60 minutes at its design power of 1,100 megawatts, a duration capability adequate for most missions." Technology phase of NERVA program was nearing completion and next step was to develop engine to flight capability, funds for which were included in FY 1969 budget request, he said. No action deferring this step could be taken without losing a major portion of capability in this field. Nuclear rocket was "a focal point for pushing forward frontiers of technology . . . [and] only major advanced propulsion program in the Nation."

  High performance of nuclear rockets had been demonstrated in nine consecutive power reactor tests. Solid base of data and underrating had been built for development of flight-rated NERVA engine. "Development of the NERVA engine at this time," Klein stressed, "would capitalize on this investment and provide a major fundamental advance in propulsion capability. Its high specific impulse will provide a broad mission versatility for the high-payload, high-energy missions . . . inevitably included in a viable space program." (Testimony)
- Dr. Norris E. Bradbury, Director of Los Alamos Scientific Laboratory, testified at Rover Program Hearing of Senate Committee on Aeronautical and Space Sciences that, since project's basic reactor performance goals had been demonstrated along with basic elements of complete engine system, major emphasis of Rover Program should shift to development of overall flight engine. LASL would continue to support NERVA program chiefly in development and evaluation of improved fuel elements and other reactor core components. "Deep space has always been known to be the true domain of nuclear energy for both power and propulsion; it is my belief that the atom will be the work horse of near space as well." (Text)
- Addressing Women's National Democratic Club in Washington, D.C., Dr. Wernher von Braun, Director of Marshall Space Flight Center, said: ". . . we must not seriously impair or hamper our progress in space because we cannot foresee immediate payoffs to offset the investment we are making." He urged that U.S. "come to grasp the unlimited opportunities and the promise of space exploration." (Text; *SBD*,

4/26/68, 321; *W Post,* 4/26/68, C3)
- ESRO announced cancellation of TD-1 and TD-2 solar astronomy satellites, which were to have been built under $20-million contract by an international consortium and launched from U.S. by Thor-Delta rockets. Italy had refused to pay its share of costs, feeling its share of work too slight to justify contribution, John L. Hess later reported in *New York Times.* Earlier U.K. had refused to contribute to proposed budget expansion of ELDO, partner with ESRO in plans for European satellite communications system [see April 16]. (Reuters, *NYT,* 4/26/68, 16; Hess, *NYT,* 4/28/68, 24)
- FAA announced allocations of $74.7 million for construction and improvement of 397 public civil airports under Federal-Aid Airport Program (FAAP) for FY 1969. Program, developed from record 773 requests for aid by public agencies, provided $67.7 million to improve 356 existing airports and $7 million to construct 41 new public airports. (FAA Release 68-28)
- FAA awarded $57,345 to McDonnell Douglas Corp., $52,663 to Western Co., and $28,000 to Bureau of Mines for additional research on use of thickened safety fuels to reduce chances and severity of post-crash fires in survivable aircraft accidents. (FAA Release T 68-15)

April 26: U.S.S.R. launched *Cosmos CCXIX*—10th Cosmos in April and 9th spacecraft in 12 days—into orbit with 1,747-km (1,085.5-mi) apogee, 225-km (139.8-mi) perigee, 104.7-min period, and 48.4° inclination. Soviet scientist Prof. Georgi Pokrovsky in *Nedelya,* Sunday supplement to *Izvestia,* predicted that interlinked satellites might some day form artificial Saturn rings around earth.

NASA Executive Secretary Dr. Edward C. Welsh said U.S.S.R.'s launch activity April 14–26 made most active 12 days in space history of any nation and "a great acceleration" of Soviet space effort. "For some time we've had indications that they're putting in an increasing rate of men and resources."

James J. Haggerty, Jr., wrote in *Journal of the Armed Forces* that U.S.S.R. satellite launches in 1968 might for the first time since 1957 exceed U.S. spacecraft orbited. Launches in Cosmos series, which included a variety of spacecraft, had continued to accelerate, he noted, with 34 Cosmos launches in 1966 and 59 in 1967.

Satellite reentered March 2, 1969. (Cohn, *W Post,* 4/27/68, A15; UPI, *NYT,* 4/27/68, 15; *J/AF,* 4/27/68, 9; GSFC *SSR,* 4/30/68; 3/15/69)

- Maj. William J. Knight (USAF) flew X-15 No. 1 to 209,600-ft altitude and 3,545 mph (mach 5.05) from Edwards AFB. Purposes of test flight were to check Saturn insulation horizon scanner and fixed ball nose. (NASA Proj Off)
- A 15-lb pig-tailed monkey, like one scheduled to orbit earth for 30 days on board Biosatellite D in 1969, had successfully completed simulated space flight fully instrumented with some two dozen separate biological sensors. Test, first joining of instrumented primate and its complete array of biological instrumentation with flight-type spacecraft, met all objectives, including 15-day medical countdown, 3-day simulated flight, and 5-day monitoring. (NASA Release 68-76; *W Post,* 4/26/68, A19)
- NASA established Aerospace Safety Research and Data Institute at LeRC to

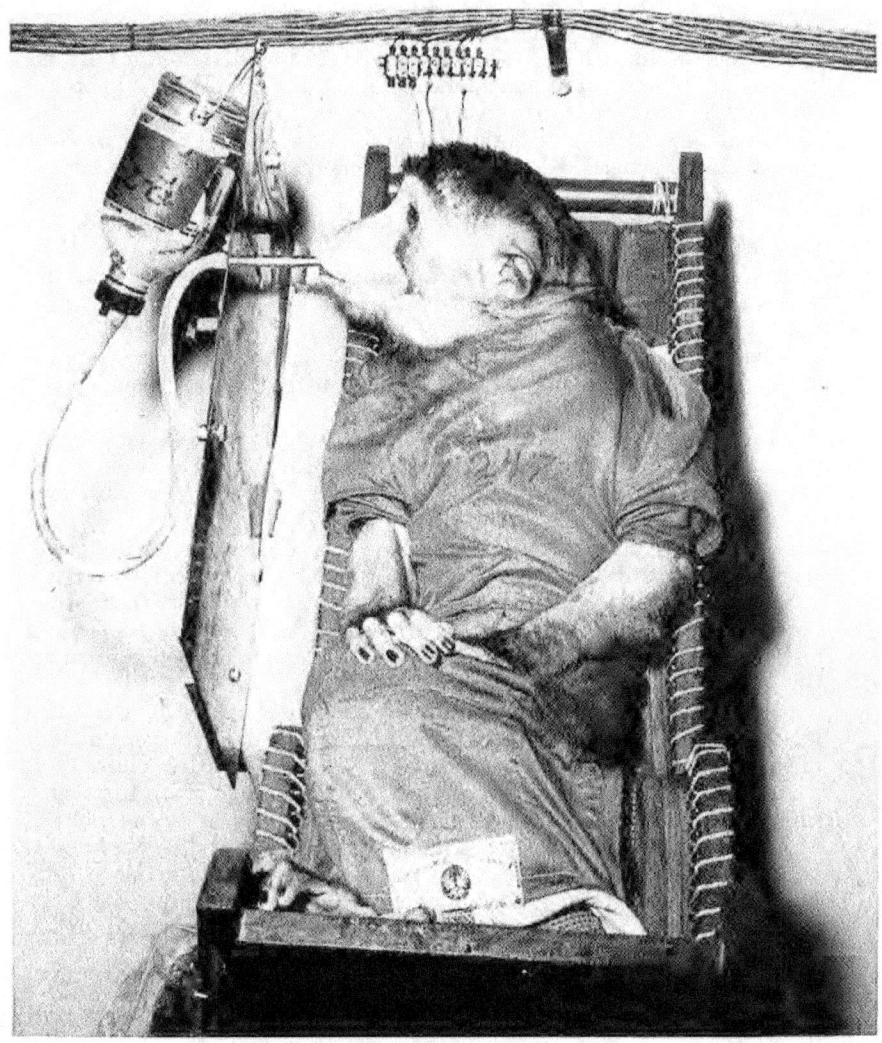

April 26: *Pig-tailed monkey was in good health after 3-day simulated test flight, with 15-day medical countdown and 5-day postflight monitoring. Monkey, identical to one scheduled for 30-day orbit of earth on NASA's Biosatellite D in 1969, had been instrumented with two dozen sensors. In photo, primate is seated in test fixture.*

maintain highest safety standards possible in national aerospace program by solving technical safety problems and providing NASA and its contractors with current information on safety data and procedures. Institute would be directed by I. Irving Pinkel, consultant on aircraft safety to USAF and FAA and former Apollo 204 accident investigator and consultant. (NASA Release 68–79; LeRC Release 68–32)

- NASA published *Constructing Inexpensive Automatic Picture-Transmission Ground Stations* (NASA SP-5079), providing instructions for building from surplus parts $500 ground station that could receive local cloud-cover pictures anywhere in the world from U.S. meteorological satellites. Booklet was available from Clearinghouse for Federal Scientific and Technical Information. (NASA Release 58-77)

April 27: NASA Administrator James E. Webb approved April 24 recommendation of Apollo Program Director M/G Samuel C. Phillips (USA, Ret.) that NASA proceed with preparation of third Saturn V Launch Vehicle for manned mission in late 1968 and retain option for another unmanned mission "if further analysis and ground testing indicate that it is the best course."

Astronauts James A. McDivitt, David R. Scott, and Russell L. Schweickart were scheduled to be launched on Saturn V in second manned Apollo space flight. First manned Apollo mission, Apollo 7 with Saturn IB booster, would carry Astronauts Walter M. Schirra, Jr., Donn F. Eisele, and R. Walter Cunningham into earth orbit in third quarter of 1968. (NASA Release 68-81; W *Star*, 4/28/68)

- NASA successfully launched 600-lb Reentry F payload by Scout booster from NASA Wallops Station to obtain inflight fundamental research data on aerodynamic heating and transition from laminar to turbulent flow in boundary layer. Payload, graphite-tipped beryllium cone 13 ft long, tapering from 0.01 in at nose to 27.3 in at base, was designed to measure heat transfer in slender cone at hypersonic speeds for comparison with ground studies. Three of Scout's four stages were used: 1st and 2nd stages fired during ascent, boosting 3rd stage and payload to 115-mi (175-km) altitude; and 3rd stage drove payload at up to 13,500 mph through earth's atmosphere. Impact occurred 800 mi downrange, northeast of Bermuda.

Reentry F experiment, sixth mission in NASA's Reentry Heating Project, was designed and directed by LaRC under sponsorship of NASA Office of Advanced Research and Technology. Payload was constructed by General Electric Co.'s Re-Entry Systems Div. (WS Release 68-9)

- Aerobee 150 MOD I sounding rocket launched from WSMR carried Naval Research Laboratory experiment to 103.2-mi (166.1-km) altitude to photograph solar corona for streamers and to photograph interplanetary dust shadows using two externally occulated coronagraphs and one solar pointing control. Rocket and instruments performed satisfactorily. (NASA Rpt SRL)

- Crash of USAF F-111A aircraft near Bowie, Tex., Oct. 19, 1967, had been caused by failure of experimental speed break—only one ever installed on F-111—USAF reported. Investigation had indicated hydraulic system tubing ruptured and flight control system was disrupted when bracket assembly failed at 1,000 mph. (AP, *W Post*, 4/27/68)

- Tass reported Moscow scientist had compared "spectrometric analysis" of cactus growing in cold areas with spectrographs of "dark areas" of Mars and concluded areas on Mars were covered with cactus-like vegetation. Other tests on cactus, scientists said, proved it could stand extremes of temperature and other conditions similar to those on Mars. (UPI, W *Star*, 4/28/68, A3)

April 29: NASA awarded $25,847,000 one-year, cost-plus-fee contract to Bendix Field Engineering Corp. for continued maintenance and operation of major portion of NASA's Manned Space Flight Network, including 11 facilities of 14-station unified S-band network for Apollo. Contract extended original two-year agreement containing three options. (NASA Release 68–82)

April 30: NASA launched Nike-Tomahawk sounding rocket from Churchill Research Range carrying Univ. of New Hampshire-Univ. of California at San Diego payload to 66.4-mi (109.7-km) altitude to measure electric field, ionospheric currents, auroral light intensity and location, and proton and electron fluxes (1–10 kev) while passing through or close to auroral display. Rocket did not perform as expected; spin rate was below that anticipated. Payload broke away from rocket and no useful data were received from experiment. Launch was third in series of four [see April 23]. (NASA Rpt SRL)

- Dr. Wernher von Braun, Director of Marshall Space Flight Center, told Senate Committee on Aeronautical and Space Sciences he was greatly concerned about future of entire space program without propulsion capability of nuclear rocket program. Nuclear propulsion was "a must for our future space needs." Failure to proceed into development phase would result in "losses of experienced personnel and cost-increase effects on the total program. . . . A one-year delay in funding could result in as much as two years delay in having an operational nuclear engine."

  Nuclear vehicle as 3rd stage on Saturn V could significantly improve payload and mission versatility, and improved capability could be utilized "to improve mission effectiveness, to increase the mission and payload reliability, and to extend the spectrum of potential missions in the late 1970's and the 1980's. Equally important . . . for high energy missions requiring the launch of two or more Saturn V's, with subsequent rendezvous in earth orbit, we will be able to reduce the number of Saturn V's needed through the utilization of a nuclear vehicle," at substantial cost savings.

  In response to questions by Sen. Howard W. Cannon (D-Nev.), Dr. von Braun said space program was "cutting edge of our technology advancements and of many advances in the applied sciences . . . [because] no other program . . . involves so many branches of technology and science." Reduction in NASA's $60-million NERVA request to $11 million recommended by House would be disastrous because to make manned space operations useful, "plenty of payload" was required. AEC funding for NERVA had been approved, but if cuts were made in NASA funding, program would be nonexistent. (Testimony; Transcript; O'Toole, *W Post,* 5/1/68, A3)

- Secretary of Defense Clark M. Clifford asked House Committee on Armed Services to restore funds cut by Senate for compromise development program for Navy F–111B aircraft. According to compromise plan, USN would continue tests on F–111B experimental models and exploratory work on alternate aircraft, VFX–1, until March 1969 and then decide whether to proceed with F–111B or to cancel program after first eight models and develop alternate. If VFX–1 were chosen, F–111Bs already produced would provide sophisticated air defense until alternate aircraft became operational in 1973. DOD had requested $425 million for 8 F–111Bs and 60 Phoenix missiles and $30 million for R&D on VFX–1. Senate instead had approved $170 million for

VFX–31. (Testimony; Sheehan, *NYT*, 5/1/68, 4; UPI, *W Star*, 5/1/68, A12)
- Univ. of Colorado physicist Dr. Edward U. Condon announced that Univ. $500,000 UFO study for USAF had been completed on schedule. Dr. Condon declined to discuss conclusions and said final report would be submitted to NAS in September. He protested May 14 *Look* magazine article, which called project fiasco, but said completion of field investigations were not related to controversy.

    Rep. J. Edward Roush (D-Ind.), citing article on House floor, questioned scientific profundity and objectivity of project and urged Congress to take over UFO investigation from USAF. (*CR*, 4/30/68, H3087; Clark, *NYT*, 5/1/68, 5)
- Republican Coordinating Committee released statement on U.S.–U.S.S.R. relations, including policy on space: "Outer space should be seen as the focus for ever increasing United States-Soviet collaboration rather than as the site of an endless series of increasingly expensive prestige races. Because our society is open, so much is known about our space program that inviting Soviet participation in the nonmilitary projects would be unlikely to endanger national security. By insisting upon reciprocal privileges we would acquire much additional knowledge about their space efforts, thus achieving a net gain for United States security. At the same time, we must not intimate that the Soviets and ourselves have an exclusive role to play in this area. We must constantly reiterate our willingness to collaborate with NATO and other Allies in space technology." (Text; UPI, *NYT*, 4/30/68, 95; Unna, *W Post*, 4/30/68, A5; *SBD*, 5/1/68, 2)
- NASA awarded General Dynamics Corp. Convair Div. $4,771,390 supplemental agreement for construction of two additional Centaur launch vehicles to be used with Atlas boosters to launch two Orbiting Astronomical Observatories (OAO) in 1960 and 1970. (NASA Release 68–83)

During April: In MIT *Technology Review*, Gen. Bernard A. Schriever (USAF, Ret.), President of B. A. Schriever Associates, and Dr. William W. Seifert, Assistant Dean of MIT School of Engineering, wrote "Air Travel Threatens to Become Hard-to-Get." Unless we "begin now to take steps to meet the demands of the future, sheer growth in population and the accompanying economic demands could so saturate our transportation system, especially the air system, that mobility could become a premium service instead of a routine accommodation." Median forecasts were that by 1980 number of domestic revenue passengers carried by airlines would quadruple, with air cargo growing 10 times. (MIT *Tech Rev*, Vol. 70, No. 6; *W Post*, 10/15/68, B4)
- House Committee on Science and Astronautics published *Survey of Views of Leading Industrial Executives on the National Space Program*. Rep. Olin E. Teague (D-Tex.), Chairman of Subcommittee on NASA Oversight, had queried 750 U.S. senior executives; 449 had replied. Majority considered $5 billion annually for NASA funding to be "about right"; 54% favored maintaining goal of manned lunar landing in this decade; 96% ranked national defense high-priority assignment; 69% felt NASA had made "much" scientific and utilitarian contribution. In context of Vietnam War, 47% favored same level of funding for space program on basis of contribution to national security. (Text)
- National Science Foundation's *Reviews of Data on Science Resources*

presented information on scientific and technical personnel in Government in 1966. Federal Government had employed 149,300 civilian scientists and engineers in October 1966, increase of 4% over December 1964 figure and annual gain of 2%—substantially below 5% annual gain in 1962–64 and in entire 1959–64 period. There were 72,500 scientists on rolls in October 1966, increase of 6% over total 68,300 reported two years earlier, and average gain of 3% per year, only half the 6% average increase in 1959–64. Approximately 13% of natural scientists and 8% of engineers in U.S. were employed in Federal Government in 1966, accounting for 9% of total U.S. employment in those occupations. DOD continued as largest employer of scientists and engineers, with total of 66,000.

Women professional scientific and technical personnel increased more than 3% between 1964 and 1966, to 32,300, 17% of total professional scientific and technical personnel in both years. (NSF Release 68–16)

May 1968

May 1: USAF launched unidentified satellite from Vandenberg AFB by Thor-Agena D booster. Satellite entered orbit with 176-mi (273.2-km) apogee, 115-mi (185.1-km) perigee, 88.6-min period, and 83.1° inclination and reentered May 15. (*Pres Rpt 68*)

- House Republican Policy Committee recommended $353-million cut in Administration's requested $4.37-million NASA FY 1969 authorization and said unless Government spending was dramatically reduced, "the cost of living may reach the moon before our astronauts." Committee urged reductions in Apollo Applications program, administrative operations, and public relations and suggested that NASA place greater emphasis on R&D programs leading to future space advances. "Stockpiling of expensive hardware that may be obsolete by the time it is finally needed" could not be justified. (Text: Sehlstedt, B *Sun*, 5/2/68, A6; *SBD*, 5/7/68, 31)

- NASA, USN, Dept. of the Interior, and General Electric Co. announced plans for Operation Tektite, 60-day study of ocean floor by four U.S. scientists isolated at 50-ft depth in Greater Lameshur Bay, Virgin Islands, in February 1969. Project, first such program undertaken jointly by Government agencies and private industry, would be longest continuous undersea study by a diving team. Previous record was 45 days.

   NASA would acquire data on human endurance; USN, on engineering, marine science, and human behavior; and Interior, on marine geology, underwater mapping, and fish life. GE would build laboratory. (Abraham, P *EB*, 5/1/68; Wilford, *NYT*, 5/2/68, 18; ARC *Astrogram*, 5/23/68, 3)

- Second stage of vehicle expected to be first manned Apollo/Saturn V space vehicle was shipped from Kennedy Space Center launch site to Mississippi Test Facility for cryogenic proof pressure test by North American Rockwell Corp. personnel. Test, scheduled for June, would further certify integrity of stage's liquid-hydrogen tank and would be conducted on all 2nd stages for manned Apollo Saturn missions. Stage would be reshipped to KSC about July 1. (MSFC Release 68–93)

- Edgar M. Cortright, former NASA Deputy Associate Administrator for Manned Space Flight, assumed duties as Director of NASA Langley Research Center. He succeeded Dr. Floyd L. Thompson, who was serving as Special Assistant to NASA Administrator James E. Webb. Cortright had served at Lewis Research Center from 1948 to 1958 and at NASA Hq. since 1958. (*Langley Researcher*, 5/17/68, 1)

- President's Office of Science and Technology released *National Atmospheric Sciences Program—Fiscal Year 1969*, report describing total $200-million Federal investment in atmospheric science in terms of priority areas and programs of 10 participating Federal agencies. Report was prepared by Interdepartmental Committee for Atmospheric

Sciences of Federal Council for Science and Technology. Planned NASA share of funding for FY 1969 was $84,573, dropping from $123,406 in FY 1968. Allocation included $50,594 for meteorological studies; $4,313 for aeronomy, including ionospheric physics atmospheric chemistry, rocket and satellite instrumentation; and $29,666 for study of planetary atmospheres.

During past year efforts had increased on weather modification and space environmental forecasting. Special effort had been made to increase support of university research groups in connection with U.S. participation in Global Atmospherics Research Program. ESSA and NSF had increased meteorology program slightly. ESSA, DOD, and NSF had increased solar-terrestrial research. (Text: OST Release, 5/1/68)

- NASA personnel changes: M/G Robert H. Curtin (USA, Ret.) was appointed Director of Facilities, reporting to Assistant Administrator for Administration. He had been Director of Civil Engineering, Office of Deputy Chief of Staff, Programs and Resources, USAF Hq.

  Ralph E. Cushman, who joined National Advisory Committee for Aeronautics in 1939, was appointed Special Assistant to NASA Assistant Administrator for Administration. (NASA Ann)

- *Oakland Tribune* editorial, "Space Race Has Big Stakes": "The orbital bomb is one chilling example of the potential military threat posed by the Soviet space program. Yet, so far, the only officially announced U.S. reaction . . . has been reports of plans for an over-the-horizon radar to lengthen the warning time against orbital devices and missiles.

  "It would be more reassuring to the nation if the Pentagon were given the authority to devote whatever resources it requires to counter the orbital bomb threat, not simply with defensive measures but with an offensive military capability of our own. . . . The ultimate prize in the space race may be national survival." (*Oakland Trib*, 5/1/68)

- USAF announced modification of F–111A flight control system to correct "binding action in the actuator mechanism." F–111As in U.S. and Thailand had been grounded briefly until "precautionary measures" had been taken. DOD could not ascertain if problem had contributed to loss of two of three F–111As in Southeast Asia during past 10 wk because wreckage had not been found. USAF said, however, F–111A's safety record was superior to that of other supersonic fighters during early flights. (Corddry, B *Sun*, 5/2/68, A3; *NYT*, 5/2/68, 15)

- Twenty nations, including U.S., U.S.S.R., and U.K., presented draft resolution in U.N. General Assembly urging "widest possible adherence" to U.S.–U.S.S.R. nuclear nonproliferation treaty and to pursue urgent negotiations on further measures to halt nuclear arms race. (deOnis, *NYT*, 5/2/68, 12)

May 2: House passed, by 262-to-105 vote, NASA FY 1969 authorization bill (H.R. 15856) of $4.031 billion, including $3.383 billion for R&D, $45 million for construction of facilities, and $602 million for administrative operations. NASA had requested $4.37 billion. House cut $142.4 million from Apollo Applications program—leaving $252.2 million, $186.4 million less than NASA had requested for orbiting workshop and lunar exploration. Administrative operations allocation was cut by $43.5 million, making total of almost $186-million reduction in $4.217-billion authorization recommended by House Committee on Science and Astronautics.

During floor debate Rep. Olin E. Teague (D-Tex.), Chairman of House Committee on Science and Astronautics' Subcommittee on Manned Space Flight, answered Apollo Applications program critics' charge that NASA's Orbiting Workshop would duplicate USAF's Manned Orbiting Laboratory, explaining that projects differed in nature and purpose. MOL objectives were to develop, operate and evaluate specialized experiments and military equipment requiring manned space operations, and DOD would draw on NASA experience in systems involved in Mercury, Gemini, and Apollo. MOL was "example of utilization by another department . . . of NASA-developed space technology." (*CR*, 5/2/68, H3229-661; Sehlstedt, B *Sun*, 5/3/68, 1; Lannan, W *Star*, 5/3/68; Griffin, *H Chron*, 5/3/68; *SBD*, 5/3/68, 15; *Aero Daily*, 5/7/68)

- Nike-Tomahawk sounding rocket launched by NASA from Churchill Research Range carried Univ. of New Hampshire-Univ. of California at San Diego payload to 161-mi (258-km) altitude. Objectives were to measure electric field, ionospheric currents, auroral light intensity and location, and proton and electron fluxes in 1- to 10-kev region while payload was passing through or close to visible auroral display. Rocket and instruments performed satisfactorily; useful data were obtained from all experiments. Launch was last in series of four [see April 30]. (NASA Rpt SRL)

- New Tanay earth station near Manila participated in U.S.-Philippines commercial satellite television inaugural with telecast between Washington, D.C., and Manila via landline to satellite earth station at Brewster Flat, Wash. Facility functioned through *Intelsat-II F-2* at 22,300-mi altitude over Pacific. (ComSatCorp Release 68-22)

- NASA Associate Administrator for Manned Space Flight, Dr. George E. Mueller, at dedication of Grissom and Chaffee Halls at Purdue Univ. said, "The pressing sociological problems besetting this nation will require a high order of technological skill to solve." Space program was contributing to "fundamental solution" of problems of poverty and human welfare by bringing "advancement in economic and technological growth." Space flight would help public understand ability of industry, science, and government to work together to mobilize resources. Honoring late astronauts Virgil I. Grissom and Roger B. Chaffee, Dr. Mueller predicted space program would "help to shape our future" and U.S. would continue "to rely upon the vision and dedication" of such men. Conquest of space would be "our most enduring memorial to these men." (Text)

- NASA Administrator James E. Webb delivered first of three McKinsey Foundation lectures, "Reflections on Government Service," before Columbia Univ. Graduate School of Business: "Our society has reached a point where its progress and even survival increasingly depend upon our ability to organize the complex and do the unusual. We cannot do the things we have to do except by employment of large aggregations of power in highly specialized forms."

   Present technological revolution was "the most decisive event of our times." Great issue of the age was whether U.S. could, within framework of acceptable institutions, "organize the use and development of advanced technology as effectively as the USSR with its totalitarian system of allocating and utilizing human and material resources." Webb

believed capabilities of U.S. system had "immense advantage over all other systems." (Text)
- U.S. military sources in Saigon said USAF F–111A aircraft had been restricted to training flights in Thailand. Three aircraft had been lost on war zone flights since being sent to Southeast Asia in mid-March. (UPI, W *Star*, 5/2/68, 1)
- Arthur E. Raymond of RAND Corp. presented Lester D. Gardner lecture at MIT, reviewing "Air Transport History and a Glimpse into the Future": "Looking back, one sees, ever since the 1920s, nothing but rapid progress in speed, range, reliability, operating altitude, carrying capacity, and volume of operations . . . but seldom if ever in human affairs does this kind of growth continue without a slowing at some point. . . . The days are gone forever when airplanes could be designed and purchased without simultaneously making provision for solving the problems introducing them will create." For "practical, utilitarian air-transport systems," he saw little advantage in speeds above those currently associated with SSTs or ranges above those associated with subsonic jets. He foresaw coupling of these speeds and ranges. He saw little advantage in larger payloads than would be carried in Boeing 747 or C–5 aircraft because of terminal congestion and high aircraft cost. (*A&A*, 7/68, 60–9)
- Donald A. Hall, designer of Lindbergh aircraft, *Spirit of St. Louis*, died in San Diego, Calif., at age 69. (*NYT*, 7/3/68, 33)

May 3: NASA launched two Aerobee 150 sounding rockets from WSMR. First carried Princeton Univ. Observatory payload to 106.4-mi (171.2-km) altitude to point two spectrographs toward hot stars in Scorpius to study their EUV radiation with 1 Å resolution and 0.3 Å resolution. Rocket and instrumentation performed satisfactorily. Second Aerobee 150, launched 30 min later, carried Columbia Univ. experiment to 2.1-mi (3.3-km) altitude to search for x-ray emission from known extragalactic objects in radio galaxy M–87 and in quasi-stellar object 3C273. Rocket performance was unsatisfactory because sustainer did not ignite. Instrumentation performance was satisfactory. (NASA Rpt SRL)
- NASA Lewis Research Center announced organizational changes: Dr. Seymour C. Himmel, Assistant Director for Launch Vehicles, was named Chief of new Special Projects Div. for jet noise and V/STOL aircraft study and to new post of Assistant Director for Aeronautics. Newell D. Sanders, Chief of Chemistry and Energy Conversion Div., would assume additional duties as Assistant Chief of Special Projects Div.

    Milton A. Beheim, Chief of Aerodynamics Branch, Advanced Systems Div., was appointed Chief of new Wind Tunnel and Flight Div.

    Edmund R. Jonash, Chief of Centaur Project Office, was named Chief of new Launch Vehicles Div., which would include previously separate Centaur, Agena, and Atlas Project Offices. William R. Dunbar would become Project Manager for Centaur. H. Warren Plohr and Edward F. Baehr continued as Agena and Atlas Project Managers. (LeRC Release 68–29)
- NASA awarded $73-million cost-plus-fixed-fee contract to Boeing Co. for technical integration and evaluation in support of Apollo program. Agreement was an addition to Boeing's previously contracted Saturn V work and could be extended as necessary. (NASA Release 68–85)
- Air Force Academy announced selection of Gen. Carl A. Spaatz (USAF,

Ret.) to receive its Thomas D. White Award, given annually for contribution to national defense and security. Gen. Spaatz was fighter pilot in World War I. During World War II he helped plan strategic bombing of Germany and commanded air forces in North Africa, U.S. Strategic Air Forces in Europe, and final bombing operations against Japan. (*Rocky Mountain News*, 5/4/68, 13)

May 4: Preliminary "pathfinder" flights could delay U.S.S.R.'s first manned lunar and planetary flights, according to Soviet scientist Vasili Parin in *Sovetskaya Rossiya*. "Experiments with animals," preceding interplanetary flight, he said, "will take many months and perhaps many years." (UPI, *NYT*, 5/4/68, 66)

- Sen. Claiborne Pell (D-R.I.) told 33rd American Assembly in Harriman, N.Y., "The proportion of government expenditures in oceanology must be brought much closer to our outer space program." Increased Federal investment in oceanology was justified by promise of economic returns, he said. Material benefits from space program were more remote. (Text; *W Post*, 5/5/68, A25)

May 5: U.K.'s *Ariel III*, launched by NASA from ETR May 5, 1967, completed successful year in orbit. It has traveled around earth 5,500 times and transmitted more than 400 million bits of usable data on upper atmosphere. All five experiments were still working well. (*Interavia-AirLetter*, 6/11/68, 9)

- Two prototypes of 20-ton DO 31 VTOL transport aircraft, capable of carrying 40 fully equipped servicemen or 80 civilian passengers, made first public appearance at Hanover Air Show, West Germany. Using combination of vectored-thrust turbofans and pods of lift engines, jet transport rose vertically and changed in air to horizontal flight in first public demonstrations. Built by Dornier Works, Munich, at $50-million cost, prototypes were financed by West German Defense Ministry, which now saw no requirement for transport. Financing would be withdrawn at close of 1968, with five years required for completion of production design. Thus far no civilian interest in production had been shown. (*NYT*, 5/7/68, 95)

May 6: Republican Coordinating Committee report charged "technology gap" existed between U.S. and U.S.S.R. Military space systems emphasis had been on passive satellites until approval of Manned Orbiting Laboratory program—delayed for years. Administration, "fearful lest new developments might provoke undesirable Soviet reactions," had failed to exploit boldly new concepts in science and technology. U.S. R&D had not been aggressively pursued, with consequential slowdown in new weapons development. (Text; AP, *NYT*, 5/27/68, 8; Golden, *P Inq*, 5/27/68)

- Astronaut Neil A. Armstrong ejected and parachuted to safety from NASA's $2.5-million Lunar Landing Research Vehicle (LLRV) while flying simulated lunar landing at Ellington AFB, Tex. Vehicle, which had reached 500-ft altitude, crashed and burned on impact. Cause of accident was unknown. (*C Trib*, 5/7/68; *B Sun*, 5/7/68; MSC *Roundup*, 5/10/68, 1)

- Dr. Edward C. Welsh, NASC Executive Secretary, defended space activity before St. Louis, Mo., section of AIAA: "There are . . . those who are trapped by the illogical proposition that if the money involved were not spent on space, it would automatically flow into projects in which they are more interested.... I do not agree. It is not an 'either/or' situation. In my judgment, if this country is great—and I know it is—it has the will, the ability, and the responsibility to handle both a vigorous space program and the social and economic problems which confront it. In fact, our competence to solve the issues of the city is greater because of the space program." (CR, 6/17/69, S7314)

May 7: U.S.S.R. successfully launched *Cosmos CCXX* into orbit with 755-km (469.1-mi) apogee, 677-km (420.7-mi) perigee, 99-min period, and 74° inclination. All equipment was functioning normally. (SBD, 5/8/68, 39; AP, NYT, 5/8/68, 93; GSFC SSR, 5/15/68)

- NASA successfully launched Canadian Black Brandt IV sounding rocket from NASA Wallops Station to 510-mi (820-km) altitude. Primary objective was to check out instrumentation to be carried later in 1968 on International Satellite for Ionospheric Studies (ISIS A), Canada's third ionosphere-probing satellite, and to confirm results of similar 1967 launch. Secondary objectives were to explore spectrum of VLF electromagnetic waves, measure electron density and temperature, and measure thermal gradients in vicinity of a skin depression. Launch was conducted when Canada's *Alouette II* was passing overhead, to permit comparison of data telemetered to earth by both vehicles. Good data were obtained. (NASA Rpt SRL; WS Release 68–10)

- Lawrence A. Hyland, Vice President and General Manager of Hughes Aircraft Co., received NAA's Robert J. Collier Trophy for 1967 on behalf of Hughes Surveyor Program Team, JPL, and other companies and organizations involved in project which citation said, had "put the eyes and hands of the United States on the Moon." Award was presented by Vice President Hubert H. Humphrey for greatest annual achievement in aeronautics or astronautics in U.S.

In ceremonies at Smithsonian Institution's Air and Space Museum, Humphrey called U.S. space program "a splendid challenge and a noble mission ... one whose practical benefits for today are exceeded only by its promise for tomorrow. "I urge every American to support the future development of our space program, and I ... shall do so with pride and vigor."

Humphrey said Nation had decided to commit resources "to venture in space for one primary reason: We believe that this mission to the far-out will produce many down-to-earth benefits for men.... In fact ... the nation that is first in science and technology has a chance to be the first to overcome some of the perplexing problems that have beset mankind since the beginning of civilization." Space research "has vastly expanded our capabilities in navigation, communication and meteorology. It has given us new products and processes in such fields as agriculture, photography, metallurgy, and oceanography." Techniques "that are going to put a man on the moon are ... exactly the techniques that we are going to need to clean up our cities ...; the systems analysis approach ... is the approach that the modern city of America is going to need if it's going to become a livable social institution.

"So maybe we've been pioneering in space only to save ourselves on earth ... maybe the nation that puts a man on the moon is the nation that will put man on his feet first right here on earth....

"I think a certain extravagance of objectives—a will to push back

May 7: *Lawrence A. Hyland (right), Vice President and General Manager of Hughes Aircraft Co., looks at Surveyor model with Vice President Hubert H. Humphrey at Smithsonian Institution. Vice President presented Hyland 1967 Robert J. Collier Trophy (background) in Smithsonian ceremonies honoring Hughes Surveyor team, Jet Propulsion Laboratory, and General Dynamics Corp. for program which "put the eyes and hands of the United States on the moon."*

the frontiers of the unknown—is the test of a vital society, a nation that intends to meet the challenge of tomorrow." (Text; AP, W Star, 5/8/68, A4; Aero Tech, 5/20/68, 19)

- U.S. patent No. 3,381,917 was awarded to Wendell F. Moore, assistant chief engineer at Bell Aerosystems Co., and Edward G. Ganczak, research associate, for Bell Pogo and Flying Chair, flying platforms on which pilot could stand or sit on fuel tank. Engine was kerosene turbojet. Both had arm pieces with which pilot directed thrust. Pilot could disembark without encumbrance, advantage useful to soldiers, policemen, or firemen. (Jones, NYT, 5/11/68, 45)

- Juan T. Trippe, founder, Chairman, and Chief Executive Officer of Pan American World Airways, Inc., announced his retirement at annual meeting in New York. Board of Directors selected President Harold E. Gray, Chairman and Chief Executive Officer, and Senior Vice President Najeeb E. Halaby, President.

    Citing Trippe's 41-yr service with PAA, *New York Times* termed him "last of the aviation pioneers" to retire. One of four who forged major U.S. trunk airlines in industry's infancy, Trippe had led PAA to be first airline to fly across Pacific, first to fly across Atlantic, first to order and fly American-made jets, first to order Boeing 747, and first to order SSTs. New Chairman Gray, hired by PAA in 1929 as its 10th pilot, had made first scheduled transatlantic flight, in 1938. He had served as President since 1964. Halaby, pilot since age 17 and former FAA Administrator, had joined airline as senior vice president in 1965. (Hammer, *NYT*, 5/8/68, 63; 5/12/68, 16)

- AEC refused comment on *Science and Citizen* report U.S. had set off 3 and U.S.S.R. 22 undisclosed underground atomic tests in 1964 through 1967, bringing total underground tests to 168 for past three years. Magazine, published by Committee for Environmental Information in St. Louis, Mo., said source of information was publication of Research Institute of Swedish National Defense and that three undisclosed U.S. tests had occurred in 1964. All subsequent U.S. tests had been reported. (*W Post*, 5/7/68, 7)

May 8: House passed by record 353-to-37 vote H.R. 17023, FY 1969 Independent Offices and HUD appropriations bill. Before floor debate, H.R. 1164, with points of order against NASA provisions waived, was passed by voice vote. As passed, H.R. 17023 provided $4.008 billion for NASA —$959.777 million below FY 1967 level and $580.677 million below FY 1968 (*CR*, H3458–502).

- ESRO successfully launched first two-stage Centaure rocket fired from Italy at Perdasdefogu, Sardinia, carrying Max Planck Institute (Landau, Germany) payload to 88.5-km (55-mi) altitude in 118 sec. (AP, *W Post*, 5/9/68)

- Arrival of Prime Minister Thanom Kittikachorn of Thailand at White House marked first transpacific satellite telecast of visiting head-of-state to his home country. Telecast also inaugurated television from U.S. to Sri Racha earth station in Southeast Asia. (ComSatCorp Release 68–24)

- USAF F–111A aircraft crashed on training flight 60 mi north of Las Vegas, Nev. Instructor and student pilot escaped injury. USAF said cause of crash was not known. F–111As had been criticized after three of six sent to Thailand were lost within weeks. (UPI, *W Post*, 5/9/68, A8; AP, *W Star*, 5/9/68, D7; UPI, *W News*, 5/9/68, 2)

- Director of Defense Research and Engineering, Dr. John S. Foster, Jr., testified before House Committee on Armed Services on FY 1969 defense research, development, test, and evaluation program. "We have a strong technical-military position today only because we built a strong research and technology base in the past. . . . Yet there are some indications that the program is eroding." The "net effect of continuing this trend will be a serious weakening of our long-term national security position." Research and technology funding in FY 1968 was "about 70% of the FY 1964 level, a critical 30% reduction." (Text; P *SB*, 6/9/68)

- Charles W. Mathews, NASA Deputy Associate Administrator for Manned Space Flight, addressed Society of Automotive Engineers Space Technology Conference in Washington, D.C.: "A major goal of the Apollo Applications Program is to accelerate the evolution of the utility of space flight using the very major capability that has been developed in the Apollo Program. . . . First, our experience to date leads to the conclusion that the utility of space flight will be greatly enhanced by the participation of men onboard the spacecraft. Second, in manned space flight our ability to maintain operations for durations considerably in excess of those now obtained is necessary for efficient operations. . . . The initial step in this approach is the establishment of a workshop in earth orbit."

Saturn IB orbiting workshop, manned after launch by three-man crew, would be followed by manned Apollo Telescope Mount, which would dock with workshop for 56-day mission to study solar phenomena—"first application of man in space to conduct advanced scientific experiments." Features and operating modes of Saturn V workshop, to follow, would be based on information gained from Saturn IB workshop. (Text)

May 9: U.S. reconnaissance satellite, orbiting at altitudes of several hundred miles, had discovered that U.S.S. *Pueblo* was no longer moored in North Korean port of Wonsan. Vessel had occupied that berth since her capture by North Koreans Jan. 23. State Dept. confirmed absence of vessel but would not discuss source of information. (Goulden, *P Inq*, 5/10/68, 2)

- In second McKinsey Foundation lecture at Columbia Univ., NASA Administrator James E. Webb discussed "Goal Setting and Feedback in Large Scale Endeavors." NASA had created "in-house technical and administrative competence" making possible "correct judgments" and thus could move "to the voter-judgment arena with confidence." NASA's "integrated system" approach had proved more effective than "independent components" approach of past in solving problems of space development. Successful working partnership of universities, industry, and government had yielded product, in usable resources, "greater than the sum of its parts." Scientific R&D expenditures in large endeavors could contribute more to economic growth in next decade than any other single factor. Maximum transfer of technology to nonspace use should be "purposefully and systematically sought." Costs of space accomplishments had been "less than three percent of the total of our federal expenditures" for first 10 yr and "less than five one thousandths of our gross national product." More than 90% went to laboratories and factories, outside NASA. (Text)

- Dr. Charles A. Berry. Director of Medical Research and Operations, MSC, was elected 1969 President of Aerospace Medical Assn. at 39th Annual Meeting in Bal Harbour, Fla. Astronaut Edwin E. Aldrin was named honorary member. (MSC *Roundup*, 5/24/68, 1)

- USAF restricted all F-111A flights in U.S. and in Southeast Asia pending investigation of crash in Nevada May 8. Five F-111As in Thailand would conduct no air strikes. (AP, *W Post*, 5/10/68, 1; AP, W *Star*, 5/10/68, A7; *NYT*, 5/11/68, 2)

May 9–10: NASA held Second Conference on Sonic Boom Research at Headquarters to review status of NASA university program on sonic boom research, survey research program at NASA centers, determine most pressing areas of research for SST, and determine most promising avenues of research on sonic boom overpressure reduction. Appraising proceedings, I. R. Schwartz, NASA OART Research Div., said that "significant progress has been made during the past year in our understanding and analysis of the generation and propagation of sonic boom. Further, it has been analytically demonstrated in the NASA program . . . that the utilization of a particular aerodynamic technique can result in large reductions in sonic boom overpressures. . . . [This] allows us to expect vast improvements in future SST conceptual designs." (NASA SP-180)

May 10: MSFC contract activity: Brown Engineering Co. received $1,007,000 one-year contract renewal for support services at MSFC Space Sciences Laboratory. A. L. Mechling Inc. was awarded $556,416 one-year contract for barge towing service on the Tennessee, Mississippi, and Ohio rivers, Gulf of Mexico, and Atlantic intercoastal waterways. (MSFC Releases 68–100, 68–101)

• Special Assistant to President for Science and Technology, Dr. Donald F. Hornig, on receipt of first Mellon Institute Award at Carnegie-Mellon Univ. for "the application of science to the betterment of mankind," discussed "A Crisis for Science." After two decades of progress, "Congress and the public ask whether we can afford it after all. We now find ourselves pulling back from the exploration of space, slowing down on the development of universities, and even holding back on health research." Scientific community, however, had "done much to alienate itself from the society which supports it." (Text; *Science*, 7/19/68, 248)

May 11: NASA's Goddard Satellite Tracking Center reported upper stage of *Molniya I-8* booster, launched April 21 by U.S.S.R., had reentered earth's atmosphere and disintegrated over Florida. (AP, *NYT*, 5/12/68, 87)

May 12: At dedication ceremony attended by 19,000, USAF officially renamed Bunker Hill AFB, Indiana, Grissom AFB in honor of Astronaut Virgil I. Grissom who died Jan. 27, 1967, in Apollo fire. (*NYT*, 5/13/68)

• Nearly fallout-free nuclear explosive appeared within U.S. grasp as result of two AEC Plowshare nuclear excavation tests Jan. 26 and March 12 at Nevada test site. Small size of explosives used had confined fallout within few hundred yards of craters, with radiation at source undetectable after three days. (O'Toole, *W Post*, 5/12/68)

May 13: Senate Committee on Aeronautical and Space Sciences approved $4.151 billion NASA FY 1969 authorization, adding $119.137 million to H.R. 15856, which had authorized $4.031 billion. Figure still was approximately 5% below $4.37 billion requested by President Johnson. Senate voted $350 million for Apollo Applications, against $253.2 million by House, and $55 million for NERVA program, as against $11.7—increasing R&D funds to $3.475 billion. Both houses agreed on allocation of $2.025 billion for Apollo program. Senate committee cut construction of facilities funds by $5.4 million to $39.6 million, but increased administrative operations by $32.4 million to $635.6 million. (NASA *LAR* VII/48; AP, *NYT*, 5/14/68; Sehlstedt, B *Sun*, 5/14/68)

• ESSA was studying solar flare prediction technique and ground observatory support for NASA's Apollo Telescope Mount flight scheduled for 1971. NASA had transferred $30,000 to ESSA for six-month investigation to enable astronauts to know when and where solar flare would occur within accuracy of 1,000 mi out of 860,000-mi-wide face of sun. ESSA also would study data required for maximum results from ATM mission and examine existing data collection networks and observing procedures. Effort would be monitored by ATM Project Office and Space Sciences Laboratory at MSFC. (MSFC Release 68–103; *SBD*, 5/15/68, 73)

- Free world's largest experimental comsat was being built by Hughes Aircraft Co. under USAF Space and Missile Systems Organization management. The 1,600-lb satellite—designed to provide testing for tactical communications between military units in the field, ships at sea, and aircraft—was scheduled for late 1968 delivery. It would be launched by Titan III–C booster into 22,300-mi-altitude orbit and would be equipped with three antenna systems. During orbital tests, standard military UHF band would be used for air-ground communications and super-high-frequency portion of X band for military satellite communications. (AFSC Release 64.68)
- National Sporting Aviation Council, formed Feb. 16, adopted official charter at first meeting held in Washington, D.C., affirming its original mission to promote progress and development of all forms of sport aviation in U.S. through National Aeronautic Assn. and worldwide through Fédération Aéronautique Internationale. (NAA *News*)
- Dr. Walter Haeussermann was selected to Fellowship in American Astronautical Society for "direct and significant contribution to the field of astronautics," as Director of Astrionics Laboratory, MSFC. (MSFC Release 68–102)
- John B. Tuke had assembled operable weather-picture receiving station at his home in Stranraer, Scotland, out of spare electronic parts, old antenna, electric motor, and tuning fork for about $480. One of first amateurs to construct homemade receiving set, Tuke was able to receive pictures from ESSA and Nimbus satellites for 15 min, from lower Spain to Arctic, on each pass. Interest in amateur stations had grown to point that NASA had published booklet of instructions for building ground stations [see April 26] and Electro-Mechanical Research Co. in College Park, Md., had begun selling packaged sets for $5,000. (Wilford, *NYT*, 5/13/68)

May 14: ComSatCorp Chairman James McCormack submitted Annual Report to shareholders' meeting at Washington, D.C. *Intelsat I, II F–2, II F–3,* and *II F–4* were reasonably loaded with commercial communications traffic. Intelsat III series was scheduled for late summer delivery and launching during fall 1968 and spring 1969. Intelsat IV series would be presented for Government approval shortly. It was hoped development could be under way before end of 1968. More than half the 40 earth stations anticipated to be in operation by 1969 were expected to be operating by end of 1968. Despite technical problems, satellite operating circuits maintained 100% reliability. Total of 48 nations were represented by Interim Communications Satellite Committee, to which ComSatCorp had submitted several U.S. proposals: relating investments of global members directly to amount of their use of system; limiting voting power to 50% maximum with substantive issues decided by two-thirds majority; ComSatCorp's continuing as Consortium manager with contractual obligations made more specific.

ComSatCorp opposed authorization by Federal Communications Commission of separate satellite system for broadcast distribution as had been proposed.

With first quarter operating income of $372,000 and investment income of $1.426 million, ComSatCorp realized net income of $1.798 million. At close of first quarter 1968, investments in communications fa-

cilities amounted to $73 million. ComSatCorp expected this to exceed $100 million at end of 1968. (Text)
- NASA's Test and Training Satellite *Tts I*, carried pickaback as secondary payload on *Pioneer VIII* and ejected into orbit Dec. 13, 1967, reentered atmosphere and was believed to have burned up over Easter Island in Pacific. Spacecraft supplied 14-station worldwide tracking network training for Apollo flights. (NASA Release 68–86; AP, P *EB*, 5/15/68; *SBD*, 5/15/68, 75)
- Dr. Harold Brown, Secretary of the Air Force, told Los Angeles Junior Chamber of Commerce that if U.S.S.R.'s long-range bomber force improved, U.S. might need new 2,000-mph F–12 interceptor rejected by Secretary of Defense Robert S. McNamara in 1967. Although U.S.S.R. had about 155 long-range bombers and U.S. had 680, USAF officials feared U.S.S.R. might build advanced manned aircraft to overpower what McNamara had acknowledged to be an obsolete conventional air shield. DOD had announced major cutbacks in existing air defense units May 13. (Text; *Omaha W–H*, 5/15/68, 21)
- Parafoil, steerable parachute being developed and tested by Notre Dame Univ. under contract with USAF's Flight Dynamics Laboratory, brought Sgt. Robert McDermott (USA) down from 35,000-ft altitude in demonstration at Wright-Patterson AFB. Parafoil had glide of nearly three feet for each foot of vertical drop, enabling jumper to steer toward target. Steerable parachutes normally used for spot landings glided only about ½ ft for each foot of fall. (AP, *NYT*, 5/15/68, 24)
- Dr. Walton L. Jones, Chief of NASA's Biotechnology and Human Research Div., Office of Advanced Research and Technology, in an interview reported study of 3,500 residents of Chicago, Denver, Dallas, and Los Angeles had indicated some were more annoyed by aircraft noise than others. Noise was more objectionable inside the house than out. The deeper people slept, the more boom it took to wake them. People could adapt to sonic boom, but data compiled thus far did not indicate to what extent. Dr. Jones' work on improving safety of aircraft seats showed "many injuries could be avoided if seats were designed to dissipate more energy." Good forward-facing seat, possibly with shoulder harness, would be better than rear-facing seats. (*CSM*, 5/14/68; *CR*, 5/28/68, E4743)
- Dr. Arthur Rudolph, Saturn V Manager at MSFC for past 4½ yr, was retiring after 38 yr in rocketry, MSFC Director, Dr. Wernher von Braun, announced. He would be succeeded by Lee B. James, Deputy Director of Apollo program in NASA Office of Manned Space Flight. (MSFC Release 60–106; AP, *NYT*, 5/15/68, 5; AP, W *Star*, 5/15/68, A2)

May 15: Dr. Edward C. Welsh, NASC Executive Secretary, told National Space Club in Washington, D.C., U.S.S.R. in 1968 had been surpassing its 1967 payload successes numerically and, "what is more significant . . . surpassing substantially the number of successful payloads launched this year by the United States. While our activity curve is moving down . . . theirs is headed up." U.S.S.R. had "orderly, persistent, and well planned space program, including a vigorous project for a manned landing on the lunar surface." U.S. had operational lead in communications, navigation, and meteorology but U.S.S.R. showed "considerable activity" in those fields. U.S. had been generally more

successful in unmanned lunar and planetary performance, but U.S.S.R. was more active. (Text)
- Fiftieth anniversary of U.S. Airmail Service was celebrated in Washington, D.C., with gathering at National Air and Space Museum of pioneer airmail pilots and presentation by Postmaster General W. Marvin Watson of commemorative 10-cent postage stamp. Watson observed that, from small beginning, "virtually all first class mail which can be effectively airlifted is now so moved and almost 80 per cent of all letter mail now travels by air." (Text; Lidman, *NYT*, 5/16/68, 30; *Aerospace*, Spring 68)
- NASA Javelin sounding rocket launched from Churchill Research Range carried Univ. of Iowa experiment to 500-mi (805-km) altitude to observe frequency-time spectra and measure relative phase and amplitude of magnetic and electric fields of VLF radio noise in 30 hz–10 khz range; measure electric field amplitude of naturally occurring radio noise from 7–70 khz; measure differential energy spectra for protons and electrons over 5 ev–50 kev energy range; and measure DC electric field perpendicular to payload spin axis. Rocket and instrument performance was satisfactory and all scientific objectives were met. (NASA Rpt SRL)
- NASA announced that sinus condition that was aggravated by pressure changes and had not responded to treatment led to grounding—perhaps permanently—of Astronaut John S. Bull. Astronaut Robert A. R. Parker had suffered spine fracture during parachute training at Williams AFB, Ariz., but injuries were not thought to be serious. (AP, *B Sun*, 5/16/68, A3; *W Star*, 5/16/68; *H Post*, 5/16/68, 8)
- NASA awarded North American Rockwell Corp. $2.575-million contract modification to continue Saturn V (S–II) "battleship" or ground test program at Santa Susana, Calif., test center through July. Modification increased total contract for S–II stage development testing and facility maintenance planning to $1.281 billion. (MSFC Release 68–109)

May 16: Iris I (*Esro II–B*) International Radiation Investigation Satellite, designed, developed, and constructed by European Space Research Organization under July 8, 1964, NASA–ESRO agreement, was successfully launched by NASA from WTR by four-stage Scout booster. Orbital parameters: apogee, 673.7 mi (1,084.2 km); perigee, 204.1 mi (328.5 km); period, 98.9 min and inclination, 97.2°. Primary NASA mission objectives were to place *Iris I* into planned polar orbit and provide tracking and telemetry support. The 164-lb, 35½-in high, 30-in-dia cylindrical satellite carried seven experiments for solar-astronomy and cosmic-ray studies representing six different organizations from U.K., France, and the Netherlands. Six experiments were operating as planned and their scientific objectives were being achieved.

Iris I was backup spacecraft to ESRO II–A, which had been launched May 29, 1967, but had failed to achieve orbit because of Scout 3rd stage malfunction. ESRO was responsible for experiment instrumentation, delivery of spacecraft to launch site, equipment and personnel necessary to mate spacecraft to launch vehicle, and spacecraft testing. NASA provided Scout launch vehicle, conducted launch operations, and supplied data and tracking acquisition support. (NASA Proj Off; NASA Release 68–75; UPI, *W Post*, 5/17/68, D13; *W News*, 5/17/68, 5; GSFC SSR, 5/31/68)

May 16

May 16: *In cooperative program, NASA launched Iris I (Esro II-B) International Radiation Investigation Satellite, for European Space Research Organization from WTR by four-stage Scout booster. In April photo, Iris I is prepared for mission to orbit seven solar-astronomy and cosmic-ray experiments from six organizations.*

• First recording of pulsar signals at short wavelengths, 2293 mc, was made by Drs. Alan Moffet, Ronald D. Ekers, and Richard M. Goldstein of Cal Tech, using 210-ft Mars dish antenna at Goldstone Tracking Station operated by JPL for NASA. Signals' energy indicated origin from natural source such as white dwarf stars or neutron stars. CP–1919, one of three pulsars observed, was weakest source ever recorded. At

2293 mc, signal was one-hundredth that of other two, though at longer wavelengths signal was one of two strongest of four known pulsars. ("CP" referred to Cambridge pulsar list; number indicated location.) Dr. Moffet placed pulsars 100 to 10,000 light years from earth, in Milky Way Galaxy. Although signal strengths varied, periods remained constant. (Cal Tech Release; Pasadena *Independent,* 5/15/68)

- President Johnson announced intention to nominate following for appointment to six-year terms on National Science Board: R. H. Bing, Chairman, Dept. of Mathematics, Univ. of Wisconsin; Harvey Brooks, Dean of Engineering and Applied Physics, Harvard Univ.; William A. Fowler, professor of physics, Cal Tech; Norman Hackerman, President, Univ. of Texas at Austin; Philip Handler, Chairman, Dept. of Chemistry, Duke Univ. Medical Center; James G. March, Dean of Social Sciences, Univ. of California at Irvine; Grover Murray, President, Texas Tech; and Frederick E. Smith, professor of zoology, Univ. of Michigan. (*PD,* 5/20/68, 802)

- In last of three McKinsey Foundation lectures NASA Administrator James E. Webb discussed "Executive Performance and Evaluation." NASA program at peak level had employed more than 400,000 full-time workers and 20,000 contractors, subcontractors, and suppliers and drawn on more than 150 universities, in addition to 8,000 NASA-supported professors, scientists, and technicians. It had worked with urban communities on problems, reclaimed wasteland, constructed massive buildings, and developed new transportation means. NASA would not be viewed in history as unique but was likely "to prove a prototype."

  As Administrator, Webb said, his purpose was to work toward environment within which NASA could be as innovative in the management of all its activities as it was in its scientific and technical work. Large-scale endeavor required executives of unusual type and had to be designed to enable executives to perform in an unusual way. (Text)

- *New York Times* editorial urged congressional passage of legislation to ban sonic booms: "While technical and budgetary difficulties have slowed development of an American supersonic aircraft, the British and French are progressing with their joint-venture model. It is important to have some protection on the law books before the booms come crashing down on the nation's ears. . . . Developers . . . say they 'assume' it will fly at supersonic speed only over water but this is too important an issue to be left to anyone's assumption." (*NYT,* 5/16/68, 46; 5/25/68, 32)

May 17: NASA Aerobee 150 sounding rocket launched by NASA from WSMR carried GSFC experiment to 102.5-mi (165-km) altitude to measure UV radiation from several early-type stars. Rocket and instruments performed satisfactorily. (NASA Rpt SRL)

- NASA announced it had notified agency elements to hold permanent employment to onboard strength at close of May 13, to minimize impact of potential manpower adjustments necessitated by cuts in FY 1969 budget. Field centers had been advised of possible further reduction by 75% of attrition May 14–June 30 (one replacement for every four losses). FY 1969 budget included positions for 32,727 permanent civil service employees. NASA installations had been requested to assess program impact of continued reduction, which could exceed 1,000 if current limitations continued. Actual supportable employment level for FY

May 17

1969 depended on final appropriation approved. (NASA Release 68–91; AP, *NYT*, 5/19/68, 80; *SBD*, 5/23/68, 114; Kluttz, *W Post*, 5/27/68, A21)

- House Committee on Interstate and Foreign Commerce approved bill to control aircraft noise levels and sonic boom. Committee, by voice vote, agreed on provisions empowering FAA to set standards for measurement and regulations for control. (H.R. 1463; *W Post*, 5/17/68, 5)
- NASA named five-man board with counsel to investigate crash of Lunar Landing Research Vehicle piloted by Astronaut Neil Armstrong at Ellington AFB, Tex., May 6. Board would determine probable cause of accident, identify and evaluate corrective actions, and evaluate implication for LLRV and lunar module design and operations. (NASA Release 68–90)
- LaRC Associate Chief of Flight Mechanics and Technology Div., John P. Campbell, had been selected by American Helicopter Society to receive its Paul E. Haueter Award for 1968, *Langley Researcher* announced. Award was given annually for significant contributions to development of vertical lift aircraft other than helicopters. Campbell was cited "for his personal contributions to and direction of NASA research programs which have provided a sound technology base for a large number of VTOL designs." (*Langley Researcher*, 5/17/68, 1)
- NASA announced organizational change effective May 19. Functions and personnel of Procurement Management Div. of Management Operations Directorate, Office of Manned Space Flight, was transferred to Office of Industry Affairs. Procurement Office would become Program Support Div. (OMSF), reporting to Director of Procurement. (NASA Ann)

May 18: Nimbus B spacecraft launched by NASA from WTR failed to enter orbit because of malfunction in booster. Thorad-Agena D booster and spacecraft were destroyed by Range Safety Officer. Satellite, third in Nimbus meteorological series, contained two 25-w SNAP–19 radioisotope electric power generators fueled with plutonium 238 to supplement solar panels and enclosed in capsules designed to withstand impact and corrosion. Debris fell into Pacific between Vandenberg AFB and San Miguel Island. (WTR Release; AP, *W Star*, 5/18/68; *P EB*, 5/18/68; AP, *W Post*, 5/19/68, A3; AP, *NYT*, 5/18/68; 5/25/68)

- NASA successfully launched Stratoscope II, Princeton Univ.'s balloon-borne telescope, from Scientific Balloon Flight Station, Palestine, Tex., to altitude of 80,000 ft. Purpose was to photograph sky from altitude above 95% of earth's atmosphere. (NASA Release 68–93)
- U.S.S.R. would conduct "further tests of the landing system of space apparatuses" from May 20 to June 30 in two Pacific areas, Tass announced. Aircraft and ships were warned not to enter 80-mi-dia circular area 1,200 mi south of Hawaii near Christmas Island and 90- by 180-mi rectangular area 1,600 mi south of Hawaii between noon and midnight local time each day. (AP, New Orleans *Times-Picayune*, 5/19/69; *SBD*, 5/21/68, 99–100)
- Entire thrust of Soviet space program pointed to circumlunar flight by cosmonauts before U.S., according to former astronaut Cdr. M. Scott Carpenter (USN). Speaking at Beloit College symposium on oceanography, he predicted NASA might try to send crew around moon in autumn

as counter to Soviet manned space spectacular. (Lewis, *B News*, 5/19/68)

- NASA, NAS, and NRC would be "taking a long close look" at applied satellites capable of producing concrete benefits to mankind, James J. Haggerty, Jr., wrote in *Journal of the Armed Forces*. Study, recommended in NAS report released Feb. 8, would focus on economic aspects. Among new areas of application proposed were: forestry-agriculture satellite to survey and identify vegetation resources through standard and infrared photography; oceanographic satellite to predict fish locations, isolate plankton areas, and improve ship routing and weather detection through use of sensors; hydrological satellite to aid in management of water resources and to facilitate weather forecasting; and geological satellite to extend process of using airborne sensors to detect minerals, oil, and gas. Group also would study economic benefits of advances in existing applied spacecraft, such as "advanced weathersat" to interrogate high-altitude weather balloons and broadcast satellite to promote "extremely broad program of education and culture."

  NAS report had set dollar values on benefits: savings of $8 billion annually in agriculture and construction alone from two-week weather forecasting; $5 million in management of municipal water supplies; and "hundreds of millions" in potential saving to shipping, fishing, and coastal engineering industries, but satellites would have to be developed at "considerable investment." (Haggerty, *J/AF*, 5/18/68, 9, 19)

- AEC Chairman, Dr. Glenn T. Seaborg, presented first Swords Into Plowshares Award for major contributions in nuclear science to Columbia Univ. in New York. Occasion marked opening of Nuclear Week, sponsored by private foundations, industry, and government, to attract young people to careers in nuclear science. (*NYT*, 5/19/68, 81)

- Fifth International Congress for Noise Abatement, meeting in London, adopted resolution urging governments to prohibit supersonic flights over their territories. Dr. Charles Wakstein, Dartmouth College scientist, said if overland SST flights were permitted, about 50 million Americans could be exposed to average of 15 sonic booms daily. Flights could cost $100 million annually in damage to homes. However, evidence thus far indicated little psychological harm to those surrounded by constant noise. (Shuster, *NYT*, 5/19/68, 3)

- NASA had named Wade St. Clair, former consultant and director on radio and television programs, Director of Special Events, Office of Public Affairs, succeeding Brian M. Duff, who had joined Urban Coalition staff. St. Clair most recently had served as Duff's deputy. (NASA Release 68–88)

- F–111A supersonic aircraft participating in air show at Holloman AFB, N. Mex., crashlanded, without injuring crew or spectators. USAF said ban on F–111 flights was still in effect but did not apply in this case since aircraft was still owned by manufacturer, General Dynamics Corp. (*W Post*, 5/19/68, 1; *W Star*, 5/19/68, A12)

- Man's "most rewarding new frontier for the next generation" might lie in ocean space, according to *New York Times* editorial. "Imminent opening of the deep-sea bed for commercial exploitation offers exciting possibilities for the enrichment of mankind . . . but unregulated rush to cash in on the wealth of the ocean floor could provoke serious international conflicts." U.S. caution, however, in approaching U.N. es-

tablishment of international control had merit because "knowledge of the potential of ocean space is still fragmentary and vital national interests are at stake." (*NYT*, 5/18/68, 30)
- President Habib Bourguiba of Tunisia visited MSC for general orientation and tour during two-day visit to Houston. (AP, *New Haven Register*, 5/17/68)

May 20: NASA Aerobee 150 MI sounding rocket launched from WSMR carried GSFC experiment to 102.6-mi (165.1-km) altitude to obtain monochromatic picture of sun using spectroheliograph and to measure solar spectral irradiance using photoelectric radiometers. Rocket and instrumentation performance was satisfactory. (NASA Rpt SRL)
- ESRO successfully launched Centaure rocket from Kiruna Range in Sweden to collect micrometeoroid data. (*SBD*, 5/23/68, 118)
- In "remote sensing" experiment by NASA and U.S. Geological Survey, pilots and scientists began two-week air-ground sweep of Southern California in search of solutions to environmental problems. Area had been selected because of variety of geographic features. Aircraft would test sensors for later use in satellite surveys of land, water, mineral, human, and energy resources. (Copley News Service, *San Diego Union*, 5/20/68)
- Australian Air Force was sending technical team to U.S. to investigate airworthiness of F-111 aircraft. Australian government had ordered 24 F-111s in 1963 for estimated $112 million. Cost had already increased to $266 million, and additional increases were expected. Delivery would be refused until Australia was satisfied mechanical faults had been eliminated. (*NY Post*, 5/20/68, 21)
- NASA and British Ministry of Technology began month-long tests of devices for predicting performance of aircraft brakes at NASA Wallops Station on runway modified to include varied surface conditions. Under direction of Walter B. Horne, LaRC, tests in which FAA and USAF also were participating, would enable engineers to establish system for runway length requirements to bring landing aircraft to safe stop. (NASA Release 68-89)

May 21: NASA announced it would launch two Mariner spacecraft, F and G, on flyby missions to Mars in 1969. Project responsibility was assigned to JPL; LeRC would be responsible for launch vehicle.

Deep Space Network stations would provide tracking and communications during launch period scheduled between mid-February and mid-April. Spacecraft would reach Mars between end of July and mid-August. Arrival dates of two spacecraft would not be less than five days apart. Atlas-Centaur combination would for first time be used as launch vehicle for planetary exploration. Spacecraft would weigh approximately 900 lb. *Mariner III* (Nov. 5, 1964) weighed 575 lb. (NASA Special Release)
- NASA announced that Dr. Mac C. Adams, Associate Administrator for Advanced Research and Technology, would resign July 1 to return to private industry. He would be succeeded by James M. Beggs, Director of Purchases and Traffic, Westinghouse Electric Corp., who would join NASA June 1. Bruce T. Lundin, Associate Director for Development, LeRC, was named NASA Deputy Associate Administrator for Advanced Research and Technology. Commenting on Dr. Adams' departure, NASA Administrator James E. Webb said, "He brought to NASA strong man-

agement and engineering competence . . . [and] gave us brilliant leadership in critical times." (NASA Release 68–92)

- USAF grounded 42 F–111As to repair defective hydraulic valve. Aircraft had been under restrictions since May 8 accident at Nellis AFB, Nev., of which defective valve was "the most probable cause." Twelve earlier model F–111As not containing defective part were not grounded. Actuator valve problem was suspected in third F–111A crash in Thailand, attributed to capsule of sealant lodged in flight control system. Modifications should be completed in June. (DOD Release 472–68; AP, NYT, 5/22/68, 5; AP, W Star, 5/22/68, A20)
- Scientists from Stanford Univ. and Univ. of California, using telescope at Lick Observatory atop Mt. Hamilton, Calif., and telescopes on Kitt Peak, Ariz., had found one of four recently discovered pulsars flashed visible wavelengths of light at half the rate of its radio pulse. Discovery was disclosed by Univ. of California at Berkeley astronomer, Dr. David Cudaback, at pulsar meeting in New York sponsored jointly by NASA Institute for Space Studies and Belfer Graduate School of Science, Yeshiva Univ. (Sullivan, NYT, 5/21/68, 14; 5/22/68, 24; 5/26/68, 12)
- Sen. Mark O. Hatfield (R-Ore.), addressing IEEE Region Six Conference in Portland, criticized order of priorities by which "survival" was superseded by "our desire to explore new technological frontiers. . . . We must rationally decide if our goal of promoting the well-being of man is better served through huge expenditures to beat the Russians to the moon or through developing methods to feed a hungry world. It is a very black mark on our sense of values and a contradiction of man's quest for progress when one part of the world competes for the moon while the other part competes for a loaf of bread." (Text)
- Arthur C. Clarke, co-author of *2001: A Space Odyssey*, told Los Altos Morning Forum proposed Federal cuts in NASA budget would curtail progress in "medicine, agriculture, weather forecasting, communication and other endeavors." NASA budget now was "smaller than the amount American women spend on cosmetics in a year." (Anderson, *San Jose Mercury*, 5/23/68)
- Soviet space experts Dr. Georgii Pokrovskii and Aleksandr Yavnel believed automatic docking of *Cosmos CCXII* and *CCXIII* April 15 had brought U.S.S.R. closer to day when it could establish large permanent earth-orbiting stations and planetary laboratories. In Moscow interviews, they claimed docking had cleared way for launching segments of large spacecraft by individual rockets, then assembling them in common orbit. (Macomber, *St. Louis G–D*, 5/21/68)

May 22: First public demonstration of satellite communications between merchant ship and shore was given before Propeller Club of Port of Baltimore, Md. Civic officials conversed via NASA's *Ats II* with captain of vessel off Valparaiso, Chile, coast. Conversation was transmitted over telephone lines to GSFC and relayed to Rosman, N.C., to satellite, and to ship. (*NYT*, 6/26/68, 86)

- Post-launch investigations at MSFC and at Rocketdyne Div. of North American Rockwell Corp., J–2 manufacturer, had revealed J–2 rocket engine malfunctions on April 4 *Apollo 6* mission had been caused by ruptures of small-diameter fuel lines feeding engine igniters. Improved designs for lines had been completed and new lines were being manufactured for extensive testing before use in next Saturn IB and Saturn V launch vehicles. (*Marshall Star*, 5/22/68, 1)

- Aerospace Industries Assn. released its 1967 Annual Report, reviewing "industry's continuing pattern of growth": 11% increase in sales over 1966, to $27.2 billion; maintenance of 1,400,000-man work force, making aerospace industry largest U.S. manufacturing employer; and $575-million increase in exports, to $2.2 billion, including 42.9% increase in civilian aircraft, to $789 million. (Text)
- NASA selected Fairchild-Hiller Corp., General Electric Co., and Lockheed Missiles & Space Co. for competitive negotiations for $5-million fixed-price contract to develop designs for basic spacecraft configuration of F and G models of Applications Technology Satellite. Firms would provide complete specifications and versions of critical spacecraft systems. (NASA Release 68-95)
- MSFC awarded third one-year contract extension, at $6.448 million, to RCA Services Co. for institutional support services at MSFC. (MSFC Release 68-112)
- NASA established Office of Facilities, merging facility functions and applicable staff elements of Office of Manned Space Flight, Office of Space Science and Applications, Office of Advanced Research and Technology, Office of Tracking and Data Acquisition, and Facilities Management Office. New office, under direction of M/G Robert H. Curtin (USAF, Ret.), would encompass planning, design, acquisition, construction, repair, alteration, maintenance, operations, utilization, and disposal of facilities including land and collateral equipment. It would report to the Assistant Administrator for Administration. (NASA Ann)

May 23: World's first passive satellite, *Echo I*, launched by NASA Aug. 12, 1960, reentered earth's atmosphere and disintegrated over southeastern Pacific. Launched as passive communications and air-density research satellite, *Echo I* also served geodesists determining accurate continental and intercontinental distances. (UPI, *W Post*, 5/24/68; *W News*, 5/24/68, 3; *Newsweek*, 5/27/68; *SBD*, 5/28/68, 138)

- USAF launched unidentified satellite from Vandenberg AFB by Thor-Burner II booster into orbit with 561-mi (902.8-km) apogee, 509-mi (819-km) perigee, 102.1-min period, and 98.9° inclination. (*Pres Rpt 68*)
- Smithsonian Institution scientists positively identified as a meteorite, fist-size black stone which damaged Denver, Colo., warehouse roof in July 1967. It was "first recovered fresh-fall" meteorite in U.S. since Bells, Tex., meteorite Sept. 9, 1961. (Cleveland *Plain Dealer*, 5/24/68)

May 24: U.S.S.R. successfully launched *Cosmos CCXXI* into orbit with 2,082-km (1,293.7-mi) apogee, 214-km (133-mi) perigee, 108.3-min period, and 48.4° inclination. (*SBD*, 5/27/68, 132; GSFC *SSR*, 5/31/68)

- First satellite telephone link between U.S. and Australia began with opening of earth station at Moree, Australia. (*W Star*, 5/26/68)
- Mathematical tools used to predict lunar motions were so imprecise that they hampered evaluation of lunar flights and use of moon as timepiece, according to two articles in *Science* by group of JPL scientists. Amplifying report, Dr. J. Derral Mulholland, JPL scientist in charge of revision of methods of prediction, said errors seemed to derive largely from inadequate emphasis on gravital influence of other planets, particularly Jupiter and Venus. JPL was preparing computer program to pre-

dict lunar position for future space flights. (*Science*, 5/24/68, 874–8; Sullivan, *NYT*, 5/24/68, 47)

- James T. Murphy, Deputy Manager in NASA Saturn V Program Office at MSFC, was named Director of MSFC's Safety Office. He would also serve as member of MSFC Flight Readiness Review Board. E. W. Neubert would continue as Chairman of MSFC Safety Board. (MSFC Release 68–114; *Marshall Star*, 5/29/68, 1)

- Before Los Angeles Junior Chamber of Commerce, Sen. Thomas H. Kuchel (R-Calif.) said, "What we need and what we lack is an administration which will allocate space dollars according to the skills and manpower and brain power available, rather than on the basis of political preference." No Federal program was more in need of public understanding than space and its relationship to balanced economy, he said. When looking at future space investment we needed to count "payoff" already gained. Spinoff from space technology was providing "enormous boon to both soldier and civilian." We needed space goals which clearly stated potential benefits of program to people, he said, "or the public will justifiably fight a single penny being thrown into space." (UPI, *San Diego Union*, 5/25/68; *CR*, 6/10/68)

- Six-month slippage in launch date of U.K.'s Skynet comsat to spring 1969 was due to "normal R&D problems," according to USAF Space and Missile Systems Organization. Philco-Ford Corp.'s Space and Re-entry Systems Div. was building two Skynets compatible with U.S. Initial Defense Communications Satellite Program for delivery to U.K. While Skynet program was linked with U.K.'s latest defense review, Defense and Technology Ministry sources said its cancellation was not expected, though U.K.'s military withdrawal from Asia would substantially affect its use and scope. (*Aero Daily*, 5/24/68)

May 25: Aerobee 150 MI sounding rocket launched by NASA from WSMR carried GSFC photometers and objective-grating wide-angle camera to 105-mi (168.9-km) altitude to measure radiation from celestial bodies in $\lambda 2800–\lambda 1300$ spectral range. Payload instrumentation performed satisfactorily; Aerobee pointing control system (ACS) malfunctioned. (NASA Rpt SRL)

- Unexpected solar flares could cause serious radiation damage to SST passengers and lead to aircraft navigation errors, according to Dr. Adolph Razdow, President of Yardney Electric Corp., Razdow Laboratories Div. Global solar watch, he said, could provide 10-min warning of strong solar flares, enabling pilot to descend to safer altitude. NASA's Solar Particle Alert Network (SPAN) could give 20-min flare warning to unprotected astronauts on moon, who could then take shelter or take antiradiation drugs. Peak year in solar flare cycle would be 1969. (UPI, *NYT*, 5/25/68, 59)

May 26: NASA appointed H. Kurt Strass, Chief of Earth Orbital Mission Office, Advanced Spacecraft Technology Div., to head newly established Office of Aerospace Safety Research Programs. New office, which would operate within Office of Advanced Research and Technology, would report to Associate Administrator for Advanced Research and Technology through Director, Space Vehicles Div., OART. (NASA Ann)

May 27: U.S.S.R. announced successful completion of rocket tests in two Pacific areas to test equipment for spacecraft landings at sea. Series, which began May 20, had been scheduled to last until June 30. (*SBD*, 5/29/68, 144)

- FAA announced Washington National Airport was first in U.S. to equip main instrument runway with color-coded centerline lights cautioning pilots entering last 3,000 ft of runway during takeoffs and landings. (FAA Release T 68-18)

- *Aerospace Daily* survey showed most experts considered military R&D funds prime candidate for budget cuts. Services had been told to slash $900 million from FY 1969 requests as their part in $6-billion reduction demanded by Congress, which exhibited little support for either Manned Orbiting Laboratory program or Sentinel ABM system.

  Federal Contract Research Center also was likely to be cut, though Pentagon sought $48 million, $3 million less than FY 1968. Government was spending estimated $17 billion annually on research and there was "rising general Congressional reluctance to approve more funds or even to provide additional money merely to keep pace with the inflation of the economy." (*Aero Daily*, 5/27/68, 26)

- Charles J. Donlan was named Acting Director, Advanced Manned Missions Program, NASA Office of Manned Space Flight, replacing Edgar M. Cortright. He would serve in addition to his duties as Deputy Associate Administrator for Manned Space Flight (Technical). (NASA Ann)

- *May 28:* In second Mars lander experiment [see April 23] NASA concluded JPL R&D program demonstrating feasibility of sending lightweight scientific capsules to Mars by successful 250-ft free-fall drop of 63-lb heat-sterilized planetary landing craft from hovering helicopter onto asphalt pavement at Goldstone Tracking Station. Capsule, which struck at 80 mph, or impact of 2,500 g, was powered by 12-cell silver-zinc battery which produced sufficient electricity after impact to extend wind gauge and to operate radio transmitter and timer aboard Mars lander. Battery was believed to be toughest yet developed. (NASA Release 68-152; W *Star*, 9/8/68, A17; Pasadena *Independent Star-News*, 9/8/68; *NYT*, 10/15/68, 34; JPL *Lab-Oratory* 7/68, 8)

- Gen. John P. McConnell, USAF Chief of Staff, was awarded National Geographic Society's Gen. Thomas D. White Space Trophy for 1967 for "effective leadership and direction of U.S. Air Force in development and utilization of aerospace vehicles." Award was presented annually to military or civilian member of USAF who made most outstanding contribution to U.S. progress in aerospace. (NGS Release; AP, *NY News*, 5/29/68, 22)

- MSFC announced it had extended for one year $3,647,603 cost-plus award-fee contract held by Management Services, Inc., for technical services at MSFC. (MSFC Release 68-116)

May 29: Command module for Apollo 7, first manned Apollo flight, left North American Rockwell Corp.'s Downey, Calif., plant enroute to Kennedy Space Center. Service module had arrived at KSC May 17. (NASA Release 68-100)

- Saturn V 2nd stage planned for first manned Apollo/Saturn V mission passed cryogenic proof pressure test at NASA Mississippi Test Facility. (MSFC Release 68-119)

- NASA announced changes in Aerospace Safety Panel personnel: Dr. Charles D. Harrington, President of Douglas United Nuclear, Inc., was elected Chairman. Dr. Eberhard F. M. Rees, Deputy Director (Technical) at MSFC and Special Assistant to Apollo Spacecraft Manager, resigned from panel to devote full time to these assignments. Newly named to panel were: Frank C. Di Luzio, President of Raynolds Electrical and Engineering Co., Inc., and Vice President of EG&G, Inc.; Gen. Orval Cook (USAF, Ret.); and Dr. Harold Agnew, Weapons Div. Leader, Los Alamos Scientific Laboratory. Dr. William A. Mrazek, Assistant Director for Engineering and Chief Engineer for Industrial Operations at MSFC, was named consultant. (NASA Release 68–98)

- NASA selected Computer Sciences Corp., Lockheed Electronics Co., Computer Applications, Inc., and Service Technology Corp. (LTV), for competitive negotiations on $1-million one-year renewable contract to provide computer support services at Electronics Research Center. (NASA Release 68–99)

May 31: U.S.S.R. successfully launched *Cosmos CCXXII* into orbit with 520-km (323.1-mi) apogee, 281-km (174.6-mi) perigee, 91.3-min period, and 70.9° inclination. Satellite reentered Oct. 11. (AP, *NYT*, 6/2/68; GSFC *SSR*, 5/31/68; 10/15/68)

- U.K. successfully launched Skylark sounding rocket from Adelaide, Australia, to altitude of 149 mi (239.8 km) after initial delays caused by weather. (Reuters, *NYT*, 6/2/68, 55)
- Roderick W. Spence, Los Alamos Scientific Laboratory, outlined advantages of nuclear rockets and offered chronological resumé of Rover program in *Science*. Thirteen years had produced reliable reactor ready for development into flyable engine. Development of complete nuclear-electric propulsion system would be difficult and expensive, "but if man wants to continue to explore space to the best of his ability, such a development seems inevitable." (*Science*, 5/31/68, 953–9)
- NASA announced LaRC and LTV Aerospace Corp. would negotiate unified contract valued in excess of $14 million to provide complete system management for Scout launch vehicle for 24-mo period beginning Nov. 1. (NASA Release 68–101)
- NASA announced extension until Feb. 13, 1969, of contract with Sperry-Rand Corp. for mission support services to Systems Reliability Directorate at GSFC. Value of extended cost-plus-award-fee agreement with two one-year options was approximately $3.5 million. Included in services were operation and maintenance of test facilities such as large space chambers, laboratory equipment, and instrumentation. (GSFC Release G–35–68)
- USN told Pratt & Whitney Div., United Aircraft Corp., it would pay approximately $180 million less than $1.2-billion contract price for 1,640 F–111 engines ordered in February 1967. Adjustment of approximately $19 million in price of engines already delivered brought difference between USN and Pratt & Whitney prices to almost $200 million. USN believed more efficient operation could produce engines at lower cost. (Kelley, W *Star*, 6/4/68, 5)
- USAF awarded Lockheed Missiles & Space Co. $1.174-million initial increment to $2.959-million cost-plus-incentive-fee contract for launch services at ETR. (DOD Release 511–68)
- In response to pressure from smaller countries, U.S. and U.S.S.R. agreed to make changes in proposed nonproliferation treaty: stronger guarantees to small countries which would benefit from peaceful use of nuclear power, more urgent efforts to end world arms race, and enforcement of

During May: Project engineer W. B. Horne, with T. J. Yager, examines effects of Convair 990 landing on grooved runway at Wallops Station. Grooves—one inch apart, one-fourth inch wide, and one-fourth inch deep—increased friction on wet runways.

U.N. charter authority against use of force generally. (Brewer, *NYT*, 6/1/68, 1)

During May: Karl G. Harr, Jr., President of AIA, commented in spring issue of *Aerospace* on preliminary report of space applications study being conducted by NASA and NAS. "Concept of funding space research as an investment has not yet found wide acceptance." Intangibility of benefits —renewed national prestige, expanding technological capability, and enormous scientific gain—was to blame. Report indicated, however, that "investment will pay far greater bonus . . . monetary value alone may run into billions." Advantages of applications program superseded dollar value. Potential contributions to "welfare of the planet" included possibilities of greater food availability, enhancement of air safety, more effective management of natural resources, higher general education level, closer relations among nations with cooperative management of earth's total resources. Opportunity to harvest these gains was present but "we must pursue . . . program aggressively" by increasing our investment. (*Aerospace*, Spring 68, 3)

- LaRC was conducting operational research program at NASA Wallops Station to evaluate runway grooving as a means of reducing aircraft hydroplaning, loss of friction during high-speed landing because of water or slush film. Aircraft landed on specially prepared runways with grooves $\frac{1}{4}$ in wide, $\frac{1}{4}$ in deep, and 1 in apart and braking effectiveness was recorded by instruments on aircraft and on ground. Research indicated grooving helped to alleviate all known phenomena which resulted in low tire-ground friction, but scientists were concerned that undesirable vibrations might be introduced into aircraft. (*Langley Researcher*, 5/17/68, 4)

- Lt. Gen. Ira C. Eaker (USAF, Ret.) wrote in *Data* he believed sonic boom problem would be eliminated, permitting transcontinental flight and realization of estimated 1,200 SSTs in service by 1990. "No federal program since World War II had ever had the thorough analysis, careful examination and close scrutiny accorded the present SST plan." (*Data*, 5/68, 8)

June 1968

June 1: U.S.S.R. launched *Cosmos CCXXIII* from Plesetsk Cosmodrome into orbit with 979-km (608.3-mi) apogee, 221-km (137.3-mi) perigee, 89.9-min period, and 72.9° inclination. Satellite performed satisfactorily and reentered June 9. (UPI, P *EB*, 6/3/68; *SBD*, 6/4/68, 168; GSFC *SSR*, 6/15/68, 7/15/68)

June 2: New York Times editorial criticized Administration's failure to support International Biological Program's proposal to conduct five-year worldwide study of spoilage and near exhaustion of earth's resources by chemical pollution. "One would think that the urgency and importance of these matters would quickly rally general support to these and related efforts to meet the worldwide challenge . . . [but] Ivan L. Bennett, Jr., Deputy Director of the Office of Science and Technology, opposed congressional appropriation of $5 million . . . for American participation . . . during fiscal 1969. The parochial priorities implied . . . must deepen concern as to whether Spaceship Earth can be saved for its proliferating cargo of fragile organisms whose increasingly ravenous demands steadily deplete and damage the ecological basis of life on this planet." (*NYT*, 6/2/68)

June 3: U.S.S.R. planned to test-fly Tu-144 supersonic aircraft during 1968, possibly within few months, Soviet Vice Minister for Air Industry Alexander Kobzarez told press at Third International Air and Space Salon in Turin, Italy. Tu-144 would carry 120 passengers up to 1,500 mph over 4,000-mi range. Soviet aircraft on display included $6.4-million, long-range, 186-passenger Il-62 jet aircraft and $2-million Mi-6 and Mi-10 heavy-duty helicopters. (*NYT*, 6/4/68, 93)

- NASA and DOD would spend estimated $1.406 billion during next decade for parachutes, aerodynamic decelerators, aerial recovery systems, planetary landing devices, and spacecraft escape systems, according to Frank A. Burnham in *Aerospace Technology*. Technology was marked by rapid growth and widespread applications, including drogue chutes for supersonic aircraft; personnel parachutes capable of being "flown" to spot landings; recovery systems for Apollo, MOL, military spacecraft, drones, and test missiles; systems to drop cargo from altitude and deposit it on ground from aircraft in flight; and air "snatch" rescue systems. (*Aero Tech*, 6/3/68, 26)

June 4: U.S.S.R. launched *Cosmos CCXXIV* from Baikonur Cosmodrome into orbit with 311-km (193.2-mi) apogee, 167-km (103.8-mi) perigee, 89.1-min period, and 51.8° inclination. Satellite functioned normally and reentered June 12. (AP, *NYT*, 6/5/68, 7; *SBD*, 6/5/68, 172; GSFC *SSR*, 6/15/68)

- Nike-Apache sounding rocket launched by NASA from WSMR carried 145-lb GSFC payload to 80.8-mi (130-km) altitude to evaluate parachute recovery system for Nike-Apache and Nike-Cajun launch vehicles.

Rocket and instrumentation performed satisfactorily; excellent data were obtained from all telemetry channels. (NASA Rpt SRL)

- MSFC awarded Boeing Co. $1,123,591 contract modification to perform failure effects analysis, components criticality determination, and reliability report on Saturn V launch vehicles through SA-510, 10th flight vehicle. (MSFC Release 68-121)
- At Glassboro (N.J.) State College commencement, President Johnson discussed developments in U.S.-Soviet relations during past year: completion of space law and space rescue treaties, new consular treaty, bilateral air agreement, and negotiations on agreement to avoid costly ABM race. He proposed International Council on Human Environment, accelerated efforts to develop global satellite communications system, conducting of Arctic and Antarctic exploration, and study of possible productive uses of rain-rich tropical forest. (PD, 6/10/68, 903; NYT, 6/5/68, 1F)
- Rep. Silvio O. Conte (R-Mass.) inserted in *Congressional Record* results of questionnaire mailed to his constituents which showed 3,857 out of 8,145 would reduce Government spending on space program. Second and third most popular candidates for cuts were public works (2,298) and poverty program (2,070). (NASA *LAR* VII/55)
- NASA announced that North American Rockwell Corp. Rocketdyne Div. injector would be used in Bell Aerosystems Co. engine of first manned lunar module. Decision on whether to use Bell or NAR injector for lunar mission had not yet been made. (MSC Release 68-41)

June 4-28: Seventh session of Legal Subcommittee of U.N. Committee on Outer Space met in Geneva, Switzerland. U.S. delegation included two NASA officials, Paul G. Dembling, General Counsel, NASA, Alternate Representative; and Daniel M. Arons, Attorney-Adviser, NASA, as adviser. (NASA Hq, 10/14/68)

June 5: USAF launched unidentified satellite from Vandenberg AFB by Titan III-B booster. Satellite entered orbit with 262-mi (421.6-km) apogee, 85-mi (126.8-km) perigee, 89.9-min period, and 110.5° inclination and reentered June 17. (*Pres Rpt 68; SBD,* 6/25/68, 272)

- Rep. Emilio Q. Daddario (D-Conn.) inserted in *Congressional Record* address he had planned to deliver to House, deploring 1969 budgetary cuts in research field, particularly 20% reduction in National Science Foundation funds. Tendency in times of stress to "jump on research as the vulnerable part of the Federal budget" came from thoughtlessness, but slashing research funds was particularly serious "in view of the administrative cutbacks in basic research funds being made by the so-called mission-oriented agencies." America's "tortuous new physical, social and security problems" could not be resolved without "new tools, new methods, new approaches" and only way to develop them was by research. (*CR,* 6/5/68, H4572)
- President Johnson, in accordance with Senate Concurrent Resolution 67 approved May 29, bade Secretary of Commerce C. R. Smith to continue coordination of national efforts in World Weather Watch program by providing forum for consultation among interested Federal agencies and departments, requiring submission of plans for following fiscal year to Congress annually on March 1, and developing arrangements to further international participation and cooperation in weather program. (*PD,* 7/8/68, 1057-8)

- NASA announced appointment of R. Emerson Harris, Supervisor of System Safety, Boeing Co., as Assistant Director of Safety (System), reporting to Bob P. Helgeson, Director of Safety. Harris also would serve as Deputy Executive Secretary to the Aerospace Safety Advisory Panel. (NASA Ann)

June 6: NASA investigation board reported that primary cause of Oct. 5, 1967, crash of T–38 jet trainer near Tallahassee, Fla., was "a jam in the lateral control system (ailerons) from unknown source." Accident had claimed life of pilot, Astronaut Clifton C. Williams. Board recommended improvement in T–38 inspection procedures. (NASA Release 68–102; UPI, *W Post*, 6/7/68, A2; Sehlstedt, B *Sun*, 6/7/68, A3)

- NASA appointed Jerome Freibaum, Assistant Executive Secretary, Office of the Administrator, to succeed L/C Wayne Mathews (USAF, Ret.) as Program Manager of ComSatCorp-related activities in Space Applications Programs Office. Col. Mathews had joined Lockheed Aircraft Corp. (NASA Ann)

- *New York Times* editorial praised U.S.–U.S.S.R. cooperation on revision of nuclear nonproliferation treaty. Textual changes "are substantively important, but their psychological impact is even greater." The way Soviet-American agreement was reached "was itself extraordinary. The two co-chairmen of the Geneva conference came to New York determined to accept no amendments, but midway through the debate . . . they realized concessions might be essential. They prepared a joint contingency plan which contained the treaty changes they both would accept if necessary." When changes had to be made, "they were able to negotiate details in a few hours. The flexibility of United States and the Soviet Union . . . should make [U.N. majority for treaty] a virtual certainty." (*NYT*, 6/6/68, 46)

June 8: NASA successfully launched two Aerobee 150 sounding rockets from WSMR. First rocket carried Naval Research Laboratory and Univ. of Maryland payload to 111.3-mi (179-km) altitude to flight-test flight design verification unit (FDVU) of high-resolution spectroheliograph planned for use on Apollo Telescope Mount-A. Second rocket carried American Science and Engineering, Inc., payload to 93.7-mi (151-km) altitude to obtain high-resolution x-ray pictures of active region of sun during solar flare and general x-ray emission of solar corona. Rocket and instrumentation performed satisfactorily, but payload of first rocket failed to separate, preventing functioning of parachute recovery system. (NASA Rpts SRL)

- Univ. of California scientist Dr. David Cudaback, in telephone interview, said rhythmic light flashes coming from one of newly discovered pulsars, which he described at NASA pulsar meeting May 21 in New York, had turned out to be rhythmic effect from his tape recorder. However, Kitt Peak National Observatory, which reported flashes at same meeting, still argued they seemed real. (Sullivan, *NYT*, 6/8/68, 32)

June 9: Cosmos CCXV, launched by U.S.S.R. April 18, had carried telescopes to observe stars without interference from earth's atmosphere, *Pravda* announced. Orbiting astronomical observatory had trained its eight mirror telescopes and one x-ray telescope on specific hot stars long enough for adequate study, partially by slowing satellite's rotation to 0.01 usual rate with magnetic anchor that fixed itself on earth's magnetic field. Data received in 150 radio sessions was being processed by computer. Satellite reentered June 30. (Anderson, *NYT*, 6/10/68, C14; Reuters, *W Post*, 6/10/68, A3; *SBD*, 6/11/68, 199; GSFC *SSR*, 7/15/68)

- Nike-Apache sounding rocket launched from Churchill Research Range carried GSFC payload to 92.7-mi (149.5-km) altitude to measure intensity and energy spectra of low energy protons helium nuclei and heavier nuclei present during Polar Cap Absorption (PCA) event. Launch was first of six scheduled during solar event. Parachute failed to deploy apparently because of technician's failure to connect flight system pyrotechniques. Extendable nose cone malfunctioned for undetermined reason. Telemetry functioned for full flight duration. (NASA Rpt SRL)

- JPL Historian R. Cargill Hall had been selected for inclusion in 1968 edition of *Outstanding Young Men of America*, compiled annually by 14-man national board of editors. (*LA Times*, 6/9/68)

- Univ. of Colorado electrical engineer, Dr. Warren L. Flock, was using radar to monitor migration of birds in study to help Government aircraft controllers predict bird migrations or establish aircraft guidance system eliminating hazards of collisions between aircraft and birds. (*NYT*, 6/9/68)

- "The seven fat years of the space effort are ended, and the seven lean years have begun," William Hines wrote in Washington *Sunday Star*. During FYs 1962–1968 Congress had granted NASA more than $30 billion. In FY 1969 agency would be limited to slightly under $4 billion. Transition from fat to lean was "completely understandable." NASA had received favored treatment because it had high-priority goal, but with attainment now in sight, NASA was no longer in priority position. "Beginning of the end" came in FY 1967 when funding fell nearly $250 million from 1966 level. Budget for FY 1968 was down $400 million from FY 1967, and cut from FY 1968 to 1969 probably would be close to half billion dollars. With current inflation added, NASA "will be lucky to have $4 billion," he said. Agency would be operating in 1969 "at an economic level lower than that of six years ago when the moon program—and the fat years—were just getting started." (*W Star*, 6/9/68, E7)

June 10: Senate passed by vote of 66 to 4 H.R. 15856, NASA FY 1969 authorization bill of $4.013 billion—$357 million less than Administration's requested $4.37 billion, $18 million less than sum voted by House, and $136.4 million less than sum recommended by Senate Committee on Aeronautical and Space Sciences. Bill provided $3.37 million for R&D, $39.6 million for construction of facilities, and $603.17 million for administrative operations. During debate Sen. Carl T. Curtis (R.-Neb.) successfully proposed amendment to bar NASA space research grants to colleges and universities that refused to allow armed forces recruiters on their campuses. (*CR*, 6/10/68, S6943–77; *W Post*, 6/11/68, A3; *B Sun*, 6/11/68, 1; *SBD*, 6/11/68, 198; AP, *NYT*, 6/12/68, 11)

- Nike-Apache sounding rocket launched by NASA from WSMR carried 115-lb Dudley Observatory payload to 95-mi (152.4-km) altitude to collect micrometeoroid particles during meteor shower and recover payload for analysis and flux determination. Rocket and instruments performed satisfactorily; good data were obtained from telemetry. Pay-

June 10

load parachuted successfully but early efforts at recovery were not successful. (NASA Rpt SRL)

- House passed H.R. 3400 giving FAA authority to set standards and make regulations for control and abatement of aircraft noise and sonic boom. Bill amended Title VI of Federal Aviation Act of 1958. (NASA *LAR VII/59*; *Science*, 6/21/68, 1321)
- NASA awarded $200,000 fixed-priced contract to Lockheed Missiles & Space Co. Space System Div. for eight-month study of earth orbital flight emergency escape device. Contract called for development of conceptual design of three-man entry vehicle for emergency escape from vehicle in earth orbit up to 300 mi. Device could be launched into orbit with future space vehicles and remain in orbit until needed. (NASA Release 68–103)
- Press reported recently released testimony at February hearings of House Committee on Appropriations' Subcommittee on Department of Defense. Air Force Chief of Staff Gen. John P. McConnell had described Foxbat as only Soviet fighter aircraft "that we cannot match." Undergoing late stages of testing, it had estimated mach 3 maximum speed, 70,000- to 75,000-ft maximum altitude, 1,600-mi combat range without external fuel tanks, and 1,400-mph cruise speed at 50,000-ft altitude. U.S. observers believed fighter-bomber version could be operational in 1969 or 1970; advanced radar-equipped interceptor version with internally carried missiles could be in squadron service by 1970 or 1971. Secretary of the Air Force, Dr. Harold Brown, had testified that U.S. was not behind U.S.S.R. in aircraft development, while Gen. McConnell had said that F–X (fighter experimental), U.S. fighter in Foxbat class, was still being designed and was "a little late." (Transcript; Corddry, B *Sun*, 6/11/68, 1)
- U.N. General Assembly's Political Committee by vote of 92-4, with 22 abstentions, endorsed draft treaty to halt spread of nuclear weapons. Chief U.S. delegate Arthur Goldberg called vote "a milestone on the road to a more peaceful and secure international order." (de Onis, *NYT*, 6/11/68, 1)
- Pratt & Whitney Div. of United Aircraft Corp. announced it had reached agreement with USN on prices covering all military aircraft engine programs during 1968–1970. Week before, USN had informed the company that top price it was prepared to pay for 1,640 jet engines built for the F–111 was $180 million less than asked by the company. (Text: *WSJ*, 6/11/68, 34; Kelly, W *Star*, 6/11/68, B5)

June 11: Cosmos CCXXV was launched from Kapustin Yar by U.S.S.R. into orbit with 519-km (322.3-mi) apogee, 248-km (154.1-mi) perigee, 92.1-min period, and 48.4° inclination. Instruments functioned normally and satellite reentered Nov. 2. (*SBD*, 6/13/68, 215; GSFC *SSR*, 6/15/68; 11/15/68)

- NASA Nike-Apache sounding rocket launched from Kiruna, Sweden, carried Max Planck Institute (Germany) experiment to 114.2-mi (183.5-km) altitude to measure micrometeoroids by sensitive microphone detectors and electron emission and to measure electrical fields in ionosphere by static voltmeter. Good data were received. Rocket and instrumentation performance was satisfactory. (NASA Rpt SRL)
- Canadian Black Brandt IV sounding rocket was launched by NASA from Barreria do Inferno Range near Natal, Brazil. It carried MSC Space

Physics Div. 80-lb South Atlantic Anomaly Probe (SAAP) experiment package containing spectrometer, magnetometers, ion chambers, and heavy ion detector to 502-mi (806-km) altitude to provide fast response capability for measuring radiation dose and dose rate at orbital altitude over South Atlantic. Project, cooperative effort of NASA and Brazilian Comissao Nacional de Atividades Espaciais (CNAE) in support of Apollo program, would use data to study dynamics of inner Van Allen belt and possible radiation hazard to astronauts on low-altitude missions. (NASA Releases 68–94, 68–97K; NASA Rpt SRL; *SBD*, 6/13/68, 213)

- XB–70 No. 1, flown by NASA test pilots Donald L. Mallick and Fitzhugh L. Fulton, Jr., reached 9,500-ft altitude and 316.5 mph to obtain inflight data for SST program. Flight ended prematurely after 71 min because of hydraulic system malfunction. (XB–70 Proj Off; NASA Release 68–105; *SBD*, 6/17/68, 229)
- NASA's HL–10 lifting-body vehicle, flown by NASA test pilot John A. Manke, successfully accomplished eighth flight after being air-launched from B–52 aircraft. During four-minute pilot-checkout flight Manke tested limit cycle characteristics and performed stability and control maneuvers. (NASA Proj Off; NASA Release 68–105; *SBD*, 6/17/68, 229)
- NASA and AEC successfully completed intermediate power run of Phoebus 2A nuclear reactor at Jackass Flats, Nev. Level of approximately 2,000 mw, highest achieved in program to date, was held about 30 sec to verify satisfactory operation of all equipment. In future test, reactor was expected to be operated at power level of 4,000–5,000 mw. (AEC/NASA Release L–131)
- U.S. patent No. 3,380,687 was granted to Edwin H. Wrench, staff scientist at Convair Div., General Dynamics Corp., and five associates for satellite dispenser designed to release cluster of satellites into orbit. Launched by rocket-powered booster and propelled by its own engine after separation, dispenser would release satellites one at a time upon radio commands from ground. (Patent Off PIO; Jones, *NYT*, 5/4/68, 53)
- President Johnson submitted to Senate nomination of GSFC Deputy Director, Dr. John W. Townsend, Jr., as Deputy Administrator of Environmental Science Services Administration. He would succeed Dr. Werner A. Baum, who had accepted presidency of Univ. of Rhode Island. Dr. Townsend, scientific administrator and ionospheric physicist, had been head of Naval Research Laboratory's Rocketsonde Branch and Deputy Science Coordinator of Project Vanguard before he joined NASA in 1958 as Chief of Space Sciences Div. (*PD*, 6/17/68, 960; ESSA Release ES 68–38; *W Star*, 6/11/68; *W Post*, 6/11/68)
- Joseph E. Karth (D-Minn.) criticized general public's "let it wait" attitude toward science and technology in Dandridge M. Cole Memorial Lecture before AIAA Greater Philadelphia Section: Poll after poll showed "strong inclination to slow down or cut back R&D—particularly the space program. . . . Indeed, sentiment has grown that drastic cuts should be made in R&D to finance a variety of requirements—Vietnam to urban renewal. . . . In April Senate turned down "by only two votes" move to reduce DOD R&D budget half million dollars on top of

committee cut of $240 million. In House, space budget was cut to $4-billion level, $1.25 billion less than 16 mo before.

Technology seemed major factor in economic and social progress and was essential to maintaining international position, but it needed to be more responsive to urgent social problems. "A strong case can be made to continue supporting vigorous R&D efforts in the U.S., although directions and priorities must change to suit our needs from time to time." (*A&A,* 7/68, 4–5)

June 12: U.S.S.R. launched *Cosmos CCXXVI* from Plesetsk. Orbital parameters: apogee, 519 km (398.9 mi); perigee, 597 km (370.9 mi); period, 96.8 min; and inclination, 81.1°. Equipment was functioning normally. (*SBD,* 6/13/68, 217; GSFC *SSR,* 6/15/68)

- X–15 No. 1, flown by NASA test pilot William H. Dana, reached 3,545 mph (mach 4.96) and 214,000-ft altitude to measure earth's horizon at 40-mi altitude. Data obtained would be used to redefine horizon as navigational reference for Apollo spacecraft on return flight from moon. (X–15 Proj Off; NASA Release 68–105; *SBD,* 6/17/68, 229)

- In what President Johnson described as "the most important international agreement in the field of disarmament since the nuclear age began," U.N. General Assembly endorsed nuclear nonproliferation treaty by vote of 95 to 4 with 21 abstentions. Making surprise appearance, President told Assembly Geneva Disarmament Conference should begin early discussions on "limitation of strategic offensive and defensive nuclear weapons systems." He promised quick U.S. action in ratifying and carrying out treaty, which provided: nuclear nations would not transfer arms or explosive devices to any nonnuclear nations or assist them in obtaining or manufacturing such items; nonnuclear nations would not receive such weapons or accept assistance in manufacturing them; nonnuclear nations would accept inspection system under International Atomic Energy Agency; research, production, and use of nuclear energy for peaceful purposes, except for development of nuclear explosives, would be made available through international body; nuclear nations would pursue negotiations for early end to nuclear arms race and for complete disarmament treaty.

   Twenty-five-year treaty would be in force when ratified by U.S., U.S.S.R., U.K., and 40 nonnuclear countries. (de Onis, *NYT,* 6/13/68, 1; Kilpatrick, *W Post,* 6/13/68, 1; *PD,* 6/17/68, 954–6)

- *June 13:* USAF Titan III–C booster launched from ETR inserted eight Initial Defense Communications Satellite Program (IDCSP) jam-resistant repeater satellites into separate, random, near-synchronous, 21,000-mi-altitude equatorial orbits. Addition of 8 comsats to 18 IDCSP comsats launched since June 16, 1966, was expected to boost average satellite-availability time between Vietnam and Hawaii from 92% to 98% for top-priority messages and from 60% to 85% for lower priority and to considerably extend expected lifetime of communications system. (*W Post,* 6/14/68; AP, *P Inq,* 6/14/68, 2; *Pres Rpt 68*)

- *Ogo V* mission, launched from ETR March 4, was adjudged successful by NASA. Initial results provided first evidence of electric fields in bow shock. Other data indicated presence of electric field discontinuities in solar wind and observations of new particle and field phenomena. Performance of three-axis stabilization system had been excellent and, except for several unexplained transmitter anomalies, performance of other subsystems was satisfactory. During first three months in orbit *Ogo V* had accumulated over 47,000 hr of experiment operations and was continuing to provide high-quality data for 21 of 24 onboard experiments. (NASA Proj Off)

- NASA Nike-Apache sounding rocket launched from Kiruna, Sweden, carried Max Planck Institute (Germany) experiment to 114.2-mi (183.5-km) altitude to measure micrometeorites and electrical fields in ionosphere. Preliminary analyses indicated good data were received. Rocket and instrumentation performance was satisfactory. (NASA Rpt SRL)

- NASA Associate Administrator for Organization and Management Harold B. Finger received AIAA's James H. Wyld Propulsion Award—including certificate and $500 honorarium—for "outstanding achievement in the development of the technology required for a nuclear flight propulsion system." Award was presented at Honors Luncheon during 4th Propulsion Joint Specialist Conference in Cleveland, Ohio. (AIAA Release)

- *Chicago Tribune* criticized "ridiculously small" congressional cuts in NASA FY 1969 authorization bill: "Of all the ways in which the government has found to spend taxpayers' money, the space program is the most expendable. . . . It is irrelevant to the war in Vietnam . . . irrelevant to the problems of poverty, crime, violence, and urban decay." It could be "scrapped in its entirety without seriously affecting any of the critical problems we face today." However, "today's space research does have a bearing on tomorrow's defense. It is better to keep the machinery running in low gear than to shut it down and then try to start it up again." (*C Trib*, 6/13/68)

- U.N. endorsement of nuclear nonproliferation treaty had "turned back the doomsday clock," said *New York Times* editorial. Broad support for it would impose political restraints on all nations and reinforce national leaders in resisting demands for nuclear arms development. Treaty put "a particular duty" on U.S. and U.S.S.R. to "move rapidly to check their own dangerous nuclear arms race and to find new ways to prevent international conflicts." (*NYT*, 6/13/68, 46)

- Sen. John S. Cooper (R-Ky.) led bipartisan drive in Senate to deny DOD FY 1969 funds to begin deployment of Sentinel ABM system. There was "no present threat to American security from a Chinese intercontinental ballistic missile attack," he said, and it was "difficult to believe they would invite the certain destruction of their country by a nuclear attack upon the U.S." when they attained the capability. Coalition hoped to postpone Sentinel deployment for at least one year by striking approximately $600 million for further R&D. (Finney, *NYT*, 6/14/68, 18; AP, *W Post*, 6/14/68, A28)

June 13–15: NASA successfully accomplished first radar tracking of asteroid Icarus, which passed close to earth once every 19 yr. Using 85-ft-dia and 210-ft-dia antennas at Goldstone Tracking Station, scientists followed Icarus' approach and June 14 flyby of earth at 3,945,000 mi and 66,215 mph, logging 15½ hr of radar contact to obtain data on asteroid's size, velocity, rotation, and composition. Preliminary results of observations by Harvard Univ. and Univ. of Arizona suggested asteroid might be composed of iron and have two- to four-hour rotation period. (NASA Release 68–106; *LA Times*, 6/16/68; Pasadena *Independent Star-News*, 6/16/68; Glendale *News-Press*, 6/18/68; Sullivan, *NYT*, 6/27/68, 41)

June 14: NAS announced establishment of 12-member Computer Science and Engineering Board chaired by Dr. Anthony G. Oettinger of Harvard Univ. to assess implications of rapid evolution of computers and their increasingly pervasive influence on individual lives and on the national welfare. (NAS Release; UPI, *NYT*, 6/15/68, 20)

- AIA President Karl G. Harr, Jr., spoke before House Ways and Means Committee in support of H.R. 17551, Administration's proposed 1968 Trade Expansion Act. "It is no accident that the U.S. supplies more than 72 percent of the transport aircraft in airline use throughout the free world. . . . World markets have been won by designing and building products that are safe, reliable, economical, easy to maintain and backed by a vast network of manufacturers' service organizations." Industry's 1967 gross sales had been $27.2 billion, of which a record $2.2 billion or 8.3% of total were exported. In 1967, industry had accounted for more than 40% of U.S. merchandise trade surplus. Annual international sales revenue had exceeded $1 billion in 10 of past 12 yr. Recent Dillon, Read & Co. forecast, Harr said, showed U.S. commercial jet transport export sales would exceed $13 billion over next decade, anticipated gain of 300% over $3.2 billion in past decade. (Text)

- USAF announced it had authorized resumption of flights by some of 42 F–111A aircraft grounded since May 8, after series of crashes. Tail control assembly rod, suspected cause of last crash, had been modified and installed in aircraft released. (AP, *W Post*, 6/15/68, A15; AP, *W Star*, 6/15/68, A10)

June 15: Rare photographic zenith tube, PZT, which provided almost exact measurement of astronomical time, had been installed at Dominion Observatory, Priddis, Alberta, Canada. It would be used in conjunction with another PZT at Royal Greenwich Observatory near Herstmonceux, U.K., to study continental drift by establishing exact measurements between two points on same latitude and observing whether distance changed with time. (Can Press, *NYT*, 6/15/68, 20)

June 16: Astronauts Joseph P. Kerwin, Vance D. Brand, and Joe H. Engle began 177-hr simulated space journey with Apollo 2TV–1 model spacecraft inside giant vacuum chamber at MSC, to prove spacecraft structure and inner pressure vessel and verify Apollo's environmental control system in temperature and vacuum extremes. (West, *H Chron*, 6/17/68; *H Post*, 6/18/68, 2; NAR *Skywriter*, 6/21/68, 1)

- NASA and France's Office National d'Etudes et de Recherches Aérospatiales (ONERA) had agreed on cooperative research project on tilt-rotors for vertical-flying aircraft. Data were expected to assist in design of rotors and propellers for V/STOL aircraft. NASA would provide rotors, hardware, and instrumentation for tests, minimum engineering support, and basic computations of structural strength; ONERA would conduct tests in Modane, France, wind tunnel at no cost to NASA. (NASA Release 68–104)

- Pearl I. Young, first woman technical employee of National Advisory Committee for Aeronautics, died at age 72 in Hampton, Va. Author of technical editing manual used throughout NASA, head of NASA technical editing staffs at LaRC and LeRC for 28 yr, and engineering teacher for 12 yr, Miss Young had retired from NASA in 1961 and at time of her death was completing a biography of Octave Chanute. (Newport News

Daily Press, 6/17/68; *W Star*, 6/18/68, B4; *Denver Post*, 6/23/68, 33)

June 17: Tests to evaluate propellant lines for J–2 engine had been successfully conducted in vacuum chamber at Arnold Engineering and Development Center, B. J. Richey reported in *Huntsville Times*. Engineers had discovered that frosting which enveloped lines and provided extra protection when engine was fired at ground temperatures failed to form in vacuum or simulated vacuum, so lines burned through. Lines leading to engine's augmented spark indicator, believed to have caused engine shutdowns and restart failure during *Apollo 6* flight April 4, had been redesigned and strengthened without affecting engine performance. (Richey, *Huntsville Times*, 6/17/68)

- Tests at ARC VTOL Static Test Facility had shown that inverted "V" fence of porous metal could prevent soil erosion around landing pads for VTOL aircraft. Fence, developed for USAF Aero Propulsion Laboratory by Northrop Corp. Norair Div., was less than 16 in high; placed between aircraft and edge of pad, it permitted some of blast to flow through and deflected some upward at 45°, diffusing downblast. (AFSC Release 83.68)

- Subcommittee on Science, Research, and Development submitted to House Committee on Science and Astronautics its report *Managing the Environment*. Subcommittee recommended development of national policy for environment, including: use of environment for benefit of all mankind; maximized productivity consistent with continued usage into very long-term future; systematic management of applied science and technology; incentives to industry, land developers, and local governments; international agreement on projects with widespread or long-term effects; assessment of new and extended applications of science; avoidance of speculative statements and emotional appeals; and increased education and information program in ecological principles. (Text)

- *Washington Post* editorial suggested House-approved bill H.R. 3400 granting FAA authorization to control aircraft noise might cause conflict of interests by instructing FAA to consider safety, "technical practicability," and "economic reasonableness" as well as noise. Many citizens would feel FAA already considered these criteria "too much." Nor could it be ignored that FAA also was charged with building U.S. SST potential "boom maker." House bill merited Senate support, however, as a first step. Its significance was "that finally the demands of the public are being heard in Congress over the roar of aircraft, airline, airport, air-traveler and air-bureaucrat interests too little concerned with noise." (*W Post*, 6/17/68, A18)

June 18: U.S.S.R. launched *Cosmos CCXXVII* from Baikonur into near-circular orbit with 271-km (168.4-mi) apogee, 202-km (125.5-mi) perigee, 89.2-min period, and 51.8° inclination. Satellite performed satisfactorily and reentered June 26. (*SBD*, 6/19/68, 244; GSFC *SSR*, 6/30/68)

- House unanimously accepted $4.013-billion Senate version of NASA FY 1969 authorization bill (H.R. 15856) [see June 10] without conference and sent to President Johnson for signature. Final bill, $357.027 million less than NASA request, reduced funds asked for Apollo program by $13.8 million, to $2.025 billion; cut Apollo Appli-

June 18

cations from requested $439.6 million to $253.2 million; and halved advanced missions funds, to $2.5 million. It increased House figure of $11.7 million for NERVA program to $55 million, still $5 million under request. Lunar and planetary exploration funds were cut from requested $107.3 million to $92.3 million, and space applications from $112.2 to $98.7 million. Final authorization figures were $3.37 billion for total R&D, $603.173 million for administrative operations ($45.027-million reduction), and $39.6 million for construction of facilities. All items were subject to appropriations bills, not yet passed. (*CR*, 6/18/68, H5052–3; Lannan, *W Star*, 6/19/68, D9; *SBD*, 6/19/68, 239; *Science*, 6/28/68, 1432)

- Apollo 7 spacecraft, which would carry three astronauts on 11-day earth-orbital mission in late 1968, was undergoing combined systems tests at KSC. Unmanned and manned altitude runs would follow, before spacecraft would be sent to Cape Kennedy to be placed on Saturn IB launch vehicle. Saturn had undergone tests and checkout on Launch Complex 34 for several months.

    NASA announced Apollo mission—which would demonstrate performance of spacecraft's command module (CM), service module (SM), crew, and support facilities—would begin with mid-morning launch, carrying commander Walter M. Schirra, Jr., CM pilot Donn F. Eisele, and LM pilot Walter Cunningham into earth orbit. During second revolution, crew would separate Apollo from rocket's upper stage and perform transposition and simulated docking maneuver similar to one to be performed on lunar mission. Reentry under manual control of crew would culminate in splashdown in Atlantic at approximately 7 am EDT on 11th day. (NASA Release 68–108; KSC Release 285–68)

- ARC reported that 90 U.S. scientists at ARC working group sessions had agreed electromagnetic systems would provide best data on moon's interior and recommended three techniques, magnetometer systems, radio-frequency systems, and radar. Magnetometer systems could measure blocking of fluctuating solar magnetic field borne by solar wind by using network of emplacements on moon's surface. Highly conductive hot lunar core would cut off field completely, while cold core would let most of field pass through. Magnetometer on lunar crawler could make similar measurements to find highly conductive masses of water, lava, or ore. Radio-frequency probes could make inexpensive subsurface surveys through spacecraft radio signals which would penetrate moon's surface and bounce back to be measured as to wave characteristics. Wave-change data could help define ore, ice deposits, or permafrost. High-frequency radar systems could study lunar surface through radar bounced off moon from earth with return signal received by earth antennas, or beamed from spacecraft in lunar orbit or from lunar crawler at lunar surface, with return signals received by spacecraft or crawler. (ARC Release 68–9)

- Opening session of 3rd Eurospace U.S.-European Conference at Munich discussed disparity between U.S. and European space funding. Total spent on Western European space research in 1967 was $300 million, 0.005% of gross national product of cooperating countries; NASA budget was 14 times that amount. West German Minister for Scientific Research, Dr. Gerhard Stoltenberg, proposed merger of ELDO and ESRO, 10% increase in total funding, and possible cooperation with U.S. in developing launch vehicle for use after 1975. International Space Research Committee Vice President Richard W. Porter urged increased cooperation with NASA, which already had concluded bilateral agreements with several countries for scientific satellite launchings, eight of which had proved successful thus far. Some 42 countries were working with U.S. on meteorological satellite programs; 11 were cooperating on communications projects. There were 20 tracking stations in 11 countries cooperating in U.S. launching projects and 395 foreign technicians, engineers, and scientists had worked in U.S. space centers. European industrial sources expressed hope system could be evolved for joint U.S.-European work on projects. (*InteraviaAirLetter*, 6/19/68, 1)

- Dr. Gardiner L. Tucker, Deputy Director of Defense Research and Engineering (Electronics & Information Systems), told Military Operations Subcommittee of House Committee on Government Operations DOD would proceed with new phase of Defense Satellite Communications System (DSCS), procuring more than three advanced synchronous satellites and developing over six new ground terminals. Funding of satellites, which would use very narrow-beam steerable antennas illuminating 1,000- to 2,000-mi-dia area of earth, would start in FY 1969. DSCS satellites would be comparable to Intelsat IV series under consideration for commercial service.

  New DOD phase represented technological advance over advanced DSCS considered in 1967 but was less ambitious in cost and size, with cost range of $100–200 million, contrasted with earlier $500 million. Current system had 25 satellites in orbit and operational, each equipped with switch to turn it off after six years in orbit. Number of terminals had grown from original two in field and seven transportable to worldwide operational system with terminals near Washington, D.C., and in New Jersey, Colorado, California, Hawaii, Guam, Australia, Korea, Okinawa, Philippines, South Vietnam, Thailand, Ethiopia, and Western Germany, plus six shipboard terminals.

  Cooperative programs with Allies in satellite communications included procurement of two synchronous Skynet satellites for U.K. as part of Initial Defense Communications Satellite Program (IDCSP) with first launch scheduled for mid-1969. Joint efforts were under way to ensure operability of U.S. and U.K. satellites and terminals, and work was under way to procure two Skynet satellites for NATO under program similar to that of U.K. First launch was planned for late 1969, with U.S. controlling satellite. U.S. had signed Memorandum of Understanding with six NATO countries establishing joint testing program utilizing *Les V* experimental satellite, launched July 1, 1967. (Transcript: DOD Release 603–68; Johnson, *Av Wk*, 6/24/68)

- Secretary of Transportation Alan S. Boyd testified before Senate Commerce Committee's Aviation Subcommittee in support of S. 3645, proposing airport development and airways systems improvement. By 1974, revenue passenger-miles flown by U.S. airlines were expected to rise to 200 billion, more than double 1967 level; hours flown by carriers, from 5.2 million in 1967 to 8.6 million, increasing 65%; and general aviation hours flown, 50%, from 21.9 million to 31.8 million. Air carrier fleet would increase from 2,272 aircraft to 3,320 and general aviation fleet would increase from 104,706 to 160,000. Aircraft handled by FAA towers would more than double, as would those handled by FAA route control centers. Increase in speed and quality of aircraft also would increase demand for use of FAA facilities.

S.3645 would authorize direct loans for potentially viable airports when reasonable private financing was unobtainable and grant up to 50% of cost for projects attributable to service by subsidized carriers. Federal assistance would be available only for development projects related to landing areas and safety facilities. DOT would be required to submit National Airport System Plan within two years. (Testimony)

- Response from U.S.S.R. and other nations to proposal for international decade of ocean exploration in 1970s had been "quite favorable," Vice President Hubert H. Humphrey, Chairman of National Council on Marine Resources and Engineering Development, said in report released by council. "I welcome the enthusiasm with which the Decade is being received. . . . We look forward to a continuation of this spirit of close collaboration, for the seas can, and must serve the interest of mankind." (Text; NCMRED Rpt, 5/68; Madden, *NYT*, 6/18/68, 16)

June 19: U.N. Security Council approved security guarantee by U.S., U.K., and U.S.S.R. to provide immediate assistance to nonnuclear nations facing nuclear attack, completing U.N. action on nonproliferation treaty endorsed by U.N. General Assembly June 12. Ten voted in favor of guarantee; France, India, Brazil, Pakistan, and Algeria abstained. (de Onis, *NYT*, 6/20/68, 10C)

- In letter to Sen. Richard B. Russell (D-Ga.), Chairman of Senate Armed Services Committee, Secretary of Defense Clark M. Clifford said that "it would be a serious mistake to eliminate construction and procurement funds in fiscal year 1969 for the deployment of the Sentinel System." Program represented 12 yr of R&D at cost of $3 billion, he said, and time had come "when we can no longer rely merely on continued research and development but should proceed with actual deployment of an operating system." Congressional decision to eliminate deployment funds would disrupt work under way and lose two years in availability of operating system which was important to U.S. security.

  Senate coalition responded with letter from Assistant Defense Secretary Paul C. Warnke acknowledging one-year delay in Chinese ICBM program and that Sentinel ballistic missile defense system had "also slipped a little." (Text; Finney, *NYT*, 6/20/68, 24; Corddry, B *Sun*, 6/20/68, A3)

June 20: USAF launched two unidentified satellites from Vandenberg AFB by Thor-Agena D booster. One entered orbit with 251-mi (403.9-km) apogee, 113-mi (181.8-km) perigee, 90.3-min period, and 85° inclination and reentered July 16. Second entered orbit with 322-mi (518.2-km) apogee, 273-mi (439.3-km) perigee, 94.1-min period, and 85.1° inclination. (*SBD*, 6/25/68, 272; *Pres Rpt 68*)

- Analysis of data returned by *Pioneer VIII* had led ARC scientists to speculate that earth's magnetic tail, extension of its magnetic envelope (magnetosphere) blown out by solar wind to resemble comet's tail, might be shorter than the 200 million mi suggested by certain theoretical calculations. When *Pioneer VII* flew through tail region at 3.5-million-mi distance from earth after Aug. 17, 1966, launch, it found long period when solar wind was completely or partially blocked out, suggesting spacecraft had observed end of organized tail region. Conditions encountered in tail area by *Pioneer VIII* at 1.75 million mi from earth after Dec. 13, 1967, launch, were similar. However, instead of smooth cylindrical structure expected at smaller distance, *Pioneer VIII* found conditions resembling turbulent wake, leading some scientists to conclude tail was shorter. Others, including Pioneer Project Scientist, Dr. John Wolfe, believed tail might include successively turbulent and smooth areas. (ARC *Astrogram,* 6/20/68; AP, *NYT,* 6/22/68, 53)

- Eastern Air Lines and McDonnell Douglas Corp. announced joint program to evaluate propeller-driven STOL aircraft for use on 300- to 500-mi intercity flights by trial of 64-passenger French Breguet 941 (McDonnell Douglas 188) aircraft. Beginning in September, performance of STOL aircraft, which could take off from 1,000-ft runway and cruise at 250 mph, would be compared with that of conventional jet aircraft on regular commercial air shuttle routes on same schedule to determine time saved by using separate runways and terminal airspace. (Hudson, *NYT,* 6/21/68; *W Post,* 6/21/68, A20; *WSJ,* 6/21/68, 28)

- U.K. withdrew from $420-million European Nuclear Research Center project because of financial difficulties resulting from devaluation of pound Nov. 18, 1967. Officials said U.K. could not afford to contribute its $93.6-million share in proposed 300-bev European nuclear accelerator. U.K. withdrawal from European Conference on Satellite Communications had been announced April 16. (*W Post,* 6/21/68, A16)

- NATO planned establishment by early 1970 of comsat network of advanced relay spacecraft in synchronous orbit 21,000 mi over Atlantic for communications between its military units in Europe and U.S. Reportedly $7.9-million contract calling for delivery in autumn 1969 of two spacecraft had already been signed. (AP, *W Post,* 6/21/68, A7)

- Commenting on reductions in NASA authorization bill, *Kansas City Times* editorial said: "We have supported the space program in the past, not as a window dressing but as an expression—and a catalyst—of the inventiveness and technical energies of the American people. We still support it, and believe it is beneficial.

  "We recognize, nonetheless, that it is but one of many costly and sometimes competing activities. Congress has decided that in a time of burdensome military expenditure, and of pressing domestic needs, the space budget is one logical place to apply the knife of economy. NASA planners will simply have to find creative ways to live with that decision." (*KC Times,* 6/20/68)

June 21: U.S.S.R. successfully launched *Cosmos CCXXVIII.* Orbital parameters: apogee, 241 km (149.8 mi); perigee, 203 km (126 mi); period, 88.9 min; and inclination, 51.6°. Satellite reentered July 3. (*SBD,* 6/24/68, 260; GSFC *SSR,* 6/30/68, 7/15/68)

- NASA's HL-10 lifting-body vehicle, piloted by Maj. Jerauld R. Gentry, successfully completed ninth flight. Purposes were to obtain stability and control data through pitch and rudder pulse maneuvers, verify yaw rate measurements, investigate pilot limit cycle, and verify predicted optimum use of landing rockets during landing flare. (NASA Proj Off)

- Nearly 1,000 scientists and educators attended "crisis" meeting called by New York Academy of Sciences to demand science be declared "disaster area" because of threatened $6-billion Federal budget cut. Federal R&D funding, after decade of average 22% annual expansion, had risen only 2.5% annually since 1964, while number of scientists had risen 20%. It was feared resulting crush would damage U.S. leadership in science and technology. MIT Provost Dr. Jerome B. Wiesner said, "If the Congress continues to do what it is now doing we'll wake up with another sputnik in a decade." Federal R&D funding had risen from $74 million in 1940 to $16.9 billion in 1968 with October 1957 "shock of Sputnik" giving greatest impetus. Cut of $6 billion would mean "we're going to cut not only into the fat, but into the flesh of lots of areas," Dr. Donald F. Hornig, President's Science Adviser, said. NASA's sustaining university program had been one of chief casualties of recent cutbacks with budget slashed from $45 to $10 million yearly, forcing drop from 1,300 to 50 training grants which had produced more than 1,000 Ph.D.s since 1961. Government officials had chided scientists for being ill-prepared for leveling of Federal support and for alleged detachment from political realities. (Reinhold, NYT, 6/21/68, 1)

June 22: Alexandru Birladeanu, Deputy Prime Minister of Romania and president of Romanian National Council of Scientific Research, visited KSC for general orientation and tour. (KSC Release 290–68)

June 24: NASA Administrator James E. Webb testified before Senate Committee on Appropriations' Subcommittee on Independent Offices that NASA was still uncertain as to exact levels at which a number of projects could be included in FY 1969 operating plan and that at House-passed $4.008-billion appropriations level NASA's aeronautical and space activities would have to be sharply curtailed. "We will have to reduce and stretch out ongoing programs and eliminate or defer the work that would have enabled us to continue the research and development . . . looking toward a future resumption of tests and missions which will soon grind to a halt." Because of reduced budget, Webb said, NASA would have to accept one-year gap in Saturn V production, discontinue production of Saturn IBs, delay initiation of NERVA development, either severely limit all proposed planetary orbiter missions or eliminate 1971 or 1973 missions, and curtail work toward using space systems for direct economic benefits.

If NASA budget were further reduced under Revenue and Expenditure Control Act of 1968, requiring $6-billion reduction in Federal expenditures in FY 1969, result could be disastrous to goals of national space program. Although NASA would do everything possible to fulfill commitment to Apollo program and would continue on smaller scale programs of greatest and most immediate national importance, further budget cuts would require complete termination of Saturn production and cancellation of production of associated manned spacecraft; elimination of Titan-Mars 1973 missions and possible elimination of Mariner-Mars 1971 missions; and further delays, curtailments, and cancellations which might require more harsh steps such as canceling orders for hardware already under contract and mothballing entire installations. (Testimony; NASA *LAR* VII/65)

- House Committee on Government Operations issued report urging immediate curtailment of Federal grants to foreign scientists, especially in developed countries, to conduct nonessential research. Committee said, despite balance of payments deficit, Government grants for foreign research projects amounted to $20 million yearly, more than $15 million in developed countries. Committee recommended limiting funds to projects urgently needed by U.S. until end of emergency. (H.R. 1578; AP, W *Star*, 6/24/68)

- NASA Aerobee 150 A sounding rocket successfully launched from NASA Wallops Station carried 300-lb payload containing two white rats to 89-mi (143.2-km) altitude in second of four experiments to study rats' behavior in artificial gravity field and determine minimum level of gravity needed by biological organisms during space flight. During five minutes of free fall, rats selected artificial gravity levels created through centrifugal action by walking along tunnel runway in extended arms of payload. Data on their movement and position were telemetered to ground stations. Payload impacted 53 mi downrange in the Atlantic; no recovery was attempted. (WS Release 68–11; NASA Release 68–112; NASA Rpt SRL)

- House passed by 269-to-42 vote H.R. 3136 authorizing study of increased use of metric system in U.S. (*CR*, 6/24/68, H5341–6; AP, *W Post*, 6/25/68)

- *Denver Post* editorial: "With the goal of the project—landing of men on the moon by 1970—so close at hand, it would be sad indeed if NASA lost some key people now because of budget and morale problems.... There are future manned flights and space experiments to consider, and these will surely be crippled at infancy if NASA's budget is cut back too severely." (*Denver Post*, 6/24/68)

June 24–27: Fourth International Symposium on Bioastronautics and the Exploration of Space was held in San Antonio, Tex., under sponsorship of AFSC's Aerospace Medical Div.

In keynote address Dr. Edward C. Welsh, Executive Secretary of NASC, noted no other program had given such impetus to technological and economic growth as national space program. "Those who oppose adequate spending on space technology are deliberately or inadvertently campaigning for a lower standard of living for our people, a declining Gross National Product for our Nation, and a secondary position in strength to that of the Soviet Union." U.S. investment in space to date, he said, "has mostly been an investment in the future, the returns of which can be lost in large measure if we lack the vision and the vigor and the desire to keep this country great by maintaining a vigorous space effort.

"Every major power and every nation eager to raise its standard of living and world influence strives to participate in space technology and space exploration. It certainly would be ironic if the United States, as the world's leader in international cooperation and the world's leader in standard of living, were to abandon or even neglect the source of such strength. I believe it might be labeled the worst mistake in history." (Text; *CR*, 6/24/68, E5775–6; *SBD*, 6/25/68, 267; *Aero Daily*, 6/25/68)

Gen. James Ferguson, AFSC Commander, said bioastronautics problems and provisions could have serious impact on mission performance, space station design, cost, and operations. Principal problems included those of crew rotation, crew size, compartment volume per crew member, radiation exposure, versatility of astronauts, station atmosphere, and prolonged weightlessness. In one study, savings from doubling crew rotation intervals from 30 to 60 days ranged from $220 million to $470 million per year, depending on altitude. Savings from increasing from 60 to 90 days were another $100 million a year. For same cost, slightly longer interval could support two more astronauts. Crew rotation intervals of two or more months should be goal. "If a future station can be expected to be useful over a period of many years and its cost can be amortized accordingly . . . efforts to achieve long crew rotation intervals have a very large potential payoff." (Text; *Aero Daily*, 6/28/68)

Dr. Robert R. Gilruth, MSC Director, described 100-man orbital workshop that could be operational by late 1970s. Proposed 615-ft-long, 1-million-lb vehicle, carrying 10,000 lb of experiments, would be launched in three separate parts by three Saturn V boosters and assembled in space. Baton-shaped station would revolve around hub in center which would serve as spaceport and zero-gravity laboratory. Crew would live and perform some work in 240-ft arm on one side of hub. On other side, 375-ft arm would house engine which would spin entire station at 3.5 rpm, creating centrifugal force to serve as artificial gravity. (Maloney, *H Post*, 6/26/68; AP, *W Star*, 6/26/68; *CSM*, 6/28/68)

Arthur C. Clark, co-author of *2001: A Space Odyssey*, suggested that most earth inhabitants could not be very objective about possibility of extraterrestrial life because they were too "geocentrically minded," still considering earth the center of the universe. "The whole history of astronomy teaches us to be cautious of any theory purporting to show that there is something special about the earth. In their various ways, the other planets may have orders of complexity as great as ours. Even the moon—which looked a promising candidate for geophysical simplicity less than a decade ago—has already begun to unleash an avalanche of surprises.

"The discovery that Jupiter is quite warm and has precisely the type of atmosphere in which life is believed to have arisen on earth may be the prelude to the most significant biological findings of this century." If we discover no trace of extraterrestrial life, he said, "even such a negative finding would give us much sounder understanding of the conditions in which living creatures are likely to evolve—and this in turn would clarify our views on the distribution of life in the universe as a whole." (Leavitt, *AF/SD*, 8/68, 59–62)

June 25: NAS Committee on SST-Sonic Boom's Subcommittee on Human Response reported that although studies indicated little cause for physiological concern, psychological impact of sonic boom would be discouraging for supersonic flight over land by present SST configurations. Report stressed, however, that although no damage to hearing or other direct physical damage was expected, indirect physiological responses could be caused by startle produced by even moderate booms. Committee's recommendations to develop "commercial SST which will be able to fly supersonically over populated areas at frequent intervals without undue annoyance to the residents" included: further development of concept of "utility" in comparing monetary and nonmonetary costs and benefits of flights; continuation of laboratory studies of booms; construction of additional boom-simulation facilities and improvement of existing ones; continuation of studies of human reaction to varied boom levels; and studies of human response during sleep

and effects of repeated awakenings. (NAS Release; Schmeck, *NYT*, 6/26/68, 41; NAS–NRC–NAE *News Report*, 6–7/68, 1)
- NASA awarded $36,271,376 three-year cost-plus-award-fee contract to RCA Service Co. for maintenance and operation of Satellite Tracking and Data Acquisition Network (STADAN) facilities at GSFC; Rosman, N.C.; and Fairbanks, Alaska. Contract carried two one-year extension options. (NASA Release 68–110)
- MSFC had brought 22 Super Loki Dart rockets from Space Data Corp. to replace larger, costlier Cajun-Dart sounding rockets in high-altitude atmospheric research at KSC. Super Loki, costing $800, in contrast to $2,800 for Cajun, would deliver similar performance in taking high-altitude wind measurements before and after Saturn launch vehicle flights. (MSFC Release 68–139)
- USN announced selection of 54 men to serve as aquanauts in 60-day Sealab III experiment in underwater living, scheduled to begin in October. Former Astronaut M. Scott Carpenter (Cdr., USN), team leader for Sealab II (Aug. 28–Sept. 26, 1965), would serve as Senior Aquanaut. Ocean floor experiments would be conducted at 620-ft depth off San Clemente Island by 40 aquanauts in 5 teams serving 12 days each. Remaining 14 men would serve as alternates and backup surface support divers. (DOD Release 579–68; *Aero Daily*, 6/26/68)
- President Johnson announced he would nominate AEC Chairman, Dr. Glenn T. Seaborg, to fill unexpired term of Dr. Samuel M. Nabrit, who had resigned from AEC Aug. 1, 1967. Term would end June 30, 1970. Dr. Seaborg's current term would expire June 30, 1968. President also would nominate James T. Ramey, AEC member since 1962, to new five-year term expiring June 30, 1969. (*PD*, 7/1/68, 1012–3)
- U.S patent No. 3,390,336 was issued to Dr. Michael J. DiToro, Vice President for Science of Cardion Electronics, for Adapticom, instrument that corrected multipath reception or time-spread responsible for fuzziness and consequent errors in high-speed radio and telephone communications and eliminated ghosts from facsimile transmission. (Jones, *NYT*, 6/15/68, 49)
- Senate by vote of 78 to 3 passed H.R. 16703, authorizing construction at military installations, including funds for ABM land acquisition and construction. Approximately $1.2 billion for Sentinel program was included in various budget requests of DOD and AEC for FY 1969. During final day of debate Sentinel system advocates had warned more of U.S.S.R. missile threat and less of Red Chinese missile threat which Sentinel had been designed to counter. (*CR*, 6/25/68, S7721; *W News*, 6/25/68, 4)

June 26: Cosmos CCXXIX was launched by U.S.S.R. from Plesetsk into orbit with 328-km (203.8-mi) apogee, 222-km (137.9-mi) perigee, 89.8-min period, and 72.9° inclination. Satellite reentered July 4. (*SBD*, 6/27/68, 286; GSFC *SSR*, 6/30/68, 7/15/68)
- Phoebus 2A nuclear rocket reactor was successfully tested by NASA and AEC at Jackass Flats, Nev. During 32-min ground test, reactor reached peak power level of approximately 4,200 mw, operating for about 12 min at above 4,000 mw. Power density exceeded that required for 75,000-lb-thrust NERVA (Nuclear Engine for Rocket Vehicle Application). Test was part of NASA/AEC nuclear rocket program. (AEC/NASA Release L–148; *W News*, 6/27/68, 3)

June 26
- NASA issued Apollo Status Summary. In preparation for first manned Apollo flight, Apollo 7 spacecraft command module (CM) and service module (SM) had been mated in KSC altitude chamber to confirm compatible operation systems. Saturn IB launch vehicle 2nd-stage engine was being modified to strengthen propellant feed lines to augmented spark igniter. Apollo/Saturn 503 mission might be first manned Saturn V flight.

  Astronauts Joseph P. Kerwin, Vance D. Brand, and Joe H. Engle had successfully completed eight-day vacuum chamber test of Apollo 2TV-1 CM and SM at MSC June 24. All tests necessary to help verify Apollo for first manned flight had been completed, with review of test data in progress.

  Last of seven scheduled verification tests of modified Apollo earth landing system had been postponed to complete analysis of all possible test conditions before drop of 13,000-lb full-scale Apollo CM from aircraft at Naval Air Facility, El Centro, Calif. Test would simulate severe landing condition using one of two drogue parachutes and two of three main parachutes. Ultimate load test of two modified Apollo drogue parachutes in reefed condition was scheduled no earlier than June 27. Repeat of unsuccessful test, it would be conducted from 13,00-lb parachute test vehicle at aircraft altitude of 46,000 ft. (Text; UPI, *W Post*, 6/22/68, A8; *SBD*, 6/25/68, 270; 6/27/68, 285)
- In GSFC tests using Omega Position Location Equipment (OPLE), track of specially equipped automobile had been located repeatedly within 1,000 ft of its actual route on Baltimore-Washington Parkway by *Ats III*, in 22,300-mi-altitude orbit. Satellite had also tracked and located boat in Chesapeake Bay, NASA calibration aircraft, and Coast and Geodetic Survey ship, *Discoverer*, in Caribbean. Meteorologists believed OPLE system, designed primarily as meteorological experiment for tracking balloons and floating buoys, might produce new data on wind circulation in atmosphere and its effect on weather. (NASA Release 68-111; AP, B *Sun*, 6/27/68)
- NASA awarded Teledyne Systems Co. $1,358,728 incentive contract to design and construct prototype airborne computer unit for Centaur launch vehicle. NASA would have option to purchase five additional units, support equipment, and spare parts for $759,872. Contract would be managed by LeRC. (LeRC Release 68-44)

June 27: NASA successfully launched four-stage Pacemaker rocket carrying 52-lb spacecraft from NASA Wallops Station to test performance of phenolic nylon charring ablation material, foamed quartz material, MOD V ablation material, and foamed Teflon material. Spacecraft reached 7,200 mph and was lowered into Atlantic by parachute after four-minute flight. Recovery helicopter retrieved payload, which would be evaluated at LaRC. (WS Release 68-12)
- House Appropriations Committee struck all funds for SST development from Administration's FY 1969 budget and asked return of $30-million carryover to Treasury. President Johnson had requested authorization of $223 million. Cut was unlikely to postpone development of project, which would be continued during FY 1969 entirely with carryover funds. First flight had been scheduled for first quarter, 1972. (*CR*, 6/27/68, H5766-7; Hoffman, *W Post*, 6/28/68, A3)

- NASA awarded one-year $1,250,000 cost-plus-award-fee contract with two one-year renewal options to LTV Service Technology Corp. for computer support services at ERC. (ERC Release 68–9)
- MSC engineers Edwin Samfield and William C. Huber were granted patent No. 3,389,877 for inflatable tether to connect orbiting spacecraft or to connect astronauts and spacecraft. Tether, which became semirigid when inflated to avoid problems of flexible tether, consisted of nylon tube with aluminum end pieces and shock-absorbing struts for attachment to spacecraft. It could be folded and stowed in end piece and expanded with compressed gas when needed. Prototypes were being constructed at MSC. (Patent Off PIO; Jones, *NYT*, 6/29/68, 37)
- Jet flying belt designed to propel wearer for minutes over multimile range at speeds from hovering to 70 mph and at varying altitudes was described by manufacturer, Bell Aerosystems Co., at Washington, D.C., press briefing. Miniature turbojet engine using kerosene fuel was designed by Williams Research Corp. for DOD Advanced Research Projects Agency. (Schmeck, *NYT*, 6/28/68, 18)
- Westinghouse Defense and Space Center engineer Paul J. Kiefer had received AIAA annual award for "outstanding contribution to aerospace sciences or technology" for overall mechanical design and development of Gemini rendezvous radar system and for development of lunar TV camera for use in Apollo series. (Westinghouse Release; AP, W *Star*, 7/5/68, A3)
- Soviet Foreign Minister Andrey A. Gromyko announced at Supreme Soviet meeting in Moscow that U.S.S.R. was ready to open discussions with U.S. on mutual limitation of antiballistic missile defense systems. U.S.S.R., he said, was anxious to sign immediately international document prohibiting use of nuclear weapons and to reach agreement on mutual restriction and subsequent reduction of strategic nuclear vehicles. (UPI, *W News*, 6/27/68, 3; Anderson, *NYT*, 6/28/68, 1)

June 28: NASA test pilots L/C Emil Sturmthal and Col. Joseph F. Cotton flew XB–70A to 39,400-ft altitude and mach 1.23 in flight from Edwards AFB. Purpose was to check exciter vane function, aeroelasticity, stability, control, and gust and canard loads and determine ground effects during load approaches. (XB–70 Proj Off)
- NASA announced probable spring 1969 launch of Nimbus B2 to replace Nimbus B experimental weather satellite intentionally destroyed after launch May 18. Primary meteorological objective would be to obtain data from advanced sensors to demonstrate infrared sounding techniques for determination of temperature profiles. Replacement mission would cost 1/3 of $61.9-million cost of Nimbus B and would eliminate critical 21-mo gap in U.S. meteorology satellite program. *Nimbus I*, launched Aug. 28, 1964, operated 26 days. *Nimbus II*, launched May 15, 1966, was still transmitting, but with inoperable tape recorders. (NASA Release 68–114)
- Third anniversary of beginning of commercial service by ComSatCorp's *Intelsat I* (*Early Bird*), launched April 6, 1965. Service had been inaugurated by President Johnson in 25-min, six-nation conference call with European officials. (ComSatCorp Release 68–31; A&A 65)
- Presidential memorandum advised agency and department heads to achieve provisions of P.L. 90–364, Revenue and Expenditure Control

June 28

Act of 1968, calling for reduction in Federal spending and lending of at least $6 billion below original 1969 fiscal estimates and restriction of hiring until Federal civilian employment was reduced to June 1966 level. (*PD*, 7/8/68, 1041)

- NASA awarded three-year, $20,126,224 cost-plus-award-fee contract to Bendix Field Engineering Corp. for maintenance and operation of Satellite Tracking and Data Acquisition Network (STADAN) facilities at GSFC; Fort Myers, Fla.; Lima, Peru; Tananarive, Malagasy Republic; Mojave, Calif.; Quito, Ecuador; and Santiago, Chile.

  NASA also awarded $27.6 million cost-plus-incentive-award-fee contract to TRW systems group for work on mission trajectory control program and Apollo spacecraft systems analysis program for MSC. (NASA Releases 68-113, 68-115)

- *Wall Street Journal* reported interview with Prof. Edwin L. Resler, Jr., Director of Cornell Univ. Graduate School of Aerospace Engineering, on possibility of reducing sonic boom created by supersonic transports to tolerable level by changing design of engines for big aircraft to slow down expansion of exhaust stream. "We can . . . reduce the shock wave effect and its consequent boom to a tolerable level so that overland flights of supersonic transports would be feasible," Prof. Resler said. (*WSJ*, 6/28/68)

- House Appropriations Committee released hearings on DOD appropriations for FY 1969 which helped explain why Vietnam war had produced strain on President Johnson's budget. USAF was now paying $2.4 million for single rescue helicopter; during World War II, each B-17 aircraft that had bombed Germany had cost $190,000 and each B-29 used over Japan had cost $635,000. Government had bought 200 World War II fighters for the $1.1 million it cost USAF for technical manuals for single type of aircraft in 1968. USN was spending $30,000 for single torpedo and $20.3 million for ammunition for battleship U.S.S. *New Jersey* to fire at Vietnam shore targets. It was requesting $51.8 million for one-year supply of aerial targets. Super Jolly helicopter which USAF was introducing in Vietnam was twice as expensive as predecessor, Jolly Green Giant, and could carry 60 passengers or 18,500-lb cargo at 195 knots. F-111 was being produced at $8 million each; C-5A, world's largest aircraft, had a unit cost of $25 million, with USAF requesting 120 aircraft. (Transcript; AP, B *Sun*, 6/30/68, 2)

June 30: Lockheed-Georgia Co. test pilot Leo J. Sullivan successfully flew C-5 Galaxy, world's largest aircraft, on 94-min first test flight from Dobbins AFB over Georgia countryside at speeds ranging from 143-mph takeoff to 230 mph and reaching 10,000-ft altitude. No attempt was made to reach maximum speed of 550 mph. C-5, powered by four TF-39 turbofan jet engines, each delivering 41,000 lb of thrust, could carry 265,000-lb payload over 2,875-mi range or 100,000-lb payload over 6,325-mi range at cruising speed of 506 mph. Military version would carry 350 fully equipped troops. USAF had ordered 58. "We like to talk about a commercial plane similar to the C-5 which could carry nearly 1,000 passengers," said Lockheed-Georgia President T. R. May, but he found idea of carrying both passengers and freight attractive. "We have preliminary plans for airplanes weighing over a million pounds. But it is fairly clear that the world is not quite ready for a

June 30: USAF-Lockheed C–5 Galaxy, world's largest aircraft, takes off from Dobbins AFB on first test flight. C–5 reached 230 mph and 10,000-foot altitude on 94-minute flight, not attempting maximum speed of 550 mph. (AFSC PHOTO)

commercial plane of this size." (AFSC *Newsreview*, 6/68, 1; Witkin, *NYT*, 7/1/68, 1; AP, W *Star*, 7/1/68, 1; *AFHF Newsletter*, 8/68, 1)

* Dr. Robert Jastrow, Director of GSFC Institute for Space Studies, reviewed Arthur C. Clarke's *The Promise of Space*. Clarke had described chronology of Apollo decision as "politics and astronautics combined" and had written: "The verdict of history may well be that the United States made the correct decision even if from dubious motives." Dr. Jastrow said Clarke "seems to betray a point of view that the primary purpose of the space program is, or should be, the exploitation of its scientific potential and the search for knowledge in the space around the earth and on the other bodies of the solar system.

"My own view is that he is mistaken. Spacecraft have yielded important scientific discoveries . . . but it seems clear to me that preservation of national security, and not scientific research per se, was the motivation for the Kennedy proposal. Kennedy acted out of a deep gut instinct, shared by the Congress and the American people, that the United States had been presented with a major challenge to which it must respond effectively or pay a heavy penalty. The decision on the expanded space program may have been accelerated by the events of the spring of 1961, but the Soviet challenge . . . was permanent."

Dr. Jastrow saw promise of space as "dollars-and-cents return in increased productivity in the U.S." To Clarke it was "the universe—or nothing." (*W Post, Book World*, 6/30/68, 1; *CR*, 7/9/68, E6290)

During June: Last of 11 JC–130 Hercules turboprop aircraft left Patrick AFB, Fla., to be replaced by larger, high-speed jet EC–135N Apollo Range Instrumentation aircraft. With complex electronic instrumentation, Air Force Eastern Test Range Hercules had supported hundreds of Cape Kennedy space and missile launches, including Atlas, Titan, Polaris, Minuteman, Saturn, Mercury, Gemini, and Apollo. When fully operational in 1961, they had replaced earlier C–54s. (AFSC *Newsreview*, 6/68, 4)

- Scientists were pressing NASA to prepare "Grand Tour" mission of successive unpowered flybys of Jupiter, Saturn, Uranus, and Neptune to take advantage of configuration which occurred once in 179 yr. Alignment of planets would permit trip time as low as 8 yr rather than 30. Next opportunity would occur during 1975–1981 period, with 1978 and 1979 regarded as best launch years. Tour was described by Lockheed Missiles & Space Co. engineer Brent W. Silver in *Journal of Spacecraft and Rockets* as "feasible and worthwhile." It would be "waste of natural resources to pass up this opportunity," he said. Studies indicated mission could be mounted with existing technology, but NASA, because of budgetary cuts, had yet to authorize it. JPL study had postulated nine-year tour including use of electric propulsion for sustained power between planets. MIT project, beginning with exploratory Jupiter probes in 1972, would cost $80 million annually over 17 yr and would use 1,000-lb spacecraft launched by Titan III–C with Centaur upper stage. (*J/Spacecraft and Rockets*, 6/68, 633–7; Wilford, *NYT*, 6/20/68, 16)

- Evert Clark in *Astronautics & Aeronautics* praised *Report to the Congress from the President of the United States: United States Aeronautics and Space Activities, 1967* [see Jan. 30]. It was "small encyclopedia, revealing information that appears nowhere else in the public record . . . a valuable addition to the shelf of the careful collector." Presidential reports, he said, served "as signposts for the road ahead as well as irreplaceable records of the recent past" and provided "only complete, official accounting of American appropriations and expenditures for military and civilian space since 1955." (*A&A*, 6/68, 6)

- Cosmonaut Valentina Nikolayeva-Tereshkova, who became first woman to fly in space when she orbited earth June 16–19, 1963, on board U.S.S.R.'s *Vostok VI*, was elected President of the Presidium of the Soviet Women's Committee. (*Moscow News*, 6/22–29/68, 1)

July 1968

July 1: Eighth anniversary of NASA's largest Center, Marshall Space Flight Center. It became operational July 1, 1960, with 4,400 employees and facilities valued at estimated $100 million. On eighth anniversary MSFC had 6,500 employees. Plant value was estimated at $400 million, with real property values accounting for $140 million and capital equipment for remaining $260 million. Achievements during first eight years included development and successful flight of Saturn I, Saturn IB, and Saturn V launch vehicles. (MSFC Release 68–143)

- McDonnell Douglas Corp. received $9,666,800 NASA contract for 10 additional Improved (Long-Tank) Delta launch vehicles for use in variety of launches, including TOS–E for ESSA in August, Intelsat III for ComSatCorp in September, IDSCP/A for DOD in May 1969, HEOS (Highly Eccentric Orbiting Satellite) for ESRO in late 1968, and ISIS–A (International Satellite for Ionospheric Studies) in late 1968. North American Rockwell Corp. was awarded $6,968,038 contract extension for material, facilities, manpower and equipment for XB–70 flight operations, and General Electric Co. was awarded $1,957,323 extension for maintenance of XB–70 engines. Both extensions covered July 1, 1968, through June 30, 1969. (NASA Release 68–116; FRC Release 19–68)

- At signing in Washington, D.C., of nuclear nonproliferation treaty, President Johnson said: "The conclusion of this treaty encourages the hope that other steps may be taken toward a peaceful world. And . . . I have described this treaty as the most important international agreement since the beginning of the nuclear age. . . . After long seasons of patient and painstaking negotiation, we have concluded just within the past five years, the limited test ban treaty, the outer space treaty, the treaty creating a nuclear-free zone in Latin America. And the march of mankind is toward the summit and not the chasm."

    Agreement had also been reached between U.S. and U.S.S.R., President Johnson announced, "to enter in the nearest future into discussions on the limitation and the reduction of both offensive strategic nuclear weapons delivery systems and systems of defense against ballistic missiles."

    At Moscow signing of treaty, Soviet Premier Alexey N. Kosygin called agreement a "major success for the cause of peace." He disclosed contents of U.S.S.R. memorandum to all nations proposing nine-point disarmament and arms control program and called on 18-nation Geneva disarmament conference to take up proposal. (*PD*, 6/8/68; Sherman, *W Star*, 7/1/68, A1; UPI, 7/1/68; Grose, *NYT*, 7/2/68, 1, 2)

- Sudden affirmative response by U.S.S.R. to President Johnson's longstanding offer for discussion of limiting missiles may have substantial meaning, said William S. White in *Washington Post*. "If this should turn out to be the case it would be ironic, indeed. It would mean that only after renouncing his office had the President been able to convince

July 1

the Russians . . . that this country was honestly prepared to make accommodations with Moscow, so long as they were realistic and enforceable accommodations to reduce a possibility of nuclear holocaust that still hangs over the world and will so hang whatever may or may not happen in the Vietnams." (*W Post,* 7/1/68, A21)

- Boeing Co. submitted SST progress report to FAA which indicated Government might have to guarantee $2 billion in production costs, depending on ultimate number of aircraft sold. Manufacturer estimated production and certification could cost $2.96 billion, assuming production of 500 aircraft. Airline advance payments could account for $905 million; suppliers, including Boeing and General Electric Co., could raise $595 million, leaving $1.5 billion in capital to be acquired from other sources. Boeing said this capital probably would not be available unless Government guaranteed repayment. Total cost of each aircraft was estimated at $41.2 million, with probable 50% advance payment by airlines required. Since first flight tests were planned for September 1972, further studies would be conducted, Boeing said, before final recommendations on SST financing were made. (Taylor, *Am Av,* 9/16/68, 22–4; AP, B *Sun,* 9/17/68, A9)

- Resignation of Dr. Mac C. Adams, NASA Associate Administrator for Advanced Research and Technology, announced May 21, became effective. He rejoined Avco Corp., where he had worked from 1955 to 1965, as Corporate Vice President and Deputy Group Executive of Government Products and Services Group. (*NYT,* 7/2/68, 63)

- NASA appointed M/G Daniel F. Callahan (USAF, Ret.), Manager of Florida Missile Operations for Chrysler Corp., to position of Deputy Director of Administration, Kennedy Space Center, vacated in October 1967 by Frederic Miller, who became Director of Installation Support. (KSC Release KSC–331–68)

- White House announced that Gen. William F. McKee (USAF, Ret.) had submitted his resignation as head of FAA, effective July 31. There was no indication of successor. (*W News,* 7/2/68, 12; *WSJ,* 7/2/68)

- New subdivision of Air Force Systems Command, Air Force Human Resources Laboratory (AFHRL), became operational at Brooks AFB, Tex., as focal point for USAF R&D effort to satisfy technology needs in human resources education, training, and management. It would also provide technical and management assistance in support of studies, analyses, development planning activities, acquisition, test evaluation, modification, or operation of aerospace systems and related equipment. (AFSC Release 93.68)

- Commenting on C–5 maiden flight, *New York Times* editorial noted: "Of the many technological advances required for yesterday's aviation breakthrough, the most important was the quantum leap in jet propulsion capabilities represented by the C–5's motors. The enormous size of the new plane forced extraordinary use of light metals . . . to keep down weight. It also posed unprecedented manufacturing problems whose brilliantly successful solution was proved by yesterday's pathbreaking flight.

    "But will the airports of this country—and the world, for that matter —be capable of meeting the challenges . . . ? By 1978, "it may be commonplace for a few enormous planes landing minutes apart to de-

posit 5,000 or 10,000 passengers on the ground almost simultaneously. . . . Now is none too soon to begin planning for handling such masses of people. . . . The vast size of the giant new planes ahead is dwarfed only by the enormity of the unprecedented problems they pose." (*NYT*, 7/1/68, 30)
- Surveyor Project Office at JPL officially closed after directing one of U.S.'s most successful space exploration programs. Program director, Howard H. Haglund, recipient of 1968 Astronautics Engineer Award, had been accepted as Alfred P. Sloan Fellow and would attend Stanford Univ. (JPL PIO; *SBD*, 7/8/68, 10)

July 2: USAF had attributed March 3 UFO reports over eastern U.S., including 70 eyewitness accounts, to reentry of booster rocket or other launching components of *Zond IV* spacecraft launched by U.S.S.R. March 2 on apparently unsuccessful mission. Despite March flurry, there had been sharp decline in UFO reports; they were reaching USAF at one-fourth the monthly rate of 1967. As of previous weekend, 156 UFO reports had been received since Jan. 1, 1968; 21 were attributed to astronomical objects, 19 to aircraft, 10 to balloons, 8 to satellites, and 22 to other known causes. There were 35 cases pending and 41 as yet unidentified. (Sullivan, *NYT*, 7/2/68, 1)
- West Germany's major aerospace companies—Messerschmitt-Bölkow, Vereinigte Flugtechnische Werke of Bremen, Hamburger Flugzeugbau, and Dornier—formed subsidiary to coordinate all long-range aircraft and space projects. They met under auspices of West German government which had been urging greater concentration of the nation's aerospace capacity. Experts termed new organization nucleus of eventual merger of the four companies to increase West German competition in world markets. (Shabecoff, *NYT*, 7/3/68, 12)
- NASA awarded contracts valued at $579,000 to Lockheed Missiles & Space Co. and $568,313 to Northrop Systems Laboratories to build and test nonflight demonstration models for Orbiting Primate Experiment, as continuation of preliminary conceptual design studies made during 1967. Research had been begun to gain better understanding of physiological changes anticipated in long manned flights. To assess effects of weightlessness on relatively high order mammal, NASA was studying experiment which might place two unrestrained rhesus monkeys in orbit and return them for detailed examination after extended period to isolate weightlessness as a variable while maintaining all other factors near normality. Postflight examinations could reveal changes resulting from absence of gravity. Orbiting Primate Experiment was part of NASA's Human Factors Systems program to provide technology required to support man in space during extended periods. (NASA Release 68-119)
- Univ. of Virginia announced it would use $100,000 NASA grant to finance construction of 40-in astrometric telescope at its observatory south of Charlottesville, Va. Additional funding would come from estate of Leander McCormick, who provided funds for its 26-in telescope built in 1882. (AP, W *Star*, 7/3/68, A20)
- U.S. patent No. 3,390,853 was issued to North American Rockwell Corp. mechanical engineer Raymond P. Wykes for inflatable drag balloon (ballute) to be released behind reentry vehicle or lifting-body vehicle at end of a cable which pulled spacecraft's wings out from its body on

reentry and slowed it down for landing. Patent No. 3,390,492 was issued to General Electric Co. engineer Edwin T. Myskowski for glass deep-submergence module in titanium alloy frame usable as laboratory or living quarters on ocean floor in anchored or mobile form. (Patent Off PIO; Jones, *NYT*, 7/6/68, 25)

- N. Whitney Matthews, Chief of GSFC's Spacecraft Technology Div., died in Alexandria, Va., at age 52. Pioneer in space research, he had been with NASA 10 yr and had helped see Goddard through planning stages. He had worked with Projects Vanguard, Ariel, and Echo and with number of Explorer programs. He had specialized in electronic and solid-state instrumentation and control circuitry. (*W Post*, 7/5/68, B8)
- In editorial critical of June 25 NAS report on sonic boom, Washington *Evening Star* said: "There comes a time when the convenience of the few and the profit of the even fewer simply have to be made secondary to the sanity of the many. That time is arriving in the sonic boom business. There is no imaginable excuse for unleashing the boom against defenseless citizens." (*W Star*, 7/2/68, 3)

July 3: President Johnson signed H.R. 15856, NASA FY 1969 Authorization Act, which had been designated P.L. 90-373 [see June 18]. (*PD*, 7/15/68, 1099; NASA *LAR* VII/71)

- *Washington Post* editorial commented on complaints of scientists about deceleration of Federal funding for R&D. Since Federal expenditures had risen every year, there would not be "much lay sympathy for scientists who complain they are not getting their annual increase of 15 per cent. . . . Rather than crying 'crisis' . . . scientists ought to accept an ongoing obligation to help public officials devise better ways of deciding how to support the level of science that the national welfare requires." (*W Post*, 7/3/68)
- Did it matter in 1968, asked *New York Times* editorial, that Italian astronomer Galileo after three centuries might be cleared of heresy by commission authorized by the Pope? "His astronomical theories and discoveries have long since been accepted; in a real sense, it is the spirit of scientific inquiry that will be 'retried' by the Vatican Tribunal.

  ". . . it still matters in 1968 that the intellectuals, the scientists and the students be granted full freedom of inquiry and participation in modern life and government. That is the meaning of Galileo, the individual and heretic, for today." (*NYT*, 6/3/68)
- Senate approved reappointment of Gen. John P. McConnell as Air Force Chief of Staff effective Aug. 1, 1968. (*CR*, 7/3/68; S8200; *W Post*, 7/4/68, 4)
- Aluminum Co. of America and Ocean Science and Engineering, Inc., announced they would invest more than $5 million in Alcoa Seaprobe project calling for construction of ship permitting search, science, and salvage work at depths to 6,000 ft and able to hoist to surface loads weighing up to 200 tons. Planned for launch by May 1970, vessel would search ocean floor by lowering streamlined sensor, carrying side-looking sonar, at end of long semirigid pipe. (*W Star*, 7/3/68, A2)
- Jet Propulsion Laboratory announced appointment of Dr. Robert V. Meghreblian, Manager of JPL Space Sciences Div., to newly established post of Deputy Assistant Laboratory Director for Technical Divisions.

Dr. Donald P. Burcham, Deputy Manager of Space Sciences Div., would succeed him. (JPL Release)

- French government announced imminent start of new atomic test series in Pacific amid indications France would attempt her first explosion of hydrogen bomb in late summer or early autumn. Bulletin warned ships to avoid danger zone around Mururoa Atoll, about 750 mi southeast of Tahiti. (*NYT*, 7/4/68, 1)
- French Armed Forces Ministry announced successful testing of two new long-range ballistic missiles during preparation for Pacific nuclear test series. First missile, sea-to-ground, two-stage, remote-controlled rocket, would be used on France's first nuclear submarine, to enter service in early 1970. (Reuters, *NYT*, 7/5/68, 13; *W Post*, 7/5/68, A27)

July 4–8: NASA launched 417-lb *Explorer XXXVIII*, Radio Astronomy Explorer (RAE–A), from WTR by three-stage Thrust-Augmented Delta booster in first of two missions to measure frequency, intensity, and source direction of radio signals from solar, galactic, and extragalactic sources.

Spacecraft entered elliptical transfer orbit, where it was spin-stabilized with 3,656.1-mi (5,884-km) apogee, 397.7-mi (640-km) perigee, 157-min period, and 59.4° inclination. Apogee motor was fired July 7, placing *Explorer XXXVIII* into planned near-circular orbit with 3,654.3-mi (5,881-km) apogee, 3,641.2-mi (5,860-km) perigee, 224.4-min period, and 59.2° inclination. On July 8, yo-yo despin mechanism reduced spin rate from 93 rpm to 2.8 rpm. As primary objective, spacecraft would measure intensity and direction of radio signals from cosmic sources in 0.5- to 10-mhz range, not normally observable from earth. Secondary objectives were to place spacecraft into circular orbit of about 3,728-mi (6,000-km) altitude and to obtain useful data during first 30 days in orbit, for detailed study of dynamic spectra and decay rates of sporadic radio bursts. Spacecraft was expected to provide first low-frequency radio map of Milky Way and additional data on low-frequency signals from Jupiter and sun.

Explorer XXXVIII was equipped with unique antenna system consisting of two antennas made of four ½-in-dia booms which could be deployed up to 750 ft each, to form X-shaped array. Configuration was to be gravity-gradient stabilized [see July 22]. Spacecraft was also equipped with damper boom, dipole antenna, and TV cameras to monitor spacecraft performance and determine source of radio signals monitored with upper array. Radio Astronomy Explorer project was managed by GSFC under OSSA direction. GSFC constructed, designed, and tested spacecraft and provided scientific instrumentation. (NASA Proj Off; NASA Release 68–109K; Schmeck, *NYT*, 6/29/68; 8; AP, *W Star*, 7/5/68; AP, *NYT*, 7/5/68, 26; *W Post*, 7/8/68, A6; 7/9/68, A7)

- *July 4:* *Explorer XXVIII*, Interplanetary Monitoring Platform launched by NASA May 29, 1965, to investigate earth's magnetosphere and study earth-sun relationships, reentered atmosphere as had been predicted at GSFC August 1967. Mrs. Barbara Lowrey of GSFC Laboratory for Theoretical Studies had found in analysis of satellite's orbit that—during perigee pass on July 4 (actually early next day in Indian Ocean reentry area)—joint effect of sun and moon would alter orbit and cause *Explorer XXVIII* to make high-angle reentry and burn up in earth's atmosphere. Computer tests had verified analysis. (GSFC *SSR*, 7/15/68; NASA Release 68–117; *Marshall Star*, 3/6/68)

July 5: U.S.S.R. successfully launched *Cosmos CCXXX* from Kapustin Yar. Orbital parameters: apogee, 544 km (338 mi); perigee, 283 km (175.8 mi); period, 92.8 min; and inclination, 48.4° Satellite reentered Nov. 2. (UPI, *NYT*, 7/6/68; *SBD*, 7/10/68, 26; GSFC *SSR*, 7/15/68; 11/15/68)

- AEC's High Energy Physics Advisory Panel report in *Science* decried cutbacks in funds for high-energy physics "one of main fronts of science" and recommended budget increase to avert decline in U.S. effort and construction of giant bubble chamber at Brookhaven Laboratory and electron-positron storage ring at Stanford Linear Accelerator (SLAC). Work on 200-bev accelerator at Weston, Ill., should continue "at highest priority," report stressed, and provision should be made to finance joint research with U.S.S.R. using present most powerful accelerator in world at Serpukhov, near Moscow.

  Lack of approval of bubble chamber and SLAC storage ring in 1968 and 1969 budgets meant "for the first time in the history of this field, U.S. physicists will be unable to make use of some of the most modern means of research." Further, there was "clear and present danger" that U.S. would lose its leadership in this fundamental field, "an ominous step" toward situation of 1930s, "when most of the major discoveries in fundamental science were made in Europe." (*Science*, 7/5/68, 11–9; Sullivan, *NYT*, 7/7/68, 17)

- JPL scientist Dr. Robert Nathan, who had devised method using computers to improve spacecraft photos of moon and Mars, planned to link computers with electron microscopes to photograph single atom. Within six months much of connection work should be done, he said, and "with luck, we could be taking pictures of atoms in a year or so." (Dighton, Glendale *News-Press*, 7/5/68, 1)

- NASA awarded 16-mo, $178,844 cost-plus-fixed-fee contract to Lockheed Missiles & Space Co. for computer software to operate NASA/RECON remote-console information retrieval system. Consoles would be installed at field centers and NASA Hq. and linked to central computer at NASA Scientific and Technical Information Facility in College Park, Md. They would provide real-time access to NASA's worldwide collection of scientific and technical documents on aerospace. Users would need no special skill. (NASA Release 68–118)

- FCC ruled that rates charged TV networks for overseas service via satellite were not excessive and that companies providing service—AT&T, RCA Communications, Inc., ITT World Communications, Inc., and Western Union International—were no longer required to place payments for services in deferred credit fund. (AP, *NYT*, 7/7/68, 10)

- Danish government announced it had banned U.S. rocket flights to probe sunspot effects at high altitudes over Greenland during 1968 because of popular apprehension which followed January crash of nuclear-armed USAF B–52 aircraft near Thule AFB. Disappointed scientists noted 1968 was peak in 11-yr sunspot cycle; 1969 would offer hardly enough sunspots for study. (*C Trib*, 7/6/68, 5)

- Sonic booms from USAF test flights were threatening prehistoric Indian cliff dwellings and natural rock formations in Arizona. Log kept at Canyon de Chelly National Monument had recorded 16 booms in April

1967, 19 in April 1968, 20 in May 1968, and 9 in June 1968. Natural Environment Panel, participating in Interagency Aircraft Noise Abatement Program under DOT, planned to place data recorders at Yellowstone, Yosemite, Bryce, and Mesa Verde national parks to extract information on which to base plea for "adjustment" from USAF. (Blumenthal, *NYT*, 7/5/68, 11)

July 5-12: High-quality weather data were moved from Suitland, Md., by wire to NASA's Mojave, Calif., relay station and beamed, for first time, to stations in the Netherlands and West Germany via NASA's *Ats III* Applications Technology Satellite. Transmissions, including cloud maps, charts, and photo-mosaics, were received "in good form," according to ESSA. WEFAX (Weather Facsimile Experiment) project was part of World Weather Watch program to develop economical worldwide weather data distribution system. Further experiments scheduled for September included relay via *Ats III* of weather data to more than 150 Automatic Picture Transmission (APT) stations in 30 countries. (ESSA Release ES-68-43, UPI, *NYT*, 7/19/68, 35; W *Star*, 7/24/68, A14)

July 6: Ninth Molniya I comsat, *Molniya I-9*, was launched by U.S.S.R. to "ensure the operation of the long-range system of . . . communication" and TV transmission to far northern and far eastern U.S.S.R., according to Tass. Orbital parameters: apogee, 39,806 km (24, 734.2 mi); perigee, 396 km (246.1 mi); period, 11 hr 9 min; and inclination, 65°. Equipment, including instruments for transmission, command, and satellite operation, was functioning normally. (AP, *NYT*, 7/9/68, 6; *SBD*, 7/10/68, 26; GSFC *SSR*, 7/15/68)

- Japanese astronomer Minori Honda of Kurashiki Astronomical Observatory, Okayama, discovered new comet south of Capella in Auriga constellation. Tokyo Astronomical Observatory said July 14 discovery had been confirmed by three American observatories. Comet was named Honda Comet No. 6. (AP, *C Trib*, 7/15/68)

- DOD released April 23-24 testimony before Senate Committee on Armed Services' Preparedness Investigating Subcommittee. Dr. John S. Foster, Jr., Director of Defense Research and Engineering, had said F-111A wings had broken off during Jan. 23 ground test—under load greater than expected in flight but less than stipulated 50%-overload safety margin—before introduction into Vietnam combat, where aircraft had operated under protective restrictions.

    General Dynamics Corp. President Frank W. Davis later termed ground testing which broke wings off USAF F-111A "normal." Tests, he indicated, were made to determine stress limitations. "We've had no failures . . . at stress simulation to be expected in combat." (Transcript; Kelly, W *Star*, 7/7/68, A3; AP, *W Post*, 7/7/68, A22; 7/8/68, A15; Corddry, B *Sun*, 7/7/68, 1)

- Washington *Evening Star* editorial praised USAF C-5 Galaxy jet aircraft and its "impressive" civilian potential; "According to Tom [T.R.] May, Lockheed's president, all the experimental evidence indicates there are virtually no engineering limitations to building strikingly larger C-5s than those scheduled," but its commercial use would cause passenger and baggage congestion. "If the Galaxy is to become a commercial plane, then, at the most, only a third of its space should be for passengers; the rest should be for cargo. . . . Although [May is] confident that bigger and bigger C–5s can be made, he doubts that the world is ready for them. We doubt it, too." (W *Star*, 7/6/68)

July 7: Melbourne, Fla., engineer Duane Brown had applied for patent on Survey Satellite (SURSAT), system of four low-cost satellites which would enable surveyors to plot boundaries, route highways, make maps, and monitor earth's crust to accuracies of a few inches. System included regional center for processing survey data and portable receiving and recording units for field use and could be operational by mid-1970's, Brown said. (UPI, W *Star*, 7/7/68, A7)

- Successful test-firing of Phoebus 2A, world's most powerful reactor, June 26 might have been catalyst needed to bring DOD into partnership with NASA and AEC in development of nuclear energy for space propulsion, Frank Macomber wrote in San Diego *Union*. Not only was USAF becoming interested in military applications for nuclear engine, so were scientists and engineers representing aerospace industry. Phoebus firing would be followed in fall by first test of smaller NERVA XE–1 nuclear engine. Both were vital phases of NASA–AEC Rover program. (SD *Union*, 7/7/68, 12)

- Sun Shipbuilding and Dry Dock Co. announced plans for new Guppy, 4,000-lb, low-cost, undersea research vehicle to be tethered to surface ship by electric cable and capable of carrying two men to 2,000-ft depth for up to 48 hr. First vehicle would be completed in March 1969. (*NYT*, 7/8/68, 66)

- France began 1968 nuclear test series with detonation of conventional atomic warhead over Mururoa Atoll in Pacific. Device was fired to test complex measuring instruments installed for tests scheduled to culminate in explosion of France's first hydrogen bomb. (UPI, *NYT*, 7/7/68, 7; *W Post*, 7/8/68, A12)

July 8: Dr. Robert C. Seamans, Jr., Special Consultant to NASA Administrator and former NASA Deputy Administrator, had been elected to board of trustees of Aerospace Corp. (Aerospace Release; *SBD*, 7/8/68, 10)

- Approximately 36 Soviet Air Force flights with more than 85 bombers had been identified off northern coasts of North America during first half of 1968, six times scale of operations reported during last half of 1967, according to Charles W. Corddry in Baltimore *Sun*. Soviet aircraft had cruised over international waters. DOD reportedly considered flights routine. (B *Sun*, 6/9/68, 1)

- In joint communique, Dr. Donald F. Hornig, Special Assistant to the President for Science and Technology, and Alexandru Birlandeanu, member of Romanian Politburo touring U.S. scientific institutions, announced agreement to broaden scientific and technological ties, including exchange of scientists and possible collaboration in atomic energy field. Romania had asked U.S. for technical and financial aid toward construction of its first nuclear power plant by 1973. (Grose, *NYT*, 7/9/68, 1)

- *New York Times* editorial on June 21 emergency meeting of scientists to protest cuts in Federal support for basic research: ". . . deep slashes in basic research funds are likely to be extremely costly in the years ahead. The fundamental lesson of the history of science is that basic research is the indispensable seed bed for all future technology, the ultimate source of the new wealth and of the improved capacity to save lives that future technology could bring. . . . Those in Congress and the Executive Branch who are now applying the axe to Government

spending would be wise to proceed as gently as possible in this small area that is so essential for the nation's future." (*NYT*, 7/8/68, 36)
- NASA Associate Administrator for Manned Space Flight, Dr. George E. Mueller, addressed joint meeting of American Institute of Aeronautics and Astronautics (AIAA) and Canadian Aeronautics and Space Institute (CASI) in Montreal: Systems engineering concept applied to management "was pioneered and developed in aerospace programs and is being increasingly applied as a powerful tool in the management of other major enterprises." In NASA most extensive application was in Apollo program. Factors unique to manned space flight had contributed to management approach, including "sheer size of Apollo program, larger in . . . lead time, money, organization and technological development than any previous program." Special feature was high reliability and safety required. And space program had been executed under scrutiny of press, public, Congress, and scientific community.

    Weight and volume budgeting were critical. High cost of flight-testing space vehicles made maximum ground testing necessary, as well as all-up (concurrent rather than sequential) flight testing. Vehicle was as complete as practicable for each flight, to obtain maximum information from minimum number of flights and provide earliest possible system readiness. Open-ended mission concept was used to accomplish as many flight objectives per vehicle as consistent with safety and mission success. Review of status throughout mission determined length of mission. Redundant, or alternate, means of operation reduced ability of single failure to endanger crew or mission. Prime design consideration in all manned space flights was safety. (Text; UPI, *H Chron*, 7/10/68)
- NASA board investigating fatal accident at North American Rockwell Corp.'s Downey, Calif., plant Oct. 5, 1967, announced it had found that laboratory employees had ignored important safety procedures. "Most probable cause" of explosion which had killed two workmen and injured 11 was "frictional or impact force created while barium-Freon TF slurry was being transferred from a laboratory container to a shipping container." Although NAR had issued safety instruction requiring barium—used in NASA sounding rocket experiments—be handled only under dry argon atmosphere, it had been washed and sieved in open air. Board recommended full recognition of chemical hazards of combining metals and chemicals such as Freon TF and upgrading of precautions, manuals, and procedures. NAR had altered procedures, would process barium only under remote control. (NASA Release 68–122; AP, *NYT*, 7/9/68, 27)
- Inauguration of direct air service between New York and Moscow had been set for "on or after July 15" by letters between U.S. Moscow Embassy and U.S.S.R. Foreign Ministry. Soviet airline Aeroflot announced Il–62 jet aircraft service would start from Moscow July 15. U.S. carrier Pan American World Airways expected to start Boeing 707 service from New York same date. Bilateral air agreement of Nov. 4, 1966, had stipulated once-weekly return flights over 4,700-mi route. May 6, 1968, agreement added intermediate stop at Montreal, Copenhagen, Stockholm, or London. (CAB Docket 6489; State Dept Release 94; AP, *NYT*, 7/9/68, 65; Ward, *B Sun*, 7/9/68, 1; AP, *W Post*, 7/9/68, A15)

July 8–9: Two major solar flares were detected within 25-hr interval by U.S. Space Disturbance Forecast Center scientists in Boulder, Colo.

First had interfered with short-wave transmissions worldwide, according to ESSA Chief of Forecast Services, Robert Doeker; second had seemed weaker. Scientists were watching for effect of cloud of electrons spawned by first solar flare, biggest and brightest since 1966. (AP, LA *Her-Exam*, 7/10/68; AP, *NYT*, 7/10/68, 17)

- Nine astronauts participated in life support training for aircraft pilots at Perrin AFB, Tex. They were second astronaut group to attend USAF course. (NASA Apollo Status Summary, 7/10/68)

July 9: British physicist Samuel Tolansky, appointed special investigator for Apollo program, had predicted discovery of industrial diamonds among 40 lb of matter Apollo spacecraft would bring back from moon. Theory was based on supposition that lunar craters had been caused by meteor impact or volcanic eruptions producing shock waves. "You can create diamonds by passing a shock wave through carbon," he said. "And there has to be carbon on the moon." (NANA, *Pasadena Independent*, 7/9/68)

- Lockheed Missiles & Space Co. scientists were studying use of small charcoal beds to remove contaminants in space capsules where pollution hazards had been found to be "more serious than those for the man on the street." Studies had isolated 150 contaminants, most of which could be extremely toxic. (*WSJ*, 7/9/68, 23)

- President Johnson transmitted Treaty on the Non-Proliferation of Nuclear Weapons to Senate for ratification and urgently recommended that Senate "move swiftly" to enhance U.S. and world security. "The treaty," he said, "does more than just prohibit the spread of nuclear weapons. It would also promote the further development of nuclear energy for peaceful purposes under safeguards." Treaty had been passed by U.N. General Assembly June 12 and opened for signature July 1. (*PD*, 7/15/68, 1090–2)

- Fixed-wing SST design aerodynamically similar to one unsuccessfully submitted to USAF by Boeing in 1957 XB–70 competition but featuring more titanium, new flight control system, and more powerful turbojets was presented to customer airlines at FAA SST program briefing. Model was undergoing wind-tunnel tests to determine its ability to exceed mach 1 without perceptible sonic boom. (Hoffman, *W Post*, 7/9/68, 1)

July 10: NASA issued Apollo Status Summary: Apollo 7 spacecraft was being prepared for unmanned altitude chamber tests at 210,000 ft for 15 hr. If successful, manned tests might be scheduled to begin July 15 with Astronauts Walter M. Schirra, Jr., Donn F. Eisele, and R. Walter Cunningham in command module. In Apollo/Saturn 503 program, combined systems tests would continue through mid-July on Lunar Module 3.

In Apollo spacecraft loading tests, drogue parachutes would be tested within several days at Naval Air Facility, with 13,000-lb test vehicle dropped from aircraft at 46,000-ft altitude, subjecting parachutes to "ultimate loads" in reefed condition before they opened fully. Drop, repeat of previous test which failed, was to complete verification test series which had begun in 1967. (Text)

- *Cosmos CCXXXI* was launched from Baikonur Cosmodrome by U.S.S.R. into orbit with 391-km (243-mi) apogee, 206-km (128-mi) perigee, 89.6-min period, and 64.9° inclination. Equipment functioned normally and spacecraft reentered July 18. (UPI, *W Star*, 7/11/68, A5;

UPI, *NYT*, 7/12/68, 7; *SBD*, 7/12/68, 41; GSFC *SSR*, 7/15/68, 7/31/68)

- Soviet Stalin Prize physicist, Prof. Andrey D. Sakharov, contributor to development of U.S.S.R. hydrogen bomb, had issued plea for full intellectual freedom, U.S.–U.S.S.R. cooperation, and worldwide rejection of "demagogic myths," in unpublished essay entitled "Thoughts About Progress, Peaceful Coexistence and Intellectual Freedom," which was circulating in Moscow. Expressing fear that world was on brink of disaster, he urged worldwide implementation of scientific method and freedom of thought in politics, economic planning and management, education, arts, and military affairs and denounced Soviet censorship. (Anderson, *NYT*, 7/11/68, 1)
- DOD formally ordered work stoppage on USN F–111B development work being conducted by General Dynamics Corp. Action followed Congressional cuts of $460 million in program. (General Dynamics PIO; *SBD*, 7/11/68, 30)
- Sen. Eugene J. McCarthy (D-Minn.), candidate for Democratic nomination for President, in position paper urged that U.S. delay deployment of Sentinel ABM system and Poseidon and Minuteman III missiles, to facilitate agreement with U.S.S.R. on defensive and offensive armament limitation. Delay would not jeopardize U.S. security, he said, since neither Chinese nuclear threat nor Soviet ABM development is "moving ahead perceptibly." Paper was prepared by Harvard Univ. chemistry professor Dr. George B. Kistiakowsky and MIT Provost, Dr. Jerome B. Wiesner. (Text; Kenworthy, *NYT*, 7/11/68, 25; *CR*, 7/11/68, S8439–42)

July 10–12: Hearings were held by Senate Foreign Relations Committee and Senate members of Joint Committee on Atomic Energy on U.S. ratification of nuclear nonproliferation treaty. Secretary of State Dean Rusk affirmed treaty would bind U.S. to no more atomic defense action than already set forth in existing treaties and by membership in U.N. Security Council.

Gen. Earle G. Wheeler, Chairman of Joint Chiefs of Staff, and Deputy Secretary of Defense Paul H. Nitze in joint testimony said U.S. would give up nothing under terms of treaty but would benefit from major step to reduce tensions. (AP, *NYT*, 7/11/68, 16; Roberts, *W Post*, 7/11/68, A11; UPI, *NYT*, 7/12/68, 4; Sherman, W *Star*, 7/12/68, A5)

July 11: USAF successfully launched *OV 1–15* and *OV 1–16* research satellites pickaback from Vandenberg AFB by Atlas-F booster. *OV 1–15* entered orbit with 1,032-mi (1,660.8-km) apogee, 94-mi (151.3-km) perigee, 103.0-min period, and 89.8° inclination and reentered Nov. 6. *OV 1–16*, nicknamed "Cannonball," was 600-lb, 23-in-dia Low Altitude Density Satellite (LOADS) launched to measure atmospheric density between 90- and 110-mi altitudes for 25–30 days. Densest satellite U.S. had orbited, *OV 1–16* had 162-lb-per-cu-ft density, which enabled it to orbit closer to earth than any previous spacecraft. Orbital parameters: apogee, 286 mi (460.3 km); perigee, 82 mi (132.0 km); period, 90.4 min; and inclination, 89.8°. *OV 1–16* reentered Aug. 19. (O'Toole, *W Post*, 7/12/68, A21; *SBD*, 7/15/68, 44; GSFC *SSR*, 7/15/68; 8/31/68; *Pres Rpt 68*)

- DOD directive that General Dynamics Corp. halt development of USN F–111B aircraft because of weight problem would not affect USAF's F–111A program or Hughes Aircraft Co. development of Phoenix air-to-air missile, which presumably would be installed in replacement aircraft, *Wall Street Journal* reported. Of 17 F–111B prototypes planned, 8 had been produced and 6 delivered (one of which had crashed). General Dynamics was uncertain how many of remaining nine would be completed. USN had originally requested 30 aircraft. (*WSJ*, 7/11/68, 29)
- Secretary of Defense Clark M. Clifford announced USN would proceed with construction of one of two advanced nuclear submarines advocated by V/A Hyman G. Rickover to combat Soviet submarine threat. Authorization was for "super high-speed" version; "quiet" electric-powered craft was still under consideration although its development had been stopped in May. Congressional committees had supported Adm. Rickover and urged development of both types. (Dale, *NYT*, 7/12/68, 1; Kelly, *W Star*, 7/12/68, A5)

July 12: Last USN flying boat, SP–5B Martin Marlin, was formally retired from active service and turned over to Smithsonian Institution at ceremony at U.S. Naval Air Station, Patuxent, Md. Aircraft would be placed in proposed National Armed Forces Museum. (*CR*, 7/18/68, E6671)
- Dr. Stephen B. Sweeney, governmental administration professor at Univ. of Pennsylvania's Wharton School and Executive Director Emeritus of Univ.'s Fels Institute of Local and State Government, and Dr. Harold Asher, manager of General Electric Co.'s TEMPO section and former Deputy Assistant Secretary of Defense, had been sworn in as consultants to NASA Administrator James E. Webb. Dr Sweeney would specialize in university affairs, public administration, and application of science and technology to urban problems. Dr. Asher would review and analyze NASA's systems for resource management. (NASA Release 68–124)

July 13: USAF C–5 Galaxy jet aircraft, flown by Lockheed-Georgia Co. test pilot Walter E. Hensleigh, completed successful 2-hr 44-min second flight with takeoff weight of 520,000 lb—believed to be 10 tons heavier than any previous aircraft takeoff weight. During ascent to 1,000 ft, crew cut each of four GE TF39 engines individually and restarted them in air. Auxiliary units also underwent cut-restart checks. (AP, *W Star*, 7/14/68, 14)
- FB–111A, bomber version of F–111, successfully completed 30-min maiden flight from Carswell AFB, Tex., reaching 20,000-ft altitude and up to 660 mph. Equipped with advanced avionics, including onboard computers enabling pilots to alter missions in flight automatically, FB–111A's design incorporated basic fuselage of USAF F–111A tactical fighter recently grounded after three crashes in Southeast Asia. (DOD Release 652–68; AP, *W Star*, 7/14/68, A2; AP, *W Post*, 7/14/68, A5)
- Team of NASA and Max Planck Institute scientists completed 28-day tour of Argentina, Chile, Netherlands Antilles, Peru, and Venezuela. They had explored potential sites for optical observation of high-altitude ionized cloud experiment proposed as cooperative project of German Ministry for Scientific Research and NASA. Release of barium vapor at 12,000- to 20,000-ft altitudes by Scout rocket launched from NASA Wallops Station was being considered. Barium cloud would be

visible from large area of Western Hemisphere. (NASA Release 68–121)

July 14: U.S. and U.S.S.R. had exchanged private messages which raised hope initial talks on limiting nuclear missiles would begin in few weeks, according to Geneva sources quoted by *Washington Post*'s Murrey Marder. Possible obstacle was Warsaw meeting of U.S.S.R. and Eastern European officials over Czechoslovakian advance toward liberalization. U.S.–U.S.S.R. accord on nuclear missile production presumably would interact on Soviet strength in Eastern Europe, weakening it as East-West tension subsided. (*W Post*, 7/15/68, A1)

- George Alexander reviewed in *Washington Post* Erik Bergaust's *Murder on Pad 34*, story of Jan. 27, 1967, Apollo fire. Book was "characterized by sloppy errors of omission and commission, innuendo and pointlessness," Alexander said. "It was good fortune, nothing else, that the various mechanical flaws and human faults that occurred in the . . . Mercury and Gemini programs did not coincide . . . as they did inside Apollo-one. Foresight tries to prevent such coincidence, but . . . not all possible coincidence can be foreseen. . . . Accidents . . . will happen. And the searching investigation conducted by the National Aeronautics and Space Administration into Apollo-one could find no evidence that the fatal fire was anything but an accident." (*Book World, W Post*, 7/14/68, 4–5)

July 15: President Johnson formally asked Senate to ratify space rescue treaty endorsed by U.N. General Assembly Dec. 19, 1967, and signed by 43 nations April 22, terming it "another step toward stable peace on this threatened earth." Astronaut assistance and return agreement looked "beyond the old divisions of history and ideology to recognize the challenge of common peril and the benefits of common action. . . . Our laws and treaties must always keep pace with our science. But the value of this Agreement goes beyond the protection it offers to those who venture into space." It also "helps protect the peace of this planet. Surely two nations who aspire to the stars can realize the common danger and act in the common interest here on earth." (Text; *W Post*, 7/16/68, A9; AP, *W Star*, 7/16/68, A8; Nordlinger, *B Sun*, 7/16/68, 1)

- Harvard College Observatory scientists Dr. George R. Huguenin and Dr. J. H. Taylor became first U.S. scientists to identify a new pulsar when they discovered HP 1506 in northern sky near Little Dipper. Pulsar, similar to four pulsars discovered in 1967 by U.K. scientists, had pulse rate of one every 0.7397 sec, each lasting 0.020 sec. Pulse rate of other four pulsars ranged from 0.25 to 1.4 sec, with each pulse lasting 0.020 sec. Harvard scientists used National Radio Astronomy Observatory antenna at Green Bank, W. Va. (Sullivan, *NYT*, 7/19/68, 20)

- NASA Administrator James E. Webb, discussing implications of FY 1969 budget reductions at AAS Symposium in Denver, Colo., said he did not find public support for space program declining. Rather, "many people who in the years following 1961 ascribed to the space program a separate, special, top priority status are now realizing, as the national leadership in the space program has understood all along, that the space program must be regarded as only one of a number of essential activities of high priority to which the country must devote substantial resources. . . . The investments made in NASA may well add greatly to the value of investments we will have to make in these other fields."

NASA was "very much in business, and it will stay in business. We are accepting the challenge of the time and will continue a hard-hitting, technically sound program aimed at the most important objectives of the future." But he described cutbacks as well as elements of strength. "We are doing all we can to avoid terminating completely such important activities as the unmanned planetary exploratory program, but it is not likely that we will be able to proceed with the Titan/Mars 1973 missions." Saturn I Workshop would be delayed and, "for a number of years to come, missions to use the manned space flight capabilities developed in the Apollo program will be very limited." Reductions to a budget already "sharply reduced will have many very serious effects on the U.S. position in aeronautics and space. They are only the most recent in a series of cutbacks and, in effect, constitute something like final ratification of a decision . . . that the United States will not, at this time, take the steps necessary to continue the advances of the recent years."

Outlining NASA's program, Webb said two flyby missions to Mars in 1969 were largely paid for. "Even at our reduced levels, I believe we can follow the 1969 missions with two orbiter missions in 1971, but will probably have to postpone for another year the start of work on the two Titan-launched orbiter and lander missions which we had hoped to fly in 1973." The 1969 missions "were initiated three or more years ago. We are approaching the end of our approved flight programs. The number of new projects started each year has sharply dwindled since 1966 and we will soon see years go by when we will have very few flights. We may see a gap of 2 years in our manned flight program after the landing on the moon, and a second gap, equally long, after the Saturn I Workshop.

"Perhaps the most fundamental decisions ahead lie in the field of large launch vehicles. Can we gap the production of Saturn V or will we have to terminate it?" Question required reexamination of uses of Titan III and of possible development of new, less costly launch vehicles.

"Especially important" in this period was continuation of broad program of advanced research for future national needs, including broad university program. (Text)

Sen. Gordon Allott (R-Colo.) told AAS Symposium: ". . . in the thirty minutes I spend talking about the space program this morning four Tiros satellites . . . will have monitored 10 cloud covers above 40 million square miles of the earth's surface. . . . During this same period, the Goddard Space Flight Center . . . will have received 340 minutes of data from 37 active satellites; and the NASA communications network around the globe will have sent and received over 750 messages dealing with information obtained from these satellites. . . .

"Already the beneficial changes wrought by man's incredible inventive genius have made their mark. When NASA launched Early Bird for Comsat three years ago, for example, it boosted the capacity of the transatlantic telephone system by 50%. The AIAA has already estimated that this new industry will be grossing $200 million by 1975. . . .

"Obviously, this is not all. Too often, it seems, those interested in giving an accounting of the $30 billion invested in the space program stop once they have demonstrated technological spin-off or . . . unmanned communication and weather satellites. The crowning achievement of the space program has really been what man himself, in space, has accomplished during this past decade. For man has ventured into the hostile, mysterious world of space." (*CR*, 7/18/68, S8901-3)

Martin Marietta Corp. planetary scientist Allan R. Barger, who was doing theoretical work on balloon-borne Venus probe, told AAS that U.S.S.R. data released after Oct. 18, 1967, Venus probe was incorrect. Soviet report had set planet's surface temperature at about 520° F and surface pressure at about 18 times that on earth. Barger said his conclusions, based on analysis of Soviet report and on data gathered by NASA's *Mariner V* space probe as it flew by Venus' upper atmosphere, set planet's surface temperature at about 890° F and pressure at 100 or more times that on earth. (*Denver Post*, 7/15/68)

- USAF's Arnold Engineering Development Center was conducting research with 5-million-w arc heater to determine temperature and pressure limitations of ablative materials used to prevent military reentry vehicles from burning up on encountering earth's atmosphere. Military reentry vehicles had to withstand conditions similar to high-speed reentry of interplanetary vehicles on return to earth, far more severe than those to be met by lunar astronauts. Data had been produced for civilian and military agencies. (AFSC Release 117.68)

- House Committee on Government Operations' Special Studies Subcommittee, chaired by Rep. Porter Hardy, Jr. (D-Va.), ordered NASA to make every effort to cut escalating costs of its June 16, 1967, contract with Boeing Co. for technical integration and evaluation in assembly of Apollo spacecraft's three modules with Saturn V launch vehicle. Contract, negotiated by NASA in drive to improve safety after Jan. 27, 1967, Apollo fire, had been listed tentatively as costing $20 million; NASA officials now placed cost of carrying work through 1968 at $73.4 million. (Transcript; UPI, *W Post*, 7/16/68, A6)

- Food, land, and raw material shortages might compel man to establish mining operations on other planets and to grow food in space stations, according to Dr. K. A. Ehricke, North American Rockwell Corp. scientist. He said farms growing food in chemicals could be established in earth-orbiting stations fertilized by chemicals produced on Mars and other planets. (AP, *NYT*, 7/15/68, 6)

- Boyd C. Myers II, NASA Deputy Associate Administrator for Operations, Office of Advanced Research and Technology, became NASA Deputy Assistant Administrator for Administration. (NASA Release 68-125; AP, *NYT*, 7/16/68, 7)

July 15-16: Commercial air service between U.S. and U.S.S.R. was inaugurated with Moscow departure July 15 of Ilyushin-62 aircraft belonging to Soviet flag carrier Aeroflot. Aircraft, carrying 97 Soviet officials and commercial passengers, landed at Kennedy International Airport in New York July 16, after 13-hr 17-min flight via Montreal (including 1 hr 35 min circling New York area because of air traffic). U.S. flag carrier, Pan American World Airways, flew two Boeing 707 aircraft from New York to Moscow's Sheremetyevo Airport via Copenhagen July 16. First carried U.S. officials and press; second carried revenue passengers. (*W Star*, 7/15/68, A11; 7/16/68, A7; Witkin, *NYT*, 7/17/68, 28)

July 16: U.S.S.R. successfully launched *Cosmos CCXXXII* into orbit with

355-km (220.6-mi) apogee, 200-km (124.3-mi) perigee, 89.4-min period, and 65.3° inclination. Spacecraft reentered July 24. (UPI, *NYT*, 7/17/68, 30; GSFC *SSR*, 7/31/68)
- Maj. William J. Knight (USAF) piloted X–15 No. 1 to 218,500-ft altitude and 3,409 mph (mach 4.74) in flight from Edwards AFB. Objective of flight, exposure and satisfactory retraction of WTR experiment, was not accomplished because abnormally low hydraulic pressure and severe vibrations prevented aircraft's reaching required altitude. (X–15 Proj Off)
- NASA Associate Administrator for Advanced Research and Technology James M. Beggs dedicated new $3.5-million Flight Control Research Facility at LaRC. Facility, connected to LaRC's data analysis and computation center, would be used for guidance and control research in support of manned flight.

  During ceremony, Center's Digital Computer Complex Group received LaRC Group Achievement Award for "outstanding performance and dedicated efforts in combining unique concepts in computer organization and operating systems" contributing to "one of the most outstanding research computer installations in the United States." (*Langley Researcher*, 7/26/68, 1, 4)
- MSC officials announced resignation of Astronaut John S. Bull (L/Cdr., USN), third astronaut to leave space program because of medical problem. Dr. Charles A. Berry, MSC Chief of Medical Programs, told news conference Astronaut Bull had rare respiratory disease for which there was no known cure and no medical name. It was characterized by chronic sinus difficulties, lung obstruction, and sensitivity to aspirin. (UPI, *W Post*, 7/17/68, A11)
- Global warning system operational since January was providing airline pilots with as much as two months, notice of reentry of spacecraft debris, which had been averaging one reentry a day. Chances of damage by fragment to aircraft, while small, would increase with operation of SSTs at 70,000- and 80,000-ft altitudes. System, outgrowth of Volunteer Flight Officer Network formed in 1963, included more than 38,000 flight crews attached to 117 airlines, which received reentry data from NORAD computers via United Air Lines communications facilities at Denver, Colo. (Sullivan, *NYT*, 7/17/68, 27)
- President Johnson informed Geneva disarmament conference that agreement was expected "shortly" on time and place of U.S.–U.S.S.R. talks to limit nuclear missile production. In message read to opening of new session of conference, President said if progress could be made on limiting strategic delivery systems, U.S. "would be prepared to consider reduction of existing systems." (Text; *W Post*, 7/17/68, A15; *NYT*, 7/17/68, 1)

July 17: Investigation of Nov. 15, 1967, X–15 accident by NASA board indicated that pilot, Maj. Michael J. Adams (USAF) who died in crash, had suffered disorientation and operated controls improperly. Mistaking roll indicator for heading indicator, he had increased heading error, causing aircraft to spin uncontrollably at mach 5 and 230,000-ft altitude and then to go into severe pitch oscillation and disintegrate at altitude above 60,000 ft. Board requested that Government report on MH–96 control system experience and recommended use of telemetry for directional readings by NASA X–15 ground control center, careful checkout of experiments and equipment for next X-15 flight, inclusion in pilot's physical examination of special tests for tendency toward vertigo, and development of additional methods to maintain proper heading under ballistic flight conditions. (FRC Release 20-68; NASA Release 68-126)

- James C. Elms, Director of ERC, discussed "the NASA Biomedical Program in Perspective" before Third Annual Meeting of Assn. for the Advancement of Medical Instrumentation in Houston. "Despite the rapid advance of biomedical techniques since World War II, the main thrust of the activities was directed toward studying sick individuals in a normal environment. The manned space program has provided the opportunity for intensive controlled study of a select group of normal and healthy individuals in an abnormal and stressful environment. By so doing, we have achieved a better definition of the range of normality of the healthy organism which, in turn, is useful in the study and understanding of disease."

  Interaction of space and medicine had led to many medical applications of aerospace hardware. Application of electron probe microanalyzer—used for chemical analyses of microelectronic circuits—to study of red blood cells had led to unexpected clues in study of blood cancer. System to monitor heart rate, respiration, and galvanic skin response was being considered for use in measuring efficiency of dental anesthetics on children and in training teachers for retarded children. Accomplishments of bioelectronics research in interdisciplinary electronics environment included remote measuring technique for eye-pointing direction, meaningful measurement of aerosol concentration and size distribution, and automatic tracking system to identify thresholds of mental alertness. (Text)

- Univ. of California physicist Dr. Edward Teller, at hearing on nuclear nonproliferation treaty before Senate Foreign Relations Committee, urged Congress to preserve option of giving nonnuclear allies control over "purely defensive" nuclear weapons systems. He was referring, he said, to system that could be exploded only over a nation's territory, one involving "time-lock" of monthly inspection by donor nation, and one which would be proof against tampering or analysis designed to develop it into offensive system. (Maffre, *W Post*, 7/18/68, A4; Sherman, *W Star*, 7/18/68, A12)

- U.K., West Germany, Italy, and the Netherlands signed agreement to cooperate in $4.8-million project to develop and produce advanced combat aircraft for their air forces. Aircraft, scheduled to enter service in 1975, would replace U.S. Lockheed F-104 Starfighter currently being used. Orders for new aircraft were expected to reach 1,000. (Reuters, *B Sun*, 7/18/68, 2)

July 18: Cosmos *CCXXXIII* was launched from Plesetsk Cosmodrome by U.S.S.R. into orbit with 1,505-km (935.2-mi) apogee, 199-km (123.6-mi) perigee, 101.9-min period, and 81.9° inclination. Satellite reentered Feb. 7, 1969. (*SBD*, 7/22/68, 32; GSFC *SSR*, 7/31/68; 2/15/69)

- Senate passed unanimously H.R. 17023, FY 1969 Independent Offices and HUD appropriations bill, including $4.008 billion for NASA. Total for NASA agreed with House-passed total, but Senate adopted committee amendments increasing funds for construction of facilities by $12.95 million and decreasing R&D funds by same amount. As passed by Sen-

July 18

ate, bill provided $3.37 billion for R&D, $34.75 million for construction of facilities, and $603.17 million for administrative operations. Senate requested conference with House on amendments. (NASA *LAR* VII/76; *CR*, S8910–38; *SBD*, 7/19/68, 71)
- House Appropriations Committee cut $550.5 million from DOD FY 1969 appropriations, including $85 million from USAF Manned Orbiting Laboratory (MOL) program. (*CR*, 7/18/68; *SBD*, 7/19/68, 71)
- Defense Communications Agency had declared operational eight satellites added to Defense Satellite Communications System (DSCS) by successful June 13 launch from ETR, final launch of Initial Defense Communications Satellite (IDSCP) Project. Total of 24 satellites were in normal use, orbiting eastward in 21,000-mi-altitude synchronous orbit. They would remain in use until 1971. (DOD Release 668–68)
- NASA had completed tests to find solution to "longitudinal oscillations" of Saturn V booster which had occurred during April 4 *Apollo 6* mission. Tests revealed that natural frequency of vehicle structure and propulsion system frequently had coincided, multiplying amplitude of oscillations. Problem would be corrected by using accumulators, small gas reservoirs, in 1st-stage liquid-oxygen prevalves to change propulsion system frequency. Minor modifications necessary to allow helium injection into prevalves were being made on 1st stages of third and sixth Saturn Vs. (NASA Release 68–128; MSFC Release 68–158)
- Ryan Vertifan, jet V/STOL aircraft designated XV–5B by NASA, was undergoing flight tests at Ryan Aeronautical Co. in San Diego before delivery to ARC for use in aeronautical research. Aircraft's counter-rotating fans submerged in wings and driven by jet exhaust provided lift for vertical takeoff, hovering, and vertical landing. XV–5B was improved version of Ryan Aeronautical Co. research aircraft built for USA; modifications and renovations, after damage from October 1966 emergency landing at Edwards AFB, were made under $1-million NASA contract. (ARC *Astrogram*, 7/18/68, 1)
- With U.S. and U.S.S.R. ready to discuss possible mutual restriction on production of strategic missiles, research and testing of advanced spectrometer designed to police agreement had been delayed because of congressional cuts in DOD funds for Arms Control and Disarmament Agency program. Device, which analyzed missile characteristics from their exhaust trails at launch sites, had been developed at cost of $574,000 after 1964 proposal by U.S. for missile agreement with U.S.S.R. Device could be manned by international inspectors positioned one mile from launch site or read by remote control through transmission cable already developed for additional $200,000. Field testing under simulated U.S.S.R. conditions had been postponed one year until summer 1969. (Oberdorfer, *W Post*, 7/18/68, G4)
- *The Security of Japan and Prospects for 1970*, study produced for Japanese Defense Agency by Security Research Council, said Japan had technical and economic resources to produce uranium and plutonium bombs and ICBM-producing capability similar to that of France. Japanese policy to date had banned construction and importation of nuclear weapons. (*W Post*, 7/18/68, A3)
- Dr. Ernest Harry Vestine, expert on geomagnetism who had joined RAND Corp. in 1957 after 20 yr with Carnegie Institution, died in Santa Monica, Calif., of heart attack at age 62. He had been an originator of

1957–58 International Geophysical Year and had served as consultant to DOD, NASA, and Dept. of Commerce. He had been one of leaders of 1933 International Polar Year expedition, which established observatory to measure earth's magnetic field. (*W Post*, 7/19/68, B6)

July 19: NASA test pilots Donald L. Mallick and Fitzhugh L. Fulton, Jr., flew XB-70A to 42,000-ft altitude and mach 1.62 in flight from Edwards AFB. Purpose of flight was to evaluate performance at variety of speeds, check exciter vane function, determine ground effects during low approach, and evaluate pilot proficiency during touch-and-go landing. (XB-70 Proj Off)

- Astronauts James A. McDivitt, David R. Scott, and Russell L. Schweickart successfully completed checkout of Apollo spacecraft cabin flight equipment provisions under simulated mission conditions at North American Rockwell Corp.'s Downey, Calif., facility. (NAR *Skywriter*, 7/26/68, 1)

- USN awarded five $1-million contracts for preliminary design and engineering work on FB-111B replacement to Grumman Aircraft Engineering Corp., LTV Aerospace Corp., General Dynamics Corp., North American Rockwell Corp., and McDonnell Douglas Aircraft Corp. Substitute for F-111B, designated VFX-1, would have vertical sweep wings and same jet engine. (AP, *NYT*, 7/20/68, 27; *W Post*, 7/21/68)

July 20: Senate by vote of 67 to 3 passed H.R. 17903, FY 1969 public works and atomic energy appropriations bill, which included $36 million for NERVA and total of $68 million for space propulsion systems. Senate also passed H.R. 13781, authorizing $15 million for sea-grant colleges and ocean exploration in FY 1969 and $15 million in FY 1970. (NASA *LAR* VII/77; *CR*, S9047, S9069–87)

July 21: USN had awarded $143.5-million contract to Newport News (Va.) Ship Building Co. for two nuclear-powered guided-missile frigates, which would bring to five USN's total atomic-powered escort vessels. (*W Post*, 7/21/68, H1)

July 22: Partial extension of *Explorer XXXVIII*'s antennas, delayed because of unexpected spacecraft oscillations and ground computer failure, was successfully conducted by NASA after series of complex maneuvers which permitted successful gravity-gradient capture and three-axis stabilization. Antenna array's four booms would be held at planned 455-ft length for at least two weeks while data was collected and then, if spacecraft performed satisfactorily, antennas would be extended to full 750-ft length. Damper boom was deployed, experiments were turned on, and all spacecraft support systems were functioning normally. Dipole antenna was deployed July 23 and satellite was declared fully operational. Spacecraft had been launched July 4. (NASA Proj Off; NASA Releases 68–123, 68–132; *W Post*, 7/18/68, D21)

- *Explorer XXXV* (IMP–E), sixth spacecraft in Interplanetary Explorer series, completed one year of operation in lunar orbit. Seven of eight onboard experiments and all spacecraft systems were 100% operational. Eighth experiment had 5% degradation in performance. Since July 19, 1967 launch, satellite had shown that positive ions from solar wind crashed directly into lunar surface and had verified existence of solar wind void directly behind moon, enabling scientists to deduce information on moon's electrical conductivity and internal temperature. (NASA Proj Off)

- Despite sharp budget cuts NASA was not contemplating layoff of Civil Service Commission personnel, columnist Jerry Kluttz reported in *Washington Post*. As Government agencies searched for ways to meet cuts ordered by Congress, NASA appeared to be only major agency with no problem of excess employees. But because of "big money problems," NASA hoped to save dollars by abolishing half of its currently vacant positions, possibly making some "selective" layoffs at MSFC and GSFC, and continuing major cutbacks in contractor personnel. NASA would reprogram its activities and transfer funds to finance CSC positions. (*W Post*, 7/22/68, A20)
- In Project Cold Flare, joint NASA–FAA–USAF program to assess radiological effects of solar activity on future SST passengers and crews, radiation-measuring flights were being flown from Eielson AFB, Alaska, near North Pole, where solar and galactic charged particles were normally concentrated, to gather data on radiological phenomena during solar flares. (*NYT*, 7/22/68, 61)

July 22–23: Arthritic growth on spine of Astronaut Michael Collins (Maj., USAF), scheduled to pilot command module on third manned Apollo mission early in 1969, had led to his being grounded. USAF surgeons successfully removed bone spur from near base of his neck but speculated convalescence might take up to four months. MSC officials declined conjecture on his future flight status. Collins had piloted two-man Gemini mission July 18–21, 1966, during which he had performed two space walks. (AP, *W Star*, 7/23/68, A4; *W Post*, 7/23/68, A5; 7/24/68)

July 23: Senate passed H.R. 18188, Dept. of Transportation appropriations bill, by vote of 82 to 2 after approving addition of $153 million for FAA to hire 3,627 air traffic controllers to relieve congestion at nation's busiest airports. (*CR*, S9226–47; AP, *W Star*, 7/24/68, A21)
- Guidance and control equipment used during *Gemini XI* reentry Sept. 15, 1966, was being flight-tested at NASA Wallops Station to set up system performance requirements for automated landing for V/STOL aircraft. Tests were part of long-range NASA research program to develop all-weather aviation electronics systems for V/STOL aircraft. (ERC Release 68–12; WS Release 68–14; *Marshall Star*, 8/14/68, 2)

July 23–24: NASA launched series of 11 sounding rockets from NASA Wallops Station between 8:19 pm July 23 and 5:55 pm July 24 to gather upper-atmosphere data for weather research. Carried on six Nike-Apache and five Nike-Cajun sounding rockets, experiments included: two Univ. of Colorado experiments to obtain vertical profile of nitric acid density, with two spheres to measure daily density change; four payloads instrumented by GCA Corp. and Univ. of Illinois to measure electron and ion density and solar radiation in ionosphere; three joint GSFC-Univ. of Michigan grenade launches to obtain temperature, pressure, and wind data; and two Univ. of Michigan payloads to measure ambient air density by tracking two small spheres as they fell from different altitudes. Experiments were expected to yield new information about interrelationship of ionosphere and neutral atmosphere between 30- and 70-mi altitudes. (NASA Release 68–134; WS Release 68–15)

July 24: Iris I (*Esro II–B*) satellite launched May 16 by NASA for ESRO completed its 1,000th revolution of earth. Two of seven scientific experiments malfunctioned in first days after launch but were turned off and turned on again weeks later. Behavior was now satisfactory. Three other experiments had been disconnected in July after anomalies were observed. By July 15, 2,200 commands had been executed by spacecraft. (*Spaceflight*, 12/68, 417)

- ComSatCorp reported $3.3 million net income (33 cents per share) for first half of 1968, of which $1.5 million (15 cents per share) was in second quarter. Earnings for first half of 1967 were $2 million (21 cents per share) and for second quarter of 1967, $859,000 (9 cents per share). (ComSatCorp Release 68–35)
- IBM physicists Dr. Peter P. Sorokin and J. P. Lankard had designed and built pulsed laser which produced 100,000-w bursts of light lasting 2.5 millionths of a second and varying in color according to commercial liquid dye used. It could be built in home workshop with materials worth $25 to $50. By changing dye used, thereby selecting new molecule, light of new frequency was produced, enabling scientists to investigate energetic properties of molecules and atoms. (Stevens, *NYT*, 7/26/68, 55)
- Joint Committee on Atomic Energy said no valid reason had been offered for DOD's May 28 halt in spending for quiet electric drive submarine and urged that all restraints on its design and construction be removed. Recommendation accompanied release of June 21 testimony in which V/A Hyman G. Rickover strongly supported submarine and criticized DOD for delaying its development [see July 11]. (Transcript; AP, *NYT*, 7/28/68, 54)

July 25: NASA was entering competitive negotiations with Informatics Inc. and Leasco Systems and Research Corp. for one-year, $4-million contract with two one-year options for operation of its Scientific and Technical Information Facility at College Park, Md. (NASA Release 68–133)

- President Johnson appointed 15-member Federal Air Quality Advisory Board—chaired by Secretary of Health, Education, and Welfare Wilbur J. Cohen—to advise Government of effective state and community air quality control programs. (*PD*, 7/29/68, 1153; *Science*, 8/2/68, 447)

July 26: NASA Aerobee 150 MI sounding rocket launched from WSMR, carried Columbia Radiation Laboratory experiment to 88.3-mi (142-km) altitude to examine x-ray polarization of Sco XR–1 in 10- to 25-kev region with x-ray polarimeter. Rocket and instruments performed satisfactorily. Experiment worked as expected, but some counter or electronic failure, or both, occurred during early part of flight. (NASA Rpt SRL)

- USAF-sponsored unidentified flying object (UFO) investigation by Univ. of Colorado concluded April 30 had become "mired in controversy," said *Science*. Its Director, Dr. Edward U. Condon, had refused to discuss situation and critics were saying project was "biased and less than diligent investigation." Chief targets for criticism were Dr. Condon and project coordinator Robert J. Low, while "most substantial" critics were James E. MacDonald, Univ. of Arizona senior physicist, and Northwestern Univ. astronomer Dr. J. Allen Hynek, USAF's chief UFO consultant, who feared Dr. Condon would recommend against further serious UFO study. Dr. Condon's supporters had noted criticism was based on newspaper quotes, on his delight in humorous UFO anecdotes, statements from project members who had been fired, and memo writ-

ten by a subordinate before project began. They did not find evidence convincing, *Science* said. (Boffey, *Science*, 7/26/68, 339–42)
- Air Force Cambridge Research Laboratories scientists John W. Salisbury and Graham R. Hunt reported in *Science* they had found hypothesis of particle-size control of albedo incompatible with hypothesis of abundant limonite on Mars. Their observations indicated that proposal that polarimetric, spectrometric, color, and albedo measurements of light and dark areas on Mars proved limonite was major soil constituent was irreconcilable with proposal that variations in size of particle could be responsible for albedo difference between light and dark areas. They showed relative albedo was reversed from blue to red for limonite samples with different-sized particles. Observations of Mars revealed no blue-red albedo reversal between areas. Although evidence was insufficient for choice between hypotheses, they believed Mars soil was most likely, for geological reasons, to be composed of silicates stained or coated with ferric oxides. (*Science*, 7/26/68, 365–6)

July 27: Aerobee 150 MI sounding rocket launched by NASA from WSMR carried MIT payload to 84.1-mi (135.3-km) altitude to obtain data on celestial locations and energy spectra of discrete x-ray sources in three regions and to search for weak, undiscovered x-ray sources using proportional counters. Rocket and instruments performed satisfactorily. (NASA Rpt SRL)
- Sen. Clinton P. Anderson (D-N. Mex.), Chairman of Senate Committee on Aeronautical and Space Sciences, inserted into *Congressional Record* his report "Legislative History of Space Nuclear Propulsion for Fiscal Year 1969," which confirmed "the continued vigorous support of the Congress for this space research and development activity" and that "appropriate agencies should proceed with the development of the NERVA–1 nuclear rocket engine." Final Congressional action on AEC and NASA FY 1969 authorizations strongly supported nuclear propulsion development, he said.

    NASA authorization had been $55 million for Nuclear Rockets Program. Joint Committee on Atomic Energy had recommended $69 million for AEC Nuclear Space Propulsion Systems and strongly recommended program proceed. AEC had requested $72 million for Project Rover, including $49 million for development of NERVA I rocket engine. Final action of both Senate and House Appropriations Committees on Independent Offices bill and AEC appropriations bill had restored funds cut by House earlier. It was clear, said Sen. Anderson, "that the view of the Congress is that the Nuclear Rockets Program for fiscal year 1969 is one of the most important and highest priority programs in NASA and should move forward as planned. . . ." Amount finally appropriated for NASA R&D for FY 1969 would not be less than $3.3703 billion, including $55 million for the Nuclear Rockets Program. "The conference committee on the AEC appropriations . . . agreed to $53 million on the AEC's program in Reactor Development—Space Propulsion Systems, a figure more than halfway between the House lower figure [$31 million] and the Senate higher figure [$68 million]." (*CR*, 7/27/68, S9582–4; NASA *LAR* VII/81)
- Army Electronics Command had produced new battery-operated nuclear clock which would gain or lose only one second every 3,000 yr. It would be used by USA, NASA, USAF, and USN primarily in aviation-elec-

tronic systems and had potential use in U.S., U.K., Canadian, and Australian digital communications systems. (AP, *NYT*, 7/28/68, 30)

July 28: JPL astronomers Dr. Richard M. Goldstein and Dr. Shalhav Zohar had located and mapped three rugged sectors on northwest face of Venus using Goldstone Tracking Station's 210-ft antenna. Beta, most clearly defined, appeared roughly circular with 150-mi dia and 17,000-sq-mi area. Two other irregular features almost as large had appeared on radar map which covered triangular area of estimated 160,000 sq mi, equal to area of northeastern U.S. Dr. Goldstein inclined to theory that prominences were mountains, but he had not yet been able to measure their heights. "We know these features are permanent," he said "because they have appeared on all our tests" for past six years. (JPL Release BB-483)

- German scientist Dr. Otto Hahn, who had won 1944 Nobel Prize for chemistry for his 1938 discovery of nuclear fission, died at Goettingen, Germany, of heart failure at age 89. He had been consistent opponent of use of atomic weapons, urging scientists to concentrate on peaceful uses of nuclear energy. (UPI, *W Post*, 7/29/68, B4)

July 28–August 3: Australian astronomers using Mills Cross antenna array at Univ. of Sydney had detected first two pulsars to be found in southern sky. They brought to nine total pulsars discovered to date. Pulse rates of once every 0.56 and 1.96 sec were similar to all others, which ranged between 0.25 and 2 sec. Cornell Univ. astronomer, Dr. Yervent Terzian, had observed that two-second pulse rate made it appear unlikely that pulsating neutron stars could account for the signals. Calculations indicated that, if they were quivering, or "ringing" like bell, it should be at much shorter intervals than those observed. Pulse rates were more compatible, he said, with proposal that they emanated from white dwarfs, "cinders" of stars that had consumed their nuclear fuel but were larger and less dense than neutron stars. (Sullivan, *NYT*, 8/8/68, 30)

July 29: Aviation Week reported anticipated $100-million cut in FY 1969 funding could halt contractors' work on USAF Manned Orbiting Laboratory (MOL) in September, with slippage of flight schedule into 1972. Boeing Co. was working on new launch vehicle at NASA's Michoud Assembly Facility under Project Scrimp. Booster would be vehicle 75 ft in dia and 80 ft high and use TRW engine based on Boeing lunar module ascent engine experience, with 2-million-lb operational thrust—although USAF had not decided to abandon Titan III–M as launch vehicle. (*Av Wk*, 7/29/68)

- House passed H.R. 18785, military construction appropriations bill for FY 1969, which included $263.3 million for Sentinel ABM facilities. It had rejected by vote of 106 to 37 motion to delete missile funds. (*CR*, H7710-35; Crowther, *B Sun*, 7/20/68, A5; AP, *NYT*, 7/30/68, 62)

- *Aviation Week* reported DOD concern over advanced U.S.S.R. tactical fighters [see June 10]—particularly new Soviet mach 2.8 MiG-23 (Foxbat)—might result in approval of long-delayed USAF and USN projects including USN VFX-1 and follow-on VFX-2 interceptors for fleet defense, USAF FX air superiority fighter, and USAF/Lockheed F-12 interceptor. USAF originally had wanted next-generation continental defense interceptor force composed entirely of F-12s but would accept mixed

July 29

force of F-12s and modified Convair F-106 interceptors designated F-106X. (*Av Wk*, 7/29/68)

- Panel of scientists testified before House Committee on Science and Astronautics symposium on unidentified flying objects (UFOs) in unanimous support of further, more detailed UFO research. Northwestern Univ. astronomer Dr. J. Allen Hynek said U.S. should seek U.N. cooperation in setting up "international clearing house" for scientifically respectable UFO reports because there was almost a total lack of quantitative data about the phenomenon. Cornell Univ. exobiologist Dr. Carl Sagan told Committee it was not inconceivable that there were other planets with civilizations and technologies more advanced than earth's, but he cautioned against a widespread UFO investigation program which would require "some harder evidence than is now present," thus being expensive.

  Computer Science Corp. mathematician and celestial mechanics specialist, Dr. Robert L. Baker, Jr., revealed space-based sensor system operated from Colorado Springs Air Defense Command Hq. had received several anomalous UFO alarms that had not been explained. Dr. James A. Harder, Univ. of California at Berkeley engineer, suggested that power which permitted UFOs to undertake their reportedly incredible maneuverings might depend on a theoretically possible "second gravitational field" interacting with electrical field in a manner corresponding to reaction between conventional electrical motors and generators. (Transcript; Lyons, *NYT*, 7/30/68, 10; Lannan, *W Star*, 7/30/68, A3)

- NASA had extended, for $29,130,524, one-year contract with Trans World Airlines, Inc., for installation support services at KSC, bringing total of cost-plus-award-fee contract to $101,017,194. (KSC Release KSC-364-68)

July 30: U.S.S.R. launched *Cosmos CCXXXIV* from Baikonur Cosmodrome into orbit with 295-km (183.3-mi) apogee, 208-km (129.2-mi) perigee, 89.5-min period, and 51.8° inclination. Equipment functioned normally. Satellite reportedly softlanded in its home territory Aug. 5. (GSFC *SSR*, 7/31/68; 8/15/68; *SBD*, 7/31/68, 129; AP, *NYT*, 7/31/68, 3; 8/7/68)

- NASA announced addition of $35,048,000 to contract with General Electric Co. for continuation of design, fabrication, and testing for four remaining Biosatellite spacecraft, associated experiment hardware, and aerospace ground equipment—bringing total value of contract to $136,662,157. (NASA Release 68-136)

- Cone-shaped instrumented payload of French Veronique rocket released at 114-mi altitude fell into Atlantic 164 mi off French Guiana and was swept away by currents, French National Space Center announced. (Reuters, *W Post*, 7/31/68, A14)

- Bernhardt L. Dorman had resigned as NASA Assistant Administrator for Industry Affairs to return to Aerojet-General Corp., NASA announced. He would be succeeded Aug. 5 by Philip N. Whittaker, Vice President of IBM Federal Systems Div. (NASA Release 68-137)

- U.S. patent No. 3,395,565 was issued to Norton Research Corp. for calibrator developed for NASA to enable crew of space vehicle, or ground control personnel in unmanned flight, to check accuracy of pressure gauges. Prototype instrument, containing quantity of helium, had been delivered to NASA. (Patent Off PIO; *NYT*, 8/10/68, 33)

- USAF Space and Missile Systems Organization issued Lockheed Aircraft Corp. $2-million initial increment to $4,131,785 cost-plus-fixed-fee contract for reentry vehicle technology and observables program. (DOD Release 706-68)

July 31: NASA issued Apollo Status Summary: Apollo 7 prime crew, Astronauts Walter M. Schirra, Jr., Donn F. Eisele, and R. Walter Cunningham, successfully completed nine-hour test in spacecraft in KSC altitude chamber July 26 at 226,000-ft simulated altitude, with cabin pressurized first with 60% oxygen and 40% nitrogen, then with 100% oxygen at 5 psi, normal orbital atmosphere. Astronauts Thomas P. Stafford, John W. Young, and Eugene A. Cernan had completed successful manned altitude test in KSC chamber July 29 with cabin pressurized with 60/40 mixture at start and oxygen to replenish atmosphere during test. Saturn IB launch vehicle stages were mated at Complex 34 with sequence malfunction tests scheduled for Aug. 1. Propellant utilization system modifications had been completed on 2nd stage.

In Apollo/Saturn 503 program, Lunar Module 3 ascent and descent stages were being modified to correct radar lock-on problem in rendezvous radar subsystem. Saturn V launch vehicle 1st and 2nd stages were mated, with pogo suppressor modification kits on 1st stage being verified. Third-stage propulsion subsystem checks were in progress.

In South Atlantic Anomaly Probe, computer analysis had confirmed radiation levels presented no hazard to low-altitude manned Apollo orbital flights. (Text)

- Nike-Tomahawk sounding rocket launched from NASA Wallops Station carried Univ. of Maryland and Johns Hopkins Univ. experiments to 186-mi (300-km) altitude to investigate role of electrons in producing day airglow. Rocket and instrumentation—which included five-wavelength filter-wheel photometer with special sunshade, 3- to 800-ev electron spectrometer, and electron retarding-potential analyzer—performed satisfactorily. Data correlating ionospheric electron density distribution with day airglow emissions in ionosphere were obtained. (NASA Rpt SRL)

- NASA announced award of $31,270,300 contract to General Electric Co. for continuance of general support services at Mississippi Test Facility through September 1969, bringing total value of basic contract to $190,810,713. (MSFC Release 68-170)

- Dr. Henry J. Reid, former Director and Senior Staff Associate at NASA Langley Research Center, died in Gloucester, Va., after heart attack at age 72. He had become LaRC Director in 1926 and retired in June 1961, after 34 yr continuous service. (AP, *W Post*, 8/1/68, B10)

During July: Ralph Kinney Bennett in *Data* scored U.S. complacency in year which might "see some of the greatest Soviet space spectaculars of the decade." In U.S., he said, early glamour of space race had faded and NASA projects were neither as ambitious nor as well-funded as they once were. In contrast, he noted, U.S.S.R. showed signs of "a new spurt of activity, new technical accomplishments above our planet, an accelerated assault on the moon and dark rumblings of advanced military uses of the threshold of space. The time for a Soviet resurgence could hardly be better.

". . . Soviets are badly in need of a great techno-propaganda feat to reassert their influence on world public opinion. . . ." Soviet hold on

European satellite nations had shown evidence of advanced erosion. "Their position in relation to the rest of the communist world is no longer clearly defined." U.S.S.R. was spending estimated $9 billion a year on space and half that sum went toward military applications of space technology. "There is an American attitude . . . of waiting until you get burnt before you shed complacency. When Sputnik I burnt us, we came back with a vengeance. Now complacency has set in again. Perhaps we will feel the fire from the Soviets in space before this year is out and react accordingly. But in space technology . . . it's a tough way to play the ball game." (*Data*, 7/68)

- Paul G. Thomas discussed "Earth-Resource Survey from Space," in *Space/Aeronautics*. Program called for: first mapping of earth to ultimate scale level of 1:25,000; first mapping of global sea state and surface temperature to help shipping, fishing, and weather forecasters; monitoring hydrological cycle to assist watershed planning and prediction of botanical responses; measuring plant vigor and impact on supply and price of earth's food; measuring geographical surface phenomena to permit geologists to home in on mineral deposits and further arguments in morphology of planet earth. Signatures and their rapid analysis were major stumbling blocks in program. "Although there are about 50 geological signatures on record and about 100 for agriculture and forestry, each represents a one-of-a-kind situation . . . and, oddly enough, there is often more variance within a species than there is between species."

    As remote sensing concepts emerged, it was "apparent that successful earth-resource surveillance will not be simply a matter of lashing together existing technology into pseudo-systems and hoping for benefits. The human effort and development funding needed to fully implement a system for earth observation from space will make the lunar landings look like a walk to the corner candy store. Thus the shift in emphasis to cost/benefit studies, new to the space age, is certainly not without foundation. If, on the other hand, an optimized system cannot—or should not—be realized, we should be willing to stumble a bit to find out." (*S/A*, 7/68, 46–54)

August 1968

August 1: Aerobee 150 sounding rocket launched by NASA from Churchill Research Range carried ARC luster sampling instrument with collection modules, camera, and electronics package to 71.3-mi (114.7-km) altitude to sample material from noctilucent clouds as luster instrument passed through them. Sampling instrument was successfully flown and recovered. Instrument malfunction caused loss of four modules, but remaining eight were recovered in excellent condition. Experiment was considered highly successful. (NASA Rpt SRL)

- Senate passed by voice vote H.R. 18785, FY 1969 military construction appropriation bill which contained $36 million for Sentinel planning, after defeating 46 to 27 amendment to cut $227.3 million in Sentinel deployment funds from its $1.7 billion FY 1969 appropriation. (CR, 8/1/68, S9965; Wilson, *W Post*, 8/2/68, A10; *SBD*, 8/2/68, 137)
- Groundbreaking ceremony marked start of construction on Hall of the National Academy of Sciences, symposia facility to be shared with National Academy of Engineering and National Research Council. An addition to present NAS structure, new building would contain 700-seat auditorium, lounge area, 40 research and administration offices, and research archive file. Design and construction funds had come from contributions to Hugh L. Dryden Memorial Fund and from other private and Government sources, including large gift on behalf of NAE. Total cost of structure was estimated at $3.8 million. Dr. Dryden, who died in December 1965, had served 10 yr as Home Secretary with NAS, 12 yr as a NAS Council member, and 21 yr as member. He had joined NACA, NASA predecessor, in 1947 and had been named Director of NACA in 1949. (NAS Release)
- President Johnson announced appointed of David D. Thomas, Deputy Administrator of FAA, as Acting Administrator, (PD, 8/5/68, 1190)
- British Railways inaugurated commercial service across English Channel with 165-ton SRN4 Hovercraft, cutting time of 30-mi crossing between Dover and Boulogne from $1\frac{1}{2}$ hr by fastest ferry to 35 min. (Lee, *NYT*, 8/2/68, 54; Wentworth, *W Post*, 8/5/68, A3)

August 2: NASA launched two boosted Arcas I sounding rockets from Resolute Bay, Canada, carrying GSFC payloads to measure D-region electron density profiles in Polar Cap ionosphere and measure simultaneously solar proton spectra. Rockets and instruments performed satisfactorily. Minimum altitude required (52.8 mi) (85 km) was obtained, allowing excellent data recovery. Peak altitudes were not determined because of lack of radar support. (NASA Rpts SRL)

- NASA was stopping work on Saturn IB vehicles 215 and 216 "to the extent possible without involving uneconomical terminations." Contractors had been instructed to terminate work on long-lead-time items being procured for Saturn V vehicles 516 and 517; halt would cause time gap should decision be reached later to resume production. Boosters

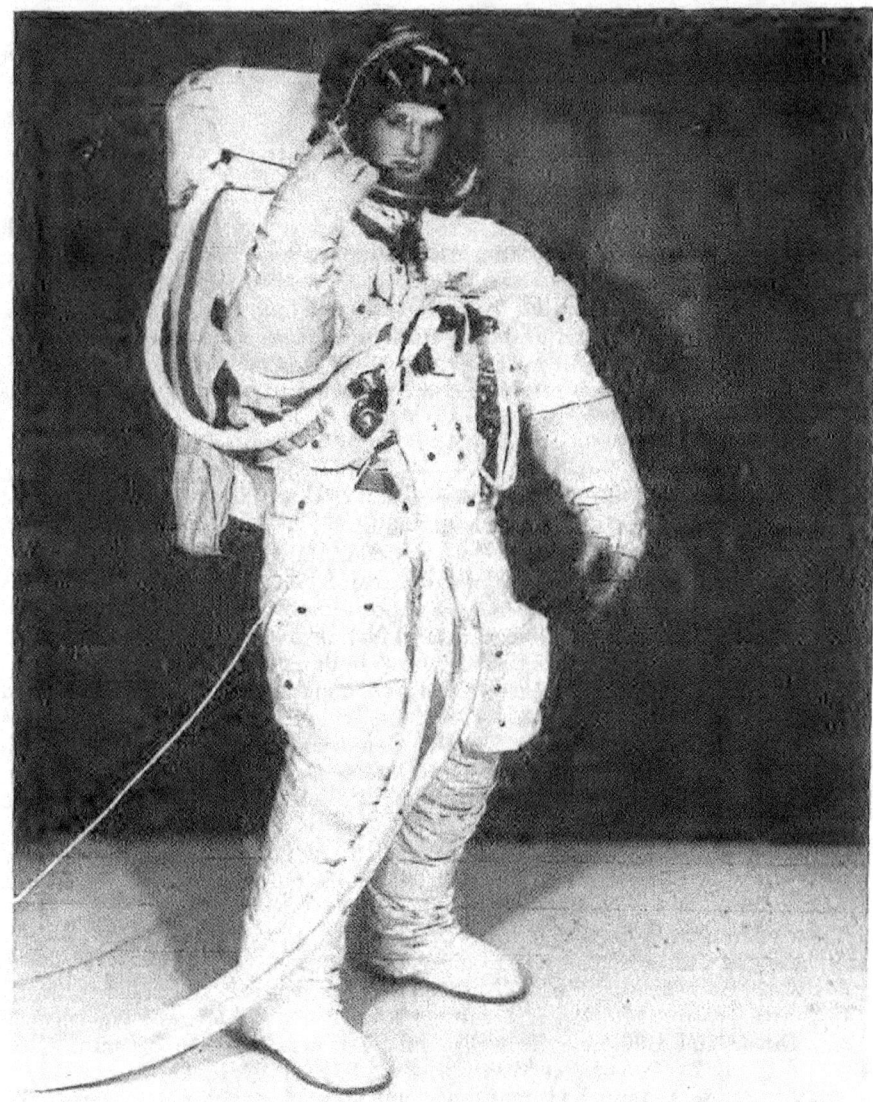

August 2: NASA introduced new fireproof Apollo spacesuits at press showing in Dover, Del. In earlier photo, test subject Ronald C. Woods wears integrated thermal micrometeorite garment (TMG) to be worn by astronauts during extravehicular activities. ILC Industries suit has back-pack life-support system developed by United Aircraft Corp.

had been requested in NASA FY 1969 budget for follow-on utilization of Apollo capacity beyond manned lunar landing. Steps were taken after Congressional cuts in budget, in anticipation of final decisions on future production. (NASA Release 68-139; *W Post*, 8/3/68, A2)

* At press showing in Dover, Del., NASA introduced fireproof Apollo space-

suits developed after Jan. 27, 1967, Apollo fire to protect astronauts against extreme heat, cold, and flash fires. Beta cloth of glass fiber was nonflammable, with 1,200° F melting point. Former spacesuit fabric would melt at 700° F and would burn if ignited. New 54-lb suits were .22 lb heavier but more flexible, consisting of inner layer of cotton underwear and cover layer of Beta cloth, nylon, and aluminized Kapton to protect against micrometeoroids and temperatures between 310° F and −250° F. Intermediate garment contained ventilation unit which, for astronauts participating in extravehicular activity, would be cooled with continuous flow of water while wearer was outside capsule and with pure oxygen inside capsule. Suits and cloth helmet to be worn after removal of plastic headwear were manufactured by ICL Industries, Inc., and cost $175,000–$200,000 each. NASA had ordered 100. (Lyons, *NYT*, 8/3/68, 3; Lannan, *W Star*, 8/3/68, A2)

- Ryan XV–5B Vertifan aircraft made first vertical and hovering flights at ARC, piloted by Ryan chief test pilot William Anderson in airworthiness tests. These were first vertical flights since extensive modifications converted aircraft from original XV–5A to new version for NASA research [see July 18]. Aircraft made three vertical lift-offs, reaching 20-ft altitude and remaining stationary in air up to one minute. (ARC *Astrogram*, 8/15/68, 3)

- Data collected for 17 mo by five Lunar Orbiters indicated rate of penetration of 0.025-mm beryllium copper surface of pressurized cells by meteoroids in near-lunar environment was approximately half the rate in near-earth environment as measured by similar detectors aboard *Explorer XVI* and *Explorer XXIII*. Each experiment used 20 pressurized-cell detectors with total exposed area of 0.186 sq m, carried aboard spacecraft in both equatorial and polar orbits at altitudes between 18.6 and 3,852.5 mi (30 and 6,200 km). Estimates of hazard near moon had ranged from less to much greater than hazard near earth. Major uncertainty had been contribution by secondary meteoroids created by impacts of primary meteoroids on moon, according to report by NASA LaRC scientists C. A. Gurtler and Gary W. Grew. Data showed no evidence of increase in hazard from such "backsplash." (*Science*, 8/2/68, 462–4)

- Univ. of Minnesota scientists J. G. Sparrow and E. P. Ney reported in *Science* results of experiment on NASA's *Oso II* Orbiting Solar Observatory designed primarily to measure zodiacal light and continuum airglow. From February to November 1965, *Oso II* had been able to pick up and distinguish between such earth light sources as lightning, gas flares in Middle East oil fields, and nighttime lights of major U.S. cities. Most significant was recognition of 10 other satellites in low earth orbit. "In principle, each satellite could have been identified." (*Science*, 8/2/68, 459–60; *W Post*, 8/2/68, A10; *SBD*, 8/5/68, 145)

- NAS and NAE would collaborate on initial study of scientific and engineering aspects of U.S. participation in International Decade of Ocean Exploration planned for 1970s, *Science* reported. (*Science*, 8/2/68, 447)

August 4: NASA had concluded agreements for cooperative scientific experiments with space and upper atmosphere research authorities of Brazil, Norway, Spain, and Sweden. Brazilian Comissao Nacional de Atividades Espaciais (CNAE) and NASA project would include four sounding

August 4

rocket launches into upper atmosphere to measure micrometeoroid flux and its variations with latitude in early autumn. Royal Norwegian Council for Scientific and Industrial Research and NASA project would observe earth's magnetic and electrical fields and charged particle environment during stages of auroral activity by launch in September and October of three barium-release ion-cloud payloads followed by instrumented payloads.

Spanish Comisión Nacional de Investigación del Espacio (CONIE) and NASA project would require launch of 24 Boosted-Dart meteorological sounding rockets to 30- to 60-km (18.6- to 37.3-mi) altitudes at two-week intervals throughout 1969 to obtain synoptic wind and temperature data on structure and circulation of upper atmosphere and wind oscillations. Swedish Space Research Committee and NASA project would study D and E ionospheric layers with instrumented payloads on four Boosted-Arcas II sounding rockets in autumn 1968.

In all projects, principal experimenters would have first rights to data which subsequently would be made available to world scientific community. Each agency would bear full cost of its agreed responsibilities without exchange of funds. (NASA Release 68–138; *SBD*, 8/5/68, 148)

August 5: U.S.S.R. was expected to make more successful launches than U.S. during 1968 for first time since 1957, with both nations emphasizing military goals, Howard Benedict of Associated Press wrote. In 1957, U.S.S.R. had launched two Sputniks, U.S. had launched none. In 1958, U.S. had launched five to U.S.S.R.'s one; thereafter, U.S. had maintained substantial edge, building to 73–46 margin in 1966. In 1967, U.S. had led by only one launch, 67–66. Through 1967, U.S. had had 400 successful launches, with 532 payloads orbited. U.S.S.R. had had 240 launches with 289 payloads. In 1968 to date, U.S.S.R. had successfully launched 40 rockets with 40 payloads; U.S. had launched 22 rockets with 31 payloads, 23 fewer than in 1967, while U.S.S.R. launches had increased by five.

Reasons for decline in U.S. launch rate were NASA budget cuts and improved instrumentation and data recovery in military launches (60% of U.S. firings). U.S. military payloads in 1968 numbered 22 to U.S.S.R.'s 24 out of total 40. Tracking experts had reported 16 of 1968 U.S.S.R. satellites on reconnaissance assignments, with 13 reentering after 8-day orbit, two after 12 days and one still aloft. Sole U.S.S.R. Fractional Orbital Bombardment System (FOBS) test, *Cosmos CCXVIII*, had been launched and returned to earth in less than one orbit April 25. Drop from nine FOBS tests in 1967 indicated U.S.S.R. might have perfected at least first stage in FOBS development.

Neither U.S. nor U.S.S.R. had launched men into space in 1968 but experts listed *Cosmos CCXII* and *CCXIII* launchings April 15 as forerunners of manned lunar mission. In 1968, U.S.S.R. had launched nine scientific satellites, three for navigation, and two each for communications, weather, and deep space probes. U.S. had made two unmanned Apollo rocket and spacecraft tests, five scientific payloads, and one moon shot. (*Huntsville Times*, 8/5/68)

- Committee on Resolutions of 1968 Republican National Convention meeting in Miami Beach, Fla., proposed platform plank on science: "In science and technology the nation must maintain leadership against increasingly challenging competition from abroad. . . . We regret that the Administration's budgetary mismanagement has forced sharp reductions in the space program. The Republican party shares the sense of urgency manifested by the scientific community concerning the exploration of outer space. We recognize that the peaceful applications of space probes in communications, health, weather, and technological advances have been beneficial to every citizen. We regard the ability to launch and deploy advanced spacecraft as a military necessity. We deplore the failure of the Johnson-Humphrey Administration to emphasize the military uses of space for America's defense." (NYT, 8/5/68, 25)

- First short-takeoff-and-landing strip for commercial aircraft in U.S. opened at La Guardia Airport in New York. Called STOLport, it was 1,095 ft long. Shortest regular runway at La Guardia was 4,000 ft. FAA said STOLport would be used for visual flying only. STOL aircraft landed and took off at 60 mph, rather than regular commercial aircraft's 105 mph. Three more stolports were to open Sept. 1 to provide taxi service between National and Dulles Airports in Washington, D.C., and Friendship Airport in Baltimore. (Shipler, NYT, 8/6/68, 1; AP, W Star, 8/6/68, A5; W Post, 8/14/68, 1)
- NASA planned flight test of relatively inexpensive general-aviation collision-avoidance system, flight test of six or more general-aviation aircraft, and series of wind-tunnel tests of Learjet at ARC under five-year program to improve light aircraft technology with FY 1969 funding of $500,000. General-aviation research program's main current emphasis was on flight-testing Piper Twin Comanche on twin-engine aircraft characteristics for comparison with measurements obtained during wind tunnel tests. ERC and LaRC work on collision-avoidance system was aimed at cooperative system costing below $1,000 commercially, in which pilot would be warned of impending collision if other aircraft was similarly equipped. LaRC was using randomly coded Doppler radar system, to be flight tested late in autumn, while ERC was developing optical system using xenon strobe light to illuminate aircraft. Light would be picked up by silicon detectors. ERC had not yet developed fully integrated system. (Am Av, 8/5/68, 6)

August 6: USAF launched unidentified satellite with Atlas-Agena D booster from ETR into orbit with 24,769-mi (39,866.8-km) apogee, 19,685-mi (31,556-km) perigee 1,436.0-min period, and 9.9° inclination. USAF's first unannounced ETR launching in five years (USAF usually launched from WTR); satellite carried payload described by USAF as "experimental." AP story in *Washington Post* said payload's primary purpose reportedly was to test improved sensing devices and other equipment designed to locate missile and other military installations and provide warning of rocket attacks from unfriendly nations. AP sources had indicated launch was from ETR so satellite could achieve high-altitude orbit where it could "dwell for a longer period over a desired area of the globe." (AP, W Post, 8/7/68, A14; SBD, 8/7/68, 156; GSFC SSR, 8/15/68; Pres Rpt 68)

- USAF launched unidentified satellite from Vandenberg AFB by Titan III–B-Agena D booster. Satellite entered orbit with 266-mi (428.1-km) apogee, 88-mi (141.6-km) perigee, 90.5-min period, and 109.5° inclination and reentered Aug. 16. (*SBD*, 8/12/68, 177; GSFC *SSR*, 8/15/68; 8/31/68; *Pres Rpt 68*)

- *Washington Post* article by George C. Wilson logged significant changes in U.S.S.R. space program: *Cosmos CCVIII* and *Cosmos CCXXVIII*, launched March 21 and June 21, had been first Soviet "spy" satellites to stay up for 12 days instead of usual 7 or 8. U.S. space experts had said they would have to watch pattern of future satellites before concluding whether purpose of longer-lived satellites was peaceful. Launches had been logged without comment in GSFC *Satellite Situation Report* but West had evidence that U.S.S.R. had recovered photographs taken by them. U.S. analysts also had noted overlapping by U.S.S.R. of reconnaissance flights, possibly to ensure backup if one failed. *Cosmos CCXXXIV* marked 100th Soviet reconnaissance launch July 30. *Cosmos CCXXXI* had been launched July 10 at time appropriate for photographing French nuclear tests in Pacific near Tahiti. (*W Post*, 8/6/68, 2)

- Republican presidential contender Richard M. Nixon told group of Southern delegates to nominating convention U.S. must restore balance of power in missile and space programs. "This Administration has allowed it to deteriorate. That is why I am opposed to any kind of program—such as defensive missiles—which allows the United States to take second place to the Soviet Union. We are going to restore our strength in the missile program."

    In the space program, he said, "We have sort of a budget problem . . . and we may have to trim it in some places. But as far as these great and important objectives that involve national defense . . . [and] prestige, we have got to be sure that the United States is first and never second in space . . . [for] any great nation, to remain great, must explore the unknown. . . . As far as I am concerned . . . there is a research gap at the present time. We aren't putting enough money into basic research to keep ahead of the Soviet Union. We are going to close that gap—and then create one where they are behind us all around the world." (*M Her*, 8/7/68, A22; Cocoa, Fla., *Today*, 8/7/68; *SBD*, 8/8/68, 162)

- In policy statement on science New York Governor Nelson B. Rockefeller, campaigning for Republican presidential nomination, had recommended that U.S. postpone manned space program after Apollo. He warned against treating space program as contest between U.S. and U.S.S.R. and advocated use of unmanned satellites for direct benefits on earth, to explore planets, and to study sun. Gov. Rockefeller said if he were President he would appoint commission "to review the whole field of government science." He recommended President make annual report to Congress on science and technology. (*SBD*, 8/6/68, 149)

- House Committee on Government Operations report, *Government Use of Satellite Communications—1968*, endorsed DOD decision to proceed with Phase II of Defense Communications Satellite System [see June 18] and urged immediate action in reprogramming and procurement. "The initial system, established partly on an experimental basis and without the benefit of more recent technological advances, needs to be supplemented and ultimately replaced." It recommended DOD planning to develop facilities and methods for emergency communications by satellite, to provide satellite services when possible instead of expanding conventional systems, and to improve ground terminals and develop new ones. (Text; *Aero Daily*, 8/6/68)

- NASA successfully completed month-long laboratory test of Biosatellite capsule which would maintain primate in space for 30 days in 1969. Fully instrumented, 15-lb, pigtailed monkey which made simulated space flight was alert and responsive when removed and was undergoing postflight examination and physiological measurements identical to those planned to follow actual flight. Test had been terminated after 28 days when biomedical instrumentation measured less urine than monkey would normally secrete.

  Biosatellite D primate flight would duplicate most routines of manned flight to provide data on effects of prolonged weightlessness. During simulated flight, monkey ate pellets, drank water, and performed two 30-min tasks each day. Test was conducted by Biosatellite experimenters at UCLA with support from NASA and General Electric Co. ARC was managing project. (NASA Release 68-142; AP, B *Sun*, 8/29/68, A4)

- USAF experiments in human reactions to extreme heat had indicated astronauts could survive reentry through earth's atmosphere if their cabin cooling system failed. In tests at Wright-Patterson AFB, Ohio, airmen were being subjected to 300°F and 400°F heat in four-foot-high oven three times daily. Some could stand up to 15 min at 300°F; limit at 400°F had been two or three minutes. (AP, *LA Times*, 8/6/68)

- Apollo 7 prime crew successfully completed water egress training in Gulf of Mexico. (NASA Apollo Status Summary, 8/7/68)

August 7: USAF launched unidentified satellite from Vandenberg AFB by Thor-Agena booster into orbit with 161-mi (257.2-km) apogee, 100-mi (160.9-km) perigee, 88.7-min period, and 82.1° inclination. Satellite reentered Aug. 27 (GSFC *SSR*, 8/15/68; 8/31/68; *Pres Rpt* 68)

- NASA Nike-Apache sounding rocket launched from NASA Wallops Station carried Rice Univ. payload to 121.6-mi (195.6-km) altitude to measure vector magnetic field in region between 90 and 130 km during electron current flow with sufficient continuity and signal strength to deduce altitude profile and absolute strength of current layer. Rocket and instrumentation performed satisfactorily. Clear magnetometer data were received during payload ascent and descent. (NASA Rpt SRL)

- NASA issued Apollo Status Summary: Apollo 7 spacecraft had been moved from altitude chamber at KSC. During week it would be mated with lunar module adapter and prepared for mechanical mate with Saturn IB launch vehicle. In Apollo Saturn 503 project, radar lock-on problem in Lunar Module 3 ascent and descent stages' rendezvous radar subsystem was being isolated. Saturn V launch vehicle 1st and 2nd stages had been mated.

  Last in series of seven water impact tests in support of first manned Apollo mission was scheduled for Aug. 13. (Text)

- AEC and U.S. Geological Survey experiments had shown that neutrons from man-made californium-252 hitting gold and silver atoms excited atoms to radioactive state; atoms then emitted characteristic gamma rays by which they were identified. With californium, irradiation equipment for technique already known could be reduced about 200,000 times, making it ideal for sea bottom probes and for devices to analyze materials on moon's surface as well as for mining and in-

August 7

dustrial purposes. Geological Survey was working with GSFC on use for study of moon. (NEA, *W News*, 8/8/68)

- *St. Louis Globe-Democrat* editorial said: "The principal criticism of the proposed cuts in the space budget is that they aren't deep enough. NASA will be left with an expenditure of nearly $4 billion in the current fiscal year. With the nation in a financial bind and the Administration pledged to prune $6 billion from the federal budget, it seems that a good deal more than now suggested should be lopped off the space program." (*St. Louis G–D*, 8/7/68)

August 8: *Explorer XXXIX* (Air Density balloon) and *Explorer XL* (*Injun V*) were successfully launched by NASA from WTR by single four-stage Scout booster. Primary mission objective was to place spacecraft in near polar orbits to extend studies of atmospheric density, geomagnetically trapped particles, and down-flux of these particles into atmosphere. Expected lifetimes were one year each.

Explorer XXXIX entered orbit with 1,548-mi (2,492.3-km) apogee, 425-mi (683-km) perigee, 117.9-min period, and 80.6° inclination. The 20.5-lb, 12-ft-dia inflatable sphere was covered with aluminum foil to reflect sunlight and radio beacons and was coated with painted white spots for temperature control. It would extend measurements of latitudinal, seasonal, and solar cycle variations in upper air density and compare previous satellite measurements for further insight into sources of atmospheric heating.

Explorer XL, 29-in-high, 30-in-dia, 153.1-lb hexagon, entered orbit with 1,573-mi (2,530-km) apogee, 424-mi (677-km) perigee, 118.3-min period, and 80.6° inclination. Satellite's primary purpose was to measure directly down-flux of charged particles into atmosphere; study geomagnetically trapped charged particles, emphasizing spectra, spatial distribution, and time variations; and correlate VLF radio emissions and measurements of low-energy positive electrons with more energetic particle measurements.

Mission was second NASA launch of two spacecraft by one Scout booster. First dual launch successfully injected *Explorer XXIV* (Air Density balloon) and *Explorer XXV* (*Injun IV*) into orbit Nov. 21, 1964. LaRC designed, developed, and constructed Air Density Explorer and Univ. of Iowa designed, developed and constructed Injun V and provided all but one of its experiments. Program was managed by LaRC under OSSA direction. (NASA Proj Off; *SBD*, 8/12/68, 177; GSFC *SSR*, 8/15/68)

- NASA announced "interim operating plan" for FY 1969 designed to achieve $3.85-billion budget. Agency would reduce staff by 1,600 and contract employees by 2,000. Purchase or construction of four Saturn boosters had been canceled [see Aug. 2], plans for 1973 instrument payloads on Mars had been substantially reduced, lunar exploration programs had been halted, and development of NERVA nuclear rocket engine would be put back until next fiscal year, at least. Apollo Applications program would be reduced $300 million from budget request of $440 million and would include only two spacecraft—Saturn I Workshop and Apollo Telescope Mount—with boosters and backup equipment. "Work toward post-Apollo lunar exploration and toward Saturn V Workshop will be limited to studies." (NASA Release 68–141)

- NASA Nike-Apache sounding rocket launched from NASA Wallops Station carried Univ. of Michigan payload to 94.7-mi (152.4-km) altitude to measure neutral atmosphere density and temperature profile in 17.4- to 63.4-mi (28- to 120-km)-altitude range, compare day measurements with night measurements by rocket to be launched Aug. 9, and flight-test and evaluate new second-generation Pitot probe system flown without ejectable nose tip. Rocket and instrumentation performed satisfactorily. Pitot tube performed well and seemed capable of reaching higher apogees, even without nose-cone tip, than old system. (NASA Rpt SRL)

- NASA announced Astronaut James A. Lovell, Jr., had replaced Astronaut Michael Collins as prime command module pilot for third manned Apollo mission. Lovell had held same position on mission's backup crew. Collins had undergone successful spinal surgery July 23 and would require three to six months recuperation. Edwin E. Aldrin, Jr., backup lunar module pilot, had been moved to command module pilot position on backup crew. Astronaut Fred W. Haise, Jr., had been assigned to backup crew as lunar module pilot. Lovell joined Astronauts Frank Borman and William A. Anders on prime crew for mission scheduled for first quarter 1969. (MSC Release; KSC Release KSC-374-68; AP, *NYT*, 8/9/68, 14; UPI, W *Star*, 8/9/68; W *Post*, 8/9/68, A7; *SBD*, 8/9/68, 169; MSC *Roundup*, 8/16/68, 1)

- GSFC team headed by Peter Minott was studying amount of energy absorbed by atmosphere through experiments with laser beam directed at *Explorer XXXVI*. Reflectors on satellite returned beam to its starting point on earth while laser detector on satellite radioed back data on the strength of its light as it passed through atmospheric conditions. NASA laser expert Dr. Henry Plotkin said, "Some day we figure that with a laser we can build a space-to-earth communications link that won't fade the way radio waves do." Laser experiments also were being conducted with *Explorer XXII, XXVII*, and *XXIX*, and two French satellites. (Kehoe, W *Post*, 8/8/68, F1)

- ComSatCorp, on behalf of INTELSAT, issued RFP for aeronautical communications satellites for two-way simplex voice communication between transoceanic aircraft and fixed earth stations. Spacecraft were to be active repeater VHF comsats with minimum five-year life for emplacement in synchronous equatorial orbit. (ComSatCorp Release 68-38)

August 8–9: Two NASA Nike-Tomahawk sounding rockets launched from NASA Wallops Station reached 178-mi (286.5-km) and 199-mi (320-km) altitudes in Univ. of Michigan experiments to obtain thermosphere structure measurements of N_2 density and temperature. Rockets and instrumentation performed satisfactorily. All scientific objectives appeared to have been met. (NASA Rpts SRL)

August 9: U.S.S.R. launched *Cosmos CCXXXV* from Baikonur Cosmodrome into orbit with 283-km (175.8-mi) apogee, 203-km (126.2-mi) perigee, 89.3-min period, and 51.8° inclination. Satellite performed satisfactorily and reentered Aug. 17. (W *Star*, 8/9/68, 1; *SBD*, 8/12/68, 177; GSFC *SSR*, 8/15/68; 8/31/68)

- NASA Nike-Apache sounding rocket launched from NASA Wallops Station carried Univ. of Michigan payload to 90.9-mi (146.3-km) altitude to measure neutral atmospheric density and temperature profile at altitudes of 12.4–74.6 mi (20–120 km), compare night measurements with day measurements by rocket launched Aug. 8, evaluate lunar position sensing device for use in resolving payload altitude, and measure wind velocities. Rocket and instrumentation performed satisfactorily; data were obtained as planned. Launch was to be last of this Pitot static payload configuration. New Pitot probe system would be used in future. (NASA Rpt SRL)

- NASA selected Nortronics Div. of Northrop Corp. and Fecker Div. of Owens-Illinois Co. for competitive negotiations for $2.5 million cost-plus-incentive-fee contract to design, build, and install 36-in aperture infrared telescope in NASA's high-altitude observatory aircraft. Aircraft, modified Convair 990 jet transport, was operated by ARC as national facility for astronomers and other scientists. Telescope would provide information on planets, comets, asteroids, solar surface, stars, nebulae, and galactic phenomena. Above interference of most of earth's atmosphere, it would gather information in infrared portion of spectrum not available from ground observatories. (NASA Release 68-144)

- Naval Research Laboratory scientists Dror Sadeh, Stephen Knowles, and Benjamin Au reported in *Science* observations made in two experiments in which an apparent decrease in frequency of radio or light waves was detected when optical path was in vicinity of a mass. Radio waves skirting earth and sun appeared to increase in wavelength, leading scientists to reason that all lights crossing portions of universe would be similarly affected by mass of objects along the way. Effect could account for part of "red shift" (shift in spectra of distant galaxies as they receded from earth), chief tool for measuring expansion rate of universe which, in turn, was used to measure age of universe. Thus, if observations by NRL scientists were confirmed, they could necessitate revision of long-standing estimates of age and expansion rate of universe. Experimenters, however, indicated their findings appeared inconsistent with results from "round-trip" experiments, such as those obtained during past year when radar impulses were bounced off Venus and Mercury. (*Science*, 8/9/68, 567–9; Sullivan, *NYT*, 8/10/68, 27)

- ComSatCorp, on behalf of INTELSAT, announced it had requested proposals for design of 120-in parabolic spacecraft antenna for experiments to evaluate feasibility of deployable, narrow-beam, parabolic antenna for use on spacecraft in synchronous, equatorial orbit in research on communications at 15 ghz. Antenna would be designed to provide 0.5° width beam. (ComSatCorp Release 68-40)

August 10: NASA's 860-lb *Ats IV* (ATS-D) (Applications Technology Satellite) was launched from ETR by Atlas-Centaur booster on mission to evaluate gravity-gradient system for spacecraft stabilization in synchronous orbit and to obtain data on experiments during first 30 days in orbit.

Centaur ignited, successfully injecting spacecraft into elliptical parking orbit with 475.8-mi (765.7-km) apogee, 115.4-mi (175.7-km) perigee, and 29.1° inclination. Centaur engines, which were to have reignited to inject spacecraft into 22,300-mi-altitude synchronous equatorial orbit west of South America, failed to ignite. Since programmer did not receive accelerometer burn indication, it did not issue command to separate and spacecraft and Centaur stage remained joined, tumbling uncontrollably, in parking orbit. Three attempts to fire apogee kick motor were unsuccessful because firing was inhibited by separation switches. Efforts would be made to operate all experiments possible, but information beyond proof of operation was not expected. Although exact rate and direction of tumble had not been determined, preliminary assessment indicated rate was too high for successful boom deployment.

Ats IV was fourth in series of seven satellites designed to improve spacecraft technology, develop long-life control systems, advance spacecraft communications, and improve long-range weather predictions. First spacecraft to be placed in earth orbit by Centaur booster, *Ats IV* was 56 in in dia, 72 in long, and would have been 251 ft across with gravity-gradient rods fully deployed. Spacecraft carried four applications technology experiments—gravity-gradient stabilization system, day-night Image Orthicon Camera (IOC), microwave communications experiment, and ion engine—and no science experiments. *Ats I* (launched Dec. 6, 1966) had exceeded its test objectives and was still operating satisfactorily. *Ats II* (launched April 5, 1967), though judged a failure because of eccentric orbit, had transmitted some useful data. *Ats III* (launched Nov. 5, 1967), last ATS spacecraft scheduled to be launched by Atlas-Agena D booster, had operated successfully and transmitted color photos of earth. ATS program was managed by GSFC under OSSA direction. (NASA Proj Off; NASA Release 68–127; UPI, *NYT*, 8/12/68, 34; AP, *P Inq*, 8/12/68; *Lewis News*, 8/16/68, 1)

- Dr. George E. Mueller, NASA Associate Administrator for Manned Space Flight, told British Interplanetary Society in London: "I believe that the exploitation of space is limited in concept and extent by the very high cost of putting payload into orbit, and the inaccessibility of objects after they have been launched. Therefore I would forecast that the next major thrust in space will be the development of an economical launch vehicle for shuttling between earth and the installations, such as the orbiting space stations which will be operating in space." Dr. Mueller said efficient earth-to-orbit transportation system would be needed to shuttle thousands of tons of material in and out of space, operating in mode similar to that of large commercial air transports and compatible with airports. Same technology could be applied to terrestrial point-to-point transport. (Text)
- New York *News* editorial criticized severe cut-backs in space funding: "Most ominous, we think, are the delays and curtailments likely to be economy-compelled in the Apollo moon program . . . too-drastic cut-backs in funds for space programs and researches would be foolish at best, and suicidal at worst, considering Russia's known efforts to develop space weapons for world blackmail. What if Queen Isabella had limited Christopher Columbus to one little ship and a skeleton crew?" (NY *News*, 8/10/68)

August 11: Photographic evidence from U.S. satellites had indicated U.S.S.R. construction of Moscow antimissile defense had been slowed down, if not halted. U.S. intelligence officials had been satisfied from photographs that Tallinn defense across northeast approaches to U.S.S.R. was against U.S. bombers, not missiles. (*W Post*, 8/11/68, 1)

August 12: NASA Nike-Apache sounding rocket launched from WSMR carried Dudley Observatory, Albany, N.Y., payload to 96.5-mi (155.2-km)

August 12

altitude to collect, recover, and evaluate micrometeoroid particles during approximate maximum of Perseid meteor shower. Rocket and instruments performed satisfactorily and telemetry data were obtained during entire flight time. Preliminary data indicated excellent experimental results. Payload was not damaged and was recovered within one hour. (NASA Rpt SRL)

- Simplified instrumentation technique using miniaturization, developed by NASA's FRC for monitoring pilots of XB-70 aircraft and wingless lifting-body vehicles, had been adapted to make possible almost instantaneous transmission of electrocardiograms taken of ambulance patients enroute to hospital doctors awaiting their arrival. In procedure taking total two minutes, data obtained from patient were flashed over ambulance radio to control center, relayed by telephone to cooperating hospital, and fed to ECG recorder. System had been successfully tested by a Los Angeles ambulance service for several months. (NASA Release 68–145; *W Post*, 8/15/68)
- *New York Times* reported interview in which General Electric Co. engineers described ARMS (Application of Remote Manipulators in Space), concept being pursued under $68,000 NASA contract to study feasibility of orbiting robots to refuel, resupply, and repair crippled spacecraft on signals from earth. They could be in operation by mid-1970s and, if successful, could extend useful lifetimes of communications and weather satellites and, possibly, could inspect or disable hostile spacecraft. One plan was to launch 600-lb robot attached to 1,000-lb tender or "home base" satellite. With repair kit orbited on distress signal from regular satellite, robot would rendezvous and repair disabled satellite directed by a ground controller by radio signals transmitted via the tender. After each repair robot would return to tender. Engineers estimated robot could remain in orbit four to five years and would have paid for itself after 12 repairs. (Wilford, *NYT*, 8/13/68, 4)

August 13: Apollo 7 spacecraft had been mechanically mated to Saturn IB launch vehicle at Complex 34, NASA announced. Apollo 7 prime or backup crew, or both, would participate in 10-hr network simulation of first six revolutions of mission and 8-hr launch simulations during week.

Spacecraft 2TV–1 command and service modules successfully completed manned checkout Aug. 9 and 10 in preparation for five-day manned thermal-vacuum test at MSC in September to help verify spacecraft in simulated space environment for Apollo earth-orbital and lunar missions. Two-minute captive firing of 6th Saturn V booster stage at MTF Aug. 13 would try out "fix" for Saturn V longitudinal oscillations —conversion of liquid oxygen prevalves into gas-filled shock absorbers. (Text; *C Trib*, 8/13/68)

- NASA issued request for quotations for design, development, fabrication, test, qualification, and delivery of actuator system for Apollo Telescope Mount (ATM), calling for completion within nine months. System, expected to weigh 20 lb, would be contained in cylinder 18 in long and 1 ft in dia. It would be used to deploy solar panels of ATM which would be placed in orbit and docked with Saturn I workshop. (MSFC Release 68–178)
- NASA Associate Administrator for Manned Flight, Dr. George E. Mueller, had told Apollo project contractors that U.S.S.R. was developing a

"large booster, larger by a factor of two, than our Saturn V," *New York Times*' John N. Wilford reported. Wilford said estimate was part of warning to Apollo team that unless they speeded their efforts, U.S.S.R. might beat U.S. to manned lunar landing. With Dr. Mueller in Europe, NASA spokesman had said he did not know basis of information. Previously, there had been no public statement indicating U.S.S.R. might be working on rocket with greater than 10-million-lb thrust. Saturn V had 7.5-million-lb thrust. (*NYT*, 8/14/68, 11)

- NAE's Aeronautics and Space Engineering Board called for review of air transport's part in U.S. transportation network, urging that DOT and FAA take systems approach to study of R&D goals while leaving R&D itself to enlarged NASA role. Board report, *Civil Aviation and Development: An Assessment of Federal Government Involvement*, recommended, "NASA's role should be expanded to involve not only flight vehicles and their propulsion systems, which have traditionally occupied its principal attention . . . but all aspects of research and development of importance to civil aeronautics."

  First project of Board organized in 1967, study said NASA work should include "development of new technologies relating to air traffic control as well as to airports and their support facilities." In addition to NASA's background in vehicle technology and growing avionics capability, "developments in space technology, including the use of satellites for communications and navigation, offer new opportunities for improving air navigation." (Text; NAE Release; Lannan, W *Star*, 8/13/68, A8; Sehlstedt, B *Sun*, 8/14/68; *A&A*, 8/68)

- U.S. patent No. 3,396,921 was issued to Francis M. Rogallo, head of low-speed vehicle branch at LaRC, for control devices to regulate pitch and roll of paraglider planned for Apollo Applications flights. Rogallo and his wife Gertrude had developed paraglider—called Rogallo wing —under NASA contract. It was free-flight vehicle with flexible wings which could be extended upon reentry to carry astronauts to a landing. (Pat Off PIO; *NYT*, 8/17/68, 37)

- DOD announced second $1-million installment on contracts for design of Navy VFX, substitute aircraft for F–111B, held by Grumman Aircraft Engineering Corp., General Dynamics Corp., LTV Aerospace Corp., McDonnell Douglas Corp., and North American Rockwell Corp. [see July 19]. Five contractors would eventually get $3 million each under contract definition phase to end Oct. 1. (DOD Press Off; *W Post*, 8/14/68, D9)

- USN had announced selection of General Dynamics Corp. and Lockheed Aircraft Corp.-LTV Aerospace Corp. team to make contract definition studies of VSX, carrier-based antisubmarine aircraft to replace 15-yr-old S2, *Wall Street Journal* reported. VSX, powered by turbofan-jets, would have twice S2's speed, 10 times its efficiency because of its digital computers which could analyze underwater sound and present data on TV-like displays. Contract could eventually reach $1 billion. (*WSJ*, 8/13/68, 2)

- In closed meeting at Civil Aeronautics Board, Washington, D.C., 75 airline and U.S. Government officials discussed possible solutions to aviation congestion crisis, including: elimination of peak-hour charter flights; adoption of minimum flight distance rule for scheduled services from New York's Kennedy International Airport; rollback to 1967 level of Kennedy schedules; limitation of aircraft movements in peak hours, including general aviation; diversion of flights to other airports; and blackout of discount fares at peak periods. Trans World Airlines President Charles C. Tillinghast, Jr., disagreed with Port of New York Authority view that basic solution to air congestion was construction of fourth New York airport. Solution, he said, was expanding Kennedy's capabilities to meet traffic demand. (Tolchin, *NYT*, 8/14/68, 30; W *Star*, 8/14/68, A18)

- *New York Times* editorial advocated priority for passenger transports over private aircraft during peak hours at major airports—New York, Washington, Chicago, Los Angeles—until longer-range measures could be provided to alleviate congestion. "Such a solution will be strongly opposed by . . . general aviation which now numbers 112,000 private planes [including] 4,000 business jets. . . . The contention will be advanced that it is unfair to bar them from airports built with public funds through a system of preference for the 2,200 commercial airliners now in service. Yet no one can argue that it is in the public interest to keep a commercial airliner with 120 passengers aboard circling for . . . hours . . . while airport tower time and runway space are occupied by planes which may carry only two or three persons. . . . Where limited airport facilities necessitate a choice, common carriers should have preference over private planes. Delay in making that choice is an invitation to disaster." (*NYT*, 8/13/68, 34)
- Washington *Evening Star* editorial praised U.S. decision to test Multiple Independently Targetable Reentry Vehicle (MIRV) despite suggestions that experimentation be suspended as token of U.S. readiness to negotiate with U.S.S.R. on international control of missile production: ". . . negotiations . . . are still in the future and will be drawn out. . . . Meanwhile, since neither party is subjecting itself to any kind of control . . . and since the technology of missiles continues a sort of explosive advance, nothing could be more foolish or more dangerous than a unilateral suspension . . . of the testing that is clearly necessary to the defense of the nation." (W *Star*, 8/13/68, A6)
- Prof. J. Hoover Mackin, one of four experts chosen by NASA to examine first rock samples to be returned from moon, died in Houston, Tex., at age 62. He had held William Stamps Farish Chair in Geology at Univ. of Texas. (AP, *W Post*, 8/14/68, B6)
- Ralph Hazlett Upson, aeronautical engineer who won James Gordon Trophy in 1913 International Balloon Race from Paris to England and American National Balloon Races in 1913, 1919, and 1921, died in Burien, Wash., at age 80. (*NYT*, 8/15/68, 35)

August 14: NAS–NRC Space Science Board urged NASA to use developing technology of fully automated systems in planetary exploration rather than manned flights, which it felt were not "essential for scientific planetary investigation at this stage." Recommendations on priorities were made in *Planetary Exploration, 1968–1975,* prepared by panel of 23 scientists who met during week of June 9 to reappraise 1965 study in light of rapid development in science and severe cuts in NASA budget.

Report called 2% for planetary exploration out of NASA FY 1969 budget "totally inadequate." It recommended diversion of funds from manned missions to instrument exploration, including biennial flights to orbit Venus and Mars until 1975; dropping capsule on Mars in 1973 to detect life signs and a major lander later, perhaps in 1975; multiple-drop probe of Venus' surface in 1975; Mercury flight in 1973; and priority Jupiter flybys in 1972 and 1973. Panel advocated pursuit of planets in economical way: use of existing Pioneer spacecraft in Venus and Jupiter missions, elimination of second spacecraft in all missions unless "clear gain . . . will result from such double launches," and use of single launch for both Venus and Mercury missions by scheduling them when planets were in alignment (same technique could be used for "Grand Tour" of major planets in 1977, when they would be aligned in space). Referring to U.S.S.R. exploration, report said, "We certainly believe we cannot abandon this broad area of space activities to our competitors."

Report also recommended strong support for radioastronomy, including development of major new observatory primarily for planetary study, and continued support of ground-based optical planetary astronomy. It proposed coordinated, informal contact with Soviet scientists on possibility of joint planning of planetary exploration. (Text; NRC Release; Lyons, *NYT*, 8/15/68, 17; O'Toole, *W Post*, 8/15/68, A1; Lannan, *W Star*, 8/15/68, C8; Carter, *Science*, 8/16/68, 671–3)

- Secretary of Transportation Alan S. Boyd said Government would limit both commercial and general aviation traffic at New York's major airports unless aviation industry imposed its own limitations. Banishment of general aviation from Kennedy International or La Guardia Airport was "an extreme possibility"; problem of "spiderweb" of connecting flights would be tackled by diverting some international flights to airports at Boston, Philadelphia, and Washington. (Tolchin, *NYT*, 8/15/68, 1)

- Taccomsat 1, world's largest synchronous orbit satellite, was scheduled for launch by USAF Titan III–C booster in February 1969, *Aerospace Daily* reported. Spacecraft would be six times as powerful as any other comsat. Initial command and control of satellite, to be positioned near Galapagos Islands for undetermined checkout period, would be accomplished by Air Force Satellite Control Facility (AFSCF). Taccomsat 1 would have 6 kw of effective radiated power. Intelsat II satellites had 50 w. Its 1 kw of raw dc solar power was 10 times that available in Intelsat II. While Intelsat II communications bandwidth was 125 mhz, Taccomsat 1 would have down-link bandwidth of 10 mhz at X-band and 500 khz at UHF frequencies. (*Aero Daily*, 8/14/68)

- Recorded aircraft highjackings had totaled 14 in past year, with 13 U.S. airliners over southern U.S. forced to land in Cuba, Andrew Wilson of *London Observer* reported. (*W Post*, 8/14/68, A3)

August 14–27: At U.N. Conference on Exploration and Peaceful Uses of Outer Space held in Vienna and attended by 500 delegates from 74 nations, Soviet Premier Alexey N. Kosygin announced U.S.S.R. and "other Socialist countries" would establish comsat network "Intersputnik" to compete with INTELSAT. Draft agreement had been submitted in New York to U.N. Secretary-General U Thant by U.S.S.R., Bulgaria, Czechoslovakia, Hungary, Poland, Romania, Mongolia, and Cuba. Vladimir Minashin, head of comsat div. of Soviet Ministry of Communications, said Intersputnik satellite would have same synchronous orbit used by the four U.S.-provided INTELSAT satellites. U.S. Dept. spokesman said U.S. had no need to join U.S.S.R.-sponsored system;

INTELSAT, with 62 member nations, was already operating successfully, handling 95% of total international telecommunications traffic. U.S. was not opposed to Soviet proposal for political reasons, but on grounds there would be economic problems in creation of two worldwide comsat systems. (*W Post*, 8/14/68, 12; O'Toole, *W Post*, 8/15/68, A21; Hamilton, *NYT*, 8/15/68, 18; 8/20/68, 18; *WSJ*, 8/15/68, 1; Sehlstedt, *B Sun*, 8/15/68, 1)

Dr. George E. Mueller, NASA Associate Administrator for Manned Space Flight, at Vienna conference discussed three results of U.S. manned space flight program to illustrate contributions to space exploration. Program had proved man could live in space and man could do useful work in space and it had created technology to make these possible. In first category, NASA, NAS–NRC Space Science Board, NSF, ARPA, DOD groups, universities, and others had conquered problems of weightlessness in space, heavy acceleration and deceleration forces, air supply and pressure required for breathing in space, psychological problems presented by isolation, and problem of radiation.

Successful termination of Mercury and Gemini programs had proved man capable of existing in space. His capability to do useful work in space had been proved by successful rendezvous and docking of spacecraft, by space photography, and by extravehicular activity. Pervasiveness of space technology had been demonstrated by range and variety of thousands of products, including Saturn V launch vehicle, world's largest flight vehicle, and integrated circuit, one of world's smallest manufactured items.

Dr. Mueller said: "Our civilization has been built upon our accumulated knowledge of the natural laws of our environment. All of our inventions have been the result of the application of these natural laws. Advances in our civilization have always followed after discovery of some one of the missing links in our chain of knowledge. . . . We know that many of the missing links . . . will be supplied . . . as we move man with his accumulation of experience and his sophisticated equipment for exploration, into a new laboratory . . . the laboratory of the moon and outward to our solar system. . . . If all of us, from all nations, are sufficiently creative, abundantly inventive, and freely adaptive, we have it within our power to improve the lives of every man, woman, and child." (Text)

U.S.S.R. Cosmonaut Aleksey A. Leonov at Vienna conference proposed Aug. 15 that Ocean of Storms, prominent feature of lunar landscape, be renamed Ocean of Gagarin in honor of Soviet Cosmonaut Yuri A. Gargarin, who made first manned earth orbit April 12, 1961. Leonov, who took man's first walk in space in 1965, read Gagarin paper which emphasized similarity between experience of Soviet astronauts and that of crews of deep-sea exploratory craft. He said all actions taken in Soviet space vehicles were tried first in underwater craft.

Large Soviet exhibit at conference showed full-scale model of original Soviet spacecraft and listed Soviet space firsts. U.S. exhibit, small because of cut in NASA funds, depicted benefits space exploration would produce for developing countries. (Hamilton, *NYT*, 8/16/68)

J. L. Blondstein of British National Industrial Space Committee told group meeting at conference Aug. 20 that production of U.S. military satellites capable of direct TV broadcasts to military units was "imminent." He said U.S. had spent between $80 million and $100 million on development but denied information was being withheld for military security reasons. DOD spokesman in Washington had said he knew of no plans for direct TV broadcasts by military satellites. (Hamilton, *NYT*, 8/21/68, 3)

W. T. Pecora, Director, U.S. Geological Survey, said Aug. 23 that worldwide volcano and earthquake monitoring network might be provided within a few years by space satellites carrying cameras and sensing devices to give up to one hour advance notice of strong earth shocks. (AP, *C Trib*, 8/23/68)

August 15: At closed meeting at Grumman Aircraft Engineering Corp. in July, NASA Associate Administrator for Manned Space Flight, Dr. George E. Mueller, had said contractors' disregard of planned delivery dates for Apollo equipment amounted to a "disease" in the project, *New York Times* reported. Apollo Program Director L/G Samuel C. Phillips was quoted as saying, "The lunar landing next year is within our grasp, but we don't have a hold of it because of the disease Dr. Mueller cited." Project was running two years behind schedule, with first manned Apollo flight expected no earlier than mid-October and earliest lunar landing in a year. Dozens of contractors in $24-million program, urged to meet tough specifications for mission safety and success at same time, were obsessed with checking and double checking all systems as result of Jan. 24, 1967, Apollo fire. Lunar module (LM) regarded as pacing item of project, had undergone only one flight test. First LM for manned flight had developed problems in rendezvous equipment. Dr. Mueller had said rate of changes in LM was three times that of Apollo command module, whose rate of changes, in turn, was four times that of Saturn V rocket. He said changes placed added burden on technicians who should be concentrating on launching operations, not on vehicle modifications. (Wilford, *NYT*, 8/15/68, 16)

• *Washington Post* editorial saw scheduled MIRV test as threat to success of missile-limitation talks between U.S. and U.S.S.R. "Perhaps it will prove possible to level off the arms race despite MIRV, although it is generally acknowledged that this weapon raises special inspection and stability issues of its own. If talks do stick, however, the Administration must be prepared to bend on MIRV. . . . No one seriously claims that there is any immediate military justification for it; deterrence works without it and the antimissile system it was designed to penetrate evidently is in low gear. It may be acceptable for MIRV to be tested in order to ease the Administration's internal tensions and electoral exigencies. . . . But it would be intolerable to let MIRV spoil the missile talks. They hold more promise of mutual security—the only genuine kind there is—than any new weapon can provide." (*W Post*, 8/15/68, A20)

August 16: NASA successfully launched *Essa VII* (TOS–E), seventh meteorological satellite in ESSA's Tiros Operational Satellite (TOS) system, from WTR by two-stage Thrust-Augmented Long-Tank Thor-Delta booster. This was first use of long-tank Thor for a NASA Delta mission, first use of two-stage vehicle for TOS spacecraft, and first operational two-burn mission for 2nd stage.

Primary NASA mission objective was to place and operate spacecraft

August 16: NASA *launched meteorological satellite* Essa VII *for Environmental Science Services Administration. Satellite relayed photo of tropical storm Shirley Aug. 19.*

in sun-synchronous orbit with local equator crossing time between 2:35 pm and 2:55 pm so that daily Advanced Vidicon Camera System (AVCS) pictures of entire globe could be obtained regularly and dependably. Satellite achieved nearly polar, sun-synchronous, circular orbit with 918-mi (1,477.9-km) apogee, 895-mi (1,440-km) perigee, 114.9-min period, and 101.7° inclination. Drift of only 0.002° per day out of sun-synchronous orbit meant it would take perhaps 15 yr to change equator crossing time by 1 hr and added considerably to expected useful lifetime.

An advanced version of cartwheel configuration, 325-lb cylindrical *Essa VII* carried two AVCS cameras for global weather coverage. Photos would be stored onboard satellite on magnetic tape until readout by ESSA's Command and Data Acquisition (CDA) stations at Fairbanks, Alaska, and Wallops Island, Va.

During first 42 orbits, spacecraft underwent orientation maneuver to place it in wheel mode and spin rate was adjusted. By Aug. 23 all spacecraft systems had been successfully programmed and excellent pictures had been read out directly, as well as sequences stored for remote readout.

ESSA financed and managed TOS system and would operate spacecraft after NASA completed checkout later in month. GSFC was responsible for procurement, launch, and initial checkout of spacecraft in orbit. *Essa I* was launched Feb. 3, 1966; *Essa II*, Feb. 28, 1966; *Essa III*, Oct. 2, 1966; *Essa IV*, Jan. 26, 1967; *Essa V*, April 20, 1967; and *Essa VI*, Nov. 10, 1967. *Essa VII* was 17th Tiros satellite launched successfully since *Tiros I*, first weather satellite, April 1, 1960. All from *Tiros III* on had equaled or exceeded designed operation lifetimes. Millionth weather satellite photo had been received May 27, 1968. (NASA Proj Off; ESSA Release ES 68–48)

- USAF attempt to launch record 12 satellites from Vandenberg AFB with single Atlas-Burner II booster failed when heat shield apparently did not separate as planned. Telemetry during launch indicated Atlas and Burner II stages performed as scheduled, but confirmation of heat shield separation was not received. USAF was still investigating launch attempt. Upper stage had been designed to eject small satellites to measure size and shape of earth, provide targets for radar experiments, and test atmospheric drag in near space. (*W Post*, 8/18/68; Boeing Release S–9805)

- NASA's Reentry VI mission—launched April 27 to obtain fundamental inflight research data on aerodynamic heating and transition from laminar to turbulent flow in boundary layer—was adjudged successful by NASA. Boosted flight trajectory was nominal and spacecraft was placed on desired trajectory with three sigma dispersion limits. Separation of spacecraft from Scout launch vehicle occurred as planned, with all systems fully operational. Maximum spacecraft velocity (19,820 fps), altitude, and reentry angle were near nominal. Telemetry records indicated flight instrumentation performance was satisfactory throughout flight. Large unexpected amount of signal attenuation occurred at altitudes below 90,000 ft, but because of high design margins, no data were lost. Transitional and fully developed heating data were obtained during reentry. (NASA Proj Off)

- DOD successfully launched first Poseidon and Minuteman III long-range missiles from ETR. USN's two-stage, solid-fuel Poseidon, designed as submarine-launched missile, weighed 65,000 lb—twice as much as Polaris it would replace—and could carry up to 10 nuclear warheads in one Multiple Independently Targetable Reentry Vehicle (MIRV) cluster. Missile traveled 1,000 mi downrange over Atlantic.

    USAF's 76,000-lb Minuteman III, more powerful version of Minuteman I and II missiles and capable of carrying three warheads, traveled 5,000 mi downrange. (DOD Release 753–68; *W Star*, 8/16/68, A9; Wilford, *NYT*, 8/17/68, 1; AP, *B Sun*, 8/17/68, 1; *W Post*, 8/17/68, 1)

- JPL scientists Paul M. Muller and William L. Sjogren, using tracking data from *Lunar Orbiter V*, had discovered mass concentrations of dense material beneath moon's surface, centered below all five circular seas on moon's near face. They had caused acceleration in speed of *Lunar Orbiter V* spacecraft, while irregular seas had only small effects on satellite's orbital velocity. Source and nature of concentrations were not known, but their presence under circular seas indicated relationship to these seas. Further analysis was expected to yield more positive in-

August 16

formation on size and depth of masses, which might aid Apollo navigation and shed light on moon's origin and evolution. (NASA Release 68–143; JPL Release 486; Sullivan, *NYT*, 8/16/68, 42; Cohn, *W Post*, 8/16/68, A3)

- U.S.S.R.'s Tu-144 supersonic aircraft had been completed and was being prepared for maiden flight, Associated Press reported. According to Moscow radio, aircraft was at an airfield, would soon start flight tests, and would become operational "in the nearest future." (AP, *NYT*, 8/17/68, 44)

- *Washington Post* editorial on NAS–NRC Space Science Board's recommendation that NASA follow manned lunar exploration with unmanned planetary exploration: "The program they project would require more funds than so far given NASA but certainly would be more economical than preparation for manned flights to other planets. Without settling the basic issues dividing the exponents of manned flights and those in favor of unmanned flights, this program would be a logical preliminary to either alternative . . . [and] seems consistent with both scientific objectives and financial realities. And if success in this program justifies going on to manned flights, that can be decided later." (*W Post*, 8/16/68, 20)

- At DOT press conference Secretary of Transportation Alan S. Boyd and Acting FAA Administrator David D. Thomas outlined proposed restrictions on air traffic at major U.S. airports to ease increasing congestion: limit aircraft movements in peak periods to 65 per hr at Kennedy International, 60 per hr at Newark and La Guardia, 130 per hr at Chicago's O'Hare, and 70 per hr at Washington National; require all aircraft operating to and from those airports to file advance flight plans; and restrict use of those airports to aircraft meeting performance and equipment requirements which would eliminate many general-aviation planes. DOT said it planned to implement its proposals only if aviation community could not agree on alternative plans. (*WSJ*, 8/19/68, 3; Lyons, *NYT*, 8/20/68, 1)

- Goodyear Tire & Rubber Co. and medical researchers were developing artificial heart of natural rubber and semirigid polyurethane with pumps and monitoring systems developed by NASA. Although rubber heart had been tried in animals, artificial heart was still experimental and was not expected to be available for use in man until 1975. (*WSJ*, 8/16/68, 4)

- USAF Space and Missile Systems Organization awarded Philco-Ford Corp. $2-million, cost-plus-incentive-fee contract for work on ejection equipment for reentry vehicles. (DOD Release 772–68)

- Howard H. Haglund, JPL Surveyor Project Manager, had been named Stanford-Sloan Fellow for 1968–69. Recipient of National Space Club Astronautics Engineer Award and NASA Exceptional Service Medal, Haglund would attend special nine-month Stanford Univ. graduate course in advanced management. (Pasadena *Star-News*, 8/16/68; Glendale *News-Press*, 8/19/68)

August 17: In letter to President Johnson, Rep. Emilio Q. Daddario (D-Conn.), Chairman of House Subcommittee on Science, Research and Development, urged reorganization of environmental science activities of Federal Government and asked that responsibility for development of weather modification programs be given to Environmental Science

Services Administration (ESSA). His committee, originator of NSF Amendments Act of 1968 which became effective July 18, was interested in implementing section which relieved NSF of that responsibility as of Sept. 1.

Rep. Daddario stated three beliefs: "1. The weather modification program involves the development of environmental understanding which is crucial to the protection of our planetary resources; it should be supported and pursued without significant hiatus. 2. In light of the nature of the program and the present administrative organization of the executive branch, it seems clear that weather modification should become part of the portfolio of the Environmental Science Services Administration. 3. At the same time, there is considerable doubt whether the Environmental Science Services Administration should remain part of the Department of Commerce; therefore consideration should be given to executive reorganization which might either relocate the Administration or give it independent status." (Text; W *Star*, 8/23/68, A4)

- *New York Times* editorial said first flight tests of Multiple Independently Targetable Reentry Vehicle (MIRV) and prototype missile carriers Poseidon and Minuteman III had taken U.S. "into a new nuclear arms race—with itself. The critical question now is whether the Soviet Union will feel sufficiently challenged to run along. Future generations undoubtedly will look back with disbelief at the way the United States again has invented, publicized and tested a deadly new weapon which, instead of improving American security, creates an added threat to it by putting the Soviet Union under pressure to produce the same weapon and aim it at the United States." (*NYT*, 8/17/68, 26)

August 18: U.S.S.R. had begun using new supersonic jet fighter, described by *Trud*, Soviet trade union paper, as "similar to a triangle, split in two by the fuselage." Aircraft, thought by U.S. sources to be MiG-23 (Foxbat), had reportedly undergone more than 1,000 hr of flight testing. (Reuters, *NYT*, 8/18/68, 50)

- U.S.S.R. celebrated Aviation Day by extolling its supersonic bombers, but no new aircraft had been unveiled—in contrast to 1967 celebration, which featured several new jets and missiles. (*W Post*, 8/19/68, A13)
- DOD attributed May 8 crash in Nevada of USAF F-111A and "possibly" three other F-111A accidents in Southeast Asia between March 28 and April 22 to break in weld holding together two pieces of metal, totaling six inches, in hydraulic fluid control valve in aircraft's tail. In early F-111s part was forged from single piece of high-grade steel but in subsequent models it was made from two pieces welded together. F-111s, grounded after May 8 crash, had resumed flying in mid-June after replacement of two-piece rod end assemblies with original one-piece forging. (DOD Press Off; Welles, *NYT*, 8/18/68; *W Post*, 8/19/68, B7)
- Soviet economist Dr. Viktor A. Cheprakov had published in *Izvestia* what appeared to be rebuttal to prognosis by Soviet nuclear physicist Prof. Andrey D. Sakharov of convergence of communism and capitalism by year 2000 [see July 22]. In "Problems of the Last Third of the Century," translated by *New York Times*, Dr. Cheprakov foresaw sharpening of struggle between the two ideologies and discussed role of science. "Science and technology are endowed with tremendous possibilities. . . . a new achievement . . . is being recorded every 25 sec-

onds. We have figures on what and how much can be extracted from the earth, on the benefits to be derived from new methods of food production . . . and on the undoubtedly vast prospects of using the resources of seas and oceans. But the realization of all these potentialities will require the victory of socialism . . . for only socialism makes it possible to funnel vast funds into scientific research and into the practical use of its results in the interest of the peoples." (Anderson, *NYT*, 8/19/68, 1)

- *Washington Post* editorial called DOT and FAA decision to deal with air traffic congestion "long overdue." Public had become aware that "saturation point had been passed early in July and those who use the air—passengers and pilots alike—have been suffering ever since. This suffering is going to go on for some time unless the FAA and the Department of Transportation move much faster in the future than they have in the last six weeks. The hope . . . that the aviation industry can solve this congestion problem within itself seems to be just wishful thinking. The airlines and the general-aviation people have not shown the slightest sign of producing constructive proposals either jointly or separately." (*W Post*, 8/18/68, B6)

August 19: Dr. Thomas O. Paine, NASA Deputy Administrator, announced elimination of lunar module (LM) operations from first manned Apollo/Saturn V flight, Apollo 8. MSF would begin planning and training for alternate, low-earth-orbit, manned command and service module (CSM) mission for December launch but no final decision on precise mission plan would be made until after Oct. 11 Apollo 7 flight results were evaluated. LM 3, delayed in checkout, would be flown in 1969 on fourth Saturn V with CSM 104 since preliminary studies had indicated many Apollo objectives scheduled for later flights could be obtained by using Apollo 8 CSM mission. Two problems in Saturn Apollo systems—vertical oscillation, or pogo effect, in 1st stage of Saturn V and rupture of small propellant lines in upper stages—had been corrected and solutions verified in extensive ground tests.

Plans and changes were discussed in detail at NASA Hq. news conference by Apollo Program Director L/G Samuel C. Phillips (USAF), who blamed checkout problems for six-week delay in Apollo schedule. "It is our view," he said, ". . . that the probable flight readiness of LM 3 is perhaps in February, but it is in the February-March time period." He said important advantage of flying Apollo 8 in 1968 was opportunity for earlier experience in operation of Saturn V and CSM than could otherwise be obtained. In response to questions Gen. Phillips said lunar landing next year was "clearly possible." He also announced reassignment of crews: redefined Apollo 8 mission would have crew of Frank Borman, James A. Lovell, Jr., and William A. Anders; crew originally assigned to this mission—James A. McDivitt, David R. Scott, and Russell L. Schweickart—would fly on Apollo 9 with LM. New backup crew for Apollo 8 was Neil A. Armstrong, Edwin E. Aldrin, Jr., and Fred W. Haise, Jr. (Text; NASA Release 68-148; AP, *C Trib*, 8/20/68; O'Toole, *W Post*, 8/20/68, A3; Wilford, *NYT*, 8/20/68, 19; Sehlstedt, *B Sun*, 8/20/68, 1; MSC *Roundup*, 8/30/68, 1)

- U.S. was negotiating with India on use of one visual and two voice channels on NASA ATS satellite to be launched by 1971 into fixed position 22,400 mi above Indian Ocean. In her pilot comsat project, India would broadcast several hours of educational TV weekly to seven areas with total 50 million population. Remaining channels would be used to broadcast NASA scientific data. India hoped eventually to transmit educational TV from comsat to community receivers in her 560,000 villages. (NASA Proj Off; NYTNS, *LA Her-Exam*, 8/19/68)

- NASA would ask industry to submit proposals on $50-million "quiet jet engine" development program, said Hal Taylor in *American Aviation*. Contractor would be selected early in 1969 with fabrication of components to begin shortly after. Development would take three years. While engine would have takeoff thrust rating of about 23,000 lb, larger than JT3D engine in Boeing 707 and McDonnell Douglas DC-8, NASA officials had emphasized technology used would be applicable to all sizes of jet engines. Allison and Pratt & Whitney had performed preliminary design contracts and NASA felt General Electric Co. had facilities and personnel to undertake the advanced technology. Initial objective was minimum noise reduction below existing turbofan engines of at least 15 pndb on takeoff and 20 pndb at landing. Once engine was developed and run successfully, technology would be turned over to industry to be incorporated into future jet engines. (*Am Av*, 8/19/68, 23)
- President Johnson signed H.R. 3136, bill authorizing three-year study of proposed increased use of metric system in U.S. by National Bureau of Standards on year-to-year basis out of already appropriated funds at cost up to $500,000 for current fiscal year. (White House PIO; *Science*, 8/23/68, 772; *PD*, 8/26/68)
- Univ. of Colorado professor Dr. George Gamow, leading theoretical physicist, astronomer, and author, died at age 64 of circulatory illness. Winner of 1956 UNESCO Kalinga Award for his nontechnical books on atomic energy and other scientific subjects, Dr. Gamow had published quantum theory of atomic nucleus in 1928, formulated Gamow-Teller selection rule for beta emissions, and been one of chief proponents of "big bang" theory of universe. (*NYT*, 8/22/68, 35; AP, *W Post*, 8/23/68, B10)

August 20: Nike-Apache sounding rocket launched by NASA from WSMR carried Dudley Observatory experiment to 95.8-mi (154.3-km) altitude to determine micrometeoroid flux in absence of an active meteoroid shower from comparison with data from Aug. 12 launch during meteoroid shower. Rocket and instruments performed satisfactorily. Payload was recovered undamaged. (NASA Rpt SRL)

- NASA issued Apollo Status Summary: Apollo 7 spacecraft would be electrically mated to Saturn IB launch vehicle at KSC Complex 34 next week. In Apollo 503 program, command module was scheduled to be mated to service module after Aug. 20. The three stages and instrument unit of Apollo 8 launch vehicle were mated at Complex 34 and undergoing tests. Modification and retests continued on Lunar Module 3 ascent and descent stages and modification and checkout on Saturn V 2nd stage for 504 mission. (Text)
- Los Angeles International Airport officials had initiated $300,000 test program to insulate 25 nearby houses against noise from jet aircraft taking off and landing 1,700 times per day. In effort to meet increasing resentment which had resulted in lawsuits against airport totaling more than $300 million in past six years, airport officials would soundproof

roofs, walls, doors, windows, and floors. Airport would also construct $170,000 "hush house" with four-inch-thick fiber glass panels where jet engines could be run at full power without disturbing nearby residents. (AP, W *Star*, 8/20/68, A11)

- Three scientists were named winners of 1967 Atoms for Peace Award: Sigvard Eklund of Sweden, Director-General of International Atomic Energy Agency (IAEA); Abdus Salam of Pakistan, Director of IAEA's International Center for Theoretical Physics in Trieste; and Henry DeWolf Smyth, U.S. Representative to IAEA and former AEC Commissioner. Presentation of gold medal and $30,000 honorarium to each would be made in New York Oct. 14. (AP, *NYT*, 8/22/68, 61)
- French test pilot André Turcat successfully piloted Anglo-French supersonic transport Concorde in 1.2-mi first formal runway test at Toulouse-Blagnac. Aircraft, equipped with Sud-Aviation fuselage and four Bristol-Siddely engines, was scheduled for fall maiden flight and 1971 production delivery. (*W Post*, 8/21/68, A16; *Av Wk*, 8/26/68)

August 21: NASA successfully launched three Nike-Apache sounding rockets from NASA Wallops Station carrying GSFC experiments to determine D-region electron density profile during solar flare. Rockets—launched into active electron density profile after flare was reported and, subsequently, into declining activity—reached 117.6-mi (189-km), 119.8-mi (192.7-km), and 119.6-mi (192-km) altitudes. All rockets and instrumentation performed satisfactorily. Good data were obtained. Radar data showed remarkable continuity for all three trajectories, indicating that rockets carried payloads to same ionospheric region. (NASA Rpts SRL)

- Nike-Tomahawk sounding rocket launched by NASA from NASA Wallops Station carried Univ. of Maryland-Geophysics Corp. of America experiment to 163-mi (262-km) altitude. Primary objectives were to evaluate capabilities and accuracies of pulse and thermal equalization probes, investigate electron energy distribution in normal daytime ionosphere, and evaluate use of wing-slope techniques with Langmuir probes. Rocket trajectory was nominal, but despin was higher than predicted. Problems were encountered in telemetry. GCA antennas broke off at deployment and no data were received. Clamshell deployed properly and exposed pulse probe as planned, but motor for boom deployment malfunctioned. No useful data were obtained. (NASA Rpt SRL)
- NASA test pilot William H. Dana flew X–15 No. 1 to 264,000-ft altitude and 3,443 mph (mach 4.71) during flight from Edwards AFB to conduct WTR experiment and check horizon scanner, sky brightness, fluidic probe, and fixed alpha ball. (X–15 Proj Off)

August 22: Cone-shaped RAM C–II spacecraft was successfully launched from NASA Wallops Station by four-stage Scout booster at 11:16 am EDT. After reentering earth's atmosphere at 17,000 mph, 264-lb spacecraft completed its eight-minute ballistic flight by impacting 725 mi downrange 150 mi northeast of Bermuda. No recovery was attempted.

Primary mission objective of RAM C–II—15-in-long, 26-in-dia spacecraft with 12-in-dia hemisphere nose—was to measure electron concentrations in flow field at discrete locations along spacecraft during reentry at medium velocity. Secondary objectives were to measure ion concentrations in flow field as function of distance from spacecraft, measure antenna performance and signal attenuation during reentry under specified conditions, and measure spacecraft environmental parameters, including accelerations, pressures and temperatures.

Launch was second in NASA's Project RAM (Radio Attenuation Measurement) to study methods for preventing loss of radio signals from reentering spacecraft. RAM C–I, launched Oct. 19, 1967, had successfully demonstrated effectiveness of water addition to flow field and use of X-band telemetry signals in maintaining communications through ionized plasma sheath. RAM program was managed by LaRC under OART direction. RAM C–II spacecraft was designed, fabricated, and tested by LaRC. (NASA Proj Off; LaRC Proj Off; NASA Release 68–146; WS Release 68–17)

- JPL researchers Dr. Arvydas J. Kliore and Dan L. Cain in *Journal of Atmospheric Sciences* reported findings similar to July 15 findings of Martin Marietta's Allan R. Barger that surface air pressure on Venus might be 75 or 100 times greater than that on earth, or four to five times greater than that reported by U.S.S.R. scientists from data supplied by *Venus IV* spacecraft. Kliore and Cain concluded that *Venus IV* either had landed on 15-mi-high Venusian peak undetected by earth radar or had stopped transmitting before it reached planet's solid surface.

  Precise radio tracking of *Mariner V*—U.S. spacecraft which had completed flyby mission by crossing Venus orbit Oct. 19, 1967, 24 hr after U.S.S.R. had reported parachute landing on Venus—had enabled Kliore and Cain to calculate position of spacecraft's radio beam, relative to Venus' center. Resulting profiles indicated Soviet probe had penetrated to radial distance of 3,774 mi from Venus' center, which point Soviets had taken to be Venus' surface. U.S. radar studies had shown Venus radius of only 3,759 mi. If radar data were accurate, *Venus IV* measurements were made at 15-mi altitude and reported conditions far less extreme than on surface. U.S. measurement of Venus surface temperature greater than 890°–900° F was consistent with estimates from passive radio astronomy and with results of radiometer experiment conducted by *Mariner II* in 1962, when surface temperature was found to be 800° F. (NASA Release 68–147; Pasadena *Independent*, 8/22/68; Parker, *Denver Post*, 8/22/68)

- NASA Administrator for Manned Space Flight, Dr. George E. Mueller, accepting 1968 Veterans of Foreign Wars Space Award in Detroit, Mich., said: "The long-term outlook for space is good. I have just returned from the first UN Conference on the Peaceful Uses of Space. There has been a world-wide increase in awareness of the values and practicality of space flight. The nations of the world have become more conscious of the economic benefits of the space program, and the public is becoming increasingly aware of the values derived from space technology. With the increase in public awareness of the importance of space technology to the future well-being of the Nation, I believe that the new Congress and the new President will and must reaffirm support for a dynamic U.S. Space Program. . . . Man will prevail in space. On that there can be no serious question. The only question is whether this Nation will prevail in space." (Text)

- *Wall Street Journal* commented on DOT plan for easing air traffic congestion: "Many people are flying 100 or 200 miles, or even shorter distances, only because reasonably rapid and reliable rail service simply is no longer available. Before building airports still farther from where people want to go—and further jamming the highways leading to them —someone should start looking at the broader picture. While the Government's new air traffic plan may represent necessary movement on a problem that demands action, it scarcely qualifies as transportation progress." (*WSJ*, 8/22/68, 14)

August 23: After meeting privately with more than 100 representatives of aviation industry, Secretary of Transportation Alan S. Boyd said Government would propose regulations in September to unsnarl traffic jam over Washington, New York, and Chicago. (Lardner, *W Post*, 8/24/68, A3)

- NASA announced Astronaut John A. Lewellyn had withdrawn from astronaut training program because of inability to progress sufficiently with jet flight training at Reese AFB, Tex. After discussions with NASA and USAF officials, he had withdrawn to consider several opportunities within NASA and elsewhere. His departure reduced total number of NASA astronauts to 52. Nine of eleven scientist-astronauts selected in 1967 were in flight training. Dr. Brian T. O'Leary had withdrawn April 23 because he disliked flying. (MSC Release 68–63; AP, *NYT*, 8/24/68, 12)

- Dr. Alan H. Barrett and Dr. William J. Wilson of MIT reported in *Science* discovery of hydroxyl radio emissions from four infrared stars in Cygnus constellation, observed from National Radio Astronomy Observatory in West Virginia. They speculated that emissions indicated definite stage in stars' evolution. (The older a star gets the hotter it becomes, until it has almost no oxygen and hydrogen.) Further studies, supported partially by NASA, would be conducted during next few years in attempt to determine how far young stars had progressed in process of evolution by observing hydroxyl emissions. (*Science*, 8/23/68, 778–9; Wilford, *NYT*, 8/25/68, 19)

August 24: France became fifth nation to explode hydrogen bomb—joining U.S., U.S.S.R., U.K., and Communist China—when she detonated two-megaton balloon-borne device 1,800 ft above Fangataufa Atoll in Pacific. French Defense Minister Pierre Messmer later told press conference device weighed 1½–2 tons and compared favorably with devices detonated by U.S. in 1956. He stressed that fallout would be minimal because of explosion's altitude and "particularly satisfactory" weather conditions. He also revealed that France's new submarine missile had become operational with final explosion July 16 and predicted that French armed forces could be equipped with nuclear weapons within several years. (AP, *W Post*, 8/25/68, A1; Hess, *NYT*, 8/28/68, 4; Fenton, *B Sun*, 8/28/68, A2)

August 25: U.S.S.R.'s leading geneticist, Prof. Nikolay P. Dubinin, told International Symposium on Genetic Effects of Space Environment in Tokyo biological experiments on various Soviet satellites beginning with *Sputnik I* had disclosed that space flight caused higher frequency of mutation, chromosome damage, and disruptions in normal cell division mechanisms. "Dynamic flight factors," he said, were responsible; doses of cosmic radiation registered were too small to account for all genetic changes observed. Symposium was held in conjunction with 12th International Congress of Genetics. American scientists also suggested "dynamic flight factors" of vibrations, acceleration, and weightlessness might be factors in genetic damage observed aboard American spacecraft. (Reinhold, *NYT*, 8/26/68)

- *St. Louis Post-Dispatch* editorial on NAS–NRC Space Science Board recommendation that NASA "embark on an ambitious program" of interplanetary flights and expand other areas: "The program would be costly, of course, but nowhere near as costly as war, and the benefits, if the goals were reached, would be incalculable. Expenditures for purposes such as this must be fitted into the schedule of national priorities. Certainly outlays to relieve the plight of our cities and to end poverty come first. But the United States is uniquely qualified to take the lead in the sort of peaceful scientific endeavor that requires both technical skills and money, and it should not lose sight of its obligation to expand the area of human knowledge." (*St. Louis P–D*, 8/25/68)

August 26: New comet of undetermined size had been discovered by amateur astronomers John Bally-Urban and Pat Clayton, who were attending Southwest Astronomical Conference at New Mexico State Univ. Using 10-in telescope, they made discovery while observing comet Honda. Smithsonian Astrophysical Observatory had confirmed sighting of comet, which had been given official name of Bally Clayton comet. (AP, *W Star*, 8/26/68, A2)

August 27: U.S.S.R. launched two Cosmos satellites. *Cosmos CCXXXVI* entered orbit with 627-km (389.6-mi) apogee, 590-km (366.6-mi) perigee, 96.8-min period, and 56° inclination. *Cosmos CCXXXVII* entered orbit with 323-km (200.7-mi) apogee, 199-km (123.6-mi) perigee, 89.7-min period, and 65.4° inclination and reentered Sept. 4. (GSFC *SSR*, 8/31/68; 9/15/68)

- NASA issued Apollo Status Summary: Integrated Systems Test had been completed on Apollo 7 command and service module (CSM), Apollo 8 CSM would undergo combined systems tests at KSC early in week. Saturn 503 had been erected on mobile launcher. (Text)
- USAF announced actions to reduce FY 1969 expenditures by $219 million in accordance with Revenue and Expenditures Control Act of 1968. Savings of $18 million would be achieved by deactivating seven F–101 squadrons earlier than planned and closing Paine Field, Wash. Early phase-out would reduce military space authorizations by 2,661 and civilian authorizations by 372 and would remove 126 aircraft from active inventory. Revised production schedules and stretch-out of F–111A and FB–111 programs would save $201 million. Production would be leveled off at combined production rate, below previously approved rates for each aircraft, causing extended force buildup, but not delaying initial deliveries. (DOD Release 800–68; Sehlstedt, B *Sun*, 8/28/68, A2; AP, *P Inq*, 8/28/68, 10)
- NASA Administrator James E. Webb conferred with West German Science Minister Gerhard Stoltenberg in Bonn on U.S.-German space projects. (AP, *Min Trib*, 8/28/68)
- USAF awarded United Aircraft Corp.'s Pratt & Whitney Div. and General Electric Co. separate letter contracts and $11-million initial obligations to develop and test prototype of high-performance afterburning turbofan engine for USAF FX air superiority fighter and USN VFX advanced fighter. Funded jointly by USAF and USN, 18-mo competitive contracts were valued at $100 million. (DOD Release 799–68)

August 28: Cosmos CCXXXVIII was launched by U.S.S.R. into orbit with

August 28

210-km (130.5-mi) apogee, 203-km (126.1-mi) perigee, 88.6-min period, and 51.7° inclination. Satellite reentered Sept. 1. (GSFC *SSR*, 8/31/68; 9/15/68)

- NASA announced award to Bendix Corp. of $1,567,550 cost-plus-incentive-fee contract for development, test, and delivery of one prototype and five flight-model ATM star trackers in 15-mo period. ATM, part of Apollo Applications program, would be launched into earth orbit and docked with Saturn I Workshop being developed by MSFC for 28- to 56-day astronaut stays in space. Astronauts would use ATM to study sun from above earth's atmosphere. (MSFC Release 68–196)

August 28–30: AIAA and American Bar Assn. held joint Impact of Aerospace Science and Technology on Law and Government Conference in Washington, D.C.

Dr. Edward C. Welsh, NASC Executive Secretary, asserted U.S. could afford multibillion dollar space program; "at least as much as President Johnson proposed this year—nearly $7,000,000,000—and perhaps an appreciable amount more." (Figure included all Government spending, not just NASA funds.) Persons advocating big cuts in space program were "doing this country a disservice." Much of high U.S. gross national product was result of prior investment in R&D. "This Nation invests more per capita in research than any other nation in the world and that is a major reason why we have a higher standard of living. The space program is the most active, economical, and productive stimulus to research and development in history. We cannot afford to slow it down . . . unless we think we can afford less investment in the elimination of poverty, less expenditure for health improvement, less income for an expanding population, and less readiness in the realm of national security." Space program cost, though large, was "not really so expensive when we figure that we are spending less than 1% of our Gross National Product for a technological capability that increases the rate at which this Nation becomes wealthier and more powerful by many times 1%." (Text; B *Sun*, 8/29/68, A5; AP, *LA Times*, 8/29/68)

Dr. Charles S. Sheldon II, Library of Congress Research Specialist, said U.S.S.R. was launching more military espionage satellites than any other spacecraft and therefore had softened complaints about U.S. espionage satellite activity. Tracing political and social implications of Soviet space program, Dr. Sheldon said U.S.S.R. accepted "importance of basic research on the cosmic scale" and vigorous exploitation of space applications for civil purposes was definite goal. In long-term outlook, U.S.S.R. talked of "future automated, cybernetics-aided industrial state . . . in which space applications, space travel, and use of extraterrestrial resources play a vital part." He also credited NASA program of international cooperation with some influence on recent Soviet shift toward increased cooperation within and without Soviet bloc, though joint U.S.–U.S.S.R. program for space exploration would probably prove inoperable even if it could be established. (AIAA Paper 68–900; AP, *NYT*, 8/30/68, 7)

M/G Jewell C. Maxwell (USAF), director of FAA SST development, said cost of designing and developing two prototype commercial SSTs would amount to $1.587 billion. Government share in total spending from 1963 through 1972 would be $1.239 billion, or 78%. Contractors,

Boeing Co. and General Electric Co., would pay $287.8 million, or 18%, and customer airlines would contribute $59.5 million, or 4%. Principal problem in financing, he said, was protection of manufacturers against catastrophic failure. This would require "some limitation of recourse of creditors . . . in event of program failure."

In addition, Gen. Maxwell forecast $2.5-billion to $5-billion market risk for period starting in 1972 and $15 billion commercial risk for period between start of sales and end of century. (AIAA Paper 68-916; AP, *NYT*, 9/1/68, 22)

August 29: NASA's *Pegasus I, II,* and *III* meteoroid technology satellites launched Feb. 16, May 25, and July 30, 1965, were turned off by ground command after more than three years successful operation. Designed for 18-mo lifetime, Pegasus satellites, by detecting meteoroids, had confirmed protective adequacy of Apollo spacecraft for manned lunar missions but indicated that spacecraft for longer missions might encounter protection design problems. Spacecraft also provided data on Van Allen belts, earth reflectivity, solar constant, orbital and gyroscopic motions of rigid bodies, degradation of surface coatings in space, thermal control systems, and lifetime of electronic components in space operations. Pegasus program had been managed by MSFC under OART direction. (NASA Releases 68-149, 68-149A, 68-149B; *Marshall Star*, 8/21/68, 1)

- NASA scientists working with Deep Space Network (DSN) had quadrupled distances over which signals from three Pioneer spacecraft currently in interplanetary space could be heard. Pioneers could view sun from every side. Improvements in DSN receivers so far had allowed return of 50% more of their data than had been planned. All three Pioneers—now 29, 116, and 182 million mi from earth—could be heard through DSN antennas, ensuring return of data until they wore out. Together they had, to date, amassed 55 mo in orbit, measuring particles and fields of sun's atmosphere and returning seven billion bits of data. Change of polarization of tracking antennas from circular to linear had doubled received signal power. (ARC Release 68-13; ARC *Astrogram*, 8/29/68, 1)

- President Johnson signed Executive Order 11424 ensuring "flight pay" to military personnel flying spacecraft and incentive pay for hazardous duty on same basis as for those flying in conventional aircraft. Order further ensured incentive pay for personnel injured in an "aviation accident" for three months without their completing performance requirements. Order included word "spacecraft" in previous Executive Order related to hazardous duty pay for "aerial flight" and "aviation accident." (*PD*, 9/2/68, 1285)

- National Pilots Assn. issued statement defending "first-come, first-served" principle of air traffic control and opposed priorities for air carriers as solution to mounting congestion at major airports. Instead, NPA recommended upgrading pilot and aircraft requirements for using New York, Chicago, and Washington terminals. It specifically advocated requiring IFR equipment including transponder for aircraft, instrument rating for pilots, and aircraft capable of maintaining specified approach speeds. (NPA Release)

- Library of Congress had acquired from former astronaut John H. Glenn, Jr., *The John Glenn Papers*, 90,000 items representing national and international response to his first manned space flight, Feb. 20, 1962. Majority of letters covering 1962–64 were from ordinary citizens, largely elementary school children. Other items included subject files, space manuals, news clippings, scrapbooks, invitations, certificates, awards, maps, and charts which Library's Information Bulletin said "reflect the contemporary need for a popular hero—one who can be held up to the young and who at the same time must undergo the rigorous pressures brought about by 'instant' fame and worldwide recognition." Papers provided "many insights into contemporary social history." Glenn had been asked for his opinion on every conceivable topic. (Sifton, LC *Info Bull,* 8/29/68)

August 30: Apollo 7 spacecraft and launch vehicle were electrically mated at KSC Launch Complex 34. (NASA Apollo Status Summary, 9/4/68)

- ComSatCorp had notified FCC of its intention to award $270,000 fixed-price contract to Hughes Aircraft Co. for installation, operation, and maintenance of transportable earth station in California to transmit, via *Intelsat II* satellite, TV coverage of Mexican Olympics to Pacific area in October. (ComSatCorp Release 68–45)

During August: In *Astronautics & Aeronautics* editorial AIAA President Floyd L. Thompson heralded U.N. Conference on Exploration and Peaceful Uses of Outer Space in Vienna, August 14–27. Against background of new developments in arms control, it would emphasize practical benefits of space exploration, especially as to how they could aid developing nations. "This action by the UN points up the continuing concern of many about the need for cooperation among nations in the exploration of space. It recognizes that space represents an important element of national power, a force for growth, particularly in advanced technology and education." (*A&A,* 8/68, 17)

September 1968

September 1: In Baltimore *Sun* William J. Perkinson warned against "A Czechoslovakia In Space." He said experts explained U.S. and other free-world nations "must insure that future Russian leaders will never be able to blackmail nations or whole continents of nations into submitting to the will of such Soviet leaders the way the leaders of the Kremlin forced Czechoslovakia to yield through the invasion of that country on the night of August 21."

By marrying technology of "spook," or hovering, satellite with that used to produce MIRV and by use of more powerful rockets, "it would be possible to launch truly 'orbiting men-of-war' that could be used far more effectively to cow any nation than the most powerful of gunboats were able to do in the days of 'gunboat diplomacy.' " Blackmail might not occur, if arms limitation treaties granted each nation right to board and inspect all heavy satellites in orbit. "But even that right would have to be backed up by means of spacecraft and spacemen capable of exercising that right. That . . . is why the invasion of Czechoslovakia is certain to spark new interest in manned space flight." (B *Sun*, 9/1/68, 3)

September 2: John Lannan in Washington *Evening Star* called GSFC "home base" for Apollo 7 astronauts. All their communications would pass through Center. Communicating with some or all lunar spacecraft would be 14 land stations, 4 ships, and 8 aircraft around the world, manned by 4,000 persons. All circuits used would be put together from GSFC switching centers. Except for astronaut's electrocardiagram and his voice, all mission data would be channeled in digital form, which meant vast bulk of detail controlling system itself, in addition to mission data from Houston or spacecraft, could be handled rapidly by computers. (W *Star*, 9/2/68, 1)

- AEC was developing plans for "nuplexes," giant agro-industrial complexes built around nuclear reactors and using advanced agricultural and industrial technologies, said AEC Chairman, Dr. Glenn T. Seaborg, in *U.S. News & World Report*. While "modest type" could be built using reactors of type under construction, eventually feasible $1-billion nuplex could sustain 100,000 farmers, laborers, and their families and feed 5 million others, while exporting fertilizer to grow food for additional 50 million people. Nuplexes would generate own electricity and pump oceans for unlimited quantities of water for irrigation and industrial uses. Seaborg saw India and Middle East as specific areas for nuplexes. (*W Post*, 8/26/68, A12; *US News*, 9/2/68, 62)

September 3: Lunar Test Article (LTA 2R)—carried onboard *Apollo 6* flight April 4 to measure vibration, acoustics, and structural integrity for lunar module—reentered earth's atmosphere and splashed into Pacific. Tracking station lost contact with debris in orbit at about 10 am

EDT but calculated its trajectory to landing area. (NASA Release 68–155; AP, B *Sun*, 9/11/68, A3)

September 4: NASA Aerobee 150 sounding rocket launched from WSMR carried National Center for Atmospheric Research experiment to 77-mi (124-km) altitude to retrieve high-intensity air sample from 26.7- to 40.4-mi (43- to 65-km) altitude, using low-temperature condenser. All cryogenic heat exchange and sampler functions performed satisfactorily; recovery parachute deployed and payload survived impact. Sample of eight moles of air was recovered between 26.8-mi (43.2-km) and 38.9-mi (62.8-km) altitudes. (NASA Rpt SRL)

- NASA issued Apollo Status Summary: Apollo 7 crew was participating in verification test of electrically mated spacecraft and launch vehicle. For Apollo 8 mission, combined systems test on Command and Service Module 103 and checkout of Apollo Saturn 503 launch vehicle would continue through week at KSC. (Text)

- FAA issued notice of proposed rule to restrict number, type, and equipment of aircraft using "high density" airports. New York's Kennedy International and La Guardia, Newark, Chicago's O'Hare, and Washington National would be allocated 80, 60, 60, 135, and 60 operations per hr from 6:00 am to midnight. Rule would set minimum airspeed of 150 knots while under air traffic approach control and require all aircraft to be equipped with radar beacon transponder with 64-code capability and two pilots while under reservations plan. Prior approved departure or arrival reservations would be required for each flight operated under IFR or from designated high-traffic airport. Public hearing on proposed rule would be held at FAA Hq. Sept. 25. (FAA Release 68–53)

- At Fort Worth, Tex., ceremonies, Australian Minister for Defence Allen Fairhall accepted first of Australia's 24 F-111C aircraft on order from General Dynamics Corp. U.S. Secretary of the Air Force, Dr. Harold Brown, told assembled dignitaries F-111 "institutes a new generation of aircraft. . . . We have discovered . . . that when compared to fixed-wing aircraft, the variable geometry wing requires far fewer compromises to obtain the desired performance." (DOD Releases 779–68, 812–68; AFSC *Newsreview*, 9/68, 1)

- Dr. George Kozmetsky, dean of Univ. of Texas College of Business and of Graduate School of Business, had been sworn in as consultant to NASA Administrator James E. Webb, NASA announced. Specialist in system analysis, organization theory, quantative methods, and system management, Dr. Kozmetsky would serve on NASA's Management Advisory Panel. (NASA Release 68–153)

September 5: Cosmos *CCXXXIX* was launched by U.S.S.R. into orbit with 262-km (162.8-mi) apogee, 198-km (123-mi) perigee, 89.1-min period, and 51.8° inclination. Satellite reentered Sept. 13. (GSFC *SSR*, 9/15/68)

- *Washington Daily News* story by London Express Service said scientists "have told the military that a new infra-red sensing device tested in a secret American reconnaissance satellite may be able, from space, to detect the elusive submarines as they lurk under water. And the invention is said to work whether the quarry is moving, hovering in midwater, or even lying silently on the bottom. . . . More tests are to be made. But it . . . seems fairly certain that the system will be put into full operation soon." (*W News*, 9/5/68, 6)

- *USAF* announced Aug. 25 discovery of new F-111A problem during fatigue test at General Dynamics-Convair plant, San Diego, Calif., would extend restrictions on aircraft. Problem was failure of steel fitting on which swing-wings pivoted. Technical officers believed crack in metal plate was caused by faulty bolt hole rather than serious design deficiency. DOD said it would take 10 to 14 days to complete analysis. (*Aero Daily*, 9/6/68; *W Post*, 9/6/68, 1; AP, *W Star*, 9/6/68, C5; Sehlstedt, *B Sun*, 9/7/68, A7)
- President Johnson issued Executive Order 11428 terminating President's Advisory Committee on Supersonic Transport, which had been established April 1, 1964. (*PD*, 9/9/68, 1309)

September 6: NASA announced it had terminated production of 60 H-1 engines for post-Apollo Saturn IB missions under contract with North American Rockwell Corp.'s Rocketdyne Div. Delivery of 32 engines to MSFC had already been completed. Of remaining 28 engines, 1 would be complete; 1, complete without firing; 6, assembled; and hardware for 20, delivered in present production state. (NASA Release 68-154; MSC *Roundup*, 9/13/68, 1)

- Charles G. Haynes, Director of NASA Inspections Div., had been assigned additional duties as Acting Director of Headquarters Administration, NASA announced. He would replace Alfred S. Hodgson, who was on extended leave. (NASA Ann, 9/6/68)
- Col. Robert L. Stevens (USAF), holder of four of nine new world speed and altitude records set in Lockheed YF-12A jet interceptor, was named technical adviser to FAA SST Program Director, M/G Jewell C. Maxwell (USAF). (FAA Release 68-32)

September 7: During visit to MSFC, Republican presidential candidate Richard M. Nixon said of space program, "I would have this clearly understood: that I consider this program as one of our national imperatives, that it must be supported at a level assuring efficient and steady progress, that the ups and downs . . . in planning, programming and financing must be brought to a halt, and that as President I will make certain our country retains leadership in this great endeavor. . . . It is an inescapable fact of our national life today that we cannot afford to do all we wish. But we must do all we can. . . . I assert my conviction that among the claimants for Federal support I consider the space program both indispensable and of major importance to our country. . . . In every area of science—in every area of knowledge—the United States must continue to probe the unknown. In terms of our long-range security and growth, the most critical gap which we could allow to develop would be a research gap." (Text; *NYT*, 9/7/68, 1; UPI, *W Post*, 9/7/68 [photo]; *B Sun*, 9/7/68)

- Max Conrad, holder of world record for straight-line distance (7,688.5 mi) in light aircraft, set June 2–4, 1959, claimed new world distance record of 4,968 mi in closed-circuit route. He landed his twin-engine Piper Aztec at Lambert Field in St. Louis, Mo., after flying 621-mi triangular route between St. Louis, Des Moines, Iowa, and Kansas City, Mo., eight times in 37 hr 50 min. (*World Almanac*, 68, 752; UPI, *W Post*, 9/9/68, C5)

September 8: France exploded second hydrogen bomb—believed to be less than two megatons—suspended above Mururoa Atoll in the Pacific. First test had been conducted Aug. 24. (AP, B *Sun*, 9/2/68, 2; *W Post*, 9/9/68, A22)

September 9: Apollo 7 prime and backup crews successfully completed egress test at KSC. Astronauts performed both aided and unaided egress from spacecraft under simulated emergency conditions. At MSC three astronauts completed five-day vacuum chamber tests in spacecraft similar to command and service modules to be flown on second manned Apollo mission and lunar flights. (NASA Apollo Status Summary, 9/10/69)

- *U.S. News & World Report* said story of how U.S. had been steadily stripping away "some of the most valuable military secrets of Soviet Russia for the past five years" was being revealed in bits and pieces. At heart of story were spy satellites orbiting earth. Powerful spaceborne cameras could photograph in detail objects less than one foot in diameter from 150-mi altitude. "Ultramodern" radar circling in outer space now penetrated cloud covers and forest vegetation to reveal hidden missile sites, tanks, and even troops. Infrared sensors could detect submarines and missile launching. Electronic sensors could monitor atomic explosions and pick up telephone messages from deep within U.S.S.R. or Communist China. President Johnson had said spy satellites alone made U.S. space investment worthwhile.

  USAF had, to date, launched more than 200 classified payloads. At least one intelligence satellite was always ready for orbit. Most stayed operational for 1 to 25 days. Even more useful, Manned Orbiting Laboratory (MOL) was yet to come. U.S.S.R. was particularly concerned about its intelligence implications. "The Russians are starting now to understand that one of their centuries-old weapons against the rest of the world—their secrecy—is rapidly being removed." (*US News*, 9/9/68, 2)

September 10: USAF launched unidentified satellite from Vandenberg AFB by Titan III–B booster into orbit with 200-mi (321.9-km) apogee, 89-mi (143.2-km) perigee, 89.1-min period, and 106° inclination. Satellite reentered Sept. 25. (*Pres Rpt 68*; GSFC *SSR*, 9/15/68, 9/30/68)

- XB–70A, flown by NASA test pilots Donald L. Mallick and Fitzhugh L. Fulton, Jr., reached 63,000-ft altitude and mach 2.54 in flight from Edwards AFB. Accomplished successfully were ILAF systems check; exciter vane tests; evaluation of stability, control, and handling qualities; and observation of duct turbulence. (NASA Proj Off)

- USAF announced successful completion on Aug. 6 of initial flight phase of C–5A jet transport, after which aircraft had been put into six week planned layup for "configuration update and instrumentation for flutter testing." It had accumulated 23 hr 48 min in flights on June 30; July 13, 20, 25, and 31; and Aug. 1 and 6. During evaluation, C–5A had flown at maximum gross weight of 557,000 lb, logged gross maximum taxi weight of 732,000 lb, and reached 250-knot top speed and 11,000-ft maximum altitude. C–5A Galaxy was designed to operate at long-range cruise speed of 440 knots with basic mission weight of 712,000 lb. (*Aero Daily*, 9/11/68)

- Astronomer Dr. Gerritt L. Verschuur, with aid of National Radio Astronomy Observatory, had measured for first time strength of Milky Way's magnetic field as it existed in far reaches of the galaxy, to help explain theories of star formation, radio wave propagation, and cosmic ray acceleration unprovable previously. (*W Post*, 9/10/68, A10)

- Col. Albert J. Wetzel (USAF, Ret.), Director of Sponsored Programs at Tulane Univ., was sworn in as consultant to NASA Administrator James E. Webb. Former technical assistant to Director of Defense Research and Engineering and former Titan ICBM program director, Col. Wetzel would advise Webb on NASA Project Management System and other technical and personnel management activities. (NASA Release 68-156)
- NASA Deep Space Network celebrated 1,000th tracking and data acquisition support pass of *Pioneer VI* spacecraft, launched Dec. 16, 1965, as first of successful series. Pass was monitored by 210-ft antenna at Goldstone, Calif., where Pioneer Tracking and Data Systems Manager, Dr. N. A. Renzetti, congratulated members of JPL Pioneer Operations Team. (JPL *Lab-Oratory*, 10/21/68, 2)
- NAA selected Sen. Warren G. Magnuson (D-Wash.) to receive its 1968 Wright Brothers Memorial Trophy for "more than two decades of dynamic leadership in developing national and international policy that has assured United States preeminence in aeronautics throughout the world and has contributed immeasurably to the health and vitality of America's economic structure." Trophy would be presented Dec. 17 at Annual Wright Brothers Memorial Dinner in Washington, D.C. (NAA Release)
- George C. Wilson in *Washington Post* wrote: "No one is willing to predict what form a new Soviet psychological sputnik would take . . . [but] it was a gut feeling among space leaders that the United States will pay the price eventually for cutting back on its space program while the Soviets move right ahead on all fronts with theirs." Even Saturn V rocket "does not look big enough to close the thrust gap. NASA Administrator James E. Webb—one of the most underrated administrators in Washington—warns that the Soviets are about to fly a rocket much bigger. . . ."

  By-product of big boosters was luxury of weight they allowed comsats. U.S. was using satellites weighing 357 lb at launch and 192 lb in orbit. U.S.S.R. was talking about 1,000-lb comsats for competing international communications system. "The Soviet Union's bigness in boosters and payloads will be an advantage in this contest between the two AT&Ts of space." (*W Post*, 9/10,/68, 14)

September 11: NASA selected Fairchild Hiller Corp. and General Electric Co. to develop spacecraft designs for Applications Technology Satellites (ATS) F and G under $4.6-million fixed-price contracts. Configuration would consist of 30-ft-dia deployable antenna and precision control system. Both spacecraft would carry advanced communications, navigation, weather, technical, and scientific experiments. They were scheduled for 1972 and 1973 synchronous orbits at 22,300-mi altitude. Each company would design and develop basic configuration for both spacecraft and provide specifications, drawings, and test procedures. (NASA Release 68-157)

- Air Force Cambridge Research Laboratories launched largest research balloon—28.7-million-cu-ft polyethylene sphere—to record 158,000-ft altitude from WSMR. Flight carried instruments for atmospheric measurements near stratopause and terminated near Needles, Calif., after 18 hr. Previous record size for polyethylene balloon was 13.5 million cu ft.

Previous balloon record, 26 million cu ft, was held by five fiber-reinforced Mylar balloons flown by AFCRL for NASA in 1966 and 1967. (OAR *Research Review*, 1–2/69, 8–9)

- Republican Presidential candidate Richard M. Nixon announced decision to request Senate postponement of approval of nuclear nonproliferation treaty. "Despite my concern over some of its provisions, I have endorsed [the treaty]. . . . I hope that it can be universally adopted. But the Soviet invasion of Czechoslovakia has seriously damaged the prospects for early ratification of the treaty."

  Nixon statement came as Republicans blocked meeting of Senate Foreign Relations Committee for second day, preventing voting on treaty approval. (Semple, Finney, *NYT*, 9/12/68, 1; Oberdorfer, *W Post*, 9/12/68, 1)

September 12: Institute for Strategic Studies in London issued report, "The Military Balance," which said U.S.S.R. had rushed construction of land-based ICBMs and now had almost as many as U.S. By end of 1968 Soviet total was expected to reach U.S. total of 1,054. (Shuster, *NYT*, 9/13/68, 9)

- Slide-wire escape way, 1,200-ft-long, from 215-ft level to ground at KSC Complex 34 was declared man-rated following tests by five dummies and five men. Slide wire was designed as last means of escape from top of Saturn IB gantry for astronauts and close-out crews if emergency should shut off other routes. (*Spaceport News*, 9/12/68, 1)

- NASA Associate Administrator for Manned Space Flight, Dr. George E. Mueller, told World Affairs Council in Pittsburgh, Pa.: "The U.S. space program was undertaken in 1958, and accelerated by three Presidents and six Congresses who considered it basic to our national strength and essential to our continued leadership of the Free World. It is . . . significant to note that this has been a bipartisan effort, with Republican as well as Democratic support. . . . Space expenditures contribute significantly to the national power of the United States in a world where military and economic security increasingly rest upon technology." U.S. space program "has been deliberately oriented toward cooperation with other countries. It is providing opportunities for foreign scientists to contribute and develop their talents and, at the same time, gives other nations a chance to share not only in the published results of space research, but in the accomplishment of these achievements as well. . . ." (Text)

- Senate adopted S.R. 391, changing assignments of minority Senators to standing committees. Measure excused Senators Edward W. Brooke (R-Mass.) and Charles H. Percy (R-Ill.) from Senate Committee on Aeronautical and Space Sciences and assigned Senators Mark O. Hatfield (R-Ore.) and Charles E. Goodell (R-N.Y.) to Committee. (NASA LAR VII/90)

September 13: Explorer XXXVIII (RAE–A), launched from WTR July 4, was adjudged successful by NASA. Satellite had completed 30 days of successful three-axis gravity-gradient-stabilized operation in orbit as of Aug. 22. Experiment instrumentation and all spacecraft support systems were operational and functioning satisfactorily. Recording of variations in galactic emission, low-frequency solar bursts, and radio signals from earth had been acquired. Some evidence had been found of enhanced noise bands at medium frequencies within magnetosphere, and strong noise enhancement had been observed in vicinity of South Atlantic anomaly. (NASA Proj Off)

- Aerobee 150 sounding rocket launched by NASA from portable launcher at WSMR carried GSFC payload to 3.1-mi (5.05-km) altitude. Objectives were to flight-qualify production lot of VAM-20 variboosters, demonstrate Aerobee rail launch feasibility, determine exiting loads on rail launcher, establish procedures for rail launching, and verify boost-phase dispersion calculations. Booster web burning time, booster pressure, acceleration, velocity, and system impact were as predicted. (NASA Rpt SRL)

  X-15 No. 1 was successfully flown by Maj. William J. Knight (USAF) to 254,100-ft altitude and 3,716 mph (mach 5.26) in test from Edwards AFB. Primary objectives were to conduct WTR experiment and check out horizon scanner, fixed alpha cone, and fluidic probe. (X-15 Proj Off)

- President Johnson selected Dr. Glenn T. Seaborg, AEC Chairman, to head U.S. delegation to 12th conference of International Atomic Energy Agency, which would open in Vienna Sept. 24. (UPI, *NYT*, 9/15/68, 34)

September 14: U.S.S.R. launched *Cosmos CCXL* into orbit with 283-km (175.8-mi) apogee, 203-km (126.1-mi) perigee, 89.3-min period, and 51.8° inclination. Satellite reentered Sept. 21. (GSFC *SSR*, 9/15/68, 9/30/68)

- While Soviet scientists believed in use of some drugs to help man adapt to long space flights, U.S. medical experts wanted to avoid use of medication in orbit because of unpredictable side effects they sometimes produced, Associated Press reported. U.S. astronauts had carried only nonbromide, antimotion sickness pills and these had been used only once, by *Gemini VIII* pilots Neil A. Armstrong and David R. Scott when stuck thruster caused spacecraft to roll and forced emergency landing. Soviet representatives at Vienna U.N. Conference on Peaceful Uses of Outer Space had indicated most U.S.S.R. cosmonauts had used depressants for sleeping in orbit and stimulants to counteract fatigue. To avoid need for drugs, NASA selected only highly qualified astronaut candidates and trained them to cope with situations they might find in space. (B *Sun*, 9/16/68, 8)

- Joint Committee on Atomic Energy called for prompt DOD action to revive work on quiet electric-drive nuclear submarine halted in May by DOD order when it decided to proceed with new high-speed nuclear submarine [see July 24]. (Transcript; Greene, *NY News*, 9/20/68, 8)

September 15-22: *Zond V* automatic space station was launched by U.S.S.R. and placed on lunar trajectory from parking orbit of another satellite to explore outer space and test spacecraft systems, Tass announced. All systems were functioning normally. Speculation in Moscow, later confirmed, was that spacecraft would attempt to circle moon. On Sept. 18 Sir Bernard Lovell, Director of U.K.'s Jodrell Bank Experimental Station, said spacecraft had passed within 1,000 mi of moon and was returning to earth and predicted U.S.S.R. would attempt to recover it. West Germany's Bochum Observatory agreed that *Zond V* was returning but insisted spacecraft had circled moon. U.S.S.R. initially denied both reports, but two days later, confirmed that *Zond V* had circled moon Sept. 18 and was continuing its flight. Tass said spacecraft had flown within 1,200 mi of moon and obtained data on physical characteristics of outer space near moon. Research mission had been completed and spacecraft was continuing to relay information to space stations.

Zond V's reentry and splashdown in Indian Ocean Sept. 21 was recorded by U.S., U.K., and West German space scientists, but U.S.S.R. made no official announcement until Sept 22. Tass then said *Zond V* had circled moon, explored space near moon, reentered at 11,000 mi per sec, softlanded by parachute in predetermined area, and was recovered by Soviet recovery ship—becoming first circumlunar spacecraft recovered on earth. "During the flight the station's systems and aggregates for maneuvering on the trajectory and for returning to the Earth were tested. Flight control systems of the station and the radio-engineering means for measuring the parameters of its trajectory ensured the successful execution of the programme.

"The successful flight of the 'Zond–5' automatic station over the Earth-Moon-Earth route, and its return to the predetermined area are an outstanding achievement of Soviet science and engineering. Another scientific-engineering problem has been solved, and broad prospects have been opened up for further research of outer space and planets of the solar system by automatic space stations which bring back research data to the Earth."

Tass announcement that atmosphere and pressure "remained within the limits of their present range" supported earlier speculation that *Zond V* carried prototype passenger cabin with atmosphere of artificial helium mixture which would be used by cosmonauts on future flights. Later Tass report quoted Soviet Prof. Leonid I. Sedov as saying spacecraft had special heat shield and was slowed down during reentry by "air resistance" and, "at a comparatively small height," by parachute.

Zond V was fifth spacecraft in Zond series and first Soviet spacecraft to land in water. *Zond I* (launched April 2, 1964) had failed in attempt to reach Venus and *Zond II* (launched Nov. 30, 1964) had suffered communications failure enroute to Mars. *Zond III* (launched July 18, 1965) had obtained photos of far side of moon; *Zond IV* (launched March 2, 1968) apparently had reached apogee comparable to lunar altitude and had burned in earth's atmosphere on reentry. (GSFC *SSR*, 9/15/68; 9/30/68; *W Post*, 9/16/68, A17; 9/19/68, A25; 9/23/68, A1; *NYT*, 9/16/68, 29; 9/21–23/68, 1; *Moscow News*, 10/5–12/68, 3)

September 16: At White House press conference NASA Administrator James E. Webb announced his resignation, effective Oct. 7, his 62nd birthday. He wished to retire to devote his time to interests in education and foreign and urban affairs after 25 yr in Government and nearly 8 yr as NASA chief. President Johnson would name Dr. Thomas O. Paine, NASA Deputy Administrator, to succeed him as Acting Administrator. Webb would remain as consultant to Dr. Paine until Jan. 20, 1969.

Webb told press: "I leave NASA well prepared and with a conservative financial structure to carry out the missions that have been approved. They will go forward. What we have not been able to do under the pressures on the budget has been to fund new missions for the 1970's. So, there is going to be a period when there will be fewer flights than would, in my view, have been important to do. . . . We planned with the President's leadership in 1961 and his recommendations to President Kennedy to develop an ability to fly six Saturn V's per year and six Saturn I's per year. . . . Under the reductions in the budget beginning in 1964 . . . we have reduced that anticipated flight schedule, or production schedule, [until] we have now cancelled the production lines on both these boosters. . . . what we will have to do is develop a new base of technology for the next generation of boosters [and] . . . the country will have to look with great care into what it is it wants to do with those boosters that will be in storage."

Answering query, Webb said, "I am not satisfied with the program. I am not satisfied that we as a nation have not been able to go forward to achieve a first position in space. What this really means is that we are going to be in a second position for some time to come. . . . I am satisfied with what we have developed in every field and that we have been flying three successful generations of spacecraft. . . . I think now the question is what will the fourth . . . be. We have a vast amount of capability. As the U.S.S.R. proceeds to fly and remains in the number one position, we will have the capability to start new programs as the need is clearly indicated. . . . We have worked up to a work force of 420,000 people. At the end of this fiscal year, we will be down to just over 200,000. So we have shown the administrative capability to build up and then to reduce without losing pace in the program." Webb said NASA had decided to man third Saturn V flight in December. "We did calculate that we needed 15 Saturn V's in program to make sure we would do the Lunar landing. We have funded eight. . . . Now we can only fly seven. So the real question is can we make the landing by number seven?"

A good many people, Webb said, have tended to use space program "as a sort of whipping boy. . . . But in essence, if it were not for the fiscal problems faced by the President and the Director of the Budget I would believe that the program would have been supported in the Congress and in the country at a higher level than it has been." Webb thought U.S. had reached parity with U.S.S.R. $2\frac{1}{2}$ yr earlier except in large boosters and spacecraft. If U.S. could have had 12 Saturns a year in late 1967 or 1968, "I think we would have forged ahead." Instead, "while we are reducing down to half to two-thirds of our program, they are still increasing." Webb said current U.S space program had created capability that could be used for major efforts in exploration if adequately funded. "Any danger to the United States that would come from the Russian program would be visible in time to use this capability to start out. But they are going to have the reality and the image of being out in front for a number of years to come. How dangerous that is, you will have to judge yourself." (Transcript; Sheehan, *NYT*, 9/17/68, 1; Kilpatrick, O'Toole, *W Post*, 9/17/68, A1; Dobbin, *B Sun*, 9/17/68, A9)

- U.S.S.R. successfully launched *Cosmos CCXLI* into orbit with 326-km (202.6-mi) apogee, 198-km (123-mi) perigee, 89.7-min period, and 65.4° inclination. Satellite reentered Sept. 24. (*C Trib*, 9/17/68; GSFC *SSR*, 9/30/68)
- NASA launched series of three sounding rockets from NASA Wallops Station. First Nike-Cajun carried GSFC experiment to 42.6-mi (68.6-km) altitude to determine ozone concentration and water vapor distribution as a function of altitude and obtain finite data during parachute de-

scent from 43.4-mi (70-km) altitude. Basic payload design proved operational. Rocket performance was good; peak altitude was 6% below predicted. Instrumentation's tone ranging was lost after payload despin and new AB signal replacing FM signal slowly deteriorated after liftoff. Radar tracked all systems from liftoff to impact. Although parachute was not fully deployed during descent and aluminum oxide hydrometer failed to project because of spring malfunction, key events occurred within two seconds of predicted times and significant data were obtained. Midair-retrieval attempt was not successful, but payload and parachute were recovered.

Second Nike-Cajun carried GSFC payload to 73.1-mi (117.6-km) altitude to obtain vertical profile of temperature, pressure density, and wind data between 21.8- and 59-mi (35- and 95-km) altitudes by detonating 19 grenades and recording their sound arrivals at ground. Rocket performed satisfactorily. Grenade timer malfunctioned and instead of ejecting 19 grenades during ascent, vehicle ejected 1 grenade after apogee during descent and 12–14 grenades which detonated at sporadic intervals during descent. Final destruct grenade destroyed payload before impact. Some significant data could be correlated from ground arrivals.

Arcas sounding rocket carried U.S. Naval Ordnance Test Station payload to 34.5-mi (55.5-km) altitude to measure ozone concentration at 18.6- to 37.5-mi (30- to 60-km) altitudes by observing solar UV light attenuation by ozone during parachute descent. Rocket performed nominally; altitude achieved was within 10,000 ft of predicted. Instruments performed satisfactorily and good data were obtained. Parachute deployed properly and payload was tracked by radar and retrieved in midair. (NASA Rpts SRL)

- Aerobee 150 sounding rocket launched by NASA from portable launcher at WSMR carried GSFC experiment to 3.14-mi (5.05-km) altitude to: (1) flight-qualify production lot of VAM–20 variboosters, (2) demonstrate feasibility of Aerobee rail launch, (3) determine exit loads, (4) establish rail-launching procedures, and (5) verify boost-phase dispersion calculations. System was ballasted with water and lead to produce mass and mass distribution of fully loaded rocket with 200-lb net payload. Rocket and instruments performed satisfactorily. Booster web burning time, booster pressure, acceleration, velocity, and system impact occurred as predicted. (NASA Rpt SRL)

- German scientists at Max Planck Institute—using U.S., U.K., and French rockets—had shown that borders of magnetic energy extending from earth could be made visible directly to persons on earth, *New York Times* reported. Release of barium clouds in space produced glowing, colored filaments outlining magnetic field. Technique showed influence of electric fields in space. Preparations were being made to release barium cloud in heart of inner radiation belt around earth before year's end. Cloud would inscribe in space magnetic lines of force which held radiation belt in place. ESRO had prepared Highly Eccentric Orbiting Satellite (HEOS) for launch by U.S. booster in experiment which would be too high for naked eye to observe. Telescopic Schmidt cameras were to be used. (*NYT*, 9/17/68, 1)

- NASA awarded $3,290,500 contract to Pittsburgh Des Moines Steel Co. for fabrication and installation of heavy steel test chambers and related systems as part of second phase in expansion of Propulsion Systems Laboratory at LeRC. (LeRC Release 68-56)

September 17: Presidium of Supreme Soviet ratified space rescue treaty signed by U.S. and U.S.S.R. April 22, according to Tass report. Presidium took care of legislative matters when Supreme Soviet was not in session. (AP, W *Star*, 9/19/68, A10)

- Dr. Thomas O. Paine, NASA Deputy Administrator whom retiring Administrator James E. Webb had said would be named Acting NASA Administrator by President Johnson, represented "new breed of scientist-administrators making their way into government," said John Lannan in Washington *Evening Star*. Unlike Webb, he was new to Government, having come up through "the scientific ranks of the sprawling General Electric Co. over a period of 19 years." As Webb's chief deputy since March, Dr. Paine had been authorizing and signing major announcements but had yet to confront Congress. He and Webb saw eye-to-eye, he had said, on agency's current program emphasis, which had been criticized as too heavily oriented to manned flight. "I am a very strong supporter of doing as much as we can in planetary areas." Dr. Paine said, "but it's one of those cases where you've got limited resources and you're trying to do all the things you can. You've got to do things in aeronautics; we've got to finish up the Apollo program; and we've got to do more in earth resources (surveying them from space) here on this planet." A lot of NASA's future "is in the questions that we'll be raising as we continue to probe the planets."

  On balancing space needs against problems of cities, Dr. Paine felt U.S. suffered from "what I would call almost a national hypochondria . . . in many ways crippling some of the forward-looking things we're able to do. . . . I feel that one of the very highest priority matters is the war on poverty and the problems of the cities. But in the meantime, we're making . . . a lot of progress in the civil rights area and really, this nation is a good deal healthier than we're giving it credit for today." He was confident of a resurgence of public interest in space once the Apollo program got underway. (W *Star*, 9/17/68, A4)

- Under Administrator James E. Webb's leadership, Don Kirkman commented in *Washington Daily News*, NASA had "put John Glenn in orbit in the Mercury capsule, brought back all the Gemini spacemen without mishap, and intended to put U.S. astronauts on the moon before the 1960s ended." Unmanned spaceships had photographed moon and Mars and probed Venus host of weather, communications, and navigation satellites had been launched. But since Jan. 27, 1967, Apollo fire and with needs of Vietnam war, "every new budget brought new blows." (*W News*, 9/17/68, 9)

- U.S. patent No. 3,402,295 was issued to Robert W. Astheimer, Vice President of Barnes Engineering Co., for process by which aircraft pilot could spot clear air turbulence (CAT) far enough ahead to avoid danger. To discover rise in temperature which marked CAT, pilot would scan ahead with radiometer to detect significant heat radiation from carbon dioxide uniformly distributed through air. (Pat Off PIO; Jones, *NYT*, 9/21/68, 45)

September 17–18: Senate Foreign Relations Committee approved nuclear nonproliferation treaty by vote of 13 to 3 after three previous meetings failed to attract majority of members. Senate Democratic leaders expressed hope treaty could be brought to vote on Senate floor before mid-October Congressional adjournment. In formal statement after Committee approval, Vice President Humphrey, Democratic Presidential candidate, called upon Republican Presidential candidate Richard M. Nixon to support ratification. U.S.S.R.'s occupation of Czechoslovakia, Humphrey said, should not deter Senate approval of "crucial" treaty. It was not agreement between Washington and Moscow, but a treaty already signed by 81 nations and developed among many signers as well as the United Nations. (Sherman, W. *Star*, 9/17/68, A5; Unna, *W. Post*, 9/18/68, 1; Finney, *NYT*, 9/18/68, 1; Furgurson, B *Sun*, 9/19/68, 1)

September 18: NASA launch from ETR of Intelsat-III F-1, first of four Intelsat III comsats scheduled for 1968–69, failed to reach orbit when three-stage Long-Tank Delta booster, on maiden flight, pitched backward and was exploded several thousand feet above Atlantic. NASA Launch Director Robert Gray said booster encountered trouble in gyroscope system controlling pitch rate 20 sec after blastoff. Launch vehicle became erratic and uncontrollable, pitching backward 1 min 2 sec into flight, and 6 sec later ETR safety officer sent signal to destroy vehicle. Delta fuel tanks had already begun to tear apart and burst into flame. Pieces of booster and satellite fell into ocean 12 mi off Florida. Experts were studying radio data to determine cause of failure, which ComSatCorp said would not have significant effect on 1968 revenue. Delay in completing cycle of four Intelsat IIIs would be felt in 1969; ComSatCorp officials termed it "major blow" to plans for early full-scale global system.

To be owned by 63-nation INTELSAT consortium, each of satellites was designed to have 1,200-voice capacity or four TV channels; completed series of four was to make coverage available around the world. First had been planned for synchronous equatorial orbit above Atlantic. (ComSatCorp PIO; ComSatCorp Release 68–46; AP, *W Post*, 9/19/68, A3; Dimond, *W Star*, 9/19/68, A21; *CSM*, 9/20/68)

- USAF launched two unidentified satellites from Vandenberg AFB by Thor-Agena D booster. One entered orbit with 243-mi (391.1-km) apogee, 111-mi (178.6-km) perigee, 90.1-min period, and 83° inclination and reentered Oct. 8. Second satellite entered orbit with 318-mi (511.8-km) apogee, 312-mi (502.1-km) perigee, 94.7-min period, and 83.2° inclination. (*Pres Rpt 68*)

- Nike-Cajun sounding rocket launched by NASA from Churchill Research Range carried GSFC payload to 12-mi (19-km) altitude to obtain variations in temperature profile and atmosphere structure. Launch was one of series scheduled for winter 1968–69 at Churchill, Point Barrow, Alaska, and NASA Wallops Station. Nineteen grenades deployed and exploded as planned. Second stage failed to ignite causing low apogee, which prevented acquisition of useful data. (NASA Rpt SRL)

- At international symposium in Washington, D.C., sponsored by NAS and GSFC, Univ. of California scientist Dr. W. Ian Axford explained new phenomenon, "polar wind," which was constantly blowing away small portions of earth's atmosphere and losing them in space. In polar regions, earth's magnetic lines of force plunged almost straight down, leaving considerable area "accessible to space." Charged particles moving rapidly could escape local entrapment by planet's field. U.S satellites had observed such particles rising toward space at supersonic speeds. They climbed vertical magnetic force lines and wound up in long tail which protruded from earth on side away from sun (distortion of earth's magnetic field by solar wind of charged particles flowing continuously from sun). Univ. of California scientist Peter M. Banks was credited with deducing existence of polar wind. (*W News*, 9/18/68, 18)

- Four NAVSATS (satellites resembling comsats) in synchronous equatorial orbit could control entire air traffic of U.S. and 18 could handle entire world's air traffic, according to TRW Systems Div., which demonstrated system during Air Force Assn. meeting in Washington, D.C. Satellites would serve as reference points for aircraft in flight, ships at sea, or mechanized ground equipment and could locate user's longitude and latitude to accuracy within 60 ft and his altitude to within 120 ft. They would operate like comsats but, instead of relaying messages from point to point, would beam signals to aircraft. By taking bearings on any two satellites, computerized unit in aircraft could pinpoint aircraft's location by radio. Militarily, system could lead to accurate all-weather bombing capability and armored column movements in poor terrain or weather, or help infantrymen in the field. For civilian air traffic control, it could overlap existing equipment and eventually phase it out. (Lannan, W *Star*, 9/18/68)

Boeing Co. confirmed it was most likely to submit conventional fixed-wing design resembling large F–4 Phantom fighter as design for U.S. SST when firm's final proposal became due at FAA, no later than Jan. 15. If approved, aircraft would start moving from paper to titanium, Boeing President T. A. Wilson told Air Force Assn. seminar. M/G Jewell C. Maxwell, FAA's SST Program Director, said aircraft could fly in first half of 1972 and be ready for passenger service as early as 1974 (1976 if Government decided to test prototype before starting production line). A. H. C. Greenwood, Assistant Managing Director for British Aircraft Corp., said Anglo-French Concorde would fly in 1968 and be in commercial operation by summer 1972. (Wilson, *W Post*, 9/19/68, D1)

- Prototype four-engine turboprop STOL aircraft built in France as Breguet III and in U.S. as McDonnell Douglas 188 was demonstrated for press in flight over Washington, D.C., metropolitan area. Eastern Airlines, McDonnell Douglas, and FAA were testing aircraft as possible replacement for jet aircraft on Eastern's crowded shuttle service between Washington, New York, and Boston. When fully developed, aircraft would carry 100 passengers at 400 mph with all-weather capability. It would take off and land on less than 800-ft runway and operate on more direct routes closed to conventional aircraft under noise-abatement rules. Airline spokesman had said aircraft could be operational in "early 1970." (Yarborough, W *Star*, 9/18/68, A1; Valentine, *W Post*, 9/19/68, B1)

- AEC Chairman, Dr. Glenn T. Seaborg, was named 1968 winner of Arches of Science Award presented annually by Pacific Science Center of Seattle, Wash., to American who had contributed to public understanding of science. Award of $25,000 would be presented Oct. 16 to Dr. Seaborg, who won Nobel Prize in chemistry in 1951 and was codiscoverer of plutonium. (*W Post*, 9/18/68, B5; AP, *NYT*, 9/18/68, 12)

- *Washington Post* commented on resignation of James E. Webb as NASA Administrator: "Virtually from scratch, with the country in a swivet over Soviet prowess, James E. Webb took over the Nation's feeble space program and drove the United States firmly into the space age. . . . he created—in the civilian space agency—the largest and most labryrinthine engineering organization in American history. . . . In his eight-year stewardship, NASA proceeded under presidential and Congressional flogging toward the goal of a manned lunar landing 'in this decade.' It achieved repeated successes and but a single important failure—the fire that claimed the lives of three astronauts last year.

  "It is characteristic of this self-effacing, organization-minded man that he should choose to retire now at his 62nd birthday, practically on the eve of the Apollo launching—a career-capping event if ever there was one. His purpose is to hand NASA over to his lieutenants before the change of administration, on the theory that the agency will thereby have a better chance of riding out next year's power transfer. The country is fortunate that Mr. Webb is in good health and determined to apply himself vigorously to further the Nation's understanding of the space age. For him this is far more than a matter of projectiles and orbits; it goes to the heart of the necessities and the aspirations of a great modern society.

  "Whereas Mr. Webb leaves NASA with its current mission adequately funded and its capabilities well formed, he does not leave it with a set mission beyond landing on the moon. This is hardly his fault. The fading American taste for competition with the Russians in space and the rising competition of other claimants for Federal funds explain NASA's uncertain estate; its budget has been cut $1.4 billion in four years. . . . it will be up to the next Administration and the next Congress to chart America's future in space. That they have a choice is the singular achievement of Jim Webb." (*W Post*, 9/18/68, A22)

September 19: House adopted conference report on H.R. 17023, FY 1969 Independent Offices and HUD Appropriations, which allocated $3.995 billion to NASA. [See Sept. 25]. (*CR*, H8970–1; NASA *LAR* VII/95)

- In 12-yr NASA-sponsored experiments with roosters spinning at 130 rpm up to 15 mo on double-deck centrifuge, Univ. of California's Davis Chronic Acceleration Laboratory gravity experts had noted possible adverse effects of long-term weightlessness on astronauts. Without steps to counter effects, "man in space for long periods will find his body decalcified," said Dr. Arthur H. Smith. "He will not be able to stand up without fracturing bones. All his muscles will shrivel up including those around his blood vessels. He will be overloaded with fat which he may not be able to use." Among methods being developed to offset disturbances to human body in space were drugs which stabilized proper balance of red cells and steroids which helped maintain calcium metabolism. Roosters were used in tests since their upright posture, heart-to-head distance, and circulatory systems were similar to man's. (Hillinger, *W Post*, 9/19/68, G11)

- NASA said it had reassigned management responsibility for Saturn I Workshop airlock module and modified lunar module ascent stage for

Apollo Telescope Mount from MSC to MSFC, to establish satisfactory balance between Apollo Applications and Apollo programs. Move placed AA program design and integration responsibilities under single NASA center. Management responsibilities encompassed systems engineering, including development test and integration, to ensure compatibility of flight hardware and ground support equipment. Airlock module, mounted on Saturn IB vehicle, would provide 65-in-dia airlock tunnel for Apollo astronauts between spacecraft and living area of workshop inside hydrogen tank of rocket's 2nd stage. Lunar module ascent stage was being modified as control station for ATM. AA program would further develop space capability in series of earth-orbital long-duration flights using Apollo hardware and other facilities. (NASA Release 68–159)

- NAS published report of NASA-sponsored two-year study of planetary astronomy by NAS–NRC Space Science Board's 16-scientist panel headed by Dr. John S. Hall, Director of Lowell Observatory, Ariz. *Planetary Astronomy: An Appraisal of Ground-Based Opportunities* recommended assignment of high priority to erection of 60-in optical telescope in mountains of northern Chile, where atmospheric conditions permitted favorable view of Mars, in time to observe planet's closest approach to earth in almost 50 yr, August 1971. Also recommended were construction of large dish antenna in U.S. for planetary radar astronomy; two large radiotelescope arrays in California and West Virginia, plus third planetary radio facility to pick up signals at short wavelengths; and 120-in infrared telescope in desert region where vapor would be less likely to interfere with infrared radiation. (Text; Wilford, *NYT*, 9/20/68, 29)
- Boeing Co. dedicated $8-million low-speed wind-tunnel complex for development of helicopters and STOL and V/STOL aircraft at its Vertol Div., Eddystone, Pa. (P *EB*, 9/19/68)

September 20: *Cosmos CCXLII* was launched by U.S.S.R. Orbital parameters: apogee, 404 km (251 mi); perigee, 269 km (167.2 mi); period, 91.2 min; and inclination, 70.9°. Satellite reentered Nov. 13. (UPI, *NYT*, 9/21/68, 14; GSFC *SSR*, 9/30/68; 11/15/68)

- NASA Aerobee 150 MI sounding rocket launched from WSMR carried NAS-Univ. of Wisconsin experiment to 107.7-mi (173-km) altitude to measure radiation from celestial objects in spectral regions 2,800–1,200 Å, 60–40Å, and 15–2Å using photometers and gas-filled proportional counters as x-ray detectors. Both x-ray and UV payloads functioned well and acquired data. (NASA Rpt SRL)
- NASA Boosted Arcas II sounding rocket launched from NASA Wallops Station carried GSFC experiment to 26.5-mi (42.7-km) altitude to determine feasibility of launching Boosted Arcas II from Arcas tube launcher and to obtain data on complete system flight. Booster pressure and acceleration were nominal throughout flight. Pressure transducer produced data until 2nd stage ignition. However, there was no drag separation after booster burnout. Separation occurred at 2nd stage ignition. Radical coning occurred during burn and predicted altitude of 71.5-mi (115-km) was not attained. (NASA Rpt SRL)
- NASA Nike-Tomahawk sounding rocket launched from Andoya, Norway, carried GSFC experiment to analyze electric fields from observed motions of neutral and ionized barium clouds during auroral condition.

Preliminary results indicated good photographic coverage was obtained from all sites. (NASA Rpt SRL)

- Apollo 7 prime crew—Astronauts Walter M. Schirra, Jr., Donn F. Eisele, and R. Walter Cunningham—held news conference at MSC. Crew saw no reason to delay Apollo 7 launch beyond scheduled lift-off on Oct. 11 but results of flight readiness test following week would determine definite plans.

  Questioned about safety, mission commander Schirra said, "when we go over the sill for launch day, we will consider that all those risks that are appropriate for this type of mission are either understood by us or that they are low enough that we have a very, very high probability. . . . We would like 10.8 days and we will do our best to do it. . . . We've had a goal that is rather a hard one to achieve, particularly one that we have to follow on when we lost three of our compatriots, and we don't want any mistakes . . . to happen again. We have not been the 'kid-around' types that we might have been in the past, we're much more serious about it, because this is a much more complicated machine and there are many, many more people involved in it. . . . I think you will find that you will see a good performance out of this total crew and we have tried very hard to make this machine work just the way it should. We have basically lived with it at the plant [and] at the Cape . . . and if somebody even takes a small component off it, we become furious and say, 'Why did you remove it?' We expect answers immediately . . . that is the way we are working."

  In answer to question, Schirra said it had taken 1 min 5 sec for all three to exit from spacecraft to white room in simulated emergency, while still being careful not to damage equipment. In real emergency, crew could probably exit in 45 sec. (Transcript)

- Astronaut Walter M. Schirra, Jr., announced he would retire as NASA astronaut after commanding Apollo 7 mission. At 45, Schirra would be world's most experienced astronaut, having flown six orbits Oct. 3, 1962, on *Sigma 7* in Mercury mission and having commanded two-man Dec. 15–16, 1965, *Gemini VI* mission, which had participated with *Gemini VII* in first rendezvous of spacecraft in space. He would remain in space program, he said, "until we effect the job we set out to do"—to land men on moon and return them. (*W Post*, 9/21/68, A7; *W Star*, 9/21/68, A2)

- Saturn V launch vehicle has passed Design Certification Review conducted over communication facilities linking NASA and contractor personnel at Washington, D.C., Houston, Tex., Cocoa Beach, Fla., and Huntsville, Ala., into one conference. Purpose was to certify overall design of rocket, including engineering modifications after two previous Saturn flights. Certification for manned mission was issued, subject to verification of data in minor areas in later, limited review. Under particular consideration were engineering changes to correct pogo, or excessive oscillations, during second Saturn V flight and anomalies experienced by J–2 engines on upper stages. Solutions had been tested to satisfaction. Major efforts on both items would continue. (MSFC Release 68–216)

- Commenting on expected Boeing Co. decision on fixed-wing design for SST, Washington *Evening Star* editorial said: "America's first fumbling rush into the field of supersonic transportation now appears destined to end not with a boom, but with a whimper . . . from the company which, after years of effort and a few hundred million dollars expenditure, has apparently decided to scrap the swing-wing concept and retreat to the drawing board. . . . It was the boldness and originality of the variable wing that won the contract for Boeing, so there would seem to be room for complaint if it now develops that the system is too heavy to be practical." (W *Star*, 9/20/68, A10)

- AEC announced selection of Dr. John Archibald Wheeler of Princeton Univ. to receive its $25,000 Enrico Fermi Award for 1968 for "his pioneering contributions to understanding nuclear fission, and to developing the technology of plutonium production reactors, and his continuing broad contributions to nuclear science." Award would be presented in ceremony Dec. 2, 26th anniversary of first sustained controlled nuclear reaction. (AEC Release L-224)

- Dr. Dinsmore Alter, astronomer, first to provide evidence that moon was not as inactive as had been thought, died in Oakland, Calif., at age 80. He had been director of Griffith Planetarium in Los Angeles from 1935 until his retirement in 1958. On Oct. 26, 1956, from Mt. Wilson, Calif., he had seen apparent mistiness at bottom of crater which later led to re-evaluation of concepts of moon. (*NYT*, 9/24/68, 47)

September 21: Nike-Tomahawk sounding rocket launched by NASA from Andoya, Norway, carried GSFC payload to analyze electric fields from observed motions of neutral and ionized barium clouds during an auroral condition. All four barium clouds were released in visible aurora as planned. Good photographic coverage was obtained from all sites. Peak altitude would be determined from triangulation of photos. (NASA Rpt SRL)

September 22: NASA Aerobee 150 sounding rocket launched from WSMR carried NRL experiment to 117.4-mi (188.9-km) altitude to obtain UV photographs of sun's disk and white light photographs of outer solar corona in conjunction with measurements made by others during total solar eclipse. Payload contained two externally occulting coronagraphs, with Univ. of Colorado solar pointing control. Solar pointing control operated properly, with exceptionally steady pointing; white light corona and streamers were recorded in both coronagraphs with excellent resolution. XUV heliograph and spectroheliograph were recorded with excellent resolution. (NASA Rpt SRL)

- Washington *Sunday Star* commented on resignation of NASA Administrator James E. Webb: "The abrupt announcement, which was notably lacking in the customary ceremonial niceties, was the final confirmation that NASA's honeymoon is definitely over. The relationship between the space agency and Congress must rank as perhaps the longest, and in many ways the happiest, official honeymoon on record. For six years, hardly a frown or a cross word came between the blossoming agency and the guardians of the nation's treasure. The amiable and loquacious Webb proved himself an able administrator who knew his way around in the bureaucratic jungle. Through Mercury and Gemini programs, NASA virtually had only to ask, and Congress was ready to give. Then came the Apollo tragedy." But Jan. 27, 1967, fire did not cause the change of heart. "It was . . . a catalyst that speeded the process. But a growing coolness was bound to develop . . . for the basic economic fact is that America simply cannot continue to meet NASA's ever-expanding economic demands—not, at least, without a critical weighing of those demands against the nation's existing domestic, diplomatic and military obligations. . . .

"Long before the tragedy that threw the timetable into a shambles, the conflict between America's technical ability to move men into space and America's ability to pay the freight could be seen by those who cared to look ahead. . . . Congressional reluctance to commit the nation to further, open-ended expenditures in manned space flight is therefore inevitable. But equally inevitable is the dismay—and in some cases the bitterness—of NASA officials. Webb and his lieutenants can take justifiable pride in what has been accomplished in the brief span of the space age. . . . They have solved problems quickly and accurately that men have never faced before. In the process, they pulled together a scientific and technical team unequalled in the peacetime history of this country. Now they are watching that massive army disband. . . .

"Man will someday go to the planets and beyond. But that day must wait until the breakdown of nationalistic barriers makes possible a truly united world space program. Meanwhile, the United States should concentrate on the development of more sophisticated machines, less costly rocket systems and closer cooperation between the military and civilian space programs to avoid wasteful duplication. America's space effort must not end with the Apollo program. It must be diverted into equally exciting and frequently more rewarding channels of unmanned space exploration." (W *Star*, 9/22/68, C1)

- Associated Press roundup of comment by space authorities on U.S.S.R.'s *Zond V* mission:

    NASA Administrator James E. Webb: "The most important demonstration of total space capability up to now by any nation."

    Rep. Olin E. Teague (D-Tex.), Chairman of House Committee on Science and Astronautics' Subcommittee on Manned Space Flight: Hoped U.S.S.R. achievement would "cause some stir in Congress. . . . The Russians again did something we have not done. We slept until the Sputnik. That may be what we're doing now."

    Sir Bernard Lovell, Director of U.K.'s Jodrell Bank Experimental Station: "I think it is a very considerable achievement and I expect that a human being will be placed in a similar spacecraft in a matter of months."

    Heinz Kaminski, Chief of West Germany's Bochum Observatory: "Still in this year or at latest in first quarter of 1969 a three-man spacecraft of the Soyuz type will circle the moon. After that [in a later flight] they will land on the moon."

    In Moscow, unidentified Western expert on U.S.S.R.'s space program said that, because of seasonal factors, best time for Soviet manned lunar circumnavigation would be no later than early November—or spring 1969. "They are now in a position to send a man around the moon and back to earth, without landing him on the moon. They might try one more unmanned shot, just to be sure, and then they could send up a man." (AP, *NYT*, 9/23/68, E8)

- In *This Week*, Erik Bergaust discussed improvements in safety made by NASA since Jan. 27, 1967, Apollo fire. Though NASA leaders in Washington and at MSC believed this was "doing the job, other experts in and out of NASA privately are expressing their doubts," he said. They felt "NASA had been slow to correct [and] . . . its solutions have been more political than practical." Experts insisted, "that while progress has been made, there have been very few significant improvements in NASA's own management. . . . reports circulating in Washington [since] the accident indicated the conditions that led to the Apollo fire may not have been corrected." (*This Week*, W Star, 9/22/68, 7)

September 23: U.S.S.R. launched *Cosmos CCXLIII* into orbit with 297-km (184.6-mi) apogee, 206-km (128-mi) perigee, 89.5-min period, and 71.3° inclination. Satellite reentered Oct. 4. (GSFC *SSR*, 9/30/68; 10/15/68)

- NASA Nike-Tomahawk sounding rocket launched from Andoya, Norway, carried GSFC experiment to analyze electric fields from observed motions of neutral and ionized barium clouds during auroral condition. All four barium clouds were released in visible aurora as planned, and good photographic coverage was obtained from all sites. (NASA Rpt SRL)

- NASA Apollo Program Mission Director William C. Schneider told news conference astronaut training for possible manned moon-orbiting mission later in 1968 was under way. "We will do the maximum the systems will allow," he said, "and the maximum you could foresee would be lunar orbit." Decision in November, after evaluation of Apollo 7 data, would determine if Apollo 8 crew—Astronauts Frank Borman, William A. Anders, and James A. Lovell, Jr.—were to attempt lunar orbit in December. NASA Deputy Administrator, Dr. Thomas O. Paine, pointed out that plans for possible manned lunar mission had been made long before U.S.S.R.'s *Zond V* circumlunar flight.

    In Oct. 11 Apollo 7 launch, Schneider said, earth orbital flight was set to run up to 10 days 19 hr at altitudes from 120 to 150 mi. Primary objectives were tests of command and service modules with crew, tests of ground control facilities, demonstration of capability for rendezvous with spent rocket section, and eight firings of service propulsion system. Schneider said he would rate Apollo 7 a success if "we get rendezvous and stay up three days to accomplish our main objectives."

    Discussing plans to follow Apollo program with lunar exploration program Dr. Paine said, if forced to choose between Apollo Applications program in earth orbit or lunar exploration, NASA "would be most reluctant to give up manned lunar exploration," as an area "man is uniquely qualified to contribute to" and one "of enormous scientific interest and importance." (Transcript; Wilford, *NYT*, 9/24/68, 26; Cohn, *W Post*, 9/24/68, A1)

- *New York Times* editorial on U.S.S.R.'s *Zond V* mission: "Until the nation [U.S.] has a better idea of the progress achieved during this long interval between American manned flights, it is premature to conclude that the Russians have a significant over-all lead in space capability and that Soviet citizens will certainly paint their flag on the lunar surface before American astronauts.

    "In any event, as the Soviet rape of Czechoslovakia has recently reminded the world so dismayingly, a nation's image and prestige depend primarily upon what it does here on earth, not on its feats in space." (*NYT*, 9/23/68, 32)

September 23

- Senate Preparedness Investigating Subcommittee of Senate Armed Services Committee recommended continuing expansion and improvement of U.S. nuclear submarine fleet to meet "serious challenge" from U.S.S.R., which was "giving major emphasis to qualitative improvement of its submarine fleet." Committee recommended U.S. continue building nuclear submarines after 1970 and proceed with high-speed, electric-drive, and advanced-design submarines. (Text; Finney, *NYT*, 9/24/68, 1)
- William E. Stoney, Jr., former Chief of MSC's Advanced Spacecraft Technology Div., became NASA Deputy Director (Engineering) of Apollo program. (NASA Release 68-151)

September 23-24: USAF halted all F-111A flights for second time in 1968 day after Sept. 23 crash at Nellis AFB, Nev., in which two pilots, including Australian trainee, escaped serious injury by triggering escape capsule. It was 11th F-111A accident since aircraft's inception. USAF was investigating crash in which aircraft had plummeted 300 ft to runway after "slow pull-up maneuver" during which pilots, on routine training mission, were unable to maintain aircraft control. (UPI, *W Post*, 9/24/68, A18; Witkin, *NYT*, 9/25/68, 1; AP, *W Post*, 9/25/68, A8; *WSJ*, 9/25/68, 1; Sehlstedt, B *Sun*, 9/25/68, 1)

September 23-25: Washington Airlines inaugurated nation's first regularly scheduled short-takeoff-and-landing STOL service linking Friendship, National, and Dulles Airports in Washington, D.C., area, 16 twin-engine Dornier flights daily. In other STOL developments, McDonnell Douglas Corp. and Eastern Airlines had started evaluation of Model 188 STOL over Eastern's shuttle routes [see Sept. 18]; MIT scientists Rene H. Miller and Robert W. Simpson in *Astronautics & Aeronautics* said series of studies of VTOL and STOL transportation systems for U.S. Northeast Corridor started in 1964 showed need for cooperation among states and Federal Government to pave way for V/STOL as essential mode of transportation; production arrangements had been announced for 32- to 36-passenger, short-field, turboprop transport by General Aircraft Corp.; Rutgers Univ. was getting Federal grant to continue studies of feasibility of low-altitude air shuttle service in New York-New Jersey-Connecticut area; and Boeing Co. had dedicated advanced R&D facilities for testing V/STOL aircraft at its Vertol Div. near Philadelphia. (Stout, *W Post*, 9/23/68, B2; Yarborough, W *Star*, 9/23/68, B1; 9/25/68, A53; *A&A*, 9/68, 28-34)

September 24: NASA doubled resolution capability of *Explorer XXXVIII* to map radio sources in space by extending each of satellite's four antennas to 600 ft by ground command. Antennas, which could be extended to 750-ft maximum, had been initially deployed to 455 ft each July 22. Satellite, launched July 4, had monitored solar radio emissions, variety of emissions across Milky Way, radio emissions apparently related to earth's magnetosphere, and possible emission from earth's radiation belt. (NASA Release 68-162; AP, W *Star*, 9/25/68, A5)

- NASA Aerobee 150 sounding rocket launched from WSMR carried Harvard College Observatory experiment to 111.9-mi (180-km) altitude to scan sun in 1,400-1,875 Å spectral region using photo-electronic detector and BB solar pointing control. Solar pointing control malfunctioned, changing signal on experiment detector. (NASA Rpt SRL)
- NASA's HL-10 lifting-body vehicle, piloted by Maj. Jerauld R. Gentry

(USAF), successfully completed 10th flight from Edwards AFB. Primary objectives were to check out vehicle systems, evaluate new nonlinear longitudinal gearing to control stick, obtain stability and control data, and evaluate modified cabin pressure system. (NASA Proj Off)

- NASA was conducting three separate studies to determine reasons for failures of May 18 launch of Nimbus B, Aug. 10 launch of *Ats IV*, and Sept. 18 launch of Intelsat III—which had brought its launch record down to 0.500 average. Before May failure, agency had compiled "almost incredible" launch record, said Thomas O'Toole in *Washington Post*. Sources close to NASA had said agency was conducting fourth investigation to determine whether there was "something systemic at the root of the trouble." One theory was that decline in NASA launch activity in 1968, with 42 attempted launches thus far compared to 69 at same time in 1967, had caused similar drop in attention to detail. Second theory was that layoffs and turnover in launch crews had cut down efficiency and introduced element of inexperience. Third theory was that NASA was suffering from "a sudden case of overconfidence brought on by its long string of launch successes."

  Seven-man board to investigate failure of first Intelsat III mission met for organization session. Board, convened by Dr. John E. Naugle, NASA Associate Administrator for Space Science and Applications, would have five observers in addition to seven voting members and would be chaired by Daniel G. Mazur, Assistant Director for Technology at GSFC. (NASA Release 68–160; *NYT*, 9/21/68, 14; *W Post*, 9/24/68, A10)

- Senate Foreign Relations Committee approved space rescue treaty providing for rescue and recovery of astronauts down in foreign lands. U.S.S.R. had given final approval along with 73 other nations. (AP, *NYT*, 9/25/68, 17; *WSJ*, 9/25/68, 1; UPI, *W Post*, 9/25/68, A6)

- Dr. Thomas O. Paine, NASA Deputy Administrator, told House Committee on Science and Astronautics' Subcommittee on Advanced Research and Technology that, while NASA was "maximizing . . . technical support of the DOD within the framework of their systems approach to new military aircraft," major changes in civil aeronautics made that situation "more complicated for NASA and the nation." Growth of air transportation, decline in development of military aircraft whose technology was directly applicable to civilian use, and ground transportation's saturation had created era in which "it is not simply enough to build aircraft which are bigger or fly faster; rather, future advances in aeronautics must also take the form of aircraft which complement the nation's overall transportation systems while operating harmoniously with the constraints imposed by urban environments."

  Basic issue facing NASA and U.S. in civil aeronautics R&D was proper role of Government in fostering its advances. "In my view, government actions to stimulate the development of advanced aircraft transportation system should be such as to leave the maximum initiative and business risk in the hands of industry. The government [might] carry promising new technological principles . . . into experimental hardware, but only to the point of demonstrating the soundness of the principle involved, and only when this demonstration is essential and would not be otherwise undertaken.

  NASA Deputy Associate Administrator for Aeronautics Charles W.

Harper told Committee accelerated research was needed not only on supersonic and hypersonic aircraft, but in development of quiet subsonic jets with steeper landing patterns, helicopters with greater safety and less noise, and STOL aircraft for interurban transport. He said air traffic control systems bogging down at major airports were based on technology developed in 1940. He doubted "continued evolutionary modernization" of this system would suffice in future. (Transcript; AP, W Star, 9/25/68, 1; Cohn, W Post, 9/25/68, D8; UPI, NYT, 9/25/68, 93)

- General Accounting Office revealed in report to Congress that Bendix Corp. had agreed to lower by $520,000 fees it would have collected under $465-million NASA contract for construction of Saturn V platform. GAO had found that target cost included overstated amounts for certain materials in relation to pricing data available to contractor before start of negotiations. (Text)

- United Press International quoted Soviet scientists interviewed by U.S.S.R. Army newspaper *Krasnaya Zvezda* (Red Star) as saying they would aim more space vehicles for sea landings because of successful recovery of *Zond V* from Indian Ocean. Before *Zond V*, all Soviet probes had been brought down in Siberia. Sea landing in warm climates would allow year-round launchings. Although Soviet ships did not recover spacecraft till day after landing, scientist N. Melnikov told newspaper, "The time and place of Zond's landing were calculated beforehand with precision. Everyone . . . knew the exact hour, minute, and even second of Zond's landing." *Krasnaya Zvezda* said even slightest deviation "would have resulted in tremendous overloads [on *Zond V*] which could have gone again into outer space." (W Post, 9/25/68, A11)

- *Washington Post* editorial praised "magnificent achievement" of U.S.S.R.'s *Zond V*: "In its own way, Zond-5 should serve to shake this Nation's complacency once again. Our national goals in space exploration are fuzzy, our hopes have been out of line with our commitments. From President Kennedy's glowing picture of Americans in space we have slipped into a program put together in fits and starts—fits brought about by a realization of how much it would cost to do it right and starts caused by a basic desire never to be second to anybody. Because of this, there is no use in seizing upon Zond-5 as a reason to go all-out to beat the Russians in a race for men on the moon. Our program . . . ought to move at its own pace. If that pace is sufficiently rapid to bring American astronauts to the moon first, fine. If it is not, so be it. The Russians will deserve the honor and praise they will win if their men make the first landing. In space exploration, it is more important to do things right than to do them first." (W Post, 9/24/68, A16)

- Washington *Evening Star* editorial commented on "Soviet Spectacular." U.S.S.R. had scored impressive advance with "boomerang-style" *Zond V*. "Congratulations are in order. But handwringing assertions that the Russians will be indisputable masters of the universe are decidedly premature. The shot demonstrated two major steps forward in the space sciences: A degree of accuracy in the guidance and navigational systems that is new for Russian spacecraft, and a re-entry system capable of withstanding the searing heat of a 25,000-mile-an-hour plunge into the atmosphere. . . . But the essential fact is that neither Russia nor the United States has yet successfully flown its third generation spacecraft —the ships that will eventually carry men to the moon. The outcome of the 'space race' . . . will depend on what happens in the Apollo missions immediately ahead and the Soviet manned flights still to come." (W Star, 9/24/68, A12)

- Television sequences showing Apollo 7 astronauts working in spacecraft as it passed Corpus Christi, Tex., and Cape Kennedy, Fla.—as well as shots of terrain more than 130 mi below and views of anything interesting in space would be telecast by three major U.S. networks "live" Oct. 12 through Oct. 19, NASA said. Pictures would be transmitted from camera in space capsule to MSC for instant showing over TV networks. Neither launch day nor splashdown day would be included in 11 am to 12 pm series. (Kirkman, W News, 9/24/68, 7)
- USN had selected Lockheed Aircraft Corp. design for its first Deep Submergence Search Vehicle. Craft would be able to descend to 20,000 ft to locate objects and recover small ones with claw-like projecting arm. It would be able to submerge for 40 hr and would have maximum speed of five knots. (NYT, 9/24/68)
- Press said WSMR Public Information Office had denied reports by Ohio newspaperman that disc-shaped practice parachute platform used in five Voyager space vehicle tests at WSMR had inspired reports of unidentified flying objects over southwestern U.S. in 1966 and 1967. Vehicle had been tested only five times, Acting Chief of Information Gabe Brillante had said; there had been many more UFO sightings during period. He said WSMR would have acknowledged tests if it had had inquiries about sightings. (UPI, W Star, 9/24/68, 4; Auerbach, W Post, 9/24/68, A5)

September 25: Senate adopted Senate-House Conference report on H.R. 17023, FY 1969 Independent Offices and HUD appropriations bill, clearing it for White House action. In addition to $500,000 for NASC and $400 million for NSF, bill agreed to by Conference included $3.995 billion for NASA—$12.95 million below amount passed earlier by House and Senate and $375.12 million below original budget request. NASA allocations were $3.37 billion for R&D, $21.8 million for construction of facilities, and $603.17 million for administrative operations. (CR, 9/25/68, S11393–402; NASA *LAR* VII/98)

- Rep. Seymour Halpern (D-N.Y.) on behalf of himself and 14 sponsors introduced H.R. 19990, Sonic Boom Damage Recovery Act of 1968, to protect public against anticipated damage from sonic boom from military and civilian aircraft. (CR, 9/25/68, H9149; NASA *LAR* VII/98)
- NASA issued Apollo Status Summary: Apollo 7 space vehicle Flight Readiness Test was under way at KSC, with prime crew—Astronauts Walter M. Schirra, Jr., Donn F. Eisele, and R. Walter Cunningham—to participate without spacesuits and with hatch open. Apollo 8 manned altitude runs had been successfully completed, with prime crew spending $13\frac{1}{2}$ hr in its spacecraft Sept. 20 and backup crew spending $10\frac{1}{2}$ hr Sept. 22. (Text)
- NASA announced it had selected North American Rockwell Corp. Rocketdyne Div. to provide injector for Apollo lunar module ascent engine under $10-million subcontract with Grumman Aircraft Engineering Corp. Bell Aerosystems Co., also under subcontract to Grumman, would

continue to provide engine hardware for assembly with the injector. (NASA Release 68–164; MSC Release 68–71; *WSJ*, 9/26/68, 3)

- AEC Chairman Dr. Glenn T. Seaborg told 12th General Conference of International Atomic Energy Agency in Vienna that U.S. planned realization of terms of nuclear nonproliferation treaty. "We shall continue to conduct, within the limitations of available funds, an active research and development program both on nuclear explosive devices . . . particularly suited for peaceful uses and on various peace applications for which nuclear explosions can be used. Concurrently, we will provide available information and data and technical advice and assistance to those nonnuclear weapon parties to the Treaty seeking such assistance." (AEC Release S–34–68)
- Sen. Henry M. Jackson (D-Wash.), Chairman of Senate Nuclear Safeguards Committee, in Senate review of implementation of limited nuclear test-ban treaty safeguards reported possibility of black-out of communications, radar, and missile systems by electromagnetic pulse (emp) emitted by nuclear explosion. If absorbed in electronic equipment, this surge of electricity could blow fuses or disrupt electronic components. Scientists had warned of insufficient data on emp and its possible effects on intercontinental and other missile systems. U.S. would spend third more on underground nuclear testing in 1968 than in 1967.

    Five-year-old treaty, Sen. Jackson said, was being observed. U.S. satellites kept watch from space "to the earth's surface" to guard against violation. (Text; AP, W *Star*, 9/25/68, A6; Finney, *NYT*, 9/26/68)
- EDP Technology, Inc., of Washington, D.C., independent concern providing technical support in computer use and technology, had announced it would purchase Cornell Aeronautical Laboratory, Buffalo, N.Y., from Cornell Univ. for $25 million. (NYT, 9/26/68)

September 26: USAF Titan III–C booster launched from ETR successfully inserted four satellites into separate earth orbits. Satellites were one Lincoln Experimental Satellite (*Les VI*), two Environmental Research Satellites (*Ers XXI*, also called *OV V–4*; and *Ers XXVIII*, also called *OV V–2*), and one Orbiting Vehicle research satellite (*OV II–5*).

Powered flight of Titan III–C was very close to planned parameters, according to preliminary figures. First transtage burn placed transtage and payload into parking orbit with 107.5-mi (173-km) apogee 90.8-mi (146.2-km) perigee, 87.7-min period, and 28.6° inclination. Second transtage burn moved stage and payload into elliptical transfer orbit with 22,241-mi (35,792.4-km) apogee, 112.7-mi (181.4-km) perigee, 631-min period, and 26.2° inclination.

Booster ejected *Ers XXVIII* (*OV V–2*) into orbit with 22,236-mi (35,777.7-km) apogee, 115-mi (185-km) perigee, 630.8-min period, and 26.4° inclination, where satellite would investigate radiation in Van Allen belts. Third transtage burn put satellite dispenser frame and remaining three satellites into final orbit with 22,201.9-mi (35,729.5-km) apogee, 22,037.4-mi (35,464.9-km) perigee, 1,428-min period, and 3° inclination.

Ers XXI (*OV V–4*), which would conduct research on heat transfer in liquids at zero gravity, was ejected into circular orbit with 22,225-mi (35,760-km) apogee, 22,220-mi (35,771.9-km) perigee, 1,435.8-min period, and 3° inclination.

OV II–5, synchronous-altitude research satellite equipped with 10 sensors to investigate radiation, entered separate orbit with 22,232-mi 35,771.2-km) apogee, 21,827-mi (35,771.2-km) perigee, and 1,418-min period.

Les VI, major payload, entered synchronous orbit with 22,236-mi (35,777.7-km) apogee, 22,233-mi (35,772.9-km) perigee, 1,435.9-min period, and 2.9° inclination. Second all solid-state UHF-band comsat, *Les VI* would transmit radio signals to test communications to aircraft, ships, and ground troops. First Lincoln Experimental Satellite, *Les V*, had been launched July 1, 1967. USAF officials reported all satellites were functioning properly. (DOD Proj. Off; UPI, *NYT*, 9/26/68, 8; AP, W *Star*, 9/26/68, A1; B *Sun*, 9/27/68, A3; GSFC *SSR*, 9/30/68; *Pres Rpt 68*)

- Univ. of Wisconsin meteorologist Verner E. Suomi said weather satellite research could reduce critical hours needed to identify tornado-producing cloud systems and warn public of threat. Photos from *Ats III* had shown it was possible to observe rapid expansion at top of tornado-producing clouds, while radar and ground observation tended to show only their "stems." With adequate facilities, Suomi saw possibility of satellite tornado forecasts presented directly to the public via television. (UPI, *W Post*, 9/26/68, K3)

- In *Washington Post* Rudy Abramson of *Los Angeles Times* described retiring NASA Administrator James E. Webb as "a huckster. A good one. Maybe the best Washington has ever seen." Even when he could talk politicians into putting up $5 billion in a year for civilian space program, "he left many of them feeling a little guilty because they had not done enough." Congressional opponents had charged he was overzealous to point of dishonesty in protecting his empire at NASA. "Space buffs were piqued because Webb continued calling the Apollo spacecraft a 'capsoole' and publicly seemed to view the astronauts with childlike adoration. But Webb had it where it counted. President Johnson ranked him with Robert S. McNamara as an administrator. Influential Congressmen sided with him against their colleagues who challenged the wisdom of the space program and the quality of the Webb management. . . .

  "If Richard Nixon is elected President, it is a dead certainty that he would not want to keep Webb on as NASA Administrator. Hubert Humphrey might also want his own man. But even if Webb could stay in the job, it's probably all for the better that he quit now. The investigation following the Apollo accident permanently soured his relations with some members of Congress. One has to suspect that Webb is leaving the space program because the country seemed less and less interested in listening to its evangelist." (*W Post*, 9/26/68, K4)

- Commenting on retirement of James E. Webb as NASA Administrator, Rep. Olin E. Teague (D-Tex.) on House floor said: "I share with Mr. Webb his continuing concern that the Soviets are determined to be the No. 1 power in space. Jim Webb has counseled us often and with great clarity on the seriousness of such a situation. We must heed Jim Webb's warning that the United States should be first in space. . . . [He] is to be congratulated for his unselfish dedication and great skill in leadership. . . . It will be difficult for any man to fill those large and capable shoes." (*CR*, 9/20/68, E8135)

September 27: In *Christian Science Monitor*, Robert C. Cowen asked, "Is

U.S. space ability eroding?" and warned, "Even if it beats the Soviet Union to the moon, the American space capability relative to that of the Soviet Union may have atrophied." At Boeing Co. Space Div. some valuable space teams were splitting up. "It will take time as well as money to buy this kind of asset back, if and when it is wanted." At JPL, officials had said personnel cutbacks now were "reaching the bone." More than any other facility, JPL "embodies American competence in this challenging field." Plight of Boeing, as an industrial contractor, and JPL, "as an outstanding government space laboratory, typify the peril the American space program faces. Loss of funds . . . means more than postponement of certain projects. . . . It threatens also at least partial loss of the competence to carry out such projects in the future." (*CSM*, 9/27/68)

- Dedication ceremonies were held for ComSatCorp's new high-capacity earth station for satellite communications at Etam, W. Va. Station would serve as major U.S. East Coast facility for sending and receiving all forms of communications via satellite between U.S. mainland, Puerto Rico and Caribbean, Europe, Latin America, and other Atlantic points. It was designed to work with INTELSAT satellites stationed in synchronous orbits 22,300 mi over Atlantic. (ComSatCorp Release; AP, B *Sun*, 9/28/68, A3)
- Senate Committee on Foreign Relations urged President Johnson to delay formal ratification of nuclear nonproliferation treaty until majority of nations "nearest to a nuclear weapons capability" had promised to join it, although it urged immediate Senate approval of treaty. (Unna, *W Post*, 9/28/68, 1; *NYT*, 9/28/68, 1)
- NASA appointed Ronald J. Philips Director of Technology Utilization Div., succeeding George J. Howick, who had resigned to join International Research and Marketing Corp. Philips, who had joined NASA in 1964 as management intern at MSC, had been principal staff assistant to NASA Administrator and Executive Assistant to Associate Administrator for Organization and Management. (NASA Release 68-166)

September 28: U.S.S.R.'s *Zond V* automatic research station had apparently suffered series of errors during reentry that could have been fatal if men had been on board, John Lannan reported in Washington *Evening Star*. According to one unidentified U.S. source, *Zond V* had entered atmosphere and "skipped" back out as planned, but then skipped along top of atmosphere and reentered "at a peculiar angle." Reentry caused g forces and heat loads that would have been fatal to crews if aboard, and spacecraft landed thousands of miles short of its intended landing area, which was probably on land in Soviet territory.

Chicago Tribune later supported theory that *Zond V*'s reentry was a disaster and reported that U.S. sources said evidence indicated spacecraft had: (1) failed to circle as close to moon as intended, and was thus unable to obtain useful photos of possible landing sites; (2) reentered earth's atmosphere at angle much steeper than planned, causing friction temperatures that would have killed cosmonauts; and (3) landed in water accidentally instead of in Soviet territory as planned. (*W Star*, 9/28/68, A1; *C Trib*, 10/1/68)

- Civilian pilot Bernie J. Dvorscak flew USAF's XV-4B Hummingbird II VTOL research aircraft on maiden flight from Dobbins AFB, Ga., for 28 min at 7,600-ft altitude and 240 mph with conventional takeoff and landing. First Hummingbird II vertical flight was scheduled for early November. Aircraft had four J85–19 lift engines mounted vertically in center fuselage and two J85–19 cruise engines mounted in nacelles, providing total thrust of 18,000 lb. (AFSC *Newsreview*, 10/68, 3)

- Senate Committee on Armed Services' Preparedness Investigating Subcommittee called for rapid development of new strategic bomber to supplement "obvious and admitted shortcomings" of F–111 aircraft. Report, *U.S. Tactical Air Power Program*, said: "During the past decade the Soviet Union has introduced at least one new type operational fighter every 2 years—a total of six, with 11 models. . . . United States does not have a single STOL airplane even in contract definition, [while] the Soviets are now flying three supersonic STOL models and also one VTOL." New U.S. fighter "designed to achieve air superiority is a vital and urgent requirement, should be optimized for its prime mission, not compromised by assigning it multipurpose roles," and should receive "highest priority." (Text; UPI, W *Star*, 9/29/68, A2)
- DOD had approved purchase by Lockheed Aircraft Corp. of $113,886,000 Government plant at Marietta, Ga., world's largest aircraft assembly area under one roof. USAF had declared it in "excess" of its needs and would sell it only to Lockheed, which had been using facility to produce C–5A cargo transports under $1.4-billion USAF contract. Lockheed would continue work on the huge aircraft and could be required to produce an additional 172 on priority basis under USAF agreement. (Wilson, W *Post*, 9/28/68, 1)

September 29: Highly sophisticated "multispectral analysis" under development at Space and Re-entry Systems Div. of Philco-Ford Corp. was enabling scientists to extract additional information from photos taken by satellites. Assigning color ratings to gradations of gray in pictures and filtering out color levels as many as 18 times, to create series of transparencies in vivid hues, allowed details previously undetectable in gray to be distinguished. Technique could be applied to all black and white photos, enabling scientists to map ocean bottom, prospect for precious metal, and explore composition of sun without leaving darkroom. (UPI, *NYT*, 9/29/68, 28)

- In *New York Times Magazine* article. "The Universe Is Not Ours Alone," Walter Sullivan recounted discovery of pulsars, "strange . . . radio pulsations coming from certain spots in the sky." Ten had been discovered to date. Radioastronomers leaned toward natural explanation, yet there were "tantalizing questions that remain unanswered." It was possible "that evidence for the existence of superior civilizations lies hidden in the archives of magnetic tape at our radio astronomy observatories." (*NYT Magazine*, 9/29/68, 40–1 ff)
- FAA released Kling Report outlining four plans for possible modernization of Washington National Airport. Four options, offered to extend National's short-haul role through 1970s, envisioned no major expansion except for 22-acre river fill. All proposed designs emphasized people-handling or terminal side of airport. Report was prepared by Vincent Kling and Associates, Philadelphia, with James Buckley, New York, and Jackson and Moreland, Boston, under $297,000 FAA contract awarded in May 1966. (FAA Release T 68–38; W *Star*, 9/29/68, A1; Eisen, W *Post*, 9/29/68, A1)

September 30: NASA Aerobee 150 MI sounding rocket launched from WSMR carried GSFC experiment to 126.5-mi (203-km) altitude to obtain solar extreme UV spectra from 40 to 390 Å and from 10 to 390 Å using BBRC–SPC 300D solar pointing control and recovery system. Despite better than expected performance, no solar spectrum was detected. (NASA Rpt SRL)

- NASA Administrator James E. Webb delivered John Diebold Lecture on Technological Change and Management at Harvard Univ. Graduate School of Business Administration. Speaking on "NASA as an Adaptive Organization," he emphasized "*how* as well as *what*" in NASA's buildup of capabilities and learning to "apply new ways of organizing and administering human and material resources."

    · Capabilities demonstrated during one week—second week of November 1967—in launching *Ats III*, *Surveyor VI*, *Saturn V*, and *Essa IV* represented "better than 90% of everything we would need to carry out almost any mission that even the most daring have placed on our space agenda for the next decade." Measurements made by experimental spacecraft plus ESSA's 24-hr-per-day operations were providing handsome returns in understanding and predicting weather. "This is one way our country says to every other country, every day, that we as a people want to use our new power over the forces of nature in a joint effort with them, with benefits to both of us, and not to threaten or to coerce them to follow some pattern laid down by us."

    Magnification capability of Ranger and Surveyor opened way "for lunar and planetary investigations of a type and scope undreamed of before we learned to use the rocket technology." Saturn V launch "demonstrated that we can have . . . the big-booster capability and the launch rate capability in which we have been behind the U.S.S.R." Utility of *Essa IV* was "far beyond any we can achieve in any other way. It works . . . all around the world, and feeds information into something like 296 stations in the United States and about 86 in 45 other countries. It is truly a working bird. . . .

    "I do not believe our Nation could have long continued as a great power if we had not built up the means to conduct operations in space. . . . I believe we would have sacrificed our chances to keep pace in the technological competition that is the crucial test of our times. . . . We would have denied to ourselves the tools and the knowledge necessary . . . [for] problems that beset us and the rest of mankind the benefits that surely will follow from the full development of space applications." (Text; *W Post*, 10/13/68, B5)

- World's largest commercial jet—360- to 490-passenger Boeing 747 with mach 0.84 to 0.90 cruising speed—was rolled from factory for first time in ceremony at Everett, Wash. The $20-million aircraft was scheduled for first flight in December 1968 and first passenger service, with Pan American World Airways, Inc., one year later. (Boeing PIO; AP, *W Post*, 9/30/68, A22; 10/1/68, A9; Witkin, *NYT*, 10/1/68, UPI, *W Star*, 10/1/68, A6)

- USAF announced it had called for proposals from eight aircraft companies for contract definition of advanced tactical fighter aircraft designated FX, highly maneuverable, single-place, twin-engine jet with initial operational capability in mid-1970s. Engine development contracts had been announced Aug. 27. Maiden flight was expected in 1972. FX would have significantly better air-to-air performance than any known fighter aircraft. (DOD Release 891-68)

- *Newsweek* commented on break-up of European Launcher Development Organization: "The seven-nation . . . [ELDO] which never really got off the pad, will be buried early in October. Efforts to get Britain to remain only made clear the reason for her withdrawal: France refused to accept Britain as a member of Euromart and the British pulled out. The Netherlands backs Britain's move in the face of De Gaulle's stubbornness, and Italy has already expressed reservations about the project. If Western Europe wants launchers for scientific space exploration in future, it will have to buy them from the U.S." (*Newsweek*, 9/30/68, 20)

- Gen. William F. McKee (USAF, Ret.), President of Schriever and McKee Associates, had been sworn in as consultant to NASA Administrator James E. Webb, NASA announced. Former FAA Administrator and former NASA Assistant Administrator for Management Development, Gen. McKee would advise NASA on management, aeronautics, and coordination with other Government agencies. (NASA Release 68-169)

- Dr. Hubertus Strughold retired as Chief Scientist of Aerospace Medical Div., at Brooks AFB, Tex. He had been associated with USAF aerospace medical program since 1947 and was known as father of space medicine. (Brooks AFB PIO; AFSC *Newsreview*, 10/68, 3)

- House, by voice vote, passed H.R. 12012 to encourage worldwide interest in U.S. developments and accomplishments in military and related aviation and equipment by authorizing Federal sponsorship of an International Aeronautical Exposition in U.S. (*CR*, 9/30/68, H9254; NASA *LAR* VII/101)

- LeRC began 18-mo program to flight-test advanced inlets and exhaust nozzles for supersonic transport engines. Tests, conducted in modified USAF F-106B jet aircraft, would study performance of nozzles and inlets in transonic speed range. (NASA Release 68-163)

During September: Bureau of the Budget released *Summer Review of the Budget*. From action taken on five regular appropriations bills to date, Congress was expected to reduce remaining appropriations by approximately $9.3 billion. On completion, further reductions required to comply with Revenue and Expenditure Control Act of 1968 would be determined by President. For portion of budget covered by required reduction under P.L. 90-364, total outlays were estimated at $6 billion below January budget estimate. Overall $6-billion reduction would be made approximately 50% in DOD and 50% in civilian agencies. NASA outlays beyond those for manned lunar landing would be held to $100-million level below that resulting from Congressional action. (Text)

- *Astronautics & Aeronautics* noted that NASA-sponsored translations of Russian historical works compiled 1963-68 were available from Clearinghouse for Federal Scientific and Technical Information, Springfield, Va. 22151. It was a credit to NASA's historians "that they have not neglected bringing to our attention the products of their opposing numbers." (Stehling, *A&A*, 9/68, 76)

- *Astronautics & Aeronautics* published letter from J. Gordon Vaeth of ESSA's National Environmental Satellite Center: With emphasis of space technology on applications and practical benefits, it was "curious that

". . . little is apparently being said or done about using communications satellites to 'carry' the mail over intercontinental distances." Facsimile techniques were available for scanning and electronic transmission of letter mail by geostationary spacecraft. "In-depth professional consideration of this satellite application is overdue" (*A&A*, 9/68, 17–20)

- P. J. Parker listed in *Spaceflight* "typical examples of space-inspired technology" which had either appeared or were expected to emerge from future space activity: low-weight, high-calorie "spacefoods" developed for mass-feeding undernourished peoples; adaptation of Saturn V air-bearing to other large items such as refrigerators; use of spacecraft automatic monitoring unit in hospitals for relaying data on patient's heart condition to central point; use of high-pressure oxygen in astronauts' suits to reduce lung collapse and to save lives of premature babies; potential cure for stuttering from examination of voice communication problems; electric switch adapted for use by paralyzed patients to operate hospital call boards and wheelchairs; and lunar exploration vehicle redesigned into walking-chair with reciprocating legs.

    Industrial applications included solar cells used to drive small outboard motorboats and "outback" telephone booster units; heat-resistant coatings for spacecraft used in furnaces, aircraft, and domestic cookers; and new precision tooling methods that worked to nearest millionth of an inch. (*Spaceflight*, 9/68, 306–7)

- National Science Foundation issued *Employment of Scientists and Engineers in the United States, 1950–1966*. Such employment had risen from 550,800 in 1950 to 1,412,500 in 1966—156%. Scientific occupations had increased by 185%, from 146,300 to 416,800, engineering had increased 146%, from 404,600 to 996,000. Chemists accounted for more than 25% of scientists in 1966, while medical scientists more than doubled rate for all scientific occupations.

    Between 1950 and 1966, number of R&D scientists and engineers rose 242%. Proportion of R&D personnel grew from 28% of all scientists and engineers in 1950 to 37% in 1966. Private industry employed 71% of total scientists and engineers; government, 16%; universities and colleges, 13%; and nonprofit institutions about 1%. Employment increased more rapidly in nonprofit sector (359% between 1950 and 1966). Government employment (excluding public educational institutions) showed lowest growth—106%. (Text)

October 1968

October 1: NASA 10th Anniversary: National Space Club presented special award to President Johnson at White House citing his legislative and executive leadership of national space program. At Space Club dinner in Washington, D.C., Sen. Clinton P. Anderson (D-N. Mex.) and Rep. George P. Miller (D-Calif.) presented retiring NASA Administrator James E. Webb award for his outstanding contributions to national space effort. In telegram tribute to Webb, President Johnson said, "The Nation is in his debt. He will be deeply missed but gratefully remembered as his dreams continue to become reality in the years ahead."

Telegram from Vice President Hubert Humphrey said, "The span of achievement which measures a decade of space progress is one which should be a source of pride to all of us. Now a new decade beckons, with new challenges, and new opportunities. And, such is the nature of the space age that we dare not become complacent about our rate of progress or the scope of our past accomplishments. It is unthinkable that we would allow ourselves to be surpassed in technology by any other nation. I have never questioned that space endeavors have contributed significantly to the strengthening and enrichment of our whole society, through a teamwork approach by private industry, our universities, and the Federal Government. By means of this program we have vitalized our economy, developed improved methods of management, stimulated our educational system, produced new goods and services, added to our store of scientific knowledge, and buttressed our national security." Telegram from Republican Presidential candidate Richard M. Nixon said, "The space program must continue to be one of our national imperatives, and it must be supported at a level assuring efficient and steady progress."

Anniversary ceremonies also included open house and annual awards presentation at MSFC Sept. 28–29. (NSC *Newsletter*; Texts; *Marshall Star*, 9/25/68, 4)

In anniversary statement issued by White House President Johnson said: ". . . not all of NASA's accomplishments have been out of this world. Satellites have given us a new look at the world's weather. . . . They have given us intercontinental television broadcasts, and broken down the technical barriers to worldwide communications. NASA has brought us advances in medical science, education, mapmaking, geology, transportation, and a host of other areas that promise a better life for us here on Earth. Its intensive research and development efforts have given us new materials, products and processes; raised our standards of reliability and advanced managerial techniques. These advances, together with the useful facilities it has built, will be lasting national assets long after the Moon landing is ancient history. . . .

"In the years ahead—as in the past decade—our foremost motive is to make men wiser and life on earth more meaningful. And on the mile-

October 1: *Ten years ago* NASA *officially came into being. In photo, President Dwight D. Eisenhower hands commissions to heads of new agency after Sept. 8, 1958, swearing in ceremonies. At President's left is Dr. T. Keith Glennan, first* NASA *Administrator. Accepting commission is first Deputy Administrator, the late Hugh L. Dryden.*

stone of this rewarding effort, we renew our dedication to the guiding principle we expressed 10 years ago at NASA's launching: that our further mastery of space may continue to be 'for the benefit of all mankind.' " (PD, 10/7/68, 1435–1436)

During its first 10 yr NASA had completed 234 major U.S. and international launches plus thousands of sounding rocket launches. Of 234 major launches, 189 were launch-vehicle successes and 174 were spacecraft or mission successes, with two missions still under evaluation. For these launches NASA had developed rockets ranging from 88,000-lb thrust to Saturn V with 7.5-million-lb thrust capable of sending nearly 100,000-lb payload to moon.

First phase of manned flight program, Project Mercury, had begun seven days after NASA was established. Six manned Mercury flights had proved man could go into space and function as pilot-engineer-experimenter for up to 34 hr weightless flight before returning to earth. Gemini program announced in 1961 had demonstrated work could be performed in orbit outside spacecraft in more than 12 hr extravehicular activity. Gemini included 52 experiments, among them 17 scientific—in astronomy, biology, geology, meteorology, and physics. Seven rendezvous techniques and nine dockings had been accomplished during Gemini. More than 2,000 hr manned spaceflight experience gained through

Mercury and Gemini had contributed heavily to Apollo, including 16 flights of unmanned Saturn launch vehicle which had confirmed Apollo engineering concepts and qualified all systems for manned missions. Although Apollo fire on Jan. 27, 1967, had delayed manned missions, it had resulted in safer spacecraft and improved suits for astronauts.

Ahead were two manned missions for 1968 and possibly five for 1969, culminating in landing of U.S. astronauts on moon. Facilities at NASA installations in U.S. were worth more than $2.5 billion and peak 35,000 staff included some of Nation's top scientists. Industrial work force had peaked at 400,000 and was dropping to 200,000. NASA had established global tracking network capable of communicating with vehicles as far away as far side of sun. Its aeronautics program conducted R&D on noise abatement, flight safety, supersonic and hypersonic aircraft, lifting-body vehicles, and V/STOL aircraft. NASA's technological advances included development of new electronic parts, alloys, adhesives, lubricants, valves, and pumps, as well as progress in miniaturization. More than 2,500 technical innovations applicable in industry, medicine, and other nonaerospace activities had resulted from 10 yr of NASA progress. In *Christian Science Monitor,* Neal Stanford said, "It is . . . science satellites and the new technology developed that are counted on to return to the tax-paying public the dividends that some say will make space the best investment man ever made." (*Marshall Star,* 9/25/68, 5; Wilford, *NYT,* 10/1/68; *CSM,* 10/5/68)

In Washington *Sunday Star* special report, "A Decade in Space," John Lannan commented on NASA's 10th anniversary: ". . . the space agency's real promise for improving the general welfare of mankind as set forth in the Space Act which brought it into being has apparently been too slow in being fulfilled. The fact that the space investment is only now starting to pay off, and at an increasingly rapid pace, is going unnoticed in the glare of present problems, past mishaps and the imminence of the Apollo venture. Where NASA's real goals lie are in the future—the near future and the near earth. Spin-off, the serendipity of technology, has long been used as a justification for the vast sums poured into space, but little beyond Teflon-coated frying pans have impinged on the tax-payer's mind." (*W Star,* 9/29/68, A1)

NASA Assistant Administrator for DOD and Interagency Affairs Jacob E. Smart wrote in *Space Digest:* "The tenth anniversary . . . marks the end of a decade of concerted effort across a broad front to advance the nation's capabilities in aeronautics and space. It has been a decade of accomplishment that has few peers in this country's history. . . . the good working relationships . . . between NASA and DOD have been of immeasurable benefit to them both, and the nation's space effort is the stronger for it. . . . with the prospect of tighter budgets likely, the maintenance of such relationships—and their improvement—assumes an increasing importance. The need to stretch the appropriations dollars to the maximum may be partly met by fresh efforts to find common ground where cooperation will produce economies as well as mutual benefits." (*Space Digest,* 10/68, 68–70)

In *Space Digest* Gen. James Ferguson, Commander of AFSC, described "A Decade of Cooperation—The Military-NASA Interface." AFSC's FY 1969 NASA support expenditures "amounted to millions of dollars and thousands of man-years in effort. The sums of money, and, more important, the human technological resources we are currently expending on the nation's space programs are of great consequence since they are the foundations of our future benefits. . . . we are acutely aware that the fate of future generations lies in space. . . . It is a sad commentary to state that technology has provided us with the means to conduct wars. But it is heartening to realize that one day technology will provide us with the means to prevent wars. (*Space Digest*, 10/68, 71–3)

- Aerobee 150 MI sounding rocket launched by NASA from WSMR carried Univ. of Colorado experiment to 108-mi (173.8-km) altitude to measure height profile of nitric oxide and nitrogen and test Mariner Mars UV spectrometer. UV spectrometer measured earth's day-glow and daylight between 1,100 Å and 4,300 Å. Inertial ACS successfully maneuvered payload until scat eye took control and caused instrument to scan through 10°. Experiment's measurements and tests were successful. (NASA Rpt SRL)

- Dr. Finn J. Larsen, DOD Deputy Director of Defense Research and Engineering, testified before House Committee on Science and Astronautics' Subcommittee on Advanced Research and Technology: "If our civil aviation is to continue its dramatic progress, the greatest single requirement is to accomplish . . . complete system engineering. The . . . planning must consider not only the aircraft in flight as a system, but also the entire problem of moving people from destination to destination," requiring "planning and resources on a considerably larger scale than are now available."

    NASA's responsibility for U.S. aeronautical research "should continue." DOD had used NASA research, augmenting it for defense as necessary, with "excellent coordination for many years." Military R&D was specialized, but much was transferable to civil aviation. Long-term goals of smokeless combustors and silent aircraft would be of mutual benefit. SAGE air defense computer system had contributed to FAA radar beacon system and new national standards. Much improved altimetry reporting came from DOD development for high-performance jet aircraft. Common digitizer was joint DOD–FAA project, as was TPX–42 airport traffic control facility. Eleven aircraft carriers had pilot "hands-off" capability for blind landing. Microwave scanning-beam landing systems were in testing. Collision avoidance, V/STOL, and cargo-handling R&D had civil application. (Text)

- Dr. Frank D. Drake, Director of Cornell Univ.'s Arecibo Ionospheric Observatory in Puerto Rico, said he had detected first distinct pattern to radio signals from two pulsars. He told radioastronomy seminar at National Radio Astronomy Observatory at Green Bank, W. Va., pulse rate could be explained only if source were star of extraordinary density spinning at incredible speed—such as neutron star. If correct, findings would be first scientific proof that hypothetical neutron stars actually existed. (Wilford, *NYT*, 10/2/68; Lannan, *W Star*, 10/2/68, A20)

- MIT physicist and radioastronomer, Dr. Bernard Burke, and teams of scientists using 140-ft "Big Dish" antenna at National Radio Astronomy Observatory began first radioastronomy test of Einstein's general theory of relativity in attempt to discover gravity's effect on universe. Among three basic tests proposed by Einstein to test his theory to account for action of all bodies under gravitational force was one to measure bending of light from distant source as it passed an energetic body like the sun. Dr. Burke's experiment measured signals from newly discovered quasars to determine amount of bending they underwent in passing the sun. If light, in form of quasar radiowaves, was bent, Einstein theory would receive added support; if it was not bent as much as he predicted, or if astronomers were unable to detect significant bending, theory would remain intact until further proof was provided. (NRAO Proj Off; Lannan, *W Star*, 10/1/68, A9)

- MSFC announced it had selected RCA for negotiation of $5.1-million cost-plus-fixed-fee contract for logistics and engineering support for Saturn ground computer systems and associated equipment. Contract would cover Oct. 1, 1968, through June 30, 1970. (MSFC Release 68–231)
- President Johnson announced resignation of Leonard H. Marks as Director of U.S. Information Agency in time to head U.S. delegation to negotiate permanent arrangements for INTELSAT at February 1969 conference in Washington, D.C. (*PD*, 10/7/68, 1433; Halloran, *W Post*, 10/2/68, A8; AP, *NYT*, 10/2/68, 23)
- Arnold W. Frutkin, since Feb. 1 Special Assistant to NASA Associate Administrator, resumed his duties as Assistant Administrator for International Affairs. (NASA PAO; NASA Ann, 10/2/68)

October 2: U.S.S.R. launched *Cosmos CCXLIV* into orbit with 158-km (98.2-mi) apogee, 140-km (87-mi) perigee, 87.4-min period, and 49.6° inclination. Satellite reentered same day. U.S. press later reported U.S. observers said launch appeared to have been 13th test of Soviet fractional orbital bombardment system (FOBS). Orbit followed pattern of previous tests identified as FOBS by U.S.—very low earth orbit with satellite reentering before completing first revolution of earth. (GSFC *SSR*, 10/15/68; UPI, *NYT*, 10/9/68, 12; *W Post*, 10/8/68, A10)

- Sen. Clinton P. Anderson (D-N. Mex.), Chairman of House Committee on Aeronautical and Space Sciences, told Senate: "Ten years from now NASA will be celebrating its 20th anniversary. I hope that the chairman of the Committee . . . at that time will be able to stand here and congratulate the agency and its people for 20 years of accomplishment and say that the United States is still first in space and in aeronautics. But unless we are vigilant and supply the agency with the needed authorizations and appropriations, that statement will not be made." (*CR*, 10/2/68, S11844)
- NASA awarded Chrysler Corp.'s Space Div. $10,545,753 cost-plus-award-fee extension to $77,877,486 contract for KSC support services. Extension, for July 1 through Dec. 31, covered manpower and material to design and sustain engineering, modification, testing, refurbishing, and launch support of KSC-designed equipment and Saturn IB launch operations. (KSC Release KSC–418–68)
- Sen. Stuart Symington (D-Mo.) said on Senate floor: "In the past ten years, money expended by the Defense Department for R&D has almost doubled, from $4 billion to about $8 billion. Yet since 1955, the United States has not produced a single modern fighter; in fact, it has produced no combat plane except the TFX series. The Navy version of that plane has already been abandoned; and the Air Force has once again found it necessary to ground their version because of technical difficulties." Despite "all those billions we have developed no air superior-

ity fighter capable of competing against a first-class air force such as the Soviets possess today." He said U.S. was losing its lead on seas as well because "our various Government branches produce arguments, whereas the Soviets produce the submarine." (Text; *NY News,* 10/3/68, 18)

- MSFC announced it had awarded American Science and Engineering, Inc., $5,413,000 addition to contract for final design, fabrication, assembly, integration, test qualification, and acceptance of prototype and flight unit x-ray spectrographic telescope for Apollo Telescope Mount. Award brought total value of contract to $11,617,471. (NASA Release 68–170; MSFC Release 68–234)
- National Center for Atmospheric Research and Information announced 10-ft-dia Global Horizontal Sounding Technique (GHOST) plastic balloon launched from Christchurch, New Zealand, Sept. 29, 1967, had broken all balloon flight-duration records by remaining aloft for one year. (AP, *St. Louis G–D,* 10/3/68)

October 3: Aurorae (*Esro I*) satellite, designed, developed, and constructed by European Space Research Organization under July 8, 1964, NASA–ESRO agreement, was successfully launched by NASA from WTR by four-stage Scout booster. Orbital parameters: apogee, 952.1 mi (1,532.2 km); perigee, 158.9 mi (255.8 km); period, 102.8 min; and inclination, 93.8°. Primary NASA mission objectives were to place *Aurorae* into planned orbit and provide tracking and telemetry support. The 185-lb cylindrical satellite carried eight experiments to study aurora borealis (Northern Lights) and other related phenomena of polar ionosphere, representing six different organizations from U.K., Denmark, Sweden, and Norway. All experiments were operating as planned, and their scientific objectives were being achieved.

Aurorae was second successful ESRO satellite launched by NASA. *Iris I,* launched May 16 to replace ESRO II–A which had failed to achieve orbit May 29, 1967, had entered planned orbit and conducted solar-astronomy and cosmic-ray studies. ESRO was responsible for experiment instrumentation, delivery of spacecraft to launch site, equipment and personnel necessary to mate spacecraft to launch vehicle, and spacecraft testing. NASA provided Scout launch vehicle, conducted launch operations, and supplied data and tracking acquisition support. (NASA Proj Off; NASA Release 68–158; AP, *W Star,* 10/4/68; GSFC *SSR,* 10/15/68)

- *Cosmos CCXLV* was launched by U.S.S.R. into orbit with 481-km (298.9-mi) apogee, 272-km (169-mi) perigee, 92-min period, and 70.9° inclination. Satellite reentered Jan. 15, 1969. (GSFC *SSR,* 10/15/68; 1/31/69)
- NASA's HL–10 lifting-body vehicle, flown by NASA test pilot John A. Manke, successfully completed eleventh flight from Edwards AFB. Vehicle with full load of fuel was carried to altitude, where successful jettison test was performed before launch from B–52 aircraft. (NASA Proj Off)
- NASA Deputy Administrator, Dr. Thomas O. Paine, presented NASA FY 1969 interim operating plan to Senate Committee on Aeronautical and Space Sciences. Although President Johnson had not yet signed appropriations bill, Bureau of the Budget had indicated NASA's share of $6 billion expenditure reduction might amount to $350 million. This

meant limitation of use of FY 1969 appropriations to $3.85 billion, Dr. Paine said. "Our actions also have had to be constrained by our current instructions within the Executive Branch to hold expenditures to a minimum in 1970 as well as in 1969, and to be prepared for the eventuality of budgetary limitations in FY 1970 even more restrictive than those in FY 1969. I am personally convinced that the nation's space program requires an increase in funding in FY 1970 and I am hopeful . . . we will succeed in establishing the need for a significant increase. Until this decision is made . . . we have no alternative but to proceed with an interim operating plan . . . which, where possible, holds open options we can exercise in FY 1970 if the budget is higher but which does not overcommit us if the FY 1970 budget is lower."

Plan retained $2.025 billion authorized for Apollo program—$14 million below budget request. It reduced authorized $253.2 million for Apollo Applications to $150 million, amount "required to work toward the important but sharply limited and deferred Apollo Applications program we now propose." This would include cessation of Saturn IB launch vehicle production after completion of 14th (Saturn 214) and discontinuation of Saturn V at completion of first 15 vehicles. Single Saturn I workshop and single Apollo Telescope Mount (ATM) would be launched in early 1970s. Authorized $5 million for advanced missions would be cut to $2.5 million, for continued studies related to manned earth-orbital and lunar missions. Authorized $136.9 million would be reduced to $132.1 million for physics and astronomy, with level of effort in supporting research and technology and data analysis approximately 10% lower than in FY 1968.

The $92.3 million authorized for lunar and planetary exploration would be cut to $75.8 million, with $6.8 million for lunar and $69 million for planetary—to support Mariner-Mars 1969 mission, reacquisition of telemetry from Mariner V, and Mariner-Mars 1971 mission. Plan also supported in FY 1969, at reduced funding level, capability to conduct Mars mission during 1973 opportunity in keeping with NAS recommendation [see Sept. 19]. Overall scope of mission would be reduced and schedule compressed. Operating plan provided for construction of two 210-ft antennas for Mars and other missions during 1970s. Launch vehicle procurement authorization of $115.7 million would be cut to $100.2 million and bioscience from $33 million to $32.7 million, which was $15 million below budget request and required slip of 6–12 mo in 21-day Biosatellite missions.

Space applications authorization of $98.7 million would be retained; program change necessitated by May 18 destruction of Nimbus B would result in launch of replacement, Nimbus B2, in spring 1969. Aeronautics R&D budget would remain at authorized $74.9 million, while FY 1969 effort in nuclear rockets would be limited and NERVA development deferred until 1970, when $7.5 million withheld in FY 1969 could be added to allow total of $39.5 million. Interim plan would provide $178.4 million for basic research, space vehicle systems, electronics systems, human factor systems, space power and electric propulsion systems, and chemical propulsion—at reduction of $21.5 million from NASA's budget request. Work in long-endurance life-support-equipment technologies would proceed as planned but in other areas research and technical development would be reduced, especially efforts in support of advanced space missions. Tracking and data acquisition budget would be reduced from $289.8 million authorized to $280 million, providing full Apollo schedule support but limiting support for spacecraft aloft and on future missions.

Construction of facilities funds had been reduced from $39.6 million authorized to $21.8 million appropriated. Operating plan would increase figure to $35.7 million by transferring funds from R&D appropriation. Facility planning and design funds were reduced from $3 million requested to $1 million. Interim operating plan anticipated transfer of $20.1 million from R&D to administrative operations, bringing total to $623.3 million instead of $603.2 million in authorization and appropriations acts and $648.2 million requested by NASA in budget. Dr. Paine emphasized that administrative operations appropriation did not cover only "administrative" expenses; it covered direct costs of operating NASA laboratories, research centers, development centers, and launch centers.

NASA Administrator James E. Webb told Committee: ". . . when you use words such as 'Congress consistently has supported the Apollo program,' you must add 'at a minimum level.' We have clearly indicated in every budget that the basis on which we were going forward with this support by Congress was one that did not take into account unusual risks and happenings and was, in effect, based on success in all these efforts." Webb said NASA Apollo funding was related to success on various operations and did not include "a return to test flight on the Saturn IB should we not be able to make the shift to the big rocket after this next flight." From 1961 to 1969, Webb said, "we have not had the funds to proceed except in a manner that would permit us, within the total budget, to do this lunar landing within this decade and on an all-up systems test basis. So the excruciatingly painful period of all-up testing on the Saturn V is yet ahead of us." (Testimony; Transcript)

- Senate by vote of 55 to 2 passed H.R. 18707, defense appropriations bill containing initial funds for deployment of "thin" missile system, already under construction. (*CR*, 10/3/68, S11951–79)
- Senate, after secret session, defeated by vote of 45 to 25 amendment by Sen. John S. Cooper (R-Ky.) to eliminate from $71.8-billion defense appropriations bill the $387.4 million requested by Administration to start deployment of Sentinel antiballistic missile system. Final action on largest defense appropriations bill in U.S. history was deferred. (*CR*, 10/2/68, S11872–85; AP, *W Star*, 10/2/68, A4; Finney, *NYT*, 10/3/68, 1; Lardner, *W Post*, 10/5/68, 1)
- NASA said spacecraft and parts of Agena 2nd stage of Nimbus B weather satellite launched unsuccessfully May 18, including two SNAP-19 nuclear power generators, had been found Sept. 30 by crew of research submarine four miles south of San Miguel Island off California coast. Pictures taken by submarine indicated three- by six-inch graphic cores of generators were intact; surrounding magnesium-thorium alloy casings were almost completely decayed since they dissolved in sea water almost immediately. Each core contained three pounds of plutonium. AEC had spent $200,000 searching for missing nuclear sources. Nimbus B had been destroyed shortly after launch when it veered off course. (NASA Release 68–171; Lannan, *W Star*, 10/4/68, A16)
- Boeing Co. said it would build one-fifth size thermal models of manned

space stations under $156,500 NASA contract to determine their feasibility for predicting temperatures in full-size earth-orbiting station. (Boeing Release S-9840)

October 4: NASA–USAF review board report said failure of Nimbus B mission May 18 had been caused by improper installation of yaw-rate gyro 90° from design position in Thorad-Agena launch vehicle. Board recommended revision of test procedures which failed to discern error and redesign of gyro mounting brackets to make improper installation impossible. Repeat mission, Nimbus B2, would be launched in spring 1969 because of flight's importance to meteorological research. (NASA Release 68–171; UPI, *H Chron,* 10/5/68)

- President Johnson signed H.R. 17023 as P.L. 90–550, Independent Offices and Dept. of Housing and Urban Development Appropriations Act, 1969, which included NASA appropriations of $3.995 billion. Conference report on bill had been adopted by Senate Sept. 25 and by House Sept. 19. (*PD,* 10/14/68, 1484)

- At NASA Pasadena (Calif.) Office third annual awards ceremony, Dr. John E. Naugle, NASA's Associate Administrator for Space Science and Applications, said: "During the next few years . . . our efforts in astronomical observations, in space applications and in planetary exploration, should receive priority. . . . In astronomy, perhaps more than in any other scientific discipline, major progress has been achieved when a new observing technique is used. In the next several years we will provide our astronomers with such new observational tools—telescopes and detectors in space above the absorbing and obscuring effect of the earth's atmosphere. . . . we envision a complementary program using both manned and automated observational platforms leading to permanent observatories in space in the next 10 to 15 years. On the other hand, OAO, OSO, and the small astronomy satellite will carry the burden of automated observation, while the Apollo Telescope Mount provides us with early experience in the contribution man can make in astronomy. Before very many years, it should be possible for us to move into astra-type systems in which we combine the best of both techniques, i.e., long-term automated instruments in orbit of a cost and complexity which justify their being serviced and maintained by man. . . .

    "Although . . . post Apollo plans are much more modest than previously announced, even at these reduced levels, it should be possible to carry out some important near-earth and extended lunar missions following the Apollo landings. However, there remains no clear picture as to the future of manned space flight beyond the use of the launch vehicles and rockets left over from Apollo. The resolution of the future goals of manned space flight must await the . . . next administration." (Text)

- NASA's *Oso IV* spacecraft (launched Oct. 18, 1967) had obtained valuable new data on three-dimensional structure of sun's atmosphere, Leo Goldberg, Robert W. Noyes, William H. Parkinson, Edmond M. Reeves, and George L. Withbroe of Harvard College Observatory's Solar Satellite Project reported in *Science.* During five weeks of experiment operation—before electronic failure of instrument detection system—more than 100 solar flares were recorded. More than 4,000 UV images in 52 different wavelengths over wide range of temperatures and heights in solar atmosphere were obtained. Most of emission lines represented had not been observed before with spatial resolution on solar disk.

New instrument was being prepared for flight on OSO-G in 1969, to have twice spatial resolution and 10 times time resolution of instrument on *Oso IV*. (*Science*, 10/4/68, 95–9)

- ComSatCorp, on behalf of INTELSAT, awarded Hughes Aircraft Co. $72-million contract for Intelsat IV advanced comsats. Hughes would deliver within 22 mo four flight spacecraft, one prototype, associated spacecraft test equipment, and necessary ground equipment. (ComSatCorp Release 68–52; *WSJ*, 10/7/68, 3)
- AFSC's Space and Missile Systems Organization awarded Philco-Ford Corp.'s Space and Re-Entry Systems Div. $7,805,000 fixed-price-incentive-fee contract for development, production, and launch of two comsats for NATO. First satellite would be launched from ETR in late 1969 by Thrust-Augmented Thor-Delta booster into 20,000-mi-altitude synchronous orbit. Second would be backup. (AFSC Release 145.68)

October 5: USAF launched unidentified satellite from Vandenberg AFB by Thor-Agena D booster into orbit with 316-mi (508.5-km) apogee, 300-mi (482.8-km) perigee, 94.5-min period, and 74.9° inclination. (*Pres Rpt 68*)

- U.S.S.R. successfully launched *Molniya I–10* to relay telephone and telegraph communications and TV programs to far northern and far eastern U.S.S.R. and to central Asia. Orbital parameters: apogee, 39,639 km (24,630.5 mi); perigee, 429 km (266.6 mi); period, 711.9 min; and inclination, 64.8°. (UPI, W *Star*, 10/7/68, A9; AP, *NYT*, 10/8/68, 2; GSFC *SSR*, 10/15/68)
- Republican Presidential candidate Richard M. Nixon issued policy statement, "The Research Gap: Crisis in American Science and Technology": U.S. was "shortchanging" its scientific community and risking research gap between U.S. effort and that of U.S.S.R. "Faced with dynamic possibilities for science, the current Administration is hobbled by the static philosophy that technological potentialities are limited. . . . This attitude is particularly perilous in the realm of defense. . . . In few areas of development is activity so intense and productive as in Soviet military research and development." While U.S.S.R. graduated twice as many scientists annually as U.S., American scientific community was "demoralized" by wavering attitudes toward R&D. "Scientific activity cannot be turned on and off like a faucet. The withdrawal of support disperses highly trained research teams, closes vital facilities, loses spinoff benefits, and disrupts development momentum. . . . The United States must end this depreciation of research and development in its order of national priorities. . . . It would be an urgent goal of my administration to devise effective means by which it could cooperate with industry and the academic community in an effort to make maximum use of scientific advances to help solve major national problems. . . . Our goal is to make the United States first again in the crucial area of research and development." (Text; Walsh, *Science*, 10/18/68, 335–7)

October 6: In Washington *Sunday Star* William Hines commented on James E. Webb's resignation as NASA Administrator: "Yes . . . there was a James Webb. He had his faults, God knows, and a peculiar style.

Most people would have done the job differently. But, on balance, it is difficult to see how anyone could have done it much better." (W *Star*, 10/6/68, C4)

October 7: U.S.S.R. launched *Cosmos CCXLVI* into orbit with 317-km (197-mi) apogee, 145-km (90.1-mi) perigee, 89.1-min period, and 65.3° inclination. Satellite reentered Oct. 12. (*InteraviaAirLetter*, 10/8/68, 11; GSFC *SSR*, 10/15/68)

- NASA Administrator James E. Webb issued order dissolving Apollo 204 Review Board established Jan. 27, 1967, to investigate accidental Apollo fire of that date. (Text)

- Resignation of James E. Webb as NASA Administrator, announced Sept. 16, became effective. Deputy Administrator, Dr. Thomas O. Paine, assumed duties as Acting Administrator. (NASA Off of Administrator; Off of Acting Administrator)

- In *National Observer*, Peter T. Chew criticized Americans as "uncertain, timid farers in space." During "19-month interregnum in manned space flight" occasioned by Jan. 27, 1967, Apollo fire, "Americans have become obsessed with the race question at home and the Vietnam War abroad. . . . If some doomsayers are to be believed, the vast U.S. space science and technology establishment put together during the last decade will be systematically dismantled once the manned Apollo landing has been accomplished because NASA has 'no clear mandate' to go on; cornfields will reclaim the great rocket and spacecraft-testing sites . . .; the solar system will become the exclusive playground of Soviet cosmonauts." Yet NASA's mandate to explore space "for the benefit of all mankind" had been set down in legislation establishing the agency and did not end with the moon. If anything, "the moon is the first stepping stone." Dr. Wernher von Braun "stands almost alone among the country's leaders in his ability to express in understandable terms just why we are going to the moon—and beyond. . . . To critics of the space program he replies, 'Man was born with an insatiable nosiness about his natural environment. . . . it seems to pay off handsomely, but often in the most unexpected way, to keep satisfying his curiosity about the world around him." (*Natl Obs*, 10/7/68)

- NASA announced it had awarded Technical Information Services Co. $4.3-million cost-plus-award-fee contract for continued operation of NASA's Scientific and Technical Information Facility at College Park, Md. Contract would extend through November 1969 with two one-year options. Current contractor was Leasco Systems and Research Corp. (NASA Release 68–173)

- *Newsweek* said NERVA project had "become one more casualty of cutbacks in the space program." Workers at Nevada test site "say only a skeleton staff will be left on the project by spring." (*Newsweek*, 10/7/68)

- Sen. Stuart Symington (D-Mo.) said on Senate floor: "I am now confident . . . serious consideration should be given to canceling the entire Air Force F–111 series. . . . If the plane is fundamentally unsound—and that would now appear to be the case—its termination would prevent the loss of additional billions of dollars—and what is more important, save the lives of many pilots." He said October report of Preparedness Investigating Subcommittee of Senate Committee on Armed Services "points up the grave security deficiencies that have resulted from the Department of Defense forcing the Air Force and Navy to put all the eggs of their aircraft development into one unfortunate basket." (Text; *CR*, 10/7/68, S12148-51; Witkin, *NYT*, 10/8/68, 18; *W Post*, 10/8/68, A11)

October 8: Antennas on NASA's *Explorer XXXVIII* (launched July 4) were each successfully extended to 750-ft maximum length and damper boom to maximum 630 ft by ground command. Satellite's antennas had been initially deployed to 455 ft each July 22 and extended to 600-ft each Sept. 24. Maximum extension completed planned antenna deployment sequence. (NASA Release 68-174; NASA Proj Off)

- Senate unanimously approved space rescue treaty, providing for rescue and return of astronauts downed on foreign soil. It had been signed by '75 nations. (*CR*, 10/8/68, S12215-6; AP, B *Sun*, 10/9/68, A10)
- NASA Associate Administrator for Manned Space Flight, Dr. George E. Mueller, addressed Ninth National Conference of United Press International Editors and Publishers in Washington, D.C.: "For the conception and construction of the equipment necessary to the safe transport of men into space and for their accomplishment of productive tasks in that new atmosphere, a new mix of professional and scientific disciplines has been created which has forced cooperation between engineers and medical doctors. Many of the technologies which are essential to our sending three men to the moon and back did not exist a few years ago. They had to be invented, adapted or developed. . . .

  "We now have the giant boosters which have released man from his atmosphere, and . . . life support systems that can maintain him in space. As a result of the cleanliness requirements of the space program we have the largest 'clean rooms' in the world—rooms which hospitals are now emulating. . . . Over 600 computers now comprise the largest and most advanced communications system in the world. The fuel cell, which had lain dormant for many years, was activated to power spacecraft in orbit. Thirty public utility companies now have a $27,000,000 program for the adaptation of the fuel cell for home power units. We had to know on a real-time basis how fast the hearts of the astronauts were beating while they were in space . . . how much oxygen they were using, and how their muscles were responding . . . so we invented another new system, biosensor to computer to data gathering equipment, and through communications network to the Manned Spacecraft Center at Houston—from 100 . . . or 800 . . . or from ¼ of a million miles out in space. And a half a dozen newly formed companies are now manufacturing these adapted space-created instruments for the use of doctors and hospitals here on earth."

  In conclusion Dr. Mueller quoted the late Dr. Hugh L. Dryden, first Deputy Administrator: "None of us knows what the final destiny of man may be—or if there is any end to his capacity for growth and adaptation. Wherever this venture leads us, we in the United States are convinced that the power to leave the Earth—to travel where we will in space—and to return at will—marks the opening of a brilliant new state in man's evolution." (Text)
- Commenting on James E. Webb's retirement, Sen. John Stennis (D-Miss.) said on Senate floor, "I have been a member of the Committee on Aeronautical and Space Sciences since about the time Mr. Webb was appointed to head NASA. I am not given unduly to praise a man. I

am not impressed by a title. I am impressed by a record. But I am certainly impressed with the fact that Mr. Webb carried out his responsibilities for NASA with an expenditure of $34 billion, in what might be called a crash program; and I have not seen any evidence of any activity of his except that clothed in the highest degree with integrity, honesty, frankness and openness in his dealings with the committee, with Congress, as well as with the public." (CR, 10/8/68, S12227-8)

- Dept. of State said it would issue visas to 35 Soviet space scientists to attend 19th Congress of International Astronautical Federation (IAF) in New York Oct. 13–19. However, it might recommend cancellation of sightseeing tour of KSC arranged with NASA by AIAA for scientists from 34 countries attending both IAF Congress and AIAA annual meeting in Philadelphia Oct. 21–25. No Soviet scientist had yet visited KSC, "apparently out of concern that the United States would ask for reciprocal rights in Russia for American scientists," said *New York Times*. Rep. Paul G. Rogers (D-Fla.), in letter to Secretary of State Dean Rusk, had said visit was inappropriate in light of U.S.S.R.'s invasion of Czechoslovakia, imprisonment of *Pueblo* crew by North Korea, Soviet aid to North Vietnam, and existence of Communist regime in Cuba. (*NYT*, 10/9/68)

October 8–10: NASA held conference at LaRC on progress of NASA research on noise alleviation of large subsonic jet aircraft. It dealt with nacelle acoustic treatment technology and application, noise generation and reduction at source, operational considerations, and subjective reaction. In introductory remarks, NASA Deputy Associate Administrator for Aeronautics Charles W. Harper said NASA support for research programs on noise had risen from less than $1 million per year to nearly $18 million in 1968, most of which was used to obtain industry support for program. (NASA SP-189)

October 9: Univ. of California at Los Angeles astronomer Dr. Kurt Riegel and graduate student Mark Jennings reported discovery of cloud of intensely cold interstellar hydrogen gas near region of galaxy where star formation was known to be taking place, about 3,000 light years from earth in direction of Milky Way. Dr. Riegel said, "The implication is that the process of star formation may in some way depend on the cooling of the gas floating around between the stars." (Getze, *LA Times*, 10/9/68)

- NASA was completing tests for USAF Cambridge Research Laboratories in which individual plastic hailstone models were dropped from 20,000- to 25,000-ft altitudes near NASA Wallops Station to study speed at which hailstones fell to earth and its effect on their size and growth rate in atmosphere. Wind-tunnel tests had confirmed that size and weight to which naturally formed hailstones would grow was related to speed they fell and thus to length of time spent in storm clouds. Shape and surface roughness affected fall speed by changing drag characteristics. Wallops test data would check tunnel results and would be applied in predicting growth of real hailstones. (NASA Release 68-172)

October 10: Aerobee 150 MI sounding rocket launched by NASA from WSMR carried Naval Research Laboratory experiment to 109.2-mi (174.7-km) altitude to obtain stellar spectra in Scorpius in 1,000–1,600 Å far UV range and photometric data on stellar fluxes. Rocket performed satisfactorily. Scientific objectives were not achieved because attitude con-

October 10

trol system (ACS) malfunctioned and vehicle failed to capture and point at desired targets. Experiment performed as expected, but because of ACS malfunction no film was advanced by spectrograph or aspect camera. Some photometric data were obtained by Geiger tube photometers. (NASA Rpt SRL)

- Australian House of Representatives, by 60–30 vote, defeated opposition Labor Party motion to condemn government for ordering 24 F–111C fighter-bombers from U.S. Crashes, delays, and cost increases had led to major criticism of government. Deliveries of aircraft were 18 mo behind schedule. Latest U.S. estimate of cost, including spares and ground equipment, was $294.63 million, about one-fourth Australian defense budget for 1968–69. There was no ceiling price on aircraft and no way Australia could cancel contract without $200-million penalty. (AP, *W Post*, 10/11/68, A11; *NYT*, 10/9/68, 12)

- NASA announced it had requested proposals by Nov. 18 for two experimental turbofan jet engines for extensive test program [see Aug. 18]. Objective was to reduce two major sources of noise—interaction of jet exhaust with outside air and noise created by fan—to produce turbofan demonstrator engine operational at noise level at least 15–20 db below those powering DC–8 and Boeing 707 aircraft. Specifications were developed at LeRC with assistance on contract from Allison Div. of General Motors Corp. and Pratt & Whitney Div. of United Aircraft Corp.; McDonnell Douglas Corp. studied feasibility of integrating quiet engine with DC–8. (NASA Release 68–175)

- Sen. John J. Sparkman (D-Ala.) on Senate floor said: "President Johnson is properly called the principal architect of America's space program. As Senator and Vice President he worked unceasingly to assure this country a role of leadership in the exploration of space. . . . Under President Johnson's leadership in the Senate the Space Act was passed in 1958, creating the National Aeronautics and Space Administration . . . setting up a charter to win for this country a preeminence in the peaceful exploration of space. NASA pulled together widely scattered efforts in space and built an organization unique in this country's history. It has contributed to the technological competence so vital to modern industrial society." (NASA *LAR* VII/108; *CR*, 10/10/68, S12423)

- FAA published report, *SRDS Program Goals, Achievements and Trends*, on 50 Systems Research and Development Service projects undertaken in FY 1968. Beacon tracking level of terminal automation would provide aircraft identity, altitude, and computed ground speed on air traffic control radarscopes. Over three years, automated radar tracking system ARTS III would be installed at 62 busiest airports. Computer-aided approach spacing (CAAS) system would give more consistent spacing of landing aircraft. R&D eventually would lead to Category III all-weather landing systems (AWLS) at major airports, permitting aircraft to land with zero ceiling and runway visual range. (FAA Release T–68–39)

October 11: Cosmos CCXLVII was launched by U.S.S.R. into orbit with 343-km (213.1-mi) apogee, 215-km (133.6-mi) perigee, 89.9-min period, and 65.4° inclination. Satellite reentered Oct. 19. (GSFC *SSR*, 10/15/68; 10/31/68)

- U.S.S.R.'s *Zond V* could be precursor to next step in flying complex unmanned missions to Venus or Mars as well as to carrying men to moon,

RAND Corp. Scientist Merton E. Davies and Cal Tech scientist Bruce C. Murray wrote in *Science*. Soviet descriptions of *Zond V* suggested U.S.S.R. might plan to send pair of spacecraft to Mars in late February or early March. One might land on Mars and cast off satellite to orbit it and relay signals to earth; other might fly by Mars and return to earth with film of Mars surface. Speculations were based on *Pravda* and *Krasnaya Zvezda* (*Red Star*) articles by Soviet Prof. A. Dmitriyev, which said "information from space" must be delivered "directly to the scientists' laboratory" free of "encumbrances and distortions of radioed signals." He said *Zond V* had successfully completed assignment of developing means and methods for returning space devices. Also, U.S.S.R. had previously sent pair of spacecraft on planetary mission and might repeat mission to take advantage of favorable Mars or Venus positions for flyby or landing attempts. (*Science*, 10/11/68, 245–6; Cohn, *W Post*, 10/11/68, A9)

- President Johnson transmitted NASA's 15th, 16th, and 17th Semiannual Reports to Congress covering July 1, 1966, to Dec. 31, 1967. President wrote, "I commend these reports to your attention. They contain, I believe, concrete evidence that NASA is moving forward, and that America is contributing mightily in the worldwide effort to conquer space for the benefit of all mankind."

  In letter accompanying 17th report, NASA Administrator James E. Webb wrote, "This period was overshadowed by the Apollo fire which took the lives of three of our astronauts. The thorough investigation of the accident and the steps that were initiated to improve safety by changes in design and procedures have previously been made matters of public record. This report shows that the same period was also one of progress in aeronautics and space as evidenced, for example, by the successful flights of Surveyor, Lunar Orbiter, and many other spacecraft. It was a difficult time for NASA, but one in which the agency showed, I believe, that it could react maturely to failure as well as success, and continue to deserve the confidence and support of the nation." (Text; NASA *LAR* VII/III)

- Republican Presidential candidate Richard M. Nixon said in TV program to viewers in Texas, New Mexico, Arkansas, and Oklahoma, Republican administration would strive to make U.S. "first in space." "I don't want the Soviet Union or any other nation to be ahead of the United States. . . . Let's emphasize the moon shot and others where we can make a direct break-through." (*W Star*, 10/12/68, A1)

- President Johnson vainly urged Senate ratification of nuclear nonproliferation treaty: "If the treaty does not go into effect soon, an increasing number of countries will see it in their national interest to go nuclear." He said if Senate found it impossible to remain in session to act on ratification, he might call special session after election. However, after consulting with President Johnson, Senate Majority Leader Mike Mansfield (D-Mont.) announced he was laying treaty aside for this session of Congress because to call it up during closing days would result in "a devisive political dispute" that could convert it into partisan issue and imperil its eventual approval. (*PD*, 10/14/68, 1481; *CR*, 10/11/68, S12685–90; Finney, *NYT*, 10/11/68, 12; Kilpatrick, *W Post*, 10/12/68, A14)

- NASA task force appointed by Assistant Administrator for University Affairs Francis B. Smith announced publication of *A Study of NASA University Programs*, containing assessment of programs and their benefit to NASA and academic community. NASA university programs had "made major contributions to aeronautics and space program. Research sponsored . . . has generated new concepts, has developed new technology, and has created unique facilities for further education and research. Over 50 percent of all experiments flown on NASA satellites have been generated by university programs. Universities have awarded at least 500 graduate degrees and provided continuing education opportunities to thousands . . . [and] university consultants have given policy, scientific, and engineering advice to NASA at all levels." (Text; NASA Release 68-177)

- NASA had asked its contractors to cut KSC personnel 10% in effort to save $40 million by July 1. Boeing Co. had already announced plans to reduce 4,400-man force to 4,000. Chrysler Corp. would keep 1,000 of its 1,200. Cutbacks were due to NASA budget cuts and affected only 2% of work force. (NASA PAO; W *Star*, 10/11/68, A3)
- Comparison of infrared images of lunar eclipses of Dec. 19, 1964, and April 13, 1968, showed thermal anomalies of lunar maria unchanged after $3\frac{1}{2}$ yr, Air Force Cambridge Research Laboratories researchers reported in *Science*. Graham R. Hunt, John W. Salisbury, and Robert K. Vincent wrote that hundreds of hot spots that cooled more slowly than surroundings were strikingly similar in images from both eclipses. One new, linear thermal anomaly had been discovered, whose close relation to lunar crustal fracture line suggested it might be of internal origin. Origin could give clue to formation of craters. (*Science*, 10/11/68, 252-4)

October 11-22: NASA's *Apollo 7* (AS-205), first manned mission in Apollo lunar landing program, was successfully launched from KSC Launch Complex 34 at 11:02 am EDT by Saturn IB booster. Primary objectives were to demonstrate command and service module (CSM) and crew performance; demonstrate crew, space vehicle, and mission support facilities; and demonstrate CSM rendezvous capability. All launch events occurred as planned and spacecraft, carrying Astronauts Walter M. Schirra, Jr. (commander), Donn F. Eisele (CM pilot), and R. Walter Cunningham (LM pilot), entered initial orbit with 177.8-mi (286.1-km) apogee, 138.2-mi (222.4-km) perigee, 89.9-min period, and 31.6° inclination. Saturn IB 2nd stage (S-IVB) manned control test was completed with excellent results, and S-IVB separated from CSM on schedule. Crew successfully transposed CSM and simulated docking by maneuvering CSM to within four or five feet of S-IVB.

President Johnson sent message to *Apollo 7* crew: "Everything in the Presidential office came to a halt as Foreign Minister Debré of France and I watched with mounting excitement the magnificent launch of the Saturn. . . ." Message was relayed from *Air Force I* as President flew from Washington to visit former President Harry S. Truman.

On second day Schirra told ground controllers crew was too busy to set up portable camera for live TV coverage because of minor but time-consuming difficulties. Astronauts had to pump waste water manually from spacecraft, Schirra and Eisele had trouble with their biomedical harnesses, Schirra had head cold symptoms, spacecraft evaporator system required maintenance, and hatch windows blurred and were bordered by mysterious "small hairs like fuzz." Crew fired SPS engine for 10 sec and 8 sec to set up rendezvous and maneuvered CSM to within 70 ft of tumbling spent 2nd stage, simulating techniques to be used on future flights if LM were to become disabled in lunar orbit. Crew took close-up photos of LM adapter attached to 2nd stage. S-IVB reentered earth's atmosphere Oct. 18 and splashed into Indian Ocean.

On third day crew, which had already accomplished half its objectives, photographed clouds and earth and continued checking out spacecraft systems. Power failure in spacecraft's AC electrical system was quickly restored, but overloading prompted officials to reschedule third SPS burn 20 hr sooner than planned. Burn positioned and sized ellipse for CM reaction control system deorbit in case of emergency and set up auxiliary gaging system test. Astronauts, all with head colds, appeared on national TV for 7 min for first time live from space. Crew displayed hand-printed signs bearing greetings from "the lovely Apollo room high atop everything."

Second live telecast for 11 min Oct. 15 showed closeups of spacecraft interior and astronauts so clear that observers could read astronauts' lips. Third TV appearance Oct. 16, which included nine-minute tour of spacecraft, won astronauts honorary membership in American Federation of Television and Radio Artists. Later, crew successfully fired SPS engine for fourth burn, demonstrating 20,500-lb-thrust engine's minimum impulse capability. As spacecraft passed over Hurricane Gladys in Gulf of Mexico, astronauts photographed storm and relayed data to Weather Bureau hurricane center. They also took pictures of "long plume" of air pollution. Crew continued spacecraft checkout, guidance and navigation procedures, and TV operations on seventh day, but fourth telecast was smudgy. Successful 67-sec fifth SPS burn—longest to date—was performed out of plane to test auxiliary gaging system and readjust ellipse for lifetime and CM reaction control system deorbit.

On ninth day astronauts showed viewers exceptionally sharp pictures of main control panels in fifth TV broadcast and performed close-order drill to demonstrate movement in weightless environment. Sixth SPS burn, essentially duplicate of fourth minimum impulse test burn, was successfully conducted. Seventh burn, on following day, adjusted time phasing for backup SM reaction-control-system deorbit burns. At 259:39 GET astronauts fired SPS engine for eighth time, to deorbit CSM for reentry. CM/SM separation, parachute deployment, and other reentry events were nominal, and spacecraft splashed down in Atlantic eight miles north of recovery ship U.S.S. *Essex* at 7:11 am EDT Oct. 22 after completing 163 revolutions. Crew was picked up by helicopter and flown to recovery ship within one hour after splashdown.

All primary *Apollo 7* mission objectives were achieved, as well as every detailed test objective and three not originally planned. Crew comfort and safety were enhanced by change in cabin atmosphere to 100% oxygen in flight, hot meals, and relatively complete freedom of motion in spacecraft. Engineering accomplishments included live TV from space and drinking water produced as by-product of fuel cells. NASA's *Ats III* applications technology satellite relayed TV pictures to Europe. Service module SPS main engine, largest thrust engine to be manually thrust-vector controlled, proved itself by accomplishing longest and shortest

October 11–22: *Bearded Apollo 7 astronauts emerge from recovery helicopter which carried them from landing point to deck of U.S.S. Essex after 11 days in orbit on first manned mission in Apollo Program. Left to right are Walter M. Schirra, Jr. (commander), Donn F. Eisele (CM pilot), and R. Walter Cunningham, (LM pilot) 4.5 million miles and 163 revolutions after launch.*

manned SPS burns, and largest number of inflight restarts. Manual tracking, navigation, and control achievements included full optical rendezvous, daylight platform realignment, optical platform alignments, pilot control of launch vehicle attitude, and orbital determination by sextant tracking of another vehicle. Mission also accomplished first digital-autopilot-controlled engine burn and first manned S-band communications.

All launch vehicle systems performed satisfactorily; spacecraft systems functioned with some minor anomalies, countered by backup subsystem, change in procedures, isolation, or careful monitoring so that no loss of systems support resulted.

Apollo 7 spacecraft had been redesigned for safety. Original two-piece side hatch had been replaced by a quick-opening, one-piece hatch. Flammability within CM had been reduced by extensive materials substitution, and systems redundancy had been expanded to reduce single failure points. Saturn IB launch vehicle carried less telemetry and instrumentation equipment, to lower weight and increase payload capability. New propellant lines to augmented spark igniter had been installed in J–2 engine to prevent failure which had occurred on *Apollo 6*.

Earlier unmanned Apollo flights had yielded all spacecraft information possible without crew on board. *Apollo 4* (launched Nov. 9, 1967) and *Apollo 5* (launched Jan. 22, 1968) had both been highly successful, completing inflight tests of all major pieces of Apollo hardware. *Apollo 6* (launched April 4), despite launch vehicle problems, had attained four of five primary objectives and had been recovered in excellent condition. Apollo program was directed by NASA Office of Manned Space Flight; MSC was responsible for Apollo spacecraft development, and KSC for launch operations. Tracking and data acquisition was managed by GSFC under overall direction of NASA Office of Tracking and Data Acquisition. (NASA Proj Off; NASA Releases 68–168K, 68–179; MSC Historical Off; *W Post*, 10/12–23/68; *B Sun*, 10/12/68; 10/23/68; *W Star*, 10/13–19/68; *PD*, 10/21/68, 1492)

October 12: Washington Post editorial noted total cost of space program to date was upwards of $20 billion: "The expenditure of sums of money like these cannot be justified in terms of the military or civilian spinoff, although both have been substantial, or of the knowledge we have gained about the atmosphere surrounding our planet. But they can be justified in terms of the national prestige that is to be won or lost in space and of the inevitable fate of man to pursue knowledge towards its ultimate end in hope of some day achieving a better understanding of what the earth and universe are all about. It is in this latter hope that we cheer the astronauts on, wish them well in their planned 163 orbits of the earth, pray for their safe return, and urge those who make the crucial decisions about the future space program to proceed with all deliberate speed to reach the goal President Kennedy put before us." (*W Post*, 10/12/68, A12)

- "NASA's 10th anniversary was celebrated with appropriate fanfare, but the 11th anniversary of the Space Age, which came during the same week, slipped by with barely a nod of recognition," James J. Haggerty, Jr., charged in *Armed Forces Journal*. "It should have been given more attention. The U.S. needs a continual reminder that we were dragged kicking and screaming into space exploration, that the U.S. space program was born only out of reaction to the accomplishments of another nation. . . .

  "The summary for 11 years of space launchings shows a total of 881 launches through the Oct. 4 anniversary date. About 400 of all the spacecraft launched still are in orbit.

  "The U.S. enjoyed a considerable quantitative lead at the end of the 11-year period. It sent into orbit 564 spacecraft, or 64% of the total, compared with 298 spacecraft, or 33%, for the USSR. . . .

  "Manned flights, of course, were confined to the U.S. and the USSR. Through the anniversary date the U.S. had launched 14 flights involving 24 astronauts and 1,993 man-hours; the USSR had made nine flights with 12 cosmonauts piling up 532 man-hours." (*AFJ*, 10/12/68, 13)

October 13: New observations from U.K.'s Jodrell Bank Experimental Station indicated distance estimates to pulsars had been 30 times too short. Observations from Australia's Molonglo Radio Observatory had identified pulsar PSR 1749–28, believed to be three times more distant than 10 others observed to date. Despite great distance, its pulses were more powerful than those of all but one other. From these observations, Jodrell's Dr. Graham Smith believed pulsars lay in distant halo, not in Milky Way. (Sullivan, *NYT*, 10/13/68, 74)

October 13–19: At 19th Congress of IAF in New York, Dr. Edward C. Welsh, NASC Executive Secretary, read message from Vice President Hubert H. Humphrey, Chairman of NASC: "We take considerable pride in the United States with the generally open nature of our space program and in the manner in which we have attempted to cooperate with other nations in space participation as well as in the dissemination of facts and theories gleaned from our space experience. . . . man has now begun to make the space far beyond the Earth's surface a part of his library of education and his scope of achievement. Now more than ever before the way of the future must be the way of nations working together to harness the forces of nature so that the peaceful pursuits of mankind may flourish."

Dr. Welsh said: "I would place high on the list of benefits those which flow from increased international cooperation in the field of space. . . . stimulated by the universal desire for knowledge [it] has brought somewhat closer together the peoples of the world. . . . as the practical applications of space become more evident in the form of weather predictions, communications, and increased knowledge of the world's limited natural resources, additional strength is added to the foundation for peace." (Text)

Daniel and Florence Guggenheim International Astronautics Award of $1,000, made annually by International Academy of Astronautics in recognition of outstanding contribution to progress of space research and exploration over five-year period, was presented Oct. 18 at IAF Congress banquet to Dr. Zdenek Svestaka of Astronomical Institute of Czechoslovak Academy of Sciences. He was Chairman of Commission on Solar Activity of the International Astronomical Union. (IAF IAA Release 34; AIAA Release 10/13/68)

Soviet scientist Prof. Leonid I. Sedov in news conference said U.S.S.R. was not in race to moon with U.S. "The question of sending astronauts to the moon at this time is not an item on our agenda. The exploration of the moon is possible, but it is not a priority." Prof. Sedov said Soviet timetable for manned expeditions would depend on next series of flight tests using Zond rocket. *Zond V* was successful, he said, "because the capsule returned safely, which was the purpose of the flight." Regarding U.S.–U.S.S.R. scientific cooperation, "its successful implementation is very much dependent on international relations." At present, he said, close relations did not exist. (*NYT*, 10/15/68, 48)

Informal meeting of international space scientists, organized by Northwestern Univ. UFO expert, Dr. J. Allen Hynek, in conjunction with IAF Congress, discussed advisability of world cooperation on UFOs. Several speakers urged international cooperation such as uniform UFO report forms, but it was agreed that no action should be taken until after appearance of Univ. of Colorado report expected later in year. (Sullivan, *NYT*, 10/16/68, 12)

Dr. Harold Masursky, U.S. Geological Survey astrogeologist, reported to IAF analysis of data from spacecraft that had orbited and crashed into or landed intact on moon had shown side facing earth was largely a basin similar to that of Pacific Ocean. It seemed more like earth than had been previously believed. Lunar interior had been molten. Deep cracks in its crust were lined with craters where molten material had erupted from below. Study of remains of giant crater in western Texas had shown features analogous to lunar craters with mountain in center. (Sullivan, *NYT*, 10/19/68, 19)

October 14: NASA successfully launched two Nike-Cajun sounding rockets two hours apart from Point Barrow, Alaska, to 80-mi (128.8-km) altitudes. Purpose of launches was to obtain data on variation of temperature, pressure, and wind profile by detonating 19 grenades per rocket at prescribed times and recording sound arrivals on ground. Data would be compared with data from two launches to be conducted from Churchill Research Range Oct. 15. Rockets and instruments performed satisfactorily; sound arrivals were recorded for all grenade ejections. Good data were anticipated. (NASA Rpts SRL)

- NASA FY 1970 budget request, totaling $4.698 billion, was submitted to Bureau of the Budget. Subsequently, Acting Administrator Thomas O. Paine met with BOB Director Charles J. Zwick, and BOB staff members held budget hearings with NASA officials. (NASA Off of Admin)

- Special committee of National Academy of Sciences had nominated Duke Univ. biochemist Dr. Philip Handler, Chairman of National Science Foundation's National Science Board, to succeed Dr. Frederick Seitz as NAS president. Election by mail ballot would be held in December. (*NYT*, 10/14/68, 33)

- *Barron's* editorial criticized concentration of space funding on manned lunar program rather than on military: "Congress in its wisdom already has curtailed funds for manned flights after the lunar landing—the so-called Apollo Applications Program—and it could usefully wield an even sharper axe. The money might far better go toward the military exploitation of space, which, for the past seven years, has suffered from dangerous neglect. Thus, out of the vast sums spent on space, at most one dollar in six has had a military bearing. In turn, with the possible exception of the Manned Orbital Laboratory . . . nearly every cent of the so-called military budget has gone for hardware with a passive or defensive aim, notably satellites for reconnaissance, communications, navigation and weather forecasting. In striking contrast, the Soviet Union has developed and tested . . . a weapon aptly known as Scrag, which can hurl a guided missile carrying a nuclear payload of 15 megatons or more into a partial orbit (hence, fractional) round the earth. To anyone in his right mind, FOBS constitutes a gross violation of the outer space treaty, which prohibits the placing of nuclear warheads in orbit. However, according to the confused legal eagles in the State Department and Pentagon (if not to some future hapless populace which finds itself on ground zero), anything less than a full orbit goes. . . . In the interest of survival, here is one balance the U.S must move swiftly to redress. . . . The first duty of government is to protect its people. Neither the Kennedy nor the Johnson Administration has honored that trust." (*Barron's*, 10/14/68, 15)

- *Christian Science Monitor* editorial listed purchases bought by $32.4 billion spent on national space program in past 10 years: manned-spacecraft program that should put men on the moon within a year; unmanned-satellite program of 234 major launches; stable of space rockets ranging from workhorse Scout with thrust of 88,000 lb, to Saturn V with 7,500,000 lb of thrust; facilities at NASA's 15 installations worth over $2.5 billion; staff of 35,000 NASA employees, including some of the nation's top scientists, physicists, and engineers; industrial work force that peaked at more than 400,000 (now down to about 200,000); global tracking network stretching around the world that can track, receive telemetry, control, and communicate with vehicles as far away as other side of sun; aeronautics program largely unpublicized but constantly growing; advances in technology that have developed new electronic parts, power sources, alloys, adhesives, lubricants, and highly reliable hardware components.

"This is just a bare-bones receipt for the American tax payers' $32 billion. But it represents an investment that is already producing a civilian 'spinoff' of incalculable value." (CR, 10/14/68, E9524)

October 15: NASA launched two Nike-Cajun sounding rockets from Churchill Research Range to obtain data on variation of temperature, pressure, and wind profile by detonating 19 grenades per rocket at prescribed times and recording sound arrivals on ground. Rockets reached 78.8-mi (126.8-km) and 77.8-mi (123.5-km) altitudes and performed nominally. Sound arrivals were recorded from all grenades. Data would be compared with data from Oct. 14 Point Barrow, Alaska, launches. (NASA Rpts SRL)

- NASA Nike-Tomahawk sounding rocket launched from NASA Wallops Station carried Univ. of Michigan payload to 211-mi (340-km) altitude to investigate role of quenching, dissociative recombination, ionospheric decay, and nonthermal electrons on airglow during decay period immediately following sunset. Rocket and instruments performed satisfactorily; peak altitude was as predicted, and 570 sec of telemetry was received. All measuring systems functioned properly and scientific objectives were met. (NASA Rpt SRL)

- Dr. Kurt Debus, Director of KSC, said Saturn IB Launch Complexes 34 and 37 would be put on standby basis until beginning of Apollo Applications program in late 1970, with reduction of 1,315 personnel. Shutdown, forced by space budget cuts, would save NASA $20 million. Some 10% of 2,400 contract work force would be affected, with 1,000 engineers and technicians continuing to work in other areas. Complexes were to have been maintained in event Saturn V launch vehicle encountered serious development problems, but Dr. Debus said this insurance was no longer needed. (KSC Release 463–68; AP, B *Sun*, 10/16/68, A7)

- U.K. Minister of State in Ministry of Technology, J. P. W. Mallalieu, told House of Commons U.K. would have to pay U.S. equivalent of $60 million in cancellation fees and other expenses connected with termination of its order for 50 F–111 aircraft Jan. 16, 1968. Order was valued at $650 million but this would have risen to $1 billion with computation of spares and 10-yr interest. (*WSJ*, 10/15/68, 17)

October 16: Aerobee 150 MI sounding rocket launched by NASA from WSMR carried Lockheed Missiles & Space Co. experiment to 96.6-mi (155-km) altitude to obtain quantitative measurements of spectrum and intensity of solar x-ray flux in 2- to 30-kev range, determine distribution on sun, and observe galactic x-ray sources. Rocket and instruments performed satisfactorily. All detection systems on payload acquired useful data. Communications, maintained with both satellite x-ray monitor and ground-based optical observer, indicated sun was in nonflaring state with increasing activity on west limb. (NASA Rpt SRL)

- Ray Cromley commented in *Washington Daily News:* "It is one of the tragedies of the moon race that earth satellite programs which could revolutionize certain aspects of agriculture, mineral exploration, TV broadcasting, navigation, weather forecasting and flood control are being squeezed unmercifully by Apollo. It is now clear that these unpublicized, unromantic programs promise unbelievably large payoffs. The U.S. return . . . has been conservatively estimated at $400 million to $1 billion for every $100 million invested after feasibility research is completed. The investments could be private, public or a mixture of both." (*W News*, 10/16/68, 31)
- U.S. and Australia signed five-year agreement in Canberra to expand scientific cooperation through exchanges of scientists and information, to participate in joint research projects, and to include scientists and institutions from other countries in some projects. (Reuters, *NYT*, 10/17/68, 14)
- MSFC issued to NAR's Rocketdyne Div. two contract modifications to extend engine production and delivery. An $8.4-million supplement was awarded for extension of J-2 engine production through April 30, 1970, because of overall stretch-out of launch vehicle production. Under extension, J-2 engine production would be cut from three engines per month to one. Contract for F-1 engine deliveries was extended through June 1970 under $4-million modification which decreased F-1 production rate from two engines per month to one. (MSFC Releases 68-246, 68-247)

October 17: NASA's *Ats IV* applications technology satellite reentered earth's atmosphere over South Atlantic southwest of St. Helena island. Spacecraft, launched Aug. 10, had remained in parking orbit, tumbling uncontrollably, when Centaur engines failed to reignite for second burn. (NASA Release 68-188)

- NASA issued summary of combined findings of Accident Board and Review Board appointed to investigate May 6 accident which destroyed Lunar Landing Research Vehicle (LLRV) at Ellington AFB. Pilot, Astronaut Neil A. Armstrong, had to eject few seconds after loss of helium pressure in propellant tanks caused premature shutdown of attitude control rocket system. Helium had been inadvertently depleted earlier than usual in flight. Armstrong incurred minor injuries. Review Board, appointed by then NASA Deputy Administrator Dr. Thomas O. Paine to study accident's possible impact on Apollo program, discovered no unfavorable effects on lunar landing project, particularly lunar module. It agreed with Accident Investigation Board in calling for improvements in design and operating practices in LLRV and urged more stringent control over such flying programs and greater attention to all NASA lunar landing simulators. (NASA Release 68-182)
- Atmospheric scientists at JPL and Ohio State Univ. announced successful high-altitude test of balloon-borne spectrometer to measure atmospheric radiation emitted in four-micron region. Data, obtained on flight from National Center for Atmospheric Research and Information station at Palestine, Tex., indicated sunlight reflected from earth would pose no significant obstacle to continuous effective operation of radiation sensors on spacecraft in earth orbit. Researchers concluded satellite-borne

instrument could probe earth's lower atmosphere for global weather prediction. Test marked step in NASA program to define experiments for manned earth-orbiting missions in Apollo Applications program. (NASA Release 68–176; *Pasadena Star-News*, 10/17/68)

October 17–18: IAF International Institute of Space Law held XIth Colloquium on the Law of Outer Space in New York. Lawyers and jurists from 17 countries presented papers on Treaty on Outer Space, Treaty on Rescue and Return of Astronauts and Space Objects, telecommunications by satellites, and next steps in space law. NASA General Counsel, Dr. Paul G. Dembling, presided. (NASA Hq Memo)

October 18: RAM C–II radio attenuation measurement mission launched Aug. 22 was adjudged successful by NASA. Good-quality measurements of electron and ion concentrations in flow field were obtained at discrete locations along the spacecraft during reentry. (NASA Proj Off)

- XB–70A, flown by NASA test pilot Fitzhugh L. Fulton, Jr., successfully reached 52,000-ft altitude and mach 2.18 in flight from Edwards AFB to evaluate ILAF-exiter vane systems, air vehicle performance, and handling qualities. (XB–70 Proj Off)

- USAF announced lifting of three-week ban on F–111A flights but reimposed severe limits on speeds and maneuvers in force before halt. Restrictions would be lifted following reinforcement of high-stress area of wing box to distribute load more evenly. Investigation of Aug. 27 ground fatigue testing failure had shown it was "due to an isolated small crack induced during manufacturing process in the metal surrounding a bolt hole." No other such imperfections had been found. USAF said Sept. 23 F–111A accident at Nellis AFB had occurred when pilot lost control because of excessive rearward shift of aircraft's center of gravity following fuel transfer to which crew had given inadequate attention. (DOD Release 947–68; Witkin, *NYT*, 10/19/68, 1; B *Sun*, 10/19/68, 4; AP, *W Post*, 10/19/68, A11; W *Star*, 10/22/68, A9)

- ComSatCorp reported net income of $5,054,000 (50 cents per share) for first nine months of 1968. Income included $1,750,000 (17 cents per share) for third quarter. (ComSatCorp Release 68–56)

- NASA announced it had released tracking ship USNS *Watertown* from priority role of reentry support for Apollo missions, thus effecting reduction in operational costs required by budgetary curtailments. Manned Space Flight Network land stations in Pacific, Apollo tracking ship *Huntsville*, and Apollo range instrumentation aircraft would serve returning Apollo spacecraft landing in preselected Pacific area. (NASA Release 68–181)

- Sen. Gordon L. Allott (R-Colo.) in letter to *Science* scored "Understanding Gap" between scientific community and Congress and taxpayers on Federal R&D funding: "We are limited to a great degree by revenue taken in by the Treasury if we are to make the financing of our national debt manageable. Within our admitted lack of expertise, coupled with an appalling lack of national goals or a system of priorities, I think we do a fair job of spreading out the federal dollar. We could do better, though, with some constructive help from the scientific community from an objective and realistic appraisal of the circumstances and of existing realities, and we could benefit from the establishment of some system, either a Joint Committee or something similar, which would view research on an overall basis, which would review national goals and aspirations and which might . . . make a stab at setting up some type of priority list." U.S. "might well benefit if . . . the scientific community would become 'involved,' would drop the cloak of mystery, and take the time to explain, not just to us in Congress, but to Mr. Taxpayer as well, just what it's all about." (*Science*, 10/18/68, 214–8)

October 19: U.S.S.R. launched *Cosmos CCXLVIII* into orbit with 543-km (337.4-mi) apogee, 473-km (293.9-mi) perigee, 94.7-min period, and 62.2° inclination. (*InteraviaAirLetter*, 10/21/68, 6; UPI, *W Post*, 10/21/68, A11; GSFC *SSR*, 10/31/68)

- USAF test pilot Maj. William J. Knight was named 1968 winner of Harmon International Aviator's Trophy as "world's outstanding pilot for exceptional individual piloting performance." He had piloted X–15 No. 2 to 4,520 mph Oct. 3, 1967. Maj. Knight held both command pilot and USAF astronaut's command wings, having piloted research aircraft to 280,000-ft altitude. (*NYT*, 10/20/68, 84; *CSM*, 10/21/68)

October 20: *Cosmos CCXLIX* was launched by U.S.S.R. into orbit with 2,158-km (1,340.9-mi) apogee, 491-km (305.1-mi) perigee, 112.1-min period, and 62.3° inclination. (AP, B *Sun*, 10/21/68, A4; GSFC *SSR*, 10/31/68)

October 21: GSFC used ruby laser to track *Explorer XXXVI* (*Geos II*) satellite during daylight, a significant milestone in development of laser satellite-tracking system. (Cambridge Research Lab PAO; NASA Release 68–219)

- ComSatCorp, on behalf of INTELSAT consortium, signed $72-million contract with Hughes Aircraft Co. for construction of Intelsat IV series of advanced comsats—four spacecraft and one prototype, with test and ground equipment. (ComSatCorp Release 68–57)

- Richard Witkin in *New York Times* quoted "reliable sources" as saying Assistant Secretary of Defense for Systems Analysis, Dr. Alain C. Enthoven, had forwarded paper to Deputy Secretary of Defense Paul H. Nitze proposing further cuts in F–111 production, including cancellation of interim bomber version. Dr. Enthoven, specialist in calculating cost effectiveness of competing weapons systems, claimed many projected F–111 missions could be performed by much cheaper aircraft such as Ling-Temco-Vought A–7. (*NYT*, 10/21/68, 25; *Business Week*, 10/26/68)

October 21–23: Tenth National Trendex Poll sponsored by Thiokol Chemical Corp. reported public support for space program was 17% higher than in 1967. Taken after successful *Apollo 7* mission, it showed 68% of U.S. public favored Apollo program to land man on moon by 1970, 21% did not, and 11% was undecided (in September 1967 poll, Apollo had 51% support, with 35% opposed and 14% undecided). Public desire for increased Government spending on space was at its highest point in five years, with 18% in favor, as against 7% in 1967. Support for program was highest among college-educated, those under 35, and men; 49% favored manned space exploration, versus 25% for instrumented program; 60% backed planetary exploration, with 30% opposed. Favored programs after Apollo were: (1) reusable space system, (2) lunar exploration, (3) manned space stations, (4) manned Mars exploration. (*CR*, 1/6/69, E64–6; *SBD*, 12/16/68, 197–8)

October 21–25: At Fifth Annual Meeting and Technical Display of AIAA in

October 19: *Test pilot Maj. William J. Knight* (USAF) *was named winner of Harmon International Aviator's trophy. In photo, he inspects connection between external tank and rocket-powered X-15 No. 2, which he flew to record 4,520 mph Oct. 3, 1967. Two tanks supplied additional engine propellant for increased speed. On Oct. 24, 1968,* NASA *test pilot William H. Dana took X-15 No. 1 on last flight of research program.*

Philadelphia, Boeing Co. Vice President John M. Swihart announced abandonment of swing-wing design for SST in favor of fixed-delta-wing aircraft with four independently mounted engines under triangular tail. Final detailed design would be given to FAA by Jan. 15, 1969, deadline. New design differed from delta-winged Anglo-French Concorde and Soviet Tu-144 in wider wing span and horizontal tail which, according to Boeing Vice President in charge of SST H. W. Withington, made possible superior control at low speeds and compensated for more drag encountered with sweep-back angle of wing. Aircraft cost would be same as swing-wing, $40 million. It would carry same number of passengers, 280 or more, at same maximum speed, 1,800 mph. (Witkin, *NYT*, 10/22/68, 77; UPI, *W Post*, 10/22/68, A11; AP, *W Star*, 10/22/68, A3)

Dr. William H. Pickering, JPL Director, received $5,000 AIAA Louis W. Hill Space Transportation Award "for devising, developing and supervising significant space and satellite programs for military and civilian agencies of the United States Government." ARC Director H. Julian Allen was named Honorary Fellow of AIAA, highest membership award given by Institute. It was presented annually to two Americans and one foreign national. Other 1968 recipients were James S. McDonnell,

Chairman of Board, McDonnell Douglas Corp., and England's Sir Frank Whittle, often called father of jet engine. (AIAA Releases; ARC Release 68-15; ARC *Astrogram*)

October 22: DOD Systems Analysis Office cost-effectiveness proposal submitted to Secretary of Defense Clark M. Clifford called for elimination from budget of new nuclear submarines requested by Adm. Hyman Rickover, reduction in funds already approved for purchase of antisubmarine weapons, and retirement of more than 20 diesel-powered submarines, said George C. Wilson in *Washington Post.* Proposal, according to sources, was to postpone high-speed submarine and cancel development of "quiet" one. Its severity "illustrates the money pinch the Pentagon finds itself in as it tries to cut billions from its fiscal 1969 budget as well as the new budget." (*W Post*, 10/22/68, A10)

October 23: USAF launched unidentified satellite from Vandenberg AFB by Thor-Burner II booster into orbit with 529-mi (838.3-km) apogee, 497-mi (799.8-km) perigee, 101.3-min period, and 99.0° inclination. (*Pres Rpt 68*)

- NASA successfully deployed 40-ft-dia parachute with predicted 10-lb-per-cubic-ft dynamic pressure at mach 3.5. Parachute was ejected from five-foot-long canister which had been propelled to 33-mi altitude by three-stage rocket launched from WSMR. Test was to determine possible use of parachute for aerodynamic deceleration in planetary entry missions. Another test in Project SHAPE (Supersonic High Altitude Parachute Experiments) was scheduled for November. (NASA Release 68-185; AP, *NYT*, 10/27/68, 66)

- *Ats IV* mission (launched Aug. 10 and reentered Oct. 17) was adjudged a failure by NASA. Satellite had remained in elliptical parking orbit instead of entering planned synchronous orbit when Centaur engines failed to reignite for second burn. Resulting highly elliptical orbit precluded meaningful return of gravity gradient data. Day-night camera operated, but attitude dynamics precluded reception other than smeared unintelligible pictures. Electrical operation of ion engines, microwave multiple access, and microwave wide band was verified. Boom camera returned good photos, including some of earth. (NASA Proj Off)

- NASA's HL-10 lifting-body vehicle, piloted by Maj. Jerauld R. Gentry (USAF), failed to climb to desired 45,000-ft altitude after air-launch from B-52 aircraft, apparently because of rocket engine malfunction. Vehicle glided to smooth 225-mph emergency landing on Rosamond Dry Lake. Flight from Edwards AFB was to have been HL-10's first powered flight. (NASA Proj Off; *LA Times*, 10/24/68)

- *Apollo 7* editorial comment:

    Washington Post: ". . . as the men in the space program go over the data on Apollo 7 and consider the alternatives of manned or unmanned flight on Apollo 8, they must not allow anyone's desire to beat the Russians, or to get around the moon by the end of 1968, or to fan public interest in the future of space exploration to enter into their calculations. Only if they are convinced that our knowledge is sufficient, our spacecraft is totally adequate, and our men are ready should they give the go to Astronauts Borman, Lovell and Anders for a Christmas trip into space." (*W Post*, 10/23/68, A24)

    Washington *Evening Star*: "To those who have made a close study of the space program, [Walter M.] Schirra is the astronaut's astronaut; the man whose ability stands out in that company of the superable. To those in the know, Schirra is the mischievous perfectionist, the naval officer who lives by the book when he isn't too busy carrying out an elaborate practical joke. But Schirra will surely be remembered by the public as the astronaut who caught cold, who growled when the alarm clock rang, and who blew up when he was pushed too far. And it may be that Schirra's greatest contribution to the space program is that he, the most superlative of the supermen, forcefully demonstrated to the world that his is completely and refreshingly human." (W *Star*, 10/23/68, A20)

Baltimore Sun: "The toting up and analysis of all the information brought home this time must be left to the teams of experts. So must the decisions as to what comes next, and the planning such decisions call for. The public is content to know that three men in a spaceship have added another brave and brilliant chapter to a history of which all of us are unreservedly proud." (B *Sun*, 10/23/68, A6)

- MSFC issued McDonnell Douglas Corp. $2,395,955-supplemental-contract agreement for qualification test program to verify capability of maintaining S–IVB stage auxiliary-propulsion-system modules for up to 90 days with propellants loaded. Award brought total value of contract to $965,568,493. (MSFC Release 68–252)
- Smithsonian Astrophysical Observatory dedicated multipurpose astronomical station at Mount Hopkins, Ariz. Station, supported by NASA Office of Tracking and Data Acquisition and Office of Advanced Research and Technology, would be site of experimental series on laser communications to be conducted by NASA and Smithsonian. (Smithsonian PAO)

October 24: X–15 No. 1, flown from Edwards AFB by NASA test pilot William H. Dana, successfully reached 255,000-ft altitude and 3,682 mph (mach 5.04) in 199th and last flight of program. Purpose was to conduct WTR experiment and check out fixed alpha cone and fluidic probe. Flight scheduled for Dec. 20 was later canceled because of adverse weather. It was not rescheduled because NASA announced completion of program [see Jan. 21 and Dec. 20]. (X–15 Proj Off; NASA Release 68–221; AP, *W Post*, 10/25/68; *SBD*, 10/29/68, 289)

- Boosted Arcas II sounding rocket launched by NASA from Kiruna, Sweden, carried Uppsala (Sweden) Ionospheric Observatory payload to 65.2-mi (104.3-km) altitude. Objectives were to measure electron density profile, distribution of positive and negative ions and secondary x-rays in D region and lower E region of ionosphere during auroral glow, quiet arc, and violent and pulsating auroral conditions and to study its effects on radio wave propagation. Rocket was launched in conjunction with three others. Rocket performance was 12% below predicted. Experimental results were successful. (NASA Rpt SRL)
- NASA was negotiating with General Electric Co. for data management system costing in excess of $750,000 for 15 mo. It would be used to monitor data from Barbados Oceanographic Meteorological Experiment (BOMEX), in which NASA would assist ESSA during 1969. Data from satellites, five to seven ships, many buoys, and from high in atmosphere to bottom of ocean would be processed by system. (NASA Release 68–251)
- Rep. Alphonzo Bell (R-Calif.) told American Astronautical Society meeting in Los Angeles: "In evaluating space spending as a budget

priority, it is vital to consider the relationship of space to defense. . . . Both Russia and the United States have advance surveillance capacity. . . . As long as the threat of nuclear war from any source continues, Russia and the United States will be producing ever more sophisticated orbital 'spies in the skies'. . . . That is why the space program of the United States never is going to be abandoned. It will always be high on the list of spending priorities. The reason is not charming, but basic. We need to be in space to protect ourselves. . . . In the somewhat more distant future the harvest of human rewards . . . now only beginning . . . will prove that space research and space applications justify a continuing high priority." (Text; *Aero Daily*, 10/29/68)

October 25: Soyuz II was successfully launched by U.S.S.R. into orbit with 229-km (142.3 mi) apogee, 191-km (118.7-mi) perigee, 88.6-min period, and 51.7° inclination. Satellite later was used in rendezvous maneuvers with manned Soyuz III [See Oct. 26–30] and reentered Oct. 28. (Lannan, W *Star*, 10/27/68, A1; *SBD*, 10/28/68, 279; GSFC *SSR*, 10/31/68)

- Secretary of Defense Clark M. Clifford announced decision to proceed with program for turbine electrical drive (quiet) submarine. He had ordered construction of high-speed nuclear-propelled attack submarine July 1. "The close re-examination . . . just completed has convinced me that costly as it is [$150 to $200 million compared with $78 million for new Sturgeon class nuclear attack submarine], there is no cheaper and effective way to achieve in equal time desired progress in noise suppression." (DOD Release 971–68)

- NASA announced that H. Julian Allen, who joined NACA in 1936, would retire as Director of Ames Research Center Nov. 15. ARC Associate Director John F. Parsons would serve as Acting Director. Leading authority on supersonic and hypersonic wind-tunnel design, Allen had originated concept of bluntness for reentry shapes—as used in Apollo spacecraft—and had received NACA's Distinguished Service Medal, NASA medal for Exceptional Scientific Achievement, AIAA Sylvanus A. Reed Award, and Air Force Assn.'s Air Power Trophy. After his retirement Allen would be available to NASA as a consultant. (NASA Release 68–183; *SBD*, 10/28/68, 272; NASA Ann)

- *New York Times* editorial commented on failure of swing-wing design for SST: "More than ever now the burden of proof is on those who urge that billions of taxpayers' dollars be spent on an American SST. The fallibility of their judgment has been demonstrated in the loss of the swing-wing gamble. Is there reason to suppose that their optimistic forecasts about the profits to be made from such an airplane are any sounder? The aerodynamics of different wing configurations is not the only thing that needs to be assessed in the current re-examination of the SST." (*NYT*, 10/25/68, 46)

- In Washington *Evening Star* Carl T. Rowan wrote: "Some disenchanted Americans shake their heads as they note the poverty, the hunger, the sickness, the ignorance that plague the earthlings about them, and they ask what logic provokes our government to ignore critical problems at hand while investing vast sums in space ventures of doubtful value. . . . we have become an 'either/or' society. . . . Even though our gross national product is now running at a fantastic level of $871 billion a year, it is absolutely inconceivable to most taxpayers that we can have guns and butter, space spectaculars and dramatic domestic change. . . . Well, no man of vision, imagination, or hope can possibly believe that we are wrong to search the darkest reaches of outer space. . . . Who can say that contributions to medicine, to weather control, to science in general, to the problems of feeding man, to national defense, and ultimately to peace may flow from the space program?" Space program was "inherently and intrinsically, justification enough for spending $340 for every man, woman, and child in America. But are we not wise to ask: what is man profited if he harness the universe and yet fail to conquer the meanness . . . the hatreds, that dog those who inhabit the earth? . . . Much of the public is not in a mood to finance anything else. So we shall be stuck with the ordeal of setting priorities where there is scant room for making choices." (W Star, 10/25/68, A15)

- MSFC announced Boeing Co. contract modifications totaling $4,652,364 for Saturn V R&D, to: install over 4,000 instrumentation and data acquisition systems in special 2nd stage structural test verification program to confirm design of lighter weight, more powerful 2nd stage for fourth Apollo/Saturn V and subsequent vehicles; perform an abort and alternate mission analysis for Apollo/Saturn vehicles 503 through 510; and perform reliability, quality, and component qualification program, special prelaunch analysis, telemetry systems, and Saturn V/Apollo operations system safety program. Total value of Boeing Saturn V systems engineering and integration contract was now $213,443,238. (MSFC Release 68–253)

- USAF's Space and Missile Systems Organization announced award of initial increments to cost-plus-fixed-fee contracts with McDonnell Douglas Corp.: $5-million increment was awarded to $9,829,177 contract for reentry vehicle developmental flight tests; $756,285 increment was awarded to $1,739,105 contract for reentry vehicle environmental components tests. (DOD Release 974–68)

- Edward J. Schmidt, Special Assistant to General Electric Co.'s Vice President for R&D, was sworn in by NASA Acting Administrator, Dr. Thomas O. Paine, as consultant to the Administrator in management operations as affected by scientific and technical information. (NASA Release 68–189)

October 26: Business Week editorial: ". . . since the tragedy on the launch pad . . . [Jan. 27, 1967] NASA and its thousands of supporting companies have done a tremendous job in rebuilding the spacecraft and in perfecting the safety and reliability of the entire Apollo system. The clear message of Apollo 7 is that NASA now has a spacecraft that can take men to the moon and back safely. This is a triumph for NASA and for U.S. science, engineering, and management. (*Bus Wk*, 10/26/68)

October 26–30: U.S.S.R. successfully launched *Soyuz III*, carrying Cosmonaut Georgy T. Beregovoy, from Baikonur Cosmodrome with "a powerful rocket-booster," Tass announced. Spacecraft entered orbit "close to the preset one," with 205-km (127.4-mi) apogee, 183-km (113.7-mi) perigee, 88.3-min period, and 51.7° inclination; all equipment was functioning normally. Launch was first manned Soviet mission since *Soyuz I* (April 23–24, 1967), in which Cosmonaut Vladimir M. Komarov was killed when spacecraft crashlanded following reentry.

Tass later announced that during first revolution *Soyuz III* "approached" to within 200 m (656 ft) of unmanned *Soyuz II* (launched

Oct. 25), initially by "an automatic system"; subsequent operations were performed manually by Beregovoy. On Oct. 27, Tass said, Beregovoy "independently oriented the ship in space and switched on the motor," to alter spacecraft's orbit; continued conducting scientific, technical, medical, and biological experiments and research; transmitted TV pictures of cabin interior; and approached *Soyuz II* for second time before it reentered Oct. 28. *Soyuz III* remained in orbit until Oct. 30, completing 94 hrs 51 min and 64 orbits, before it softlanded "with the use of aerodynamics," in a preset area in Soviet territory. (Lannan, W *Star*, 10/27/68, A1; O'Toole, *W Post*, 10/27/68, A1; Kamm, *NYT*, 10/27/68; *SBD*, 10/28/68, 279; 10/31/68, 297; AP, *W Post*, 10/28/68, A1; GSFC *SSR*, 10/31/68)

October 27: Lightweight plastic foam invented by ARC scientists Dr. John A. Parker and Salvatore Riccitiello showed promise for industrial fire protection, particularly fuel fires. Extremely light polyurethane with additives formed tough, protective char layer when exposed to flame, while simultaneously releasing fire-extinguishing gases which helped to quench flame. Used to fill airspaces within structures, foam would prevent oxygen from reaching and feeding a fire. Demonstrations had shown it suitable for fire protection in aircraft, spacecraft, homes, and other structures. Other possible uses included automobiles, boats, trains, oil refineries, paint and chemical processing, and laboratories. Foam was resistant to heat flow, making it an excellent insulator. (NASA Release 68-187)

- In *New York Times* John N. Wilford said some NASA Hq. officials were "hesitant to approve a lunar orbit mission out of fear of being criticized for taking undue risks by skipping preliminary test flights. They are worried about the spacecraft's electrical system, which developed some minor 'bugs' during Apollo 7, and the propulsion system, even though the on-board rocket apparently performed well in eight firings during Apollo 7." If lunar mission was decided on, it would probably be launched Dec. 21 when moon's position to earth would require minimum midcourse rocket firing maneuvers for landing and light conditions would give good view of potential lunar landing site. First astronauts on moon were expected to stay less than 24 hr, to demonstrate it could be done. In time astronauts would make many return trips and would roam moon's surface in "moon buggies." Day might come when people would establish lunar colonies. (*NYT*, 10/28/68, 12E)

- Dr. Lise Meitner, nuclear physicist who was for 30 yr scientific partner of Dr. Otto Hahn, Nobel Prize winning discoverer of nuclear fission, died in Cambridge, England, at age 89. She had been forced to leave her work with Dr. Hahn and flee Nazi Germany's antisemitism in March 1938, nine months before he announced results of experiments which indicated atom could be split. Dr. Meitner was credited with having laid much of theoretical groundwork for atomic bomb. Though it was she who named the phenomenon "nuclear fission," she took pains to disassociate her work from the bomb itself. (*NYT*, 10/28/68, 1)

October 28: NASA outlined six steps which would lead to final decision during week of Nov. 11 on next Apollo manned mission. Apollo 8, scheduled for December, was planned as manned earth-orbital mission on Saturn V vehicle. Because of *Apollo 7* success, NASA was considering

October 28

alternative mission possibilities: earth-orbital mission deeper into space, circumlunar flyby, and lunar orbit.

Steps—laid out by Associate Administrator for Manned Space Flight, Dr. George E. Mueller, and Apollo Program Director, L/G Samuel C. Phillips—were: detailed analysis and review of *Apollo 7* results to determine any necessary spacecraft changes; final certification of solutions to *Apollo 6* problems; certification of strengthened Saturn V 2nd and 3rd stage fuel lines and elimination of pogo in 1st stage; completion of ground tests before Apollo 8 command and service module (CSM) certification for lunar flight; completion of flight computer programs for deep space and lunar missions; rehearsal of CSM operations tests with mathematical models and delivery of Apollo 8 CM computer program; and completion of design certification reviews of launch vehicle and spacecraft subsystems.

Dr. Thomas O. Paine, NASA Acting Administrator, said, "The final decision on whether to send Apollo 8 around the Moon will be made after a thorough assessment of the total risks involved and the total gains to be realized in this next step toward a manned lunar landing. We will fly the most advanced mission for which we are fully prepared that does not unduly risk the safety of the crew." (NASA Release 68-190; UPI, *NYT*, 10/29/68, 14)

- Washington *Evening Star* editorial: "The Russians are going to the moon just as fast as their technology will carry them. . . . It is, of course, impossible to judge what lies ahead or guess what problems either nation may encounter before the lunar landings are carried out. But if all goes well it looks as though both nations might be ready to go in about a year. We would have no objection at all if a way could be found to 'fix' the race, and an agreement reached to make the landings literally simultaneous. It would be one way of assuring that neither nation would pursue the goal of national prestige to the point of tragedy." (*W Star*, 10/25/68, A10)

October 29: ESSA said "very minor" solar flare reported at 7:18 am EST was accompanied by large radio burst that could interfere with radio communication. It was too small to affect U.S.S.R. Cosmonaut Georgy T. Beregovoy in orbit. (UPI, *W Post*, 10/30/68, A23)

- Army Map Service technicians were building 22- by 14-ft hand-carved model of landing site astronauts would see on approaching lunar "target area," to assist NASA in simulating manned landings on moon. Model, part of lunar module simulator (LMS), would be constructed from high-fidelity lunar relief map made from *Orbiter IV* and *V* photography. (DOD Release 966-68)

- NASA announced retirement, effective Nov. 1, of Werner R. Kuers, Director of Marshall Space Flight Center's Manufacturing Engineering Laboratory since 1961. (MSFC Release 68-257; *Marshall Star*, 10/30/68, 1)

- MSFC awarded Boeing Co. $1,404,548 contract modification to predict and evaluate orbital heating effects of liquid-hydrogen boil-off, supply thermal criteria and profiles related to Saturn V 2nd stage, assist with Saturn V preflight reviews, and provide configuration accounting. Award brought total contract to $212,128,585. (MSFC Release 68-256)

October 30: Award of $70,000 Nobel Prize in physics to Univ. of California at Berkeley Prof. Luis W. Alvarez and in chemistry to Yale Univ.

Prof. Lars Onsager meant U.S. had won all three Nobel science categories for 1968, as it had in 1946. Awards in medicine and physiology [see Oct. 16] went to U.S. geneticists. Dr. Alvarez was cited for "decisive contributions" in early 1960s to physics of subatomic particles and techniques for their detection. Dr. Onsager was honored for findings published in 1931 and sometimes regarded as fourth law of thermodynamics, "the reciprocity relations of Onsager," which could determine interrelation between voltage and temperature as electric current flowed through metal wire. Awards would be presented in Stockholm Dec. 10. (Lannan, *W Star*, 10/30/68, A1; Lee, *NYT*, 10/31/68, 1; O'Toole, *W Post*, 10/31/68, A25)

October 31: U.S.S.R. launched two Cosmos satellites. *Cosmos CCL* entered orbit with 845-km (525.1-mi) apogee, 735-km (467.8-mi) perigee, 100.6-min period, and 74° inclination. *Cosmos CCLI* entered orbit with 226-km (140.4-mi) apogee, 170 km (105.6-mi) perigee, 88.3-min period, and 64.7° inclination and reentered Nov. 18. Both satellites functioned normally. (GSFC *SSR*, 10/31/68; 11/30/68; *SBD*, 11/4/68, 12)

- Dr. William H. Pickering, JPL Director, and Dr. Lee A. DuBridge, Cal Tech President, presided at unveiling of historical marker at JPL commemorating test-firing of rocket engine Oct. 31, 1936, by students of Cal Tech's Guggenheim Aeronautical Laboratory under the late Dr. Theodore von Kármán. With firing, Cal Tech had become first university actively to sponsor rocket research. Its work had gained Government sponsorship and later had led to establishment of JPL. (JPL Release 492; Diebold, *LA Times*, 11/1/68)

- NAS President Dr. Frederick Seitz announced William W. Rubey, professor of geology and geophysics at Univ. of California at Los Angles had been named Director of Lunar Science Institute, Houston, Tex. NAS had accepted interim responsibility for operation of Institute until consortium of universities could be formed to assume its direction. Formation of Institute had been announced by President Johnson March 1, 1968, to provide base for academic scientists participating in lunar exploration program, working in Lunar Receiving Laboratory, or using other facilities of Manned Spacecraft Center devoted to study of the moon. It was to serve also as center for analysis and study of lunar data obtained from NASA unmanned missions. (NASA Release 68-191; NAS Release)

October 31–November 1: Soviet academician and aerospace scientist, Prof. Leonid I. Sedov, visited Univ. of Tennessee Space Institute during AIAA-sponsored tour of U.S. At press conference he said U.S.S.R. would reach moon from orbital station but this was not crux of Soviet effort in space. "Other planets are just as important." Zond-type satellites would circumnavigate other planets and return. In question and answer period following lecture he said U.S.S.R. would not conduct manned lunar space operation within following six months.

In lecture, Prof. Sedov said it was "obvious that space technology and the associated research have a pronounced stimulating effect on the development of the technological fields . . . essential for large-scale progress—particularly in the development of automatic control systems; in radio engineering, television, and telemetry; in computer technology; in the preparation of new materials and new devices; in miniaturization and minimum-weight design; in problems associated with accuracy and reliability of automatic systems. . . ." It was clear that "space technology has become pivotal in modern industry in the broadest sense of the word."

Dr. G. G. Chernyi, Moscow Univ. professor, also participated in seminar. (Transcripts; *Aero Daily*, 11/7/68, 29)

During October: Soviet Science in the News, Electro-Optical Systems, Inc., publication, said review of Soviet technical press indicated U.S.S.R. would attempt to orbit manned space station within the year and that it possessed "well-devised and thoroughly realizable designs." First "rooms" of station would comprise Cosmos or Proton booster joined with Soyuz spacecraft. Additional rooms would combine solid and inflatable elements like polyethylene. Tests of water recovery systems in Pacific indicated broadening of Soviet techniques. Six vessels had been completed for ocean recoveries of spacecraft. Conclusion of Soviet scientists that weightlessness had adverse effect on human skeletal composition seemed to indicate space station would use artificial gravity. 'Rotation of space station of from 40 to 60 meters in diameter would generate sufficient artificial gravity to allow large number of scientists to work in space." (*SSN*, 10/68, 1; *Aero Daily*, 10/16/68)

- Dr. Robert C. Seamans, Jr., former NASA Associate Administrator (1960–67) and now MIT professor and consultant to the NASA Administrator, was nominated as AIAA President for 1969. (*A&A*, 10/68, 106)

November 1968

November 1: U.S.S.R. launched *Cosmos CCLII* from Baikonur Cosmodrone into orbit with 2,148-km (1,334.7-mi) apogee, 531-km (330-mi) perigee, 112.4-min period, and 62.3° inclination. (W *Star*, 11/1/68; *SBD*, 11/4/68, 12; GSFC *SSR*, 11/15/68)

- NASA XB–70, flown by NASA test pilots Fitzhugh L. Fulton, Jr., and Emil Sturmthal, reached 41,000-ft altitude and mach 1.62 in flight from Edwards AFB to obtain stability and control data and to test ILAF. (XB–70 Proj Off)

November 2: President Johnson presented NASA Distinguished Service Medal, NASA's highest award, to recently retired NASA Administrator James E. Webb at ceremony in Johnson City, Tex. He also awarded cluster to NASA Exceptional Service Medal held by *Apollo 7* commander Walter M. Schirra, Jr., and Exceptional Service Medals to *Apollo 7* Astronauts R. Walter Cunningham and Donn F. Eisele.

President said U.S. was "ready to take that first great step out into the solar system and on to the surface of the nearest of the many mysterious worlds that surround us in space." Noting that *Apollo 7* had logged more than 780 man-hours in space—more than had been logged "in all Soviet manned flights to date"—and had accomplished 56 mission objectives, as many "in this one flight as were accomplished in the first five manned flights of the Gemini spacecraft," he said: "This is not important as either a game or a contest. But it is important because the United States of America must be first in technology if it is to continue its position in the world. I believe today, as I did when we had our original hearings that created the Space Administration, that the United States must be first."

President read citation, presenting Medal to Webb for "outstanding leadership of America's space program from 1961 to 1968. . . . More than any other individual he deserves the credit for the great achievements of the United States in the first decade of space, and for helping man to reach outward toward the stars."

Webb responded: "The citation and medal . . . should, in my view, be converted into some kind of holographic substance so it could be divided into thousands of parts . . . and each part should really go to an outstanding person in NASA, in our scientific group, working in our universities, and in the great industrial organizations of this country that have really done the work." (Transcript; Citation; *PD*, 11/8/68, 1568–71)

- In Prague newspaper *Mlaba Fronta*, Czechoslovak Academy of Sciences' astronomer Dr. L. Krivsky said "very dangerous" radiation from solar radio storm might have forced premature ending of U.S.S.R.'s *Soyuz III* mission Oct. 30. He implied, said *New York Times*, that U.S.S.R. had either been unaware or had failed to consider radio storm forecast for late October. (*NYT*, 11/3/68, 35)

November 2: *President Johnson reads citation before presenting* NASA's *Distinguished Service Medal to James E. Webb, who retired as* NASA *Administrator Oct. 7. Medal was given for outstanding leadership of America's space program.*

- *Business Week* commented on results of "two bad decisions by agencies of the federal government." Boeing Co. was scrapping swing-wing concept on SST in favor of fixed-wing and "word seeped out of the Pentagon that a real fight has developed over whether to cut back production of . . . F–111." How were such mistakes to be prevented in future? "One lesson that emerges . . . is that the government must learn to avoid premature commitment to any huge-scale project. . . . Another lesson is that in such major decisions, an independent, technologically competent judgment should be brought to bear on the issue. On the F–111, the President's Science Advisory Council did not even look at the design features of the aircraft. On the SST, a Special Presidential Advisory Committee was set up, but it was chaired by [then Secretary of Defense Robert S.] McNamara and was dominated by top Administration officials. Such changes in procedures may not wipe out all mistakes, but they could greatly reduce the chances of astronomically costly blunders." (*Bus Wk*, 11/2/68)

November 3: USAF launched unidentified satellite from Vandenberg AFB by Thor-Agena D booster into orbit with 177-mi (284.8-km) apogee, 108-mi (173.8-km) perigee, 88.8-min period, and 82.1° inclination. Satellite reentered Nov. 23. (*Pres Rpt 68*)

November 4: President Johnson released *Noise—Sound Without Value*, report of Federal Council for Science and Technology task force, and challenged industry, universities, and public authorities to attack noise in environment from many sources. He directed Federal departments and agencies to undertake or expand noise abatement programs. Among recommendations endorsed by President, report said: NASA should complete studies of community response to airport noise, in addition to HEW, DOT, and HUD studies of effects; NASA and DOT should continue air transport noise abatement research; DOD and NASA should continue to study and set standards for noise in special situations; and DOT should develop sonic-boom-control standards. (Text; *PD*, 11/11/68, 1575–6)

- *New York Times* editorial commented on award of all 1968 Nobel Prizes in science and medicine to U.S. citizens: ". . . there are real and important roots of American scientific prowess which need to be understood and fostered so that future achievement may match or excel that of the past. This country's hospitality to refugees from political tyranny and to those seeking to better themselves economically has brought rich rewards particularly in science and technology. The nation's huge investment in education has permitted able young people to develop their talents. Generous Government support of basic research has given the nation's scientists the tools and the material security needed for the realization of their potential excellence. The abundant returns from these policies provide good reason for maintaining them so that American science can continue to flower." (*NYT*, 11/4/68, 46)

- In *Aviation Week & Space Technology* editorial, Robert Hotz said: "The element of sharp competition between the U.S. and the USSR in manned space flight has unquestionably forced progress at a much more rapid pace than if either nation were going it alone. . . . Much has already been learned by these competitors from each other. It is a pity that the Soviets' obsession with secrecy bars them from a more fruitful international exchange of technology. . . .

  "Events of the past month have put the space race into high gear again. With a lunar window opening in December, it is a strong possibility that both the U.S. and the USSR can launch a manned circumlunar mission as another step towards the ultimate lunar landing. With a Soyuz and an Apollo carrying the Hammer and Sickle and the Stars and Stripes, respectively, around the moon at about the same time, it would require a phlegmatic world indeed not to get excited at these extraterrestrial Olympics." (*Av Wk*, 11/4/68, 11)

November 5: Republican candidate Richard M. Nixon was elected President of United States, with 302 electoral votes from 32 states and popular vote of 30,041,582, or 43.42%. Democratic candidate Hubert H. Humphrey polled 191 electoral votes from 14 states and 29,817,385 popular votes, or 43.10%. American Independent Party candidate George C. Wallace won 45 electoral votes and 9,242,950 popular votes, or 13.36%. While space had not been major issue in election, candidates

had made statements in support of space program in campaign speeches. (NASA EH; *CQ Weekly Rpt*, 11/8/68, 3071)
- Soviet Academy of Sciences President Mstislav V. Keldysh told Moscow news conference *Soyuz III* spacecraft flown by Cosmonaut Georgy T. Beregovoy Oct. 30 was intended only for earth orbit. He said U.S.S.R. might send animal on lunar mission before sending human to ascertain that cosmic radiation was not too dangerous. (AP, B *Sun*, 11/6/68, A2)
- USAF awarded separate $3,941,500 contracts to Westinghouse Electric Corp. and Hughes Aircraft Co. for 20-mo competition to develop new attack radar system for ZF-15A (formerly FX) advanced air superiority fighter aircraft. Awards were initial obligations of contracts which would total $22 million during FY 1969 and FY 1970. Winner would be selected after flight tests and evaluation of both radar prototypes. (DOD Release 1006-68; *WSJ*, 11/6/68, 13)

November 6: USAF launched unidentified satellite from Vandenberg AFB by Titan III-B booster into orbit with 249-mi (400.7-km) apogee, 90-mi (144.8-km) perigee, 89.8-min period, and 106° inclination. Satellite reentered Nov. 20. (*Pres Rpt 68*)
- National Radio Astronomy Observatory astronomers at Green Bank, W. Va., disclosed discovery of first pair of pulsars, near Crab Nebula, 6,000 light yr from earth. Through association with the decayed star, they might provide clue as to pulsars' identity. (Cohn, *W Post*, 11/7/68, A4)
- With task of designing equipment for U.S. space program largely over and because of cuts in NASA spending, hundreds of scientists and engineers were losing their jobs or getting out "while the getting is good," said Peter H. Prugh in *Wall Street Journal*. Boeing Co. was laying off several hundred at New Orleans and Cape Kennedy; its Huntsville work force was down from 4,600 in 1966 to 3,000, with more cuts coming. Chrysler Corp. had cut employment at New Orleans from 3,300 to 1,500 and most of its 900 Cape Kennedy employees faced layoffs or shifts to other cities. Huntsville office of Alabama State Employment Service said area employment had declined 3,500 in past year with biggest drop in aerospace field.

    Space scientists and engineers were finding even mundane jobs difficult to land because of their specialized skills and relatively high salary demands. Exodus was worrying space experts, "who fret that a new emphasis on U.S. space efforts or new military needs would leave companies hard pressed to fill the rows of desks being vacated now." (*WSJ*, 11/6/68, 1)
- AFSC Commander, Gen. James Ferguson, addressing Fourth Biennial Guidance Test Symposium at Holloman AFB, N. Mex., cited missile guidance needs and said U.S.S.R. was "working night and day to upset the status quo. There are a number of possible advances or even breakthroughs that would give them decided advantages over us. We would be most unwise to let them take a lead in technology through our lack of decisive effort. We must, at all times, maintain a technical momentum in order for our nation to maintain adequate strength across the entire spectrum of deterrence." (Text)
- NASA announced appointment of Dr. Mathias P. Siebel as Director of MSFC's Manufacturing Engineering Laboratory, replacing W. R. Kuers,

who retired Nov. 1. Dr. Siebel had been Deputy Director of Laboratory since going to MSFC in 1965. (MSFC Release 68–262)

- Ham, first chimpanzee successfully launched on space flight [Jan. 31, 1961], and now 11 yr old, was among five great apes at National Zoo who reacted positively to tuberculosis tests, according to zoo veterinarian, Dr. Clinton W. Gray. Animals were under treatment and expected to be fit for exhibiting again within 60 to 90 days. (Schaden, W Star, 11/6/68, B2; Elsberg, W News, 2/14/69, 5)

November 7: NASA Nike-Apache sounding rocket launched from NASA Wallops Station carried GSFC payload to 64.4-mi (103.6-km) altitude to determine absolute value of positive ion concentration in D and E regions of ionosphere. Quadrupole mass spectrometer for measuring relative abundance of positive ions malfunctioned and, consequently, launch of two supporting rockets was postponed. Vehicle performance and trajectory were good and experimental related events functioned well and on time. (NASA Rpt SRL)

- NASA Nike-Apache sounding rocket launched from Thumba Equatorial Rocket Launching Station carried Physical Research Laboratory experiment to 93.2-mi (150-km) altitude to measure absolute flux and energy spectrum of x-rays from sources in constellations Scorpius, Taurus, and Centaurus and time variation of x-ray fluxes from Scorpius and Centaurus sources. Experiment also would survey southern sky for undiscovered x-ray sources. Rocket and instrumentation performed satisfactorily. Good x-ray data were reported. (NASA Rpt SRL)

- ARC *Astrogram* reported successful completion of first in series of studies by Ames Biomedical Research Branch in which primates had been restrained in chairs for 98 days to provide information on calcium metabolism and bone mineralization when normal weight load on bones was altered as in weightlessness in space. Results showed that with application of weight loads on certain bones loss of calcium in urine was within normal limits; without load, control animal's calcium loss was elevated and lasted throughout experiment. X-rays showed normal bone mineralization in vertebrae and bones of loaded pig-tailed monkey, while unloaded animal sustained mineral loss. Studies would determine methods for prevention of bone changes and improve safety and efficiency of manned space flights. (ARC *Astrogram*, 11/7/68, 2)

- NASA released Delta launch vehicle for Nov. 8 launch of Pioneer D after completion of "corrective actions" to prevent repetition of vehicle's first flight failure Sept. 18. Delta No. 59, carrying Intelsat III-A (Intelsat III/F–1), had been destroyed shortly after liftoff when vehicle began breaking up. (NASA Release 68–195)

- U.S.S.R. celebrated 51st anniversary of Bolshevist Revolution in Moscow with missile display which included no new weapons. (AP, W Star, 11/7/68, A3)

November 8: NASA's *Pioneer IX* (Pioneer D), fourth in series of five spacecraft designed to provide continuing measurements over solar cycle at widely separated points in interplanetary space, was successfully launched from ETR by Thrust-Augmented Improved Thor-Delta (DSV–3E) booster into orbit around sun. Orbital parameters: aphelion, 0.99 astronomical unit (au), or 92.04 million mi (148.10 million km); perihelion, 0.75 au, or 69.71 million mi (112.19 million km); period, 297.55 days; and inclination 0.09°.

November 8

Test and Training Satellite *Tetr II* (TETR–B) carried pickaback on 2nd stage, was successfully ejected after 3rd-stage burnout and entered orbit around earth with 582.3-mi (937-km) apogee, 231.3-mi (373-km) perigee, 97.9-min period, and 32.8° inclination. S-band transponder was operating properly. *Tetr II* was follow-on *Tetr I* (formerly designated *Tts I*; launched pickaback on *Pioneer VIII* Dec. 13, 1967), which was highly successful in testing Apollo communications network.

Primary mission objective of 147-lb, drum-shaped *Pioneer IX* was to collect scientific data on electromagnetic and plasma properties of interplanetary medium for period covering six or more passages of solar activity centers. As secondary mission, *Pioneer IX* would: (1) acquire data when highly significant solar event occurred; (2) refine primary determinations of earth and moon masses, the astronomical unit, and osculating elements of earth's orbit; (3) provide synoptic study of solar-interplanetary relations; and (4) provide target for checkout of Manned Space Flight Network equipment and training of operations personnel by launching Test and Training Satellite as secondary payload. *Pioneer IX* separation, boom deployment, and first solar orientation occurred as planned and all eight experiments were operating properly and returning good data.

Pioneer VI (launched Dec. 16, 1965), *Pioneer VII* (launched Aug. 17, 1966), and *Pioneer VIII* (launched Dec. 16, 1967) were all successful and were continuing to transmit excellent data. Pioneer program was managed by ARC under OSSA direction. (NASA Proj Off; NASA Release 68–192; UPI, *W News*, 11/8/68; *W Post*, 11/9/68, A10; AP, *LA Times*, 11/9/68; Lannan, W *Star*, 11/11/68, A17; *SBD*, 11/12/68, 46; GSFC, *SSR*, 11/15/68)

* Bureau of the Budget issued tentative allowance of $3.623 billion for NASA's FY 1970 budget request. (NASA Off of Admin)
* American Nuclear Society held panel session in Washington, D.C., on "The U.S. Space Program: Achievements and Objectives."

Dr. Edward C. Welsh, NASC Executive Secretary, declared: "We must step up the rate at which we tap the vast potential of nuclear energy for the space activities of tomorrow. . . . if we do not make greater use of nuclear energy, we will neglect our mission of learning rapidly more and more about the solar system in which we live and about the planet where we reside. In the field of propulsion, chemical rockets, both liquid and solid, can be vastly improved when combined with the products of nuclear technology." Combining nuclear stage with Saturn V "will greatly increase that rocket's power of achievement. . . . Not only will we have the vast power of the atom at our command, but it will be compact, self-contained, long lived, highly maneuverable, and virtually independent of its surrounding environment. . . . Atomic energy will enable the space effort to reach for the infinite." (Text)

NASA Associate Administrator for Advanced Research and Technology James M. Beggs discussed "Research and Technology for the Future": "The difference between success and failure of [NASA] missions . . . lies in our knowledge of the flight sciences and our skill for applying this knowledge to the development and operation of space vehicles. . . . A natural characteristic of technology is its multiapplicability; an improvement in guidance or communication equipment, for example, may find many uses in space missions as well as non-aerospace applications. A key to making this process productive . . . is a continuing research program . . . that has a good balance between the effort in the scientific and engineering disciplines and in the technologies needed to explore the unknown." (Text)

Dr. Wernher von Braun, MSFC Director, said: "The first practical application of space electric power systems, which have been under development over the past 10 years, may well be found in our second generation orbital space station program. Consistent with our present estimates of station initial power requirements, and allowing for growth, a zirconium hydride reactor coupled with a thermoelectric conversion system is being studied for application on such a space station." (Text; Reuters, B *Sun*, 11/14/68, A10)

- MSFC announced Boeing Co. had been issued $239,000 contract for 10-mo study defining two-stage derivative of Saturn V launch vehicle. With 1st (S–IC) and 3rd (S–IVB) stages and instrument unit of Saturn V, vehicle could place up to 158,000 lb in low earth orbit. Varying the number of F–1 engines in S–IC could tailor vehicle to specific missions. Five-engine configuration could put into orbit Saturn I Workshop with airlock and multiple docking adapter, plus Apollo Telescope Mount and Apollo CSM and three-man crew. Three Saturn IB vehicles would be required to do same job. Vehicle could resupply space stations and could be used for synchronous orbits and unmanned lunar and planetary flights at major savings over three-stage Saturn V. Two-stage version was called "Intermediate 20." With Centaur 3rd stage, vehicle could send about 15,000 lb to Jupiter or Saturn.

  MSFC also had signed $22,826,736 contract modification with North American Rockwell Corp.'s Rocketdyne Div. for continued production support of J–2 engines used on Saturn IB and Saturn V boosters. Modifications would improve engines' versatility. (MSFC Releases 68–264, 68–266)

- NAS–NRC Space Science Board issued *Physics of the Earth in Space—A Program of Research: 1968–1975*, report of NASA-supported study by 31 scientists at Woods Hole, Mass., Aug. 11–24. Report, dated October and fourth by Board to provide guidance for NASA's programs in space physics, said results of decade of research by artificial satellites were "revolutionary; few of the concepts of the early 1950's have survived without major revision and totally unexpected discoveries have provided fundamentally new theoretical challenges." And "results of today's space research on the physics of the Earth in space become the engineering design data of tomorrow's civilian and defense applications programs."

  Report defined program of satellite, space-probe, and sounding rocket missions for concerted attack on questions of fundamental physical mechanisms of sun-earth system, in contrast to past decade's exploratory surveys. It emphasized coordinated investigations, new experimental techniques, and major observation effort during 1974–1975 low solar activity. Recommendations included continued NASA support for balloon, aircraft, and ground-based observations and of advanced development of spacecraft instruments; better means of data handling and adequate support for data analysis; and restoration of NASA program of predoctoral traineeship grants to 1966 level. (Text)

- Astronomers reported in *Science* conclusion neither NASA's *Mariner V* (launched June 14, 1967, for flyby of planet Venus) nor U.S.S.R.'s *Venus IV* (launched June 12, 1967) had reported atmospheric conditions near level of mean surface of planet. Von R. Eshleman and Gunnar Fjeldbo of Stanford Univ., John D. Anderson and Arvydas J. Kliore of JPL, and Rolf B. Dyce of Arecibo Ionospheric Observatory (Puerto Rico) had made new determination of radius of planet, based on concurrent ranging from earth to *Mariner V* near encounter and to surface of Venus. Extrapolations of measurements had given surface values for mid-latitudes of close to 100 atmospheres pressure and $700°K$ temperature (within $100°$), rather than Soviet values of 19 ± 2 atmospheres and $544° \pm 10°K$. Soviet probe apparently was not designed to work through such thick atmosphere. Simple ambiguity (times two) in *Venus IV* altimeter reading could explain supposition that probe reached Venus surface, "since this would bring all other data into excellent agreement." (*Science*, 11/8/68, 661–5)

- At press conference, inventor-scientist Stanford R. Ovshinsky described production of electronic devices—including desktop computers; flat, tubeless TV sets that could be hung on walls; amd missile guidance systems impervious to destruction by man-made radiation. Devices were made of amorphous materials whose electrical properties differed from transistor materials. Balance of energy forces within amorphous glasses was such that application of voltage of right minimum strength made material switch from insulator to conductor. (Stevens, *NYT*, 11/11/68, 1)

- American Telephone & Telegraph Co. said it had asked FCC to authorize its acquisition of 70 additional satellite communications circuits from ComSatCorp. Purchase would boost AT&T circuits to 396. (*WSJ*, 11/8/68, 5)

November 9: Apollo 7 commander Walter M. Schirra, Jr., received from Italian Ambassador to U.S. Egidio Ortona gold medallion awarded by Assn. of Man in Space, group of Italian scientists and jurists, at Washington, D.C., party in his honor. (Billington, *W Star*, 11/11/68, B7)

November 10: On nationwide "Meet the Press" TV interview, Astronaut Walter M. Schirra, Jr., commented on NASA budget cuts: "We've built up a fantastic technology [but] talented people are starting to leave. . . . We should let it be known that we are in this for the future, not just one flight." Cost of manned missions was justified in quest for knowledge not only outward, but earthward, too. Astronauts were "looking at portions of the earth that had never been documented before. A crew can see something and respond to it, on earth or the moon." Fellow *Apollo 7* crewman R. Walter Cunningham said never had Soviet crewmen "functioned in the same operational conditions as we." U.S.S.R. was putting fewer higher trained persons in orbit, "mainly as biological specimens." (AP, *W Post*, 11/11/68, A2; AP, *B Sun*, 11/11/68, A5)

- NASA announced it soon would begin series of test flights at LaRC of XC–142 tilt-wing VTOL aircraft on loan from USAF, to determine operational problems in airport terminal areas during poor visibility. XC–142, for which Ling-Temco-Vought, Inc., was prime contractor, was propeller-driven and powered by four GE turboshaft engines. NASA also was testing Ryan Aeronautical Co.'s XB–5A, which it had modified as XV-5B, and Hawker Siddeley P-1127 vectored jet VTOL aircraft. (NASA Release 68-194)

- In *New York Times* Walter Sullivan described "The Sun-Spot Menace to Astronauts." *Apollo 7* and U.S.S.R.'s *Soyuz III* served as reminders that sunspots were reaching their 11-yr peak. If eruptions were particularly severe, protons were hurled out at almost speed of light. These could penetrate spacecraft. While *Apollo 7* astronauts were never in danger, Soviet spacecraft placed in orbits reaching north beyond latitude 51° might "nudge zone" within which protons ejected by sun "rain fiercely on the atmosphere." Major flare had occurred Oct. 30 just after *Soyuz III* returned to earth. If astronauts had been in orbit —particularly if they had been outside spacecraft—they could have been subjected to hazardous radiation. Many warnings preceded this event. For moon journey it should be possible to postpone or cut short flight if sun looked ominous. In any miscalculation, radiation exposure to astronauts inside spacecraft would be severe only during most intense outbursts. However, on prolonged journeys to other planets there would be no escape. "It may therefore be necessary to design the spacecraft so that a portion of its interior will be shielded from such radiation." (*NYT*, 11/10/68, 7E)

November 10–17: *Zond VI* automatic space station was successfully launched by U.S.S.R. and placed on lunar trajectory from parking orbit of another satellite to explore outer space and test spacecraft systems, Tass announced. All equipment was functioning normally. Speculation, later confirmed, was that spacecraft would attempt to circle moon on same route taken by *Zond V* Sept. 15–21. On Nov. 14 Tass announced that *Zond VI* had circled moon at minimum distance of 2,420 km (1,503.8 mi) and had conducted studies of physical characteristics of near lunar space before continuing its journey back to earth.

Zond VI reentered and softlanded in a predetermined area in Central Asia Nov. 17. Unlike *Zond V*, which had plunged directly through upper atmosphere, *Zond VI* skipped across outer layers of atmosphere to reduce its reentry speed and then resumed its descent with aerodynamic forces. Announcing recovery, Tass said *Zond VI* had for first time tested a "more complex and promising method of the return of spacecraft from interplanetary trajectories—the method of controlled descent with the use of aerodynamical lifting force (aerodynamical quality) of the descending craft. . . .

"The braking of the descending apparatus in the atmosphere . . . was effected along a trajectory with two immersions in the atmosphere. During the first immersion . . . the second cosmic speed . . . [11 km per sec, 24,607 mph] was reduced to 7.6 kilometers a second (17,000 mph) through aerodynamical braking. In doing so, the descending apparatus . . . was oriented through the onboard control system in such a way that it, passing through the dense layers of the atmosphere, left them and next continued along the ballistic trajectory until the second immersion . . . [in which] the further lowering of the descending apparatus was also effected along the trajectory of controlled descent with the use of aerodynamical qualities which ensured its return to the Earth in the pre-set district."

Zond VI was sixth spacecraft in Zond series [see Sept. 15–21]. (AP,

W *Star*, 11/11/68, 1; *NYT*, 11/15/68, 8; GSFC *SSR*, 11/15/68, 11/30/68; Winters, B *Sun*, 11/19/68, 1; Kamm, *NYT*, 11/19/68, 1; *SBD*, 11/19/68, 71–2)

November 12: NASA Acting Administrator, Dr. Thomas O. Paine, announced at NASA Hq. news briefing that Apollo 8—carrying commander Frank Borman, CM pilot James A. Lovell, Jr., and LM pilot William A. Anders—would be launched from ETR Dec. 21 on open-ended lunar orbital mission of at least six days. Spacecraft would circle moon 10 times at 70-mi altitude before returning to earth. Dr. Paine explained: "After a careful and thorough examination of all of the systems and risks involved, we have concluded that we are now ready to fly the most advanced mission for our Apollo 8 launch in December, the orbit around the moon. . . .

"We have reached this conclusion after a long series of intensive investigations of the status of our program, the flight hardware, ground support equipment, status of our training."

L/G Samuel C. Phillips, Apollo Program Director, cited two categories of new risks with a lunar orbital mission: "In the one . . . the spacecraft propulsion system must operate properly in order to propel the spacecraft back out of lunar orbit and on its way back to earth. And the other category of risks are those that are inherent in being some three days away from the earth as opposed to . . . between a half an hour and three hours which the crew is away from the earth in a low earth orbital mission." Although increased reliance would have to be placed on dependability of life support and electric power systems and mission would have to take on additional risks, Gen. Phillips said, "The progression of risk between the Apollo 7 mission which we have flown and the Apollo 8 mission which we have designed is a normal progression of risks in a logically stepped development, flight test program." (Transcript; NASA Release 68–199; Schmeck, *NYT*, 11/13/68, 1; O'Toole, *W Post*, 11/13/68, 1; Sehlstedt, B *Sun*, 11/13/68, 1)

Engineer Michael O'Hagan, manager of space and military systems in government contracts dept. of U.K.'s Standard Telephone & Cables Co., told Conservative Party seminar in London U.S. was "actually buying land with mineral rights" in other countries after using satellites to discover its location. Hawker Siddeley Group Ltd. scientist K. C. C. Pardoe said countries could use satellites to spy on rival nations' crops and decide best time for marketing. (Reuters, *W Post*, 11/13/68, A25)

- FAA issued 1968 edition of *National Airport Plan*, annual assessment of civil airport needs for commercial and private flying. U.S. would require 808 new airports—including 22 for airline traffic and 748 for general aviation—during next five years, to relieve congestion and accommodate growth. Plan included recommendation for 25 STOLports in heavily congested areas. (FAA Release 68–74)

- Office of Secretary of Defense issued list of 100 companies and subsidiaries awarded largest dollar volume of military prime contracts of $10,000 or more in FY 1968. Total of these was $26.2 billion, 1.9% above FY 1967. U.S. companies received $38.8 billion, 1% less than in FY 1967. No. 1 on list was General Dynamics Corp., F–111 manufacturer, with $2.24 billion in contracts; No. 2, Lockheed Aircraft Corp., $1.87 billion; No. 3, General Electric Co., $1.49 billion. McDonnell

Douglas Corp. fell from first place in 1967 to fifth with $1.1 billion. (Text; Wilson, *W Post*, 11/19/68, D7; *WSJ*, 11/19/68, 27)

- New York State Supreme Court Justice Abraham N. Geller issued temporary order blocking sale of Cornell Aeronautical Laboratory to EDP Technology, Inc., Washington, D.C., for $25 million. Order had been requested by State Attorney General Louis J. Lefkowitz, who alleged alteration of Laboratory from public purposes to profit-making organization would be a "major change detrimental to the national interest, to the quality and effectiveness of the laboratory and to the . . . public." New York State had sought to purchase facility but was reportedly unable to match EDP Technology, Inc., offer. (AP, *NYT*, 11/13/68, 11)

November 12–14: Twelve nations attending Third European Space Conference in Bonn decided to work toward creation of single European Space Authority, in effort to end dependence on U.S. space research. Committee was appointed to draft convention for new body encompassing 6-nation ELDO, 10-nation ESRO, and 12-nation CETS and functioning somewhat like NASA. Member nations would be free to choose programs they would support—which could mean small group would work on launchers as well as working with number of other nations on space applications and research.

Day preceding conference, ELDO meeting had resolved, with U.K. abstaining, to proceed with launcher development on scale designed to hold down costs. U.K. had proposed European nations abandon project, rely on U.S. boosters, and concentrate on space applications, particularly communications. U.K. agreed, however, to fulfill commitment to support launcher program until 1971. (*W Post*, 11/15/68, A20; Greenberg, *Science*, 12/6/68, 1108–9)

November 13: U.S.S.R. launched *Cosmos CCLIII* from Plesetsk Cosmodrome into orbit with 337-km (209.4-mi) apogee, 216-km (134.2-mi) perigee, 90-min period, and 65.4° inclination. Equipment functioned normally and satellite reentered Nov. 18. (*SBD*, 11/14/68, 53; GSFC *SSR*, 11/15/68; 11/30/68)

- NASA's HL-10 lifting-body vehicle, flown by NASA test pilot John A. Manke, completed first powered flight after air launch from B-52 aircraft near Lancaster, Calif., to demonstrate successful operation of XLR-11 rocket engine and to investigate effects of engine operation on basic stability and control of wingless vehicle. Manke said craft climbed and maneuvered in "marvelous fashion." Increased performance expected in successive flights would simulate lifting-body spacecraft during terminal portion of flight.

Manke ignited two of four chambers in HL-10's rocket engine and climbed from 35,000 to 43,250 ft, reaching top speed of 610 mph (mach 0.8) during 184-sec engine burn. Eventually speeds of 1,000 mph and altitudes to 80,000 ft were expected. Oct. 23 HL-10 flight had ended with early shutdown of rocket engine. (NASA Proj Off; NASA Release 68–198; FRC Release 26–68; AP, *B Sun*, 11/14/68, A10; LATNS, *W Post*, 11/14/68, A22)

- NASA announced it had assigned Astronauts Thomas P. Stafford, John W. Young, and Eugene A. Cernan as prime crew for Apollo 10 mission, scheduled for second quarter of 1969 as second manned flight of lunar module. Backup crew would be Astronauts L. Gordon Cooper,

November 13: *John A. Manke,* NASA *research pilot at* FRC, *stands before wingless HL–10 lifting-body vehicle which he flew in first powered flight, after air launch from B–52 aircraft. Manke reached 610 mph in* NASA–USAF *program to evaluate possible forerunners of reusable spacecraft. Eventual 1,000-mph speed was expected.*

Donn F. Eisele, and Edgar D. Mitchell. Flight crew support team was Astronauts Joseph H. Engle, James B. Irwin, and Charles M. Duke, Jr. Mission probabilities ranged from earth orbital operations to lunar orbital flight, with separation and docking of command and service module and lunar module. (NASA Release 68–201; AP, W *Star,* 11/14/68, A7; *W Post,* 11/14/68, A12; AP, *NYT,* 11/14/68, 21; Sehlstedt, B *Sun,* 11/14/68, 1)

- Dr. John E. Naugle, NASA Associate Administrator for Space Science and Applications, addressed International Meeting of American Nuclear Society in Washington, D.C. In past decade NASA had developed "impressive capability" in "competent and creative people who make up our government-university-industry team." It was "prepared to accomplish any goal in space exploration which the new administration may establish." NASA future included "broad, balanced, Planetary Program" emphasizing Mars but examining other planets like Mercury and Jupiter; major lunar exploration program in 1970's; astronomy program covering optical astronomy, x-ray and gamma ray fields, and low-frequency

- Dr. Abe Silverstein, Director of Lewis Research Center, was announced winner of Rockefeller Public Service Award in science category. He had supervised plans for rocket installation on Lunar Orbiter and Surveyor spacecraft.

    Dr. Silverstein had been first Director of Space Flight Programs in NASA Hq. in 1958. Under his leadership first U.S. man-in-space program, Project Mercury, had been planned and groundwork laid for Gemini and Apollo programs. He had joined NACA at Langley Aeronautical Laboratory in 1929 and helped design and later was in charge of Full-Scale Wind Tunnel. He was transferred to Lewis Laboratory in 1943, where he was responsible for conception, design, and construction of first U.S. supersonic propulsion wind tunnels. After serving in Washington, D.C., three years he returned to LeRC as Director in 1961.

    Leonard C. Meeker, State Dept. legal adviser, won award in foreign affairs and international administration for his work on 1967 space law treaty. (*W Post*, 11/14/68, B7; LERC Biog, 4/67; *Lewis News*, 11/22/68, 1)

- *New York Times* editorial, "After the Lunar Landing," said first priority would probably go to study of moon's surface and resources. "Beyond this exploration, the moon—once accessible to men—will immediately become a laboratory world for expanding knowledge in astronomy, physics, chemistry, geology and a host of other sciences. For both purposes . . . it will be desirable to create one or more permanent manned communities on the moon as soon as possible. . . . As the pioneers in space, the United States and the Soviet Union have the responsibility now to begin organization of the vast international effort that must follow the coming triumph of human courage and human ingenuity. And the United Nations, of course, is the body whose flag must fly over future lunar settlements." (*NYT*, 11/13/68, 40)

- Purdue Univ. Prof. James E. Etzel said in Evanston, Ill., interview that technique for processing sewage sludge by bombarding it with ionizing radiation from chemical emitter of gamma rays could save $1 billion a year if used by all U.S. cities. Process was pioneered by Etzel and Gordon S. Born of Purdue in cooperation with Jerome Stein, director of R&D for Chicago Sanitary District. Plan would cut solid-waste-processing costs by more than half. Treated sludge would be completely disinfected, odorless, and compressed to $1/3$ its volume. (Randal, *W Star*, 11/14/68, A5)

- USAF awarded $8.5 million addition to contract with Northrop Corp. for F–5 aircraft. (*WSJ*, 11/13/68, 7)

November 14: NASA announced it had authorized JPL to proceed with construction of two Mariner spacecraft for 1971 Mars orbit with funds allocated under approved FY 1969 NASA appropriations bill. In combined mission to assist in establishing touchdown sites for 1973 Mars lander mission, spacecraft would be launched by Atlas-Centaur boosters into orbit around Mars to examine Mars polar cap, provide high-resolution coverage of selected areas, and permit oblique views of broad areas of Mars' surface and, possibly, its moons, Phobos and Deimos. Each spacecraft would complete trip from earth to Mars in six months, with May 1971 launch and November 1971 arrival, and would orbit Mars for three months or more. (NASA Release 68–196; B *Sun,* 11/14/68, A10; *W Post,* 11/14/68, D16)

- Washington *Evening Star* editorial commented on scheduled December launch of Apollo 8: ". . . this promises to be one Christmas when the thoughts of all . . . will contain more than visions of sugarplums, of laden stockings, of gifts about to be received and bills about to come due. It is, in fact, just possible that NASA will succeed in putting that missing ingredient back into the yule season, and that more prayers will be offered this Christmas than at any time in the past 2,000 years." (W *Star,* 11/14/68, A14)

- At hearing of Congressional Joint Economic Committee's Subcommittee on Economy in Government, A. E. Fitzgerald, Deputy for Management Systems in USAF financial office, said ultimate cost to DOD of 58 C–5A cargo aircraft might be "100% above the original estimate" of $1.279 billion for Lockheed Aircraft Corp. airframes. General Electric Co. estimate of $459 million for engines was not expected to double. Increase resulted from rise in manufacturing costs, higher subcontracting prices, and rising administrative costs, rather than gross original underestimate, although there was probably some original underestimate by Lockheed. DOD was weighing all factors before deciding action on option for 62 additional C–5As which would expire Jan. 31, 1969. (Crowther, B *Sun,* 11/14/68, A12; AP, *NYT,* 11/14/68, 10; Porter, *W Post,* 11/14/68, A1)

- Soviet aviation experts told *Pravda* they had successfully tested "ornithopter," aircraft which flew by waving its wings like a bird. Craft was said to have "withstood all aerodynamic tests" and to have greater lifting power than ordinary aircraft. *Pravda* said test "opens unheard of prospects." (UPI, *P Inq,* 11/15/68)

November 15: NASA Aerobee 150 MI sounding rocket launched from WSMR carried Princeton Univ. Observatory experiment to 108-mi (174-km) altitude to obtain UV radiation of bright star in constellation Cassiopeia, using gyro-stabilized spectrograph, ACS, and recovery system. Rocket and instruments performed satisfactorily. ACS stabilized rocket on target at star. Spectra were obtained on three exposures and payload was recovered in excellent condition. (NASA Rpt SRL)

- NASA released plans for lunar landing experiments. First U.S. astronauts to land on moon in 1969 would place three scientific experiments on lunar surface instead of more complex Apollo Lunar Surface Experiments Package (ALSEP) originally planned. Change was necessitated by uncertainties in workload required to deploy ALSEP by astronauts in pressurized suits on moon's surface. Mission's primary objective would be to prove Apollo system by achieving successful moon landing and safe return to earth. During first landing, two astronauts would leave spacecraft for up to three hours on moon's surface, making observations and photographing area in vicinity of landed spacecraft, collecting soil and rock samples, and deploying experiments.

Scientific and medical data would be obtained on expenditure of astronauts' energy, monitoring ability to perform in vacuum, extreme temperature, and $1/6$ gravity for planning of longer, more complex missions. Experiments were (1) passive seismometer (self-contained 100-lb seismic station with earth-moon communications link, designed to last up to one year), (2) 70-lb laser ranging retro-reflector, and (3) 1-lb solar-wind-composition experiment. They would provide data on internal activity of moon; would provide data to improve measurement of earth-moon distance, fluctuation of earth's rotation rate, and variations in gravitational constant G, as well as to test theory of intercontinental drift by direct measurements from different continents; and would entrap noble gases (helium, neon, krypton, xenon) from solar wind for analysis.

In second lunar landing mission, astronauts would deploy full geophysical station or ALSEP and conduct detailed field geology investigation. (NASA Release 68-200)

- U.S.S.R.'s *Zond V* automatic space station [see Sept. 15–21] had carried first living organisms—wine flies, turtles, meal worms, plants, bacteria, and seeds—around moon and returned them safely to earth, *Pravda* announced. Turtles had lost about 10% of their body weight but remained very active and showed no loss of appetite. Preliminary blood tests indicated no substantial deviation from laboratory control animals, but analysis of test turtles 21 days after flight revealed excess glycogen and iron in their livers. (AP, W *Star*, 11/15/68, A4; B *Sun*, 11/16/68, A4)

- NASA Honor Awards ceremony was held at Washington, D.C., with Dr. Thomas O. Paine, Acting Administrator, making presentations. Alexander H. Flax, Assistant Secretary of the Air Force (Research and Development), was main speaker.

   Distinguished Service Medal, NASA's highest honor, was presented to Secretary Flax; Edmond C. Buckley, former NASA Associate Administrator for Tracking and Data Acquisition; Paul G. Dembling, NASA General Counsel; and Abe Silverstein, Director of Lewis Research Center.

   NASA Exceptional Scientific Achievement Medal went to LeRC's G. Mervin Ault, Edmond E. Bisson, and John C. Evvard; JPL's Richard M. Goldstein, Alan Rembaum, Lewis D. Kaplan, and Conway W. Snyder; MSFC's Otto A. Hoberg and Hans H. Hosenthien; LaRC's Mark R. Nichols; and ARC's William A. Page and John A. Parker.

   Exceptional Service Medal recipients were: Mac C. Adams, Walter Boone, Richard L. Callaghan, Arnold W. Frutkin, Alfred S. Hodgson, Mildred V. Morris, and Boyd C. Myers, NASA Hq.; Robert M. Crane (posthumous), William R. Schindlar, and Michael J. Vaccaro, ARC; Robert J. Darcey, Herman E. LaGow, and Robert J. McCaffery, GSFC; Philip Donely, Paul F. Fuhrmeister, and Harry H. Hamilton, LaRC; Robert C. Duncan, ERC; Fred H. Felberg and Alvin R. Luedecke, JPL; Arthur Rudolph, MSFC; and Hubert Ray Stanley, Wallops Station.

   Group Achievement Award went to *Apollo 7* Flight Operations Team, Instrumentation Ships Team, Mariner Occultation Experiment Team, OGO Project Team, Sonic Boom Investigating Team, and Surveyor Team. (Program; MSFC Release 68-267; Marshall *Star*, 11/13/68, 1; Pasadena *Independent*, 11/15/68; ARC *Astrogram*, 11/21/68, 1)

- Dr. R. G. McIver, head of aeromedical research at Holloman AFB, N. Mex., said tests on chimpanzees had shown astronaut experiencing sudden decompression outside spacecraft might live at least three minutes in total vacuum, giving his companions time to get him inside pressur-

November 15

ized cabin. Survival of astronauts in ruptured space cabin would depend on presence of one man wearing space suit and thus able to place other crewmen in suits after they lost consciousness. Earlier, scientists had predicted sudden death for astronauts experiencing space vacuum decompression. Revised estimates of life expectancy were based on simulated space capsule experiments with 150 chimpanzees. (AP, B *Sun*, 11/16/68, A4)

- *Science* editorial on NASA administration: "In terms of numbers of dollars or of men, NASA has not been our largest national undertaking, but in terms of complexity, rate of growth, and technological sophistication it has been unique. . . . Keeping all of [its] parts—often working right at the edge of technological knowledge and capacity—finely tuned and in close harmony has been an organization achievement of high order. . . . Ever since the space program began to take shape there has been talk of technological spin-offs. It may turn out that the most valuable spin-off of all will be human rather than technological: better knowledge of how to plan, coordinate, and monitor the multitudinous and varied activities of the organizations required to accomplish great social undertakings." (Wolfle, *Science*, 11/15/68, 753)

- In *Washington Post* George C. Wilson said Secretary of Defense Clark M. Clifford's office had recommended cutback in USAF's order for FB–111, bomber version of TFX, from 253 to 90, a saving of $1.5 billion. While DOD was under pressure to reduce spending by $3 billion, said Wilson, argument "goes to the heart of the strategic question of unmanned missiles vs. manned bombers." (*W Post*, 11/15/68, 1)

- Formal establishment of Science and Public Policy Studies Group temporarily housed at MIT became effective. Group was designed as focus of interest and information for scholars, universities, and government officials engaged in teaching and research in science and public policy. Any university with active teaching and research program in science and public policy could become affiliate. Some 50 had indicated interest. Funding for two-year period was from university contributions and matching Sloan Foundation grant. Prof. Eugene B. Skolnikoff of MIT would serve as chairman. (Group Ann)

- Dr. Frank D. Drake, Director of Cornell Univ.'s Arecibo Ionospheric Observatory (Puerto Rico), told meeting of Council for the Advancement of Science Writing at Evanston, Ill., that pulsar discovered Nov. 14 near Crab Nebula was almost certainly a neutron star. It was second discovery during week [see Nov. 6] and was located by Arecibo radio dish telescope. (Randal, *W Post*, 11/16/68, A5; Sullivan, *NYT*, 11/20/68, 31)

- Edward Wenk, Jr., Executive Secretary of Marine Resources Council, told new American Oceanic Organization in Washington, D.C., "We are ready for the next step in Federal management of marine affairs—creation of an independent civilian operating agency." Chairman of new group, Rep. George E. Shipley (D-Ill.), member of House Appropriations Committee, said, "I think that we will see the time that this country will spend as much in oceanography as we are spending in the space program." (Corrigan, *W Post*, 11/17/68, K1)

November 16: U.S.S.R.'s *Proton IV*, largest unmanned scientific satellite to date, was launched into orbit with 472-km (293.3-mi) apogee, 248-km (154.1-mi) perigee, 91.7-min period, and 51.5° inclination. Satellite weighed 17 metric tons (37,478 lb), including 12.5-metric ton (27,557.5-lb) scientific payload to continue studies of cosmic rays. Equipment was functioning normally. (AP, W Star, 11/17/68, A18; Kamm, NYT, 11/17/68, 1; SBD, 11/19/68, 72–3; GSFC SSR, 11/30/68)

- Apollo 8 prime crew—Astronauts Frank Borman, James A. Lovell, Jr., and William A. Anders—held press conference at MSC on Apollo 8 flight plan. Commander Borman explained that an important feature of mission's flight plan was establishment in advance of decision points when crew could decide whether to continue mission as planned or return to earth: ". . . the first one [decision point] will be whether we commit to translunar injection [from parking orbit] or not. . . . But even if we do commit . . . we have regular abort times along the course to the moon, so that we could—in case of a system problem . . . stop at any time and come back. Then we finally reach a point where it would be swifter to just go on around the moon than it would be to try to abort. The next great point is before we burn lunar orbit injection." Once in lunar orbit crew would receive abort data regularly and have option to return to earth every two hours. "The mission, if all works well, will be a relatively simple one," Borman said. ". . . we designed Apollo, we said we were going to the moon, and . . . finally when we get down to examining the details and saying we are really going, people start getting a little queasy about it. But I have no hesitancy about the hardware." (Transcript; O'Toole, W Post, 11/17/68, A8)

November 17: Boeing Co. in Washington, D.C., displayed scale model of new fixed-wing design for SST with downward-bent wings inboard toward fuselage and vertical tail placed well forward of horizontal tail. SST Program Director M/G Jewell C. Maxwell said in interview, "I think we have a much greater feeling of confidence than we have had in some time. We now have a design in hand that seems able to do the job." Boeing would submit new design to Government in mid-January but, said *New York Times*, "there is some nervousness about the receptivity of the new Congress for a budget request that will probably come close to $300 million for the fiscal year ending June 1970." Current estimates of overall SST cost were $1.5 billion. (NYT, 11/18/68, 93)

- In *New York Times* Walter Sullivan said nuclear specialists who met in Stockholm during April and June at invitation of International Institute for Peace and Conflict Research had confirmed effectiveness of new method of distinguishing man-made explosion from natural earthquake at thousands of miles by comparing magnitude of seismic event in waves crossing earth's surface with magnitude of "body waves" from same event that had passed through earth's depths. Analysis by scientists from U.K., Canada, and U.S.S.R. had shown strength of surface waves related to body waves was consistently less in bomb explosions than in earthquakes. (NYT, 11/17/68, 1)

November 18: Bullet fragment in brain of holdup victim Joseph Barrios was successfully repositioned into brain membrane by whirling patient in ARC centrifuge, subjecting him to force of 6 g. Fragment had moved from critical central ventricle to lateral ventricle when doctors at O'Connor Hospital in San Jose, Calif., somersaulted Barrios on rotating chair, but fragment could move again. Force of centrifuge—normally used by NASA for space research—imbedded fragment in membrane of left ventricle, where it was expected to remain safely encapsulated by scar tissue. Dec. 16 x-rays showed fragment had not moved since centrifuge treatment. Barrios was sent home from hospital Dec. 17. (Brody, *NYT*, 11/21/68, 37; AP, *W Post*, 11/21/68, A3; ARC *Astrogram*, 12/8/68, 1; ARC Medical Services Branch)

- Astronaut Walter M. Schirra was announced winner of Kitty Hawk Memorial Trophy for distinguished achievement in aeronautics. Trophy would be presented at annual Wright Brothers memorial banquet in Beverly Hills, Calif., Dec. 6. (UPI, *W Star*, 11/19/68, A2)
- NASA submitted reclama to Bureau of the Budget's tentative FY 1970 budget allowance, for total $4.074 billion. Subsequently, Acting Administrator Thomas O. Paine met with BOB Director Charles J. Zwick. (NASA Off of Admin)
- USAF denied decision had been made to cut back or cancel FB-111 production [see Nov. 15], *Wall Street Journal* said, but reports persisted that fewer than 253 would be purchased because of their high cost and indifferent performance. *Aerospace Daily* said its sources reported recommended cutback of more than 50% was "essentially accurate" and put figure at "about 100" to give USAF Unit Equipment force of 90 aircraft. (*WSJ*, 11/18/68, 12; *Aero Daily*, 11/18/68)

November 18–19: Conference on Pavement Grooving and Traction Studies was held at LaRC on results of NASA research program into hydroplaning and effectiveness of runway and highway grooving in minimizing accidents on wet pavement. Among papers presented, joint NASA-British Ministry of Technology study of tests at Wallops Station showed 1- by ¼- by ¼-inch grooving "at least doubled the friction coefficient" of aircraft runway surfaces. Studies by LaRC, California Div. of Highways, and other state highway departments showed grooving of highway pavement was effective in preventing wet-weather accidents. (NASA SP-5073)

Air Transport Assn. of America reported that 15- to 19-mo use at three major airports had convinced airlines that runway surface grooving was "an effective aid in overcoming hydroplaning" during wet landings. Airline operation evaluation had "dispelled earlier fears that grooving might damage runways or aircraft." (Text; *NYT*, 11/24/68, 84)

November 18–21: During Geneva conference of IAF on Basic Environmental Problems of Man in Space, Dr. Boris B. Yegorov, only physician to travel in space (launched in U.S.S.R.'s *Voskhod I* Oct. 12, 1964), told press conference he believed planetary flights would be achieved by 1987. U.S.S.R. was experimenting with 8- to 10-volt electrical shocks administered to improve circulation and stimulate muscles of cosmonauts to enable them to endure strain of returning to earth after prolonged period in confined space capsule. He claimed problem of weightlessness was same for short or long flights and urged immediate planning for flights to galaxies outside solar system.

Dr. Walton W. Jones, NASA Director of Biotechnology and Human Research, told conference U.S. could keep astronauts in earth orbit a year within 10 yr if U.S. would allocate sufficient resources to project. ARC experiments had shown rats subjected to gravitational pull 4.7 times above normal had increased life expectancy. He said after meeting that NASA planned experiments with rats at zero gravitation to determine if aging process was suspended while man was free from gravitational pull.

Dr. J. F. Kubis of Fordham Univ. emphasized importance of psychological factors in selecting space crews. On long flight, "lack of privacy, crowding, and continuous exposure to interaction will become sources of aggravating stress." He recommended no more than one in crew with dominant personality.

Soviet scientist Dr. N. N. Gurovsky said condition of two dogs after 22-day orbital flight (in *Cosmos CX* Feb. 22–March 16, 1966) had aroused forebodings on effect of year-long flights on astronauts. Loss of calcium in bones would make them prone to breaking. Both dogs had developed liver and intestinal ailments including edemas. Human beings could expect same effects. Earlier reports had noted dogs suffered muscular reduction, dehydration, and confusion in adjusting to walking.

Cal Tech Prof. Fritz Zwicky said some apparently unexplained deaths on earth each year could be due to blows from meteors or heavy nucleons in cosmic rays. (Hamilton, *NYT*, 11/20/68, 31; 11/21/68, 18; AP, B *Sun*, 11/22/68, 1)

November 19: NASA launched series of three sounding rockets from NASA Wallops Station to conduct acoustic grenade experiments. Nike-Cajun carried GSFC experiment to 72.6-mi (122.7-km) altitude to obtain vertical profile of temperature, pressure, density, and wind data between 21.8- and 59-mi (35- and 95-km) altitudes. Sound arrivals of 18 exploding grenades, ejected at programmed altitudes, were recorded and shock waves were measured by two rocketborne pressure transducers. Launch was correlated with similar grenade experiments launched from Churchill Research Range and Point Barrow, Alaska.

Nike-Apache carried GSFC payload to 124.3-mi (200-km) altitude to define D-region electron-density profile in conjunction with measurements by Nike-Apache launched later. Good data were obtained, but radar failed to track vehicle.

Second Nike-Apache, carrying Univ. of Michigan experiment, reached 80-mi (148-km) altitude. Objective was to obtain neutral particle parameters of pressure, temperature, and density between 24.8- and 62.1-mi (40- and 100-km) altitude with Pitot static tube and falling sphere. All key experimental events occurred as planned. (NASA Rpts SRL)

- Vice President Hubert H. Humphrey spent five hours in briefings at KSC, ascending to spacecraft level of Apollo 8 moon rocket, performing simulated rendezvous and formation flying inside mock spacecraft, and lunching with NASA officials and Apollo 8 Astronauts Frank Borman, James A. Lovell, Jr., and William A. Anders. During visit, he said: "The space program has contributed greatly to the structure of our nation. . . . I have felt it has been underfunded at times and we will pay the price. I feel it is a wise investment on the part of Congress and the public. I'll do what I can as a private citizen and a former chairman of the space council to see it is funded." (KSC Release KSC–496–68; AP, W *Star*, 11/20/68, A8; AP, W *Post*, 11/20/68, C1; *Today*, 11/20/68, 1)

- Soviet Prof. Georgy I. Petrov in *Izvestia* said successful recovery of

Zond VI opened way "for creation of spacecraft able to get to the moon, Mars, Venus, and other planets and return to earth." Such experiments, he said, "will allow us in the near future to create long-lasting orbital stations, moon laboratories manned by scientific personnel. . . . The passenger ships will differ from the present spacecraft, but now we are witnessing their birth." In another dispatch Tass said basic problem of reentry by spacecraft at "second cosmic speed," about 25,000 mph, had been solved by *Zond VI*. (UPI, W *Star*, 11/20/68, B11)

• In answer to queries [see Nov. 16] DOD issued statement on increased cost of C–5A, world's largest aircraft. "At the beginning of this program over three years ago, the Air Force estimated that the cost of development and production of the first 58 airplanes would be $2.3 billion. The corresponding estimate for the 120 airplanes ultimately contemplated was $3.1 billion. Current estimates, including economic escalation and all other factors, are $3.25 billion and $4.3 billion, indicating increases of 41% and 39% respectively." Additional costs had resulted from increased labor and materials costs due to Vietnam war and "unprecedented demand for civilian aircraft," introduction of new technology, and modifications to overcome technical difficulties. (Text; *WSJ*, 11/20/18, 2; AP, *W Post*, 11/20/68, A3; Kelly, W *Star*, 11/20/68, A11)

November 20: NASA successfully launched two Nike-Cajun sounding rockets to 74.6-mi (120-km) and 70.6-mi (113.6-km) altitudes to obtain data on variation of temperature, pressure, and wind profile by detonating grenades at prescribed times and recording sound arrivals on ground. Rockets and instruments performed satisfactorily; sound arrivals were recorded from 19 grenades on each rocket. Data would be compared with data from launches from NASA Wallops Station and Point Barrow, Alaska. (NASA Rpts SRL)

• NASA Acting Administrator Dr. Thomas O. Paine in speech before AIA Fall Conference in Phoenix, Ariz., said, "Today the United States stand at the crossroads." In FY 1969 NASA was operating close to "breakpoint" level. Below $4-billion budget level, NASA could no longer "hold together our hard-won capabilities and utilize them effectively in critical programs; some of them would have to be dropped entirely." Budget above $4 billion would "not only allow us to . . . carry forward major programs, but might permit . . . modest investments aimed at reducing costs of future space activities."

Pointing to "long-term results of the tremendous technological effort . . . mounted during World War II" and "still being exploited today," Dr. Paine urged that as "we tackle grave social ills, we've got to continue to forge ahead in other areas. We must worry about how we create new wealth as much as we do about how we better distribute today's wealth. It would be an international tragedy if America were to turn back now from its forward thrust in space at the end of an astonishingly productive first decade."

James M. Beggs, NASA Associate Administrator for Advanced Research and Technology, discussed three major aerospace needs: increased aeronautical research, low-cost boosters, and use of nuclear energy in space. In aircraft technology, NASA was "increasing . . . effort by increases in old and new aeronautical disciplines in V/STOL technology, and in noise reduction." Pressing need for boosters as reliable as existing boosters but far more economical, "means a fresh approach to the entire concept of boosters and a critical examination of each step . . . from the drawing board to the end of the boost operation." NASA was recommending that NERVA engine development be continued with engine and stage ready for a mission as early as 1977. (Texts)

- In BBC TV interview Jodrell Bank Experimental Station Director, Sir Bernard Lovell, said U.S. Apollo 8 plan for manned lunar orbit was, "On a scientific basis . . . wasteful and silly. . . . We've reached the stage with automatic landings when it's not necessary to risk human life to get information about the moon. Within a few years this information could be obtained by automatic, unmanned instruments." Sir Bernard said he was full of admiration for project but added, there was "a dangerous element of deadline beating in it."

  NASA Associate Administrator for Manned Space Flight, Dr. George E. Mueller, commented, "The purpose is not scientific but to take an important step in developing the capability of landing men on the moon." In telephone interview from Huntsville, Ala., he said, "We are taking no undue risks."

  Apollo 8 crewman William A. Anders told news conference at KSC, "We are flying primarily an operational mission and we strongly feel that a manned platform in lunar orbit with the operational equipment we have can add significant bits and pieces to scientific knowledge. . . . We think it will be a real boon for future Apollo flights to have the photographic, navigation, tracking, mapping, and other knowledge we'll bring back." (AP, B *Sun*, 11/21/68, 1; *NYT*, 11/21/68, 19; O'Toole, *W Post*, 11/21/68, A3)

- Survey by AIA predicted decline in aerospace industry employment from 1,431 million in March to 1,400 million by December because of continuing decline in space program and leveling off of employment requirements in aircraft production and R&D programs. Employment was expected to remain at December level through March 1969. Aircraft production and R&D employees were expected to decline from 854,000 to 833,000, or by 2.5%. Missiles and space employment was expected to drop from 517,000 to 507,000, or 1.9%. Scientists and engineers would continue to form 16% of aerospace employment, with technicians making up 6% of aerospace industry employment. (AIA Release 68-56)

- Soviet trade union newspaper *Trud* said mass production of powerful intercontinental rockets had started at U.S.S.R.'s "Rocket City" because "the conquest of space is expanding." Device made at the unidentified city "will help our scientists discover new mysteries of the universe for the good of man." (UPI, *NYT*, 11/22/68, 22)

- GSFC awarded $3,127,001 one-year extension of cost-plus-award-fee contract to Fairchild-Hiller Corp. for scientific and engineering support services at GSFC. (GSFC Release G-53-68)

November 21: Cosmos CCLIV was successfully launched by U.S.S.R. from Plesetsk Cosmodrome. Satellite entered orbit with 332-km (206.3-mi) apogee, 210-km (130.5-mi) perigee, 89.8-min period, and 65.4° inclination and reentered Nov. 29. (UPI, *NYT*, 11/22/68, 22; GSFC *SSR*, 11/30/68; *SBD*, 12/2/68, 129)

- NASA Aerobee 150 A sounding rocket successfully launched from NASA

Wallops Station carried 300-lb payload containing two white rats to 101-mi (162.5-km) altitude in third of four experiments to study rats' behavior in artificial gravity field and determine minimum level of gravity needed by biological organisms during space flight. During five minutes of free fall, rats selected artificial gravity levels created through centrifugal action by walking along tunnel runway in extended arms of payload. Data on their position and movement were telemetered to ground stations. Payload impacted 69 mi downrange in the Atlantic. (WS Release 68-21)

- NASA Aerobee 150 MI sounding rocket launched from WSMR carried Univ. of Colorado experiment to 121-mi (193-km) altitude to take high-resolution spectra of carbon-2 resonance doublet at wavelength 1,334 and 1,336 Å; to retake, for comparison, high-resolution spectra of oxygen-1 resonance triplet at 1,302, 1,304, and 1,306 Å; and to measure total intensities of these lines. Rocket and instrumentation performed satisfactorily. Analysis of experiment results awaited recovery and processing of flight film. (NASA Rpt SRL)

- National Science Foundation released *Research and Development in Industry, 1966*. Total 1966 industrial R&D expenditure was $15.5 billion, with industry accounting for 70% of nation's R&D performance. Federal Government R&D performed in industry reached $8.3 billion, 53% of industrial total. In January 1967, 163,900—30% of total—industrial scientists and engineers were engaged in R&D directly supported by Federal Government. In 1966, NASA and DOD financed work of 89% of R&D scientists and engineers employed by industry on Federal projects, with 54,000 working on NASA projects, an increase of 50% over 1963. DOD-financed scientists and engineers decreased 23% in same period. (Text)

- At awards dinner of National Institute of Social Sciences in New York, Charles A. Lindbergh said, "My hope that aviation would cause better relationships between our earth's peoples gave way to realization that the airplane's primary significance lay in its power of destruction. . . . My fascination with the rocket as a space-exploring vehicle has been replaced by my alarm at its ability to wipe out our civilization overnight." Lindbergh and his wife, Anne Morrow, received gold medals for "distinguished service to humanity" for their work in conservation. (UPI, W *Star*, 11/22/68, A2)

- Washington *Evening Star* said U.K. and Australian governments had drawn up joint plans for largest telescope in Southern Hemisphere, 150-in optical telescope at Siding Spring Mountain Observatory of Australian National Univ. It would cost $10.5 million. (W *Star*, 11/21/68, A6)

- ComSatCorp announced it had requested proposals for construction of two new earth stations near Talkeetna, Alaska, and Apra Heights, Guam. (ComSatCorp Release 68-63)

- Dept. of Interior released *Noise and the Sonic Boom in Relation to Man*, report of 12-scientist study group appointed by Interior Secretary Stewart L. Udall. Study found boom effects not harmful to hearing but estimated regular commercial overland SST flight would produce possible $80 million worth of damage claims annually from owners of buildings. Although value of time saved by SST passengers might be $3 billion annually, number of persons "gravely annoyed" probably would be larger than that of SST passengers. Widespread use of SSTs would introduce large quantities of water vapor into atmosphere, which might affect weather.

Group found urgent necessity for prompt decision on SST program and recommended: permitting only controlled experimental SST flights over populated areas while full determination of effects was being made; making immediate large-scale experiments with existing supersonic aircraft simulating intercity SST operations; and establishing Presidential committee to hold public hearings in all regions of U.S. likely to be affected by sonic boom. (Text; Hornig, W *Star*, 11/22/68, A4; Rich, *W Post*, 11/22/68, A2; Sehlstedt, B *Sun*, 11/22/68, A5)

November 21–22: NASA launched two Nike-Cajun sounding rockets from Point Barrow, Alaska, carrying GSFC experiments to obtain vertical profile of temperature, pressure, density, and wind data between 21.8- and 59-mi (35- and 95-km) altitudes by detonating grenades at prescribed times and recording sound arrivals on ground. Rockets and instruments performed satisfactorily; 19 grenades on each rocket exploded as planned. (NASA Rpts SRL)

November 22: NASA Aerobee 150 A sounding rocket launched from NASA Wallops Station carried Univ. of Kentucky experiment to 101.7-mi (163.6-km) altitude to determine gravity preference of small white rat when subjected to artificial gravity field [see Nov. 21]. Rocket and instrumentation performed satisfactorily. Good data were obtained. (NASA Rpt SRL)

- JPL announced its astronomers had determined asteroid Icarus was about half mile in diameter and rotated every $2\frac{1}{2}$ hr, from data received during three-day series of seven microwave probes. Icarus was clocked at speeds from 36,000 mph to 1,450 mph at 4-million-mi distance—closest approach it had made to earth in 19 yr. A 450,000-w transmitter on 85-ft antenna at Goldstone Tracking Station in Mohave desert beamed radar waves at 2,388-mc frequency. Reflected echoes were received by 210-ft antenna 14 mi away. Average radar round trip to Icarus was 43 sec. JPL radar astronomer, Dr. Richard M. Goldstein, said indications were that Icarus was "rough, even jagged, and perhaps shaped like a peach stone." Radar reflections were unable to indicate whether its surface was stony or metallic. If metallic, Dr. Goldstein said, its radius might be as small as 300 m; if stony, 600 m, which fixed Icarus' diameter at 600 to 1,200 m, with 900 m a probable figure. (NASA Release 68–197; AP, W *Star*, 11/20/68, A8; Goldstein, *Science*, 11/22/68, 903–4)

- Electronic device used to monitor heart pumping performance of astronauts was being tested as a "potentially sensitive indicator of early transplant rejections," MSC's Dr. Charles A. Berry and Dr. Lawrence F. Dietlein told symposium of medical space scientists from 15 nations at World Health Organization in Geneva. "The device, if successful, would be ideally suited for this use." Dr. Dietlein pointed out that since no needles were used there would be no risk of infection and no risk of upsetting the delicate immunological balance of patients. (AP, *Today*, 11/24/68, 12A)

- NAS in *The Mathematical Sciences: A Report* said that before World War II U.S. was consumer of mathematical talent but now was "universally recognized as the leading producer." Graduate education in

November 22

mathematical sciences at major U.S. centers was "far superior to that in all but two or three centers in the rest of the world," but shortage of college teachers was likely to worsen. Report recommended increased Federal support for basic research, including NASA programs, computer science, applied mathematics, graduate apprenticeships, and faculty improvement. (Text; Schwartz, *NYT*, 11/24/68, 74)

- Dr. Donald F. Hornig, President Johnson's Special Assistant for Science and Technology, would join Eastman Kodak Co. in "executive capacity" in early January, *Science* reported. He would also become professor of chemistry at Univ. of Rochester. (*Science*, 11/22/68, 881)
- U.S. and Romania signed agreements to exchange information on peaceful uses of atomic energy, scientific delegations, and unclassified technical literature and films. Romanian graduate students would be assigned to U.S. laboratories and universities. (Cohn, *W Post*, 11/21/68; E1; AP, *NYT*, 11/24/68, 43)

November 23: First Lady, Mrs. Lyndon B. Johnson, and daughter Mrs. Lynda Robb paid their first visit to KSC and participated in simulated moon landing in lunar landing module replica. Dr. Thomas O. Paine, NASA Acting Administrator, presented model of Apollo lunar landing module for display in Lyndon B. Johnson Presidential Library. He said model would bear plaque listing "a few of the many contributions to our space program of Lyndon B. Johnson—as Senate Majority Leader, as Chairman of the Senate Aeronautics and Space Sciences Committee, as Chairman of the National Aeronautics and Space Council when he was Vice President, and then as President of the United States." Dr. Paine told Mrs. Johnson, "We trust that this model of the lunar landing module will signify to the many visitors . . . the President's vision and leadership that has carried this nation outward into the new ocean of space." (KSC Release KSC–500–68; Shelton, W *Star*, 11/25/68, E3; Blair, *NYT*, 11/24/68, 41)

November 24: Cornell Univ. scientists reported pulsar lying in or near Crab Nebula with fastest pulse rate of pulsars discovered thus far was slowing pulse tempo at rate of one part in 2,000 a year. Discovery was made with 1,000-ft-dia antenna at Arecibo (Puerto Rico) Observatory. Walter Sullivan in *New York Times* said discovery encouraged view that astronomers "are penetrating an entire new realm of physics . . . physics of superdense matter (in the form of so-called neutron stars), magnetic fields far beyond anything observable in the laboratory and superpowerful gravity." (*NYT*, 11/25/68, 53)

- *New York Times* article said at least eight nations, including U.S. and U.S.S.R., were building astronomical observatories in Chile, which had been termed ideal site because of its latitude, near 30° South. Association of Universities for Research in Astronomy, Inc. (AURA), dependency of NSF, had invested $19 million thus far in observatory of Cerro Tololo. Other groups were European Southern Observatory (ESO, consortium of West Germany, France, Holland, Belgium, Sweden, and Denmark) and Soviet Astronomical Mission. A 36-in-telescope at U.S. observatory already had photographed powerful ray-emitting star whose existence previously had been only suspected. (*NYT*, 11/24/68, 27)
- USAF launched experimental Advanced Ballistic Reentry System (ABRES) vehicle from Vandenberg AFB. (AP, *W Post*, 11/25/68, 9)

November 25: Four foreign firms who handled all Argentine overseas telephone calls and telecommunications, with multimillion-dollar investment, had been notified by lower echelons of Argentine government that its national carrier ENTEL would be exclusive international carrier and would reserve all communications satellite channels for itself, Washington *Evening Star* reported. Argentina was to link up with ComSatCorp and INTELSAT on completion of earth station at Balcarce in June 1969. Firms—U.S.-based ITT World Communications, Inc.; Transradio, which had RCA hookup; Western Telegraph of U.K.; and Italcable of Italy—would propose mixed government-carrier company to own and operate Argentine earth station. (O'Leary, W *Star*, 11/25/68, A15)

- C–130 troop and supply carrier, used as pick-up aircraft for satellites dropping special reconnaissance photos into sea, had developed serious wing cracks which would cost USAF $11 million to repair, Bob Horton reported in Washington *Evening Star*. USAF had been reinforcing wings but would have to build entire new wing on 400 C–130 models B through E to ensure long service. Model C–130A was not affected. (W *Star*, 11/25/68, A3)

- *New York Times* editorial commented on NAS report on status and needs of American mathematics [see Nov. 22]: ". . . this era—often called the age of the computer—is really the time of the most widespread and fruitful application of mathematics ever known. . . . The cost of supporting American mathematics is so slight in relation to the vast potential benefits that even the most economy-minded Congressman should be chary of using his ax" on Federal support. (*NYT*, 11/25/68, 46)

- Harold T. Luskin, Director of Apollo Applications in NASA Office of Manned Space Flight, died in Bethesda, Md., of respiratory illness. He had joined NASA in March 1968 and had become Apollo Applications Director in May. He was past president of AIAA and had been associated with Douglas Aircraft Co. for 20 yr and Lockheed Aircraft Corp. for 9 yr. He had held engineering and management positions in connection with Agena space vehicle, USAF Manned Orbiting Laboratory, X–3 supersonic research aircraft, and DC–8. (NASA Ann, 11/26/68; *W Post*, 11/27/68, C3; *Marshall Star*, 11/27/68, 1; W *Star*, 11/27/68, B4)

- In *Aviation Week & Space Technology* editorial Robert Hotz said, "The national space program is approaching a critical watershed at about the same time the Nixon Administration will assume stewardship. . . . It is obvious now that space technology has demonstrated its ability to provide a wide range of benefits for better living on earth. It would be a national tragedy, if this country failed to capitalize on the substantial investment it has already made in developing space technology. It will fall to the Nixon Administration to make the critical decisions during the next few years that will determine the future course of the national space program." (*Av Wk*, 11/25/68, 11)

- Dr. Paul Allman Siple, polar explorer and geographer who had accompanied Adm. Richard E. Byrd to Antarctica and science adviser to U.S. Army 1946–1963 and since 1967, died at Arlington, Va., at age 59. (*NYT*, 11/27/68, 47)

November 26: Dr. George E. Mueller, NASA Associate Administrator for Manned Space Flight, told National Space Club in Washington, D.C.,

November 26

unique conditions in space suggested use of orbiting space stations for certain manufacturing processes. "For instance, liquid floating in a weightless environment takes the shape of a perfect sphere. Thus, it is conceivable that metal ball bearings could be manufactured in space to tolerances impossible on earth, yet at a cost, including transportation, less than we can now achieve. Perfect bearings would reduce friction and noise levels to the vanishing point. Free-fall casting techniques could be utilized to cast large flawless optical blanks for telescopes and by proper combinations of spinning and electrostatic forces we should be able to shape the surface as well."

Stable foams for mixtures of liquefied materials and gases, impossible to produce satisfactorily on earth, could be produced in weightlessness, resulting in "a steel foam almost as light as balsa wood with many of the properties of solid steel." Composite materials like steel of different densities and properties and glass also could be produced. (Text; Schmeck, *NYT*, 11/27/68, 48; Sehlstedt, B *Sun*, 11/27/68, A6)

- Dr. Homer E. Newell, NASA Associate Administrator, said in speech before annual convention of National Council for Social Studies at GSFC: ". . . because of the length of time between investment in basic scientific research and important use of the results thereof, support by Congress and the public of basic research is hesitating and often outright skeptical. The development and application of technology to the quick solution of current problems is better understood and more readily supported. Yet the fact is that technology to be used in the solution of a practical problem rests ultimately upon the results of basic research performed years, and often many years, ago." Experience proved, he said, "that the effort to select the basic research to support in terms of predicted usefulness would over and over again preclude support of the research that in time would prove to be the most valuable." (Text)

- NASA and Univ. of Texas dedicated world's third largest telescope, 107-in, 150-ton instrument at McDonald Observatory, Mount Locke, Tex. Observatory Director, Dr. Harlan J. Smith, said it had been booked a year in advance. The $5-million observatory would emphasize studies of moon and other planets. Government scientists hoped for information to help them design planetary spacecraft. (UPI, *NYT*, 11/27/68, 26)

- French Prime Minister Maurice Couve de Murville announced France's 1969 atomic test program would be canceled and credits would be reduced for Concorde supersonic aircraft as part of austerity program to save French franc from devaluation. French contribution to Concorde development would be cut by $12 million. It was not known how much move might delay program, with first prototype scheduled to fly in late December. (AP, W *Star*, 11/26/68, 1; Louchheim, W *Post*, 11/27/68, A1; Tanner, *NYT*, 11/27/68, 1)

November 27: NASA FY 1970 budget request of $3.878 billion was agreed on in meeting between President Johnson and NASA Acting Administrator Thomas O. Paine. (NASA Off of Admin)

- NASA awarded Allis-Chalmers $3,500,000 contract to flight-qualify Multimission Fuel Cell Assembly, an improved fuel cell electrical power system for Apollo Applications (AA) program. System had been developed under three previous NASA contracts since 1962. Allis-Chalmers would produce two assemblies for use in qualification program and two for delivery to MSC. (NASA Procurement Off; MCS *Roundup*, 12/6/68, 1)

- U.K. became first nuclear power to ratify nuclear nonproliferation treaty. At Washington, D.C., ceremony, British Chargé d'Affaires Edward E. Tompkins handed instruments of ratification to Director William C. Foster of U.S. Arms Control and Disarmament Agency. (*NYT*, 11/28/68, 9)
- Soviet journal *Aviatsiya i Kosmonavtika (Aviation and Cosmonautics)* said Soviet scientists had concluded "basis of a linked system for providing man with vital necessities on board spaceship will be the cultivation of higher plants. . . . Scientists believe that artificial soil could be used for space plant growing." Once spacecraft left earth gravitation field, "plants will be fixed in special holders and sprayed with concentrated solutions containing all necessary substances." (UPI, *NYT*, 11/29/68, 22)
- USN's Sealab III was carried by barge to San Clemente Island, Calif. It would be lowered 600 ft to Pacific Ocean bottom to serve as working and living quarters for five teams of 8 to 10 men setting up underwater trolley line, building dry and lighted hut on sea floor, starting lobster farm, and training porpoises and sea lions to fetch and carry. Sealab III, submerged to three times depths of Sealab I and Sealab II, was final experiment in Sealab program. (AP, W *Star*, 11/28/68, A36)

November 28: NASA announced Mexican cities Gomez Palacios and Torreon had been saved from flooding in wake of Hurricane Naomi when Automatic Picture Transmission (APT) station at Mexico City relayed data from *Essa VI* satellite indicating weather was clearing. Mexican authorities thus were saved from opening dangerously filled dam which would have inundated the cities. (NASA Release 68–203; UPI, *NYT*, 12/1/68, 77)

- Canadian scientist I. A. Stewart, Operations Manager of Churchill Research Range, said at first International Aerospace Exposition in Montreal he believed range, operated jointly by U.S. and Canada, could be converted for $1 million to accommodate satellite launching pad. Canada could thus rise to ranks of space-age power, he said. Of more than 200 on staff, only one was from U.S. (Can Press, *NYT*, 12/1/68, 13)

November 29: U.S.S.R. launched *Cosmos CCLV* from Plesetsk Cosmodrome into orbit with 317-km (197-mi) apogee, 211-km (131.1-mi) perigee, 89.6-min period, and 65.4° inclination. Spacecraft reentered Dec. 7. (GSFC *SSR*, 11/30/68; 12/15/68; *SBD*, 12/2/68, 129)

- MSFC announced it had requested proposals from 11 aerospace companies for six-month design and definition study for dual mode lunar roving vehicle (LRV) capable of transporting astronauts on lunar surface and of performing automated, long-range scientific traverses across moon under remote control from earth.

    Vehicle was to permit manned sorties of up to 6 mi from landed spacecraft, with total round trip of more than 18 mi. After astronauts left moon, LRV would be placed in remote control mode for automated long-range (600 or more mi) geological and geophysical trips for one year. It would collect up to 200 lb of lunar samples and measure terrain, then rendezvous with manned spacecraft for return of samples to earth. (MSFC Release 68–274; *SBD*, 12/3/68, 134; *Marshall Star*, 12/4/68, 1)

November 29

- FAA announced it had issued RFP for collection and analysis of information on engineering, economic, and operational aspects of proposed construction of airports on offshore water sites, including floating airports and those to be built on fill or piles or in areas protected by dikes. (FAA Release 68–75; *NYT*, 12/2/68, 94)
- World's largest vacuum telescope, scheduled for spring 1969 completion at Sunspot, N. Mex., could provide method of predicting solar flares, Associated Press reported. Housed in 135-ft concrete needle atop mountain ridge 4,000 ft above WSMR, at Sacramento Peak Observatory, telescope was brainchild of USAF astronomer Dr. Richard Dunn, who told AP, "Prediction of solar flares would give us time to warn astronauts working outside the radiation shielding of a spacecraft to take cover and allow us to predict periods of radio communication interference." Project cost $3.5 million. (AP, *NYT*, 12/1/68, 65)

November 30: U.S.S.R launched *Cosmos CCLVI* into orbit with 1,223-km (759.9-mi) apogee, 1,195-km (742.6-mi) perigee, 109.4-min period, and 74° inclination. (GSFC *SSR*, 11/30/68)

- European Launcher Development Organization (ELDO) attempt to place 550-lb Italian ELDO F–7 into polar orbit failed when technicians lost contact with payload shortly after launch. Satellite launched from Woomera Rocket Range by booster with U.K. Blue Streak 1st stage, French Coralie 2nd stage, and West German Astris 3rd stage stopped transmitting after abbreviated seven-second 3rd-stage burn and could not be tracked. (Reuters, *NYT*, 12/1/68, 11)

During November: Milton Alberstadt reviewed 25-yr history of MSFC's Michoud Assembly Facility. U.S. Government built plant (dedicated Oct. 24, 1943) on Michoud sugar mill site during World War II to assemble Curtiss C–46 Caravan transports. After C–46 contract was canceled, "Flying Dutchman," air-sea rescue boat carried under belly of Boeing B–17 bomber, was developed. Michoud plant was reactivated during Korean War by Chrysler Corp. to produce 810-hp engines for Patton tanks. It was selected by NASA in 1961 as site to build Saturn rocket to put man on moon. (*Boeing* Magazine, 11/68, 8–9)

- In *Air Force and Space Digest* interview General Electric Co. Vice President Gerhard Neumann, head of GE Aircraft Engine Group, said postponement of essential R&D during past several years had "mortgaged" nation's techological future. He saw hazards in current DOD contracting policies which kept industry from taking "reasonable risks" because they imposed implacable performance guarantees. He blamed parsimonious funding on Vietnam War requirements and urged lost ground be regained as soon as priorities permitted. In era of R&D austerity, Neumann advocated "lot more" work in advanced-engine research by NASA. (*AF/SD*, 11/68, 58)
- Secretary of Defense Clark M. Clifford wrote in *Air Force and Space Digest:* "Not too many years ago, the War and Navy Departments were concerned almost exclusively with men and simple machines. Defense industries were regarded as mere munitions-makers. . . . We now have a military-industrial team with unique resources of experience, engineering talent, management and problem-solving capacities, [and it] must be used to help find the answers to complex domestic problems as it has found the answers to complex weapon systems. Those answers can be put to good use by our cities and our states, by our

schools, by large and small businesses alike. The nation will be the better and the stronger." (*AF/SD*, 11/68, 76-7)
- In *Air Force and Space Digest* Capt. Gerald T. Rudolph (USAF) of AFSC Space and Missiles Systems Organization scored lack of progress in adopting systems technology to solve "nation's mounting social problems." He found two major reasons for lag: "the public does not entirely understand [it], largely because the aerospace industry has been unable to define it adequately and explain how it is used. And . . . American society has always been reluctant to accept the kind of centralized authority required to implement systems solutions, especially at the community and regional levels."

    Every effort should be made, he said, to surmount obstacles because "it is the judgment of many experts that systems technology will prove to be the most valuable parcel of knowledge to come out of present day space technology." (*AF/SD*, 11/68, 79-81)
- J. S. Butz, Jr., in *Air Force and Space Digest* article "The Men Behind Soviet Aircraft Design" wrote: "Top Russian designers are exuberant realists; and they operate under a 'prototype' system of development that is probably the most competitive and technically stimulating in the world. This system is similar to the one employed in the United States until the mid-1950s and the odds appear strong that the U.S. will return to such an approach in the 1970s, rejecting the strong dependence on 'systems analysis' and 'cost-effectiveness' studies that characterized the 1960s." (*AF/SD*, 11/68, 62-7)

December 1968

December 1: Tenth anniversary of Antarctic Treaty signed by 12 nations to promote scientific research and to bar military activity in the area. (A&A, 1915–1960)

December 2: Retirement of H. Julian Allen as Director of Ames Research Center, announced Oct. 25, would be deferred, NASA announced. Allen would be Special Assistant to Associate Director, Office of Advanced Research and Technology, and would serve as Acting Director of ARC for indefinite period. ARC Associate Director James F. Parsons, named to be Acting Director after Allen's retirement, had become seriously ill. Parsons died March 2, 1969. (NASA Ann, 12/2/68; ARC Memo from Director to Staff, 3/3/69)

- U.S.S.R. published *Zond VI* photo of 70-mi-wide lunar crater carrying name honoring geneticist Nikolay I. Vavilov, who had been sent to Siberia in 1939 for opposing ideas of Trofim D. Lysenko, and his brother Sergei I. Vavilov, who had later become president of Soviet Academy of Sciences. Nikolay Vavilov died in Siberia in 1943; his brother died in 1951. Soviet name for crater, Brothers Vavilov, had not yet been accredited by International Astronomical Union. (*NYT*, 12/3/68, 36)

- President Johnson presented 1968 Enrico Fermi Award in White House ceremony to Dr. John Archibald Wheeler, Princeton Univ. physicist, for "pioneering contributions to understanding nuclear fission and to developing the technology of plutonium production reactors and his continuing broad contributions to nuclear science." Award carried gold medal, citation, and $25,000. (*PD*, 12/9/68, 1956; *W Post*, 12/3/68, A2; *W Star*, 12/3/68, A6)

- DOD announced issuance of $4,400,000 initial increment to $31,132,689 cost-plus-incentive-fee USAF contract with Lockheed Missiles and Space Co. for launch support services at WTR. (DOD Release 1060–68)

- U.K., West Germany, and the Netherlands had decided to pool secrets and build centrifuge separation plant to produce cheap enriched-uranium reactor fuel, Don Cook reported in *Washington Post*. In 1962 agreement with U.S., the countries had contracted to keep experimentation in centrifuge secret from each other. In addition to U.S., U.S.S.R. and Communist China produced enriched uranium by gaseous-diffusion process. (*W Post*, 12/2/68, A3)

December 3: U.S.S.R. launched *Cosmos CCLVII* into orbit with 438-km (272.2-mi) apogee, 269-km (167.2-mi) perigee, 91.6-min period, and 70.9° inclination. Satellite reentered March 5, 1969. (AP, *B Sun*, 12/4/68; Reuters, *C Trib*, 12/4/68; GSFC *SSR*, 12/15/68; 3/15/69)

- XB–70A experimental supersonic bomber, piloted by NASA test pilots Fitzhugh L. Fulton, Jr., and Donald L. Mallick, reached mach 1.64 and 39,400-ft altitude during 1-hr 58-min flight from Edwards AFB to test exiter vane integrated loads and frequency and handling qualities. All

298

primary objectives were accomplished, although cabin pressure problem necessitated manual operation of system. (XB-70 Proj Off)

- President Johnson proclaimed Agreement on the Rescue of Astronauts, the Return of Astronauts, and the Return of Space Objects in effect with deposit in three capitals of instruments of ratification by U.S., U.S.S.R., and U.K. Ireland and Nepal had previously ratified treaty, which had been signed by 75 countries. Proclamation noted new agreement was essentially humanitarian. Its provisions "carry forward the goals of international cooperation to which the United States has been dedicated since enactment of the National Aeronautics and Space Act of 1958 and to which the United States renewed its commitment in the outer space treaty of 1967." (PD, 12/9/68, 1658; AP, NYT, 12/5/68, 13; NASA Off of General Council)

- President-elect Richard M. Nixon named Dr. Lee A. DuBridge, retiring President of Cal Tech, as his Science Adviser and announced formation of task force under Dr. H. Guyford Stever, President of Carnegie-Mellon Univ., to make recommendations in general science field. Task force headed by Dr. Charles Townes, Nobel Prize winning physics professor at Univ. of California at Berkeley, would recommend in space field.

  Dr. DuBridge described as a "very critical matter" loss of momentum in U.S. space program because of appropriations cuts and said reducing activities in basic and applied research damaged nation's future.

  Later, National Science Board Chairman, Dr. Philip Handler of Duke Univ., said of Dr. DuBridge, "He has one rare attribute . . . wisdom, an unusual commodity. He is knowledgeable about the role of technology in our society and has a deep belief in the importance of fundamental research to the health and welfare of the country. And he is usually understanding of the problems of the private universities." Horner, W Star, 12/3/68; Cohn, W Post, 12/3/68, A7; SBD, 12/4/68, 141; W Star, 12/4/68, A13)

- President Johnson presented Harmon International Aviator's Trophy to Maj. William J. Knight (USAF) for piloting X-15 rocket research aircraft to unofficial record speed of 4,520 mph (mach 6.70) Oct. 3, 1967. During White House ceremony President said, "What we learned from the X-15 program will enable us to improve on all our aircraft. The information gained from Major Knight's 'flying laboratory' will make the airplanes of the future safer, faster, and more efficient." (NASA PAO; W Post, 12/4/68, A3; SBD, 12/5/68, 151; PD, 12/9/68, 1660)

- Gen. Jacob E. Smart (USAF, Ret.), NASA Assistant Administrator for DOD and Interagency Affairs, addressed Armed Forces Management Assn. on NASA Interagency Coordination: ". . . I believe it would be appropriate for the President or the Congress to charge NASA with exercising the lead in a national R&D program that is designed to learn how and to what degree science, engineering, and technology can be applied to help the Departments of Interior, Agriculture, Commerce, Navy, and others to fulfill their responsibilities for development and management of natural resources. . . . National decision on whether we undertake development of operational space systems must await outcome of R&D. Issues will be more clearly identified and understood if the President and the Congress lend their interest and public support to a first class R&D program." (Text)

December 3

- FAA adopted rule proposed Sept. 4 to limit number and type of aircraft operations at five "high-density" airports in New York, Washington, and Chicago and included supplemental carriers in same category as scheduled carriers. Final rule, effective April 27, 1969, dropped proposed requirement that all aircraft operating instrument flight rule (IFR) at high-density airports have minimum two-pilot crew and be able to maintain 150-knot minimum airspeed. (FAA Release 68–76; W Star, 12/3/68, A1; Eisen, W Post, 12/3/68, A1; Witkin, NYT, 12/4/68, 1)
- Cal Tech said Mt. Wilson and Mt. Palomar observatories were keeping close watch for developing sunspots that might threaten Apollo 8 astronauts with dangerous x-ray radiation. (Reuters, W Post, 12/5/68, H9)

December 4: USAF launched unidentified satellite from Vandenberg AFB by Titan III–B booster into orbit with 454-mi (730.6-km) apogee, 94-mi (151.3-km) perigee, 89.3-min period, and 106.2° inclination. Satellite reentered Dec. 12. (*Pres Rpt 68*)

- Nike-Apache sounding rocket launched by NASA from WSMR carried Dudley Observatory payload to 95.8-mi (153.2-km) altitude to determine particle flux during quiet period preceding Geminid meteor shower (Dec. 13–14). Launch was first in series of two; second would be launched Dec. 14. Rocket and instruments performed satisfactorily. All collection trap functions occurred as programmed. Payload was tracked visually during descent and sucessfully recovered. (NASA Rpt SRL)
- Eugen Sänger Medal of German Society of Aeronautics and Astronautics was awarded to U.S. X–15 research aircraft team in Bonn, West Germany, for X–15 program's contribution to advancement of space flight. Medal honored German rocketry leader, Dr. Eugen Sänger, first to define potential of hypersonic rocket aircraft, who died in 1964. John V. Becker of LaRC accepted award on behalf of NASA–USN–USAF team. (NASA Release 68–206)
- Dr. Abe Silverstein, Lewis Research Center Director, received one of six $10,000, 1968 Rockefeller Public Service Awards—highest privately sustained honor for Federal career service—in Washington, D.C., ceremonies. (Causey, W Post, 12/5/68, A22)
- In letter to National Council on Marine Resources and Development, President-elect Richard M. Nixon said his administration would "consider an integrated program in oceanology a first priority." (AP, NYT, 12/8/68, 74)

December 5–30: NASA successfully launched ESRO's 240-lb *Heos–A* Highly Eccentric Orbiting Satellite from ETR by Thrust-Augmented Thor-Delta (DSV–3E) booster. Drum-shaped satellite entered orbit with 138,831.2-mi (223,579.4-km) apogee, 263-mi (423-km) perigee, 112-hr 29-min period, and 28.3° inclination.

Primary NASA objective was to place *Heos–A* into earth orbit which would permit successful achievement of ESRO scientific objectives and to provide spacecraft tracking and telemetry support. Eight onboard experiments designed by nine scientific groups in Belgium, France, West Germany, Italy, and U.K. would investigate interplanetary magnetic fields and study solar and cosmic-ray particles outside magnetosphere during period of maximum solar activity. By Dec. 30 seven experiments had been turned on and operated satisfactorily. Eighth experiment would be activated two months after liftoff. Spacecraft control had shifted to ESRO Operations Center, Darmstadt, Germany.

Heos–A was third ESRO mission orbited by NASA, first ESRO mission on Delta launch vehicle, and first NASA–ESRO reimbursable mission, in which ESRO would pay U.S. for launch vehicle hardware and certain launch costs. (NASA Proj Off; NASA Release 68–204; Wilford, *NYT*, 12/6/68; GSFC *SSR*, 12/15/68)

December 5: Interview with Dr. Donald F. Hornig, Science Adviser to President Johnson, appeared in *New York Times*. Dr. Hornig said problem facing his successor, Dr. Lee A. DuBridge, was finding funds for science and technology, to preserve "the vitality, the creativity and the entrepreneurial surge of the establishment." Next science adviser would have to orient new administration to give science proper place in overall scheme. One of America's great strengths was recognition that science, like economics, was part of everything.

New administration would have to set course for space program after manned lunar landing. It would have to decide how to improve nation's health care, deal with environmental pollution, and face technological problems of keeping world free of nuclear war. Particularly important task would be advancing vitality of basic science enterprise—research devoted to quest for knowledge with no immediate applications in mind. (Schmeck, *NYT*, 12/5/68, 28)

- *Wall Street Journal* editorial said major task of new Presidential Science Adviser, Dr. Lee A. DuBridge, would be "to try to help instill some semblance of order in the Government's massive involvement in science." While DOD and NASA had been biggest science spenders, other agencies were extensively involved. "Because there was such a limited effort to set priorities in the first place, it's hard for anyone to know for sure which programs the nation now can best afford to slow down or abandon. . . . [Dr.] DuBridge now must try to plan the future course of Federal science, somehow reconciling all the competing pressures from scientists, universities, industry, Congress, maybe even taxpayers." (*WSJ*, 12/5/68)

- NASA Deputy Administrator for Space Sciences and Applications Oran W. Nicks described applications of Surveyor and Lunar Orbiter techniques to Mars exploration before AIAA meeting in Washington, D.C. Two 1971 Mariner Orbiter missions would provide new tools for survey of dynamic Mars, arriving when most striking seasonal changes were evident in Southern Hemisphere. Their combined orbits and life expectancies would allow observations of dynamic changes in clouds and surface features over several months. In 1973, two additional orbiters would survey Mars from different orbits during different seasons, with support of landing mission as prime requirement. Also planned for 1973 was survivable landing spacecraft to make local measurements of environment, photograph surface and topography, and analyze soil.

"Burning question of immediate concern to you and me will be addressed by use of our new tools," Nicks said: "Is there life elsewhere? Has life existed on nearby planets and disappeared for any reason? Can nearby planets be made suitable for life?' Together, orbiters and landers form a powerful team for the study of Mars and for seeking answers to these questions. Together, they will continue to extend our capabilities in what is probably the most challenging, open-ended arena for expansion of science and technology in the decade ahead." (Text)

- NASA announced plans to use Titan III–D/Centaur in dual launches of 6,000-lb spacecraft consisting of Surveyor-type soft landers mated to Mariner 1971 class Mars orbiters in mid-1973, in Project Viking, to obtain scientific information on life on Mars. Launched 10 days apart, spacecraft would arrive about seven months later, when orbiter propulsion systems would place orbiters and landers in Mars orbit. After reconnaissance, landers would be detached and softland on Mars. (NASA Release 68–207; Pasadena *Independent*, 12/6/68)

- LeRC announced NASA-developed artificial-heart control system delivered to Cleveland Clinic, Cleveland, Ohio, would aid research into ways man-made ventricle could assist damaged natural heart. Later, system could be modified to control artificial replacement heart. Electronic R-wave detector sorted heartbeat from background noise in electrocardiogram to synchronize beat in artificial heart with natural heart and gave synchronization signal to programmer, which could adjust pumping of artificial ventricle to needs of body. (LeRC Release 68–80)

- ComSatCorp announced it had placed in full-time commercial operation two new earth station facilities at Jamesburg, Calif., and Paumalu, Hawaii, to handle U.S.-Pacific area communications via *Intelsat–II F–4* (Intelsat II–D) satellite. (ComSatCorp Release 68–66)

- Scientists at Univ. of California at Los Angeles, testing to determine if life could exist in hostile environment of Venus, said they had grown algae in atmosphere of 100% carbon dioxide. They said Venus had more than 90%. More complex plants had not survived, reported Dr. Willard F. Libby, Nobel Prize winner in chemistry, Dr. Joseph Seckbach of Israel, and Dr. Irene Aegerter of Switzerland. (AP, *NYT*, 12/8/68, 35)

- DOD announced USAF scientists, in study conducted by Air Force Cambridge Research Laboratories, would fly KC–135 flying ionospheric laboratory around auroral oval in Arctic during seven-day series of first flights attempted along this route. Purpose was to obtain data on spatial extent and intensity of optical and radio auroras in midwinter and define relationships between them. USAF was interested in influence of auroras on radio communication and navigation and on radar surveillance. AFCRL scientists hoped, eventually, to predict occurrence and severity of auroral activity. (DOD Release 1065–68)

- *Washington Post* editorial commented on FAA's limitation of operations at five major airports [see Dec. 3]: "It is unfortunate, of course, that the capacity of airports at New York, Chicago and Washington have not kept up with the growth of the Nation's air fleet. But airports do have limits, and it is growing increasingly clear that the only solution to congestion problems is to separate commercial and noncommercial traffic. It would make far more sense for those who feel they have been unfairly treated by the FAA's action to campaign for more reliever airports around large cities (and the taxes on airport users to pay for them) than to attempt . . . to overturn the FAA decision." (*W Post*, 12/5/68, A20)

- USN announced authorization of $22 million to Electric Boat Div. of General Dynamics Corp. for planning and procurement of materials and equipment for "quiet" nuclear-powered submarine. Award brought total allocation thus far to $30.5 million. Initial $8.5 million was for design. (UPI, *W Post*, 12/5/68, H7)

December 6: Aerobee 150 MI sounding rocket launched by NASA from WSMR carried American Science and Engineering, Inc., payload to 92.6-mi (149-km) altitude to collect data on celestial x-ray sources in 1- to 20-kev range. Rocket and instruments performed satisfactorily. (NASA Rpt SRL)

- NASA announced first successful orbital operation of two low-thrust space engines during five separate tests in two-month lifetime of *Ats IV* spacecraft. Ion, or electrical engines, producing thrust of less than 20 micropounds, had been fired for total 23 hr and performed perfectly. Ion engines for future spacecraft were ideal systems for countering gravitational attraction between spacecraft, moon and sun, and elliptical equatorial section—to keep satellite stationary for conducting communications, meteorological, and navigational experiments. They had controllable thrust level and direction, higher fuel efficiency, and longer fuel life than Chemically propelled engines. Reservoir of $^1/_{10}$ lb cesium could keep synchronous satellite stationary for more than three years. Before test, propulsion experts had expected ion engine firings might create radio frequency interference. Test showed no interference. Two more ion engines were planned for testing aboard ATS–E, scheduled for August 1969 launch. (NASA Release 68–205)

- In major organizational changes at Marshall Space Flight Center, Dr. W. R. Lucas, Director of Propulsion and Vehicle Engineering Laboratory, was named Director for Program Development, effective Dec. 16; Dave Newby was named Director of Center Operations; and Dr. Ernst Stuhlinger, Director of Space Sciences Laboratory, was named Associate Director for Science. All were new positions. MSFC Research and Development Operations would become Science and Engineering, with Hermann K. Weidner remaining Director; Industrial Operations would become Program Management under same director, E. F. O'Connor.

  Dr. Wernher von Braun, MSFC Director, said major function of new Program Development organization would be to "harden" complete package plans for new programs, such as space station. Director for Program Development would "help chart the course for this Center in the post-Apollo period, keeping in mind our Apollo program requirements and AAP obligations." (MSFC Release 68–276; *Marshall Star*, 12/11/68, 1)

- *Apollo 7* Astronaut Walter M. Schirra, Jr., received Kitty Hawk Memorial Trophy for distinguished achievement in aeronautics at annual Wright Brothers memorial banquet in Beverly Hills, Calif. Banquet commemorated 65th anniversary of flight of *Kitty Hawk*. (UPI, W *Star*, 12/7/68, A2)

- ComSatCorp filed application for authority to construct earth station for satellite communications in Alaska, 90 mi north of Anchorage near Talkeetna Village. (ComSatCorp Release 68–68)

- In *Washington Post*, Thomas O'Toole said Howard R. Hughes organization's holding company, Hughes Nevada Operations, had requested public disclosure of alleged DOD study by Harvard Univ.'s Dr. George Kistiakowsky and MIT's Dr. Jerome Wiesner and James Killian. Company claimed *Special Report on Underground Testing* concluded there was no reason for continued underground atomic tests. DOD spokesman denied existence of study, while Science Adviser to President Johnson, Dr. Donald F. Hornig, said only that he had met with top physicists in November to discuss safety of underground testing. "We may make a public report on it. The subject seems to be of enough importance to warrant a public airing." (*W Post*, 12/6/68, 87)

December 7–10: NASA's *Oao II* (OAO–A2) Orbiting Astronomical Observatory was successfully launched from ETR by two-stage Atlas-Centaur booster. Orbital parameters: apogee, 485.7 mi (781.6 km); perigee, 479.2 mi (771.2 km); period, 100.4 min; and inclination, 35°. Heaviest and most complex automated spacecraft ever developed by U.S., *Oao II* was 7 ft wide and 10 ft high, weighed 4,400 lb, carried 11 telescopes, and contained 238,000 separate parts.

Primary mission objective was to demonstrate flight operation to support two experiments provided by Smithsonian Astrophysical Observatory and Univ. of Wisconsin for at least 30 days. Secondary objective was to obtain scientific data over a range of star magnitudes and wavelengths for at least 50 hr. Experiments would observe interstellar dust and extremely young hot stars in UV portion of spectrum not visible to human eye or earth-based observatories. Satellite would be able to collect six hours of UV data per day—twice as much as had been obtained in 15 yr from 40 sounding rocket launches. Through its complex ground-command spacecraft-attitude system *Oao II* would be aimed at individual objects in space with precision never before attained by an orbiting satellite. Information from experiments would be radioed to earth as digital data for analysis by experimenters.

By Dec. 10 all spacecraft equipment had been turned on and operated satisfactorily: satellite had been placed in its sunbathing mode with its solar panels oriented toward sun; its six star trackers had been activated; three of the trackers had been locked on to preplanned guide stars; and satellite had established three-axis stabilization. Smithsonian experiment initial power had been turned on. Wisconsin Experiment Package would be turned on Dec. 11.

Oao II was second in series of four spacecraft in NASA's OAO program to obtain precise astronomical observations of celestial objects above earth's atmosphere. *Oao I* had been launched into almost perfect orbit April 8, 1966, but had failed because of power supply system malfunction and probable high voltage arcing in star tracker. OAO program was managed by GSFC under OSSA direction. (NASA Proj Off; NASA Release 68–186K; KSC Release KSC–68; UPI, W *Star*, 12/8/68, A5; O'Toole, *W Post*, 12/8/68; AP, W *Star*, 12/9/68, A6; *SBD*, 12/10/68, 172; Sehlstedt, B *Sun*, 12/12/68, A5)

December 7: Apollo 8 Astronauts James A. Lovell, Jr., William A. Anders, and Frank Borman—scheduled to begin lunar orbital mission Dec. 21—held preflight press conference at MSC. Borman, comparing mission to "a combat tour in Vietnam," said: "The risks we take are acceptable ones. . . . We have to accept them if we believe it's worthwhile. If I ever feel it's not worthwhile, I'll quit."

Describing the magnificence of space flight, Borman said, "When you're finally up at the moon and looking back at the world, the nationalities blend. . . . You get the concept that this really is one world and wonder why . . . we can't live together like decent people." Anders said crew "might use the opportunity [of being in lunar orbit Christ-

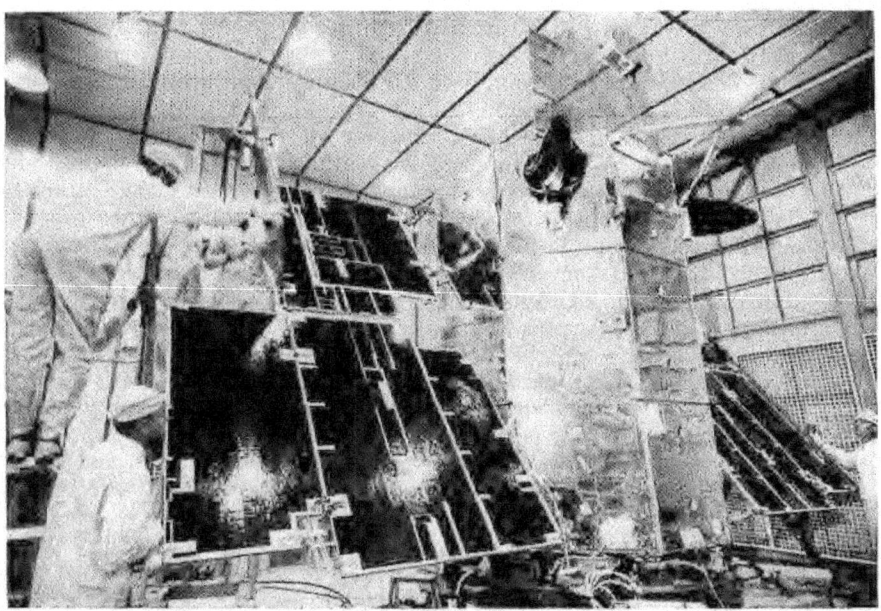

December 7–10: NASA's Oao–II 21-foot-wide, 4,400-pound Orbiting Astronomical Observatory carried 11 telescopes into orbit to seek answers to questions on origin, evolution, and future of universe. Photo was made in final flight qualification testing.

mas Day] to express to all peoples of the world our purposes and the benefits we can expect from space exploration." (Lannan, W Star, 12/8/68, A4; AP, KC Star, 12/8/68; AP, M Her, 12/8/68)

December 8: Chief Test Pilot Joseph S. Algranti successfully ejected from lunar landing training vehicle about four minutes into planned six-minute flight, when large lateral-control oscillation developed as he descended from maximum altitude of 550 ft. He ejected at 200 ft and landed by parachute while $1.8-million vehicle crashed and burned several hundred feet away. Flight was 14th for this LLTV. Astronaut Walter M. Schirra, Jr., was named chairman of board to investigate accident. (MSC Special Releases; W Post, 12/9/68, 1; SBD, 12/10/68, 172)

• In *New York Times*, Walter Sullivan said President-elect Richard M. Nixon's entourage had discussed creation of Cabinet post for science and technology. "The science adviser, in the view of the Nixon entourage, has been unable to streamline the machinery for making science policy. Science, and the problems relating to it, has outgrown its old boundaries. The big problems are interdisciplinary. Their solution requires expertise in many fields, and the cooperation of many departments of government. This has led to a proliferation of interdepartmental committees." However, "because the creation of a Cabinet post would have major repercussions within the scientific establishment, it is unlikely that such a step will be taken until there has been an extensive study, lasting perhaps a year or more." (NYT, 12/8/68, 3E)

December 9: NASA's HL–10 lifting body vehicle, piloted by Maj. Jerauld R. Gentry (USAF), successfully completed second rocket-powered flight from Edwards AFB reaching mach 0.8 (550 mph) at altitude approaching 50,000 ft. (NASA Proj Off; AP, B *Sun,* 12/10/68, 5)

- At White House Dinner honoring members of space program, President Johnson presented Presidential Medal of Freedom, Nation's highest civilian award, to retired NASA Administrator, James E. Webb, "a most distinguished public administrator, . . . a farsighted and forceful leader of this Nation in the pioneer exploration of outer space, opening new frontiers of discovery and progress for the American people."

  President told 23 Apollo astronauts, NASA officials, Government leaders in space program, and heads of rocket-building firms, "I asked you to come here tonight in the twilight of this administration, so I could pay the respect and the honor and the affection that I felt for the man who has directed your efforts and directed them so well, and so that I could express my personal admiration and respect for you."

  In response, Webb said, "The challenge of space is large and so is NASA. In all such human endeavors, organized institutional efforts are essential, and we know, in the words of Emerson, that they are 'the lengthened shadow of one man.' We in NASA know, Mr. President, that you are the man of which our civilian space effort, conducted for the benefit of mankind, is the lengthened shadow."

  At predinner ceremony, document for White House Treaty Room was signed by *Apollo 7* astronauts Walter M. Schirra, Jr., R. Walter Cunningham, and Donn F. Eisele; *Apollo 8* Astronauts William A. Anders, Frank Borman, and James A. Lovell, Jr.; and pioneer aviator Charles A. Lindbergh. Also in Treaty Room were commemorations of May 21, 1963, visit to White House by Astronaut L. Gordon Cooper following 22-orbit mission in *Faith VII* May 15–16 and of June 17, 1965, award of Exceptional Service Medal to late Astronaut Edward H. White II, and Astronaut James A. McDivitt following first U.S. space walk during June 3–7 *Gemini IV* mission.

  During exchange of dinner toasts President Johnson said that "in the hundreds of laws on which I have answered the rollcall, the bills that I have sponsored or cosponsored or amended or defeated, there is not a single one that gives more pride than the Space Act." Responding, Webb cited three generations of spacecraft in 10 yr, "put to use . . . in every major field," and said he had "strongly held view that in the kind of world we live in, our Nation needs this kind of success in this kind of endeavor." (*PD,* 12/16/68, 1689–91; McCardle, *W Post,* 12/10/68, D1; Shelton, *W Star,* 12/10/68, B6; *NYT,* 12/10/68, 86)

- World Meteorological Organization, U.N. agency managing World Weather Watch project, said work was well advanced, according to UPI. System would be fully operational in 1971 and save world economy estimated $17 billion annually in losses caused by unexpected weather changes. System called for 29,000 observations daily, of which 24,000 were already being made. Additional 2,500 by 1971 would bring implementation level to 91%. World centers at Melbourne, Moscow, and Washington would be computerized by 1969, increasing daily output of 134 charts to 223 by 1971. Additionally, 21 regional centers issuing 1,191 charts daily would increase output to 1,830 charts by 1971 and, eventually, would be linked to global communications system. Observa-

December 9: Apollo 7 and 8 *flight crews sign memorial document for White House Treaty Room. Signers are (left to right): Astronauts R. Walter Cunningham, Donn F. Eisele, Walter M. Schirra, Jr. (Apollo 7); William A. Anders, James A. Lovell, Jr., and Frank Borman (Apollo 8). Standing are Charles A. Lindbergh (who also signed), Mrs. Johnson, President Lyndon B. Johnson, retired NASA Administrator James E. Webb, and Vice President Hubert H. Humphrey.*

tions would be made by land surface stations 300 mi apart throughout world with exception of desert areas. Further data would be passed on by weather ships, upper-air sea stations, aircraft, and meteorological satellites. (W *Star*, 12/9/68, A8)

December 9–12: New York Times and Washington *Evening Star* published contents of unreleased draft report by Presidential task force appointed in 1967 to formulate national communications policy dealing with rapid technological changes and providing for adequate Government supervision. Report would recommend reorganization of U.S. communications industry to include Government-sponsored monopoly to transmit all international communications—including ComSatCorp satellites and ground stations; AT&T undersea cables; and terminals and switching stations of "record" carriers ITT World Communications, Inc., RCA Communications, Inc., and Western Union International, Inc. If ComSatCorp became single international "entity," committee's recommendations would preclude it from becoming owner and manager of domestic satellite communications system for which the report would propose pilot program.

Committee claimed single entity could make more balanced investment choice on whether to lay more cables or launch satellites and would eliminate need for duplicate transmission facilities. Conclusions were challenged in dissenting footnote to report by Dr. Edward C. Welsh, Executive Secretary of NASC, who said merger would inhibit development of satellite technology and reduce technological competition between cables and satellites that could result in lower rates.

Report also would recommend informal merger of postal and telegraph services, with Western Union Telegraph Co. permitted to operate in post offices; relaxation of FCC restrictions on cable TV to protect broadcasters; and Government sponsorship of experimental program to test TV's usefulness in assisting minority groups. (Finney, *NYT*, 12/9/68, 1; 12/10/68, 1; Aug., W *Star*, 12/12/68, A13)

December 10: Cosmos *CCLVIII* was successfully launched by U.S.S.R. into orbit with 302-km (187.6-mi) apogee, 206-km (128-mi) perigee, 89.5-min period, and 64.9° inclination. Satellite reentered Dec. 18. (UPI, W *Star*, 12/10/68, A6; AP, *NYT*, 12/11/68, 6; GSFC *SSR*, 12/15/68; 12/31/68)

- Javelin sounding rocket launched by NASA from Churchill Research Range carried Syracuse Univ. Research Corp. payload to 520-mi (837-km) altitude to observe ionization levels of helium in exposure. Data would be related to aspects of sun and earth relative to payload. Rocket and instruments performed satisfactorily. (NASA Rpt SRL)

- NASA announced award of $1,046,123 contract to Thiokol Chemical Corp.'s Elkton, Md., Div. for development of more powerful solid-fuel rocket for automated missions, including possible use as 3rd stage for Centaur and Delta. New 3rd stage was expected to be ready for use in 1971. (NASA Release 68-210)

- NASA announced joint 1969 project with German Federal Ministry for Scientific Research (BMWF) to photograph earth's magnetic lines of force high in space would involve release of barium vapor by NASA Scout rocket at 20,000-mi altitude. Scientists would use special optical equipment to view resultant glowing ionized cloud along magnetic field lines and visually map electromagnetic forces acting in barium area. Agreement on project called for BMWF to provide payload, two ground observer stations, and data analysis. NASA would furnish rocket, conduct launch from NASA Wallops Station, and provide tracking and communications services. (NASA Release 68-211)

- MSC had awarded $16.4 million, one-year extension to cost-plus-award-fee contract with Lockheed Electronics Co. Div. of Lockheed Aircraft Corp., NASA announced. Award was for general electronic, instrumentation, and engineering support services, bringing total contract value to $46.4 million. (NASA Release 68-209)

- *Washington Daily News* editorial commented on statement by physicist Dr. Ralph E. Lapp warning of possible dangers in Apollo 8 mission [see Dec. 14]: "There are perhaps sound reasons involving national prestige for trying to be the first nation to send men into a moon orbit. But surely no such reasons are compelling enough to cut corners on safety. The technical arguments advanced by Dr. Lapp are far too complex to be resolved by laymen. But after the tragic fire that took the lives of three of our Apollo spacemen two years ago it should not be necessary to urge that the National Aeronautics and Space Administration exercise all due prudence—even at the risk of losing the race around the moon." (*W News*, 12/10/68, 24)

- Secretary of Defense Clark M. Clifford announced DOD had completed Project 693 (a $3-billion cutback in FY 1969 expenditures required by Revenue and Expenditure Control Act of 1968) including $85-million reduction in MOL program. (DOD Release 1083–68; *SBD*, 12/11/68, 176)

- Soviet Finance Minister Vasily Garbuzov told semiannual budget meeting of Supreme Soviet in Moscow that nominal U.S.S.R. defense spending in 1969 would increase 6%, from 16.7 billion to 17.7 billion rubles. Institute of Strategic Studies in London had estimated before Aug. 20 invasion of Czechoslovakia that real Soviet military spending for 1968 would be approximately equivalent to $50 billion. Thus, announced increases would bring total 1969 budget to some $53 billion, as against $80 billion for U.S., which had allocated $30 billion for costs of Vietnam War. However, Soviet figure did not include military R&D and investment in defense industries.

  Budget also disclosed continued expansion of government support for scientific research. (Shabad, *NYT*, 12/11/68, 1; Shub, *W Post*, 12/11/68, A1)

December 11: President-elect Richard M. Nixon introduced his Cabinet on nationwide TV from Washington, D.C. Among appointments, William P. Rogers, Washington attorney and Eisenhower Administration Attorney General, would serve as Secretary of State; Rep. Melvin R. Laird (R-Wis.), as Secretary of Defense; and Massachusetts Gov. John A. Volpe, as Secretary of Transportation. (Herbers, *NYT*, 12/12/68)

- Apollo 8 crew, wearing their spacesuits, participated in final 2 hr 45 min of countdown rehearsal for Dec. 21 launch. Spacecraft had completed four-day rehearsal fully fueled Dec. 10. (AP, B *Sun*, 12/11/68, A9; UPI, W *Star*, 12/11/68, C4)

- NASA was unsuccessful in second Project SHAPE (Supersonic High Altitude Parachute Experiments) attempt at WSMR when parachute was ejected prematurely from five-foot-long canister after three-stage rocket had successfully propelled payload to 33-mi altitude. First test Oct. 23 had been successful. (NASA Release 68–216)

- Fédération Aéronautique Internationale (FAI) had established Yuri Gagarin gold medal honoring cosmonaut who became first man in space April 12, 1961, during Soviet *Vostok I* mission, *Space Business Daily* reported. Medal would be awarded annually to pilot contributing best performance of year in peaceful exploration of space. (*SBD*, 12/11/68, 178)

- Cal Tech's Dr. Maarten Schmidt received Rumford Premium, nation's oldest science award for "the most important discovery or useful improvement . . . on heat and on light" at AAAS dinner in Boston, Mass. Award, established in 1796 by Benjamin Thompson, Count Rumford, consisted of medal and $5,000. Dr. Schmidt had determined intense radio emission of quasars indicated they were moving away from earth at speeds up to 149,000 mps, or about 80% of speed of light. (AP, W *Star*, 12/12/68, A2)

December 12: USAF launched two unidentified satellites from Vandenberg AFB by Thor-Agena D booster. One entered orbit with 148-mi (238.3-km) apogee, 109-mi (175.4-km) perigee, 88.6-min period, and 81.0° inclination and reentered Dec. 28. Second satellite entered orbit with 916-mi (1,474.1-km) apogee, 862-mi (1,387.2-km) perigee, 114.4-min period, and 80.3° inclination. (*Pres Rpt 68*)

- Two Nike-Cajun sounding rockets launched by NASA from Churchill Research Range carried GSFC experiments to 75-mi (120-km) and 78-mi

(124.8-km) altitudes to obtain temperature, pressure, density, and wind variation data in atmosphere between 21.8- and 59-mi (35- and 95-km) altitudes by detonating grenades at various altitudes and recording their sound arrival on ground. Grenades exploded as planned. Launches were coordinated with series at Point Barrow, Alaska. (NASA Rpts SRL)

December 13: Explorer XXXIX (Air Density Explorer C) and *Explorer XL (Injun V)*, launched as dual payload Aug. 8, were adjudged successful by NASA. *Explorer XXXIX* balloon had been ejected and fully inflated. *Explorer XL* had despun, extended booms, and turned on experiments satisfactorily. Malfunction in spacecraft's power subsystem had caused loss of one-third of power generated by solar panels, reducing quantity but not affecting quality of data. Satellite would enter full sunlight Sept. 18, 1969, permitting originally planned data-acquisition rate. (NASA Proj Off)

- NASA announced agency and DOD had agreed jointly to make computer programs available to industry, educational institutions, scientific and technical organizations, and others through Computer Softwear Management and Information Center (COSMIC). Established in 1966 under NASA contract at Univ. of Georgia, COSMIC listed some 350 computer programs for sale at fraction of cost. More than 100 DOD programs, with additional NASA programs, would be added during 1969.

  Part of NASA Office of Technology Utilization effort to encourage secondary use of aerospace R&D results, COSMIC offered design information for electronic printed-circuit boards and programs for such jobs as inventory control, accounting, data control, stress analysis, equipment checkout, and structural testing. (NASA Release 68–212)

- Cal Tech trustees announced choice of Dr. Harold Brown, Air Force Secretary, to replace Dr. Lee A. DuBridge, who resigned effective Jan. 20 to become Science Adviser to President-elect Richard M. Nixon. (UPI, *W Star*, 12/14/68, A3; *W Post*, 12/15/68, A6)

- Naval Ship Systems Command had awarded $13.5-million contract to Todd Shipyards Corp. for first of nine oceanographic ships of radically new design. Prototype would provide USN with its first catamaran-style hull. Diesel-propelled, 246-ft-long ship would be designated GOR–16 (for general ocean research). (*NYT*, 12/13/68, 86)

December 14: NASA's *Oao II*, launched Dec. 7, photographed three unnamed stars in Draco Constellation 2,000 light years from earth, near Vega between Big and Little Dippers—to make first UV photos of stars. Photos were taken by telescopes of Smithsonian Astrophysical Observatory experiments aboard satellite and used UV radiation which did not penetrate earth's atmosphere. They gave GSFC astronomers enough information to reconstruct fairly detailed images of stars, GSFC spokesman said. (OSSA; UPI, *NYT*, 12/15/68, *SBD*, 12/17/68, 208)

- U.S.S.R. launched *Cosmos CCLIX* from Kapustin Yar into orbit with 1,331-km (827-mi) apogee, 213-km (132.4-mi) perigee, 100.3-min period, and 48.4° inclination. Satellite reentered May 5, 1969. (GSFC SSR, 12/15/68, 210; *SBD*, 12/17/68, 210; 3/15/69)

- Nike-Apache sounding rocket launched by NASA from WSMR carried Dudley Observatory payload to 98.9-mi (159.2-km) altitude to collect micrometeoroids during peak of annual meteor shower. Particles would be studied to determine chemical composition, size distribution, numerical density in upper atmosphere, and crystal structure.

Control experiment had been launched successfully Dec. 5. Although two doors failed to operate properly, scientific objectives were accomplished and payload was recovered. (NASA Rpt SRL)

- *New Republic* reported interview with Dr. Ralph E. Lapp, physicist and assistant laboratory director of World War II Manhattan Project. Dr. Lapp said: "We are pushing our luck, gambling that everything will work perfectly" on Apollo 8 mission. He advised delay in program so that Apollo 9, configured to have relief-capability, would be on pad ready for launch should Apollo 8 "run into trouble." Asked if U.S. could afford to delay Apollo 8, Dr. Lapp said, "The basic factor is not really technical. We are racing the Russians to the moon. A lot of people in NASA and in industry are hoping that a successful Apollo-8 orbiting of the moon—or even circumnavigation—will build up public support for an invigorated manned space program. It's just one of the weighty techno-decisions facing Mr. Nixon. He is committed to funding out the Apollo program—but post-Apollo programs await his decision." (AP, W *Star*, 12/9/68; *W Post*, 12/9/68, 3; *New Republic*, 12/14/68, 16–9)

- Deep sea drilling between U.S. and Africa by scientists of Joint Oceanographic Institutions for Deep Earth Sampling (JOIDES) had confirmed that sea floor was spreading and pushing Europe and America farther apart, John Lannan said in Washington *Evening Star*. Evidence, brought to surface in samples, or cores, extracted from earth's bottom, showed fluid internal mass of molten rock under ocean was still welling up along mid-Atlantic Ridge. Cores indicated "this newly formed crust must spread slowly aside, like a giant conveyor belt, continuously accumulating sediment . . . throughout millions of years as it moves," scientists said. (W *Star*, 12/14/68, A1)

December 15–19: NASA successfully launched *Essa VIII* (TOS–F), eighth meteorological satellite in ESSA's Tiros Operational Satellite (TOS) system, from WTR by two-stage, Thrust-Augmented, Long-Tank Thor-Delta booster. Primary NASA mission objective was to provide global cloud coverage on regular, daily basis with six-month nominal and three-month minimum lifetime. Satellite achieved nearly polar, sun-synchronous circular orbit with 903.4-mi (1,453.6-km) apogee, 874.6-mi (1,407.2-km) perigee, 114.6-min period, and 101.9° inclination.

An advanced version of cartwheel configuration, 300-lb *Essa VIII* carried two Automatic Picture Transmission (APT) cameras which would photograph earth's cloud cover and immediately transmit pictures to local APT stations in 52 nations. During first 22 orbits spacecraft underwent orientation maneuver to place it in wheel mode and spin rate was adjusted by 0.2 rpm. By Dec. 19 all spacecraft systems had been successfully programmed and excellent pictures had been read out.

ESSA financed and managed TOS system and would operate spacecraft after NASA completed checkout later in month. GSFC was responsible for procurement, launch, and initial checkout of spacecraft in orbit. *Essa VIII* was 18th Tiros satellite launched successfully since *Tiros I*, first weather satellite, April 1, 1960. All from *Tiros III* on had equaled or exceeded designed operation lifetimes. Most recent ESSA satellite launched was *Essa VII*, launched Nov. 10. (NASA Proj Off; ESSA Release ES 68–67; *SBD*, 12/17/68, 210)

December 15: Apollo 8 launch crew began lengthy countdown on time at

December 15

KSC at 7:00 pm EST, for launch scheduled for 7:51 am EST Dec. 21. (*W Post*, 12/16/68, A12; *W Star*, 12/16/68, A5; *W News*, 12/16/68, 3)

- NASA announced millions of home TV viewers in U.S., Europe, and Japan would see live pictures taken by *Apollo 8* crew with cigar-box-size camera similar to that carried on *Apollo 7* mission. They would be beamed to earth from spacecraft six times during *Apollo 8* mission, including twice while spacecraft was in moon orbit. Manned Space Flight Network stations near Madrid, Spain, and Goldstone, Calif., would convert slow-scan signal into TV picture. Still photos would be taken of TV monitor during live transmission and pictures released in Los Angeles, Madrid, and Canberra. (NASA Release 68–214)

- Scientific team headed by Nobel Prize winning physicist, Dr. Charles H. Townes, announced discovery of ammonia molecules in direction of Sagittarius toward Milky Way center, 30,000 light years away from earth. Finding was made through spectrographs at radioastronomy observatory of Univ. of California at Berkeley during studies supported by NASA, Office of Naval Research, and NSF. Ammonia was considered chemical ancestor of organic compounds and necessary step toward origin of life. Team said discovery "marks the first time that a relatively complex molecular compound has been definitely identified in the vast regions between the stars." It would spur intensified search for additional combinations of life-essential elements detected in space.

  Former MIT provost and inventor of maser, which led to development of laser, Dr. Townes had been named Dec. 3 to head President-elect Richard M. Nixon's task force to make recommendations on space program. Report would be published in *Physical Review Letters* of American Institute of Physics. (UPI, *NYT*, 12/16/68, 93; AP, *W Post*, 12/16/68, A1)

- In Washington *Sunday Star* William Hines said NASA announcement of plans for 1974 Mars landing marked revival "of a project that was sacrificed last year on the altar of the great God Apollo whose manned moon program was gobbling up all the funds at NASA's disposal. It also marked the first positive action by NASA in two years directed to starting rather than terminating activities." Its timing, one month after presidential election, seemed "to provide a tantalizing clue to the Nixonian philosophy about space." Hines said it was likely new administration would heed post-Apollo views of NAS which had recommended more attention to scientific exploration of space with instruments. (W *Star*, 12/15/68, F4)

- In *Washington Post* Thomas O'Toole said poisoning from overexposure to beryllium powder had resulted in 800 known deaths in U.S. during past 15 yr. Beryllium disease expert, Dr. Harriet L. Hardy of Massachusetts General Hospital, estimated total beryllium cases at 2,500, "about three times as many as we've heard about." New uses for lightweight, heat-resistant metal had pushed production to 150,000 lb per year. Battelle Memorial Institute estimated output would grow at 20% rate per year for next five years; NAS estimated 1979 production at six times 1969's. Beryllium was being used by Lockheed Aircraft Corp. for wheel brakes in C–5A transport and heat shields for Poseidon missile. Boeing Co. used it for new Minuteman missile shield. In past three years estimated $25 million had been spent on beryllium rocket research. O'Toole said one scientist claimed test firing in California had so contaminated site that electrician working there developed beryllium poisoning. Neither NASA nor USAF planned to abandon testing beryllium rockets, however, and beryllium use had "kicked off a lively debate inside the Federal Government." (*W Post*, 12/15/68, A1)

December 16: U.S.S.R. launched *Cosmos CCLX* into orbit with 39,576-km (24,591.3-mi) apogee, 511-km (317.5-mi) perigee, 712.3-min period, and 64.9° inclination. (AP, B *Sun*, 12/18/68, 10; *SBD*, 12/18/68, 211; GSFC *SSR*, 12/31/68)

- NASA announced decision to terminate work on two Biosatellites scheduled for 21-day missions beginning in 1971. Contract with General Electric Co. Reentry Systems Div. would be revised to retain only work on two 30-day primate-experiment Biosatellites scheduled to begin in 1969. Funding for later missions had been reduced and efficient planning made more difficult. Possibilities for experiments with greater flexibility in early 1970s had been indicated by success of smaller satellites and by studies. (NASA Release 68–215)

- FRC said two reports—one by Dr. Eldon E. Kordes, NASA Senior Staff Scientist, to American Society of Mechanical Engineers and one by Chief XB–70 Pilot for FRC, Fitzhugh L. Fulton, Jr., to Flight Safety Foundation—indicated NASA–USAF XB–70 test program was providing valuable information for operation of large supersonic commercial and military aircraft. Results from XB–70's participation in national sonic boom research program had shown methods used to predict overpressure and extent of boom were generally adequate for aircraft of this size and speed under "standard day" conditions but indicated unusual weather conditions and other factors could affect these predictions. XB–70 program was attempting to define intensity of turbulence at higher altitudes. System to improve stability, lessen turbulence-induced accelerations, and improve passenger riding qualities while lengthening aircraft's fatigue life was being tested.

  New methods of presenting flight and engine information to pilot had evolved from XB–70 program, including digital form for more precise readout. Special warning systems prevented exceeding operating limits and attitude indicator with changeable sensitivity for smoother flight control had been evaluated. XB–70 flight had demonstrated need for further research in basic stability and control characteristics of its class of aircraft. Actual cross-country flight experience had been gained with trip from Edwards AFB, Calif., to Carswell AFB, Tex., and return. Both reports emphasized that, although XB–70 was not passenger aircraft, it was similar in size and performance characteristics to proposed SSTs and was only operational aircraft approaching SST size and speed. (FRC Release 28–68)

- Apollo 8 astronauts were pronounced in perfect health by NASA Director of Medical Operations, Dr. Charles A. Berry, after three-hour physical examination as countdown proceeded toward Dec. 21 launch for moon flight. "The crew is in real fine spirits," he added. (AP, B *Sun*, 12/17/68; AP, W *Star*, 12/17/68)

- President Johnson proclaimed Dec. 17 Wright Brothers Day, commemorating aircraft flights made by Orville and Wilbur Wright 65 yr earlier: "Their first journey was shorter than the floor of the giant C–5 cargo ship that was test flown earlier this year. But those brief flights . . . on December 17, 1903, launched the air age. They changed mankind's way of life." (*PD*, 12/23/68, 1718–9; *NYT*, 12/17/68, 16)

December 17: In interview published in *New York Times*, Dr. Lee A. DuBridge, science adviser-designate to President-elect Richard M. Nixon, said problem facing Government was not "Shall Government support and use science and technology, but how shall it do it. What are the priorities. . . ." Fundamental to use of science was knowledge; therefore, "one must have in any modern society a very important and extensive free basic research enterprise establishment, largely in universities, so that new areas of knowledge will be explored." Basic research budget of country should increase at least 10% annually for next few years. When it came to using this knowledge, "the Government can, and somebody must, direct, set up the goals."

On space program, Dr. DuBridge said that "the astonishing discoveries" made by launching instruments into space justified further exploration from scientific point of view. When man entered picture as "another piece of the instrumentation that is needed for the exploration," first effect "is that the costs get large. . . . Apollo landing of a man on the moon is vastly more expensive . . . than the Surveyor landing. . . . By the same token, the information returned will be much greater, too. However, as the technology of the spacecraft improves, and our instruments need to get more complex, heavier, or longer lasting, there may very well be a time when putting a man up will actually be cheaper than trying to use automated instruments." (Sullivan, *NYT*, 12/17/68, 1).

- At Wright Brothers Memorial Dinner, Washington, D.C., sponsored by NAA, Sen. Warren G. Magnuson (D-Wash.) received Wright Brothers Memorial Trophy for "assuring United States preeminence in aeronautics throughout the world." (NAA PAO)
- At Washington, D.C., ceremony, Adm. Thomas H. Moorer, Chief of Naval Operations, USN, presented Distinguished Service Medal to Astronaut Walter M. Schirra, Jr. (Capt. USN), for space deeds "exceeded by no one afloat or airborne." (AP, *W Star*, 12/18/68, B2; *W Post*, 12/18/68, A3)
- Dr. Philip Handler, National Science Board Chairman and only nominee to succeed Dr. Frederick Seitz as NAS president in July 1969, said in interview he would urge Federal program of "bloc grants" to U.S. universities in 1969 to support science. Legislation would be introduced in new Congress, with "something like" $500-million price tag in first year, "just for starters." Universities, "completely dependent upon science project grants," had had funds for individual projects cut off leaving "numerous employees for whom they have no salaries." (Cohn, *W Post*, 12/18/68, A32)
- DOD announced USN had selected Grumman Aircraft Engineering Corp. and McDonnell Douglas Corp. to continue to contract definition phase for F-14A aircraft, formerly called VFX-1. Selection of contractor was scheduled for January 1969. (DOD Release 1109–68; *WSJ*, 12/18/68; UPI, *W Star*, 12/18/68, A6)

December 18–20: Intelsat–III F–2 was successfully launched by NASA for ComSatCorp on behalf of International Communications Satellite Consortium. Launch was from ETR by three-stage, Thrust-Augmented, Long-Tank Delta booster. The 632-lb, cylindrical satellite entered elliptical transfer orbit with 22,590-mi (36,355.1-km) apogee, 161.9-mi (260.5-km) perigee, 642.9-min period, and 28.8° inclination. All systems were functioning normally. On Dec. 20 apogee motor was fired to kick satellite into planned near-synchronous orbit over Atlantic at 51° west longitude with 22,328.2-mi (35,933.8-km) apogee, 21,833.4-mi (33,137.5-km) perigee, 0.79° inclination, and 3.25° per day eastward orbital drift.

Intelsat–III F–2, first successful launch in Intelsat III series, was backup to Intelsat–III F–1 (Intelsat III–A) which had been destroyed minutes after launch Sept. 18 when launch vehicle began to break up. Satellite was scheduled to begin commercial service Jan. 2, 1969 [see Dec. 28], handling up to 1,200 voice circuits or four TV channels. By Jan. 29, 1969, all Atlantic area service except NASA Apollo traffic would be transferred to *Intelsat–III F–2* from other Atlantic comsats. *Intelsat I (Early Bird)* and *Intelsat–II F–3 (Atlantic II)*. Etam, W. Va., earth station would become prime East Coast terminal and Andover, Me., station would be removed from service preparatory to its use as prime terminal for Intelsat–III F–4. NASA Apollo communications would be maintained through *Intelsat–II F–3* with 42-ft terminal antenna at Andover. (NASA Proj Off; ComSatCorp Release 68–69; Stevens, *NYT*, 12/19/68, 1; AP, *W Post*, 12/19/68, A3)

December 18: NASA announced appointment of William C. Schneider, Apollo Mission Director, as Director of Apollo Applications, succeeding late Harold T. Luskin, who died Nov. 25. George H. Hage, Deputy Director of Apollo Program, would be Acting Apollo Mission Director in addition to his present duties. (NASA Release 68–217)

- Apollo 8 astronauts heading for moon would be "in far less hazardous position" than they would have been as crew for Columbus, NASA Director of Manned Space Flight Safety Jerome F. Lederer said in speech before Wings Club in New York. "Columbus did not know where he was going, how far it was, nor where he had been after his return. With Apollo, there is no such lack of information." Nevertheless, mission would "involve risks of great magnitude and probably risks that have not been foreseen.

"Apollo 8 has 5,600,000 parts and one and a half million systems, subsystems and assemblies. Even if all functioned with 99.9 per cent reliability, we could expect 5,600 defects. Hence, the striving for perfection and the use of redundancy." (Text; *NYT*, 12/19/68, 56)

- Aerospace sales reached record high of $30.1 billion in 1968, an increase of almost $3 billion over 1967, Aerospace Industries Assn. President Karl G. Harr, Jr., told Washington, D.C., meeting of Aviation/Space Writers Assn. Commercial aerospace sales increased 39%, to record $6.4 billion; aerospace exports rose 32%, to $3 billion.

Military space programs in 1968 rose 3%, from $1.088 billion in 1967 to $1.121 billion, reported AIA's "1968 Aerospace Industry Review and Forecast," which Harr released. Nonmilitary space sales declined 3.7%, from $4.202 billion in 1967 to $4.047 billion in 1968. Sales of products and services for use of aerospace technology in nonaerospace areas—such as marine science, water desalination, crime control, and rapid transit—increased from $2.579 billion to $2.726 billion.

Harr predicted slight decline in total aerospace sales to about $29.6 billion during 1969 because of 25% drop in jet transport sales before deliveries of new high-capacity aircraft; continuing increase in helicopter, executive, and utility aircraft sales; modest increase in defense and nonaerospace sales; and decline in civil space sales. He noted that in third quarter of 1968 backlog of Government aerospace orders was less than that of other customers for first time since before World War II. (Text: AIA Release 68–60; W Star, 12/19/68, A19)

- USAF flew 11 newspapermen on simulated combat missions in F–111A from Nellis AFB, Nev., to demonstrate aircraft's systems. In *Washington Post*, George C. Wilson said decision to allow newsmen in cockpits of controversial plane for first time evidenced USAF's conviction "that the F–11 program is at a crucial juncture as the Nixon Administration gets ready to take office." (*W Post*, 12/19/68, A8)
- Dr. Anatoli A. Logunov, Director of Institute of High Energy Physics near Serpukhov, 60 mi south of Moscow, said in *Izvestia* that Institute's 1,000-yd-dia, 70-bev, proton accelerator had joined scientists elsewhere in search for quark. Quark was hypothetical particle thought to be elementary building block of all matter and to carry electrical charge one-third to two-thirds that of electron charge. (*NYT*, 12/20/68, 3)
- AIAA announced Dr. Charles P. Sonnett, Chief of ARC's Space Science Div., would receive Space Science Award, including $500 honorarium, "for his personal contribution as planner, leader and individual experimenter in major space science vehicle programs which have contributed to the field of space physics." He had worked in magnetospheric physics and nuclear physics and was currently concentrating on interplanetary physics. He had been principal investigator on several NASA experiments and ALSEP. Award would be presented at AIAA 7th Aerospace Science Meeting in New York Jan. 20–22, 1969.

    Dr. Stanley G. Hooker, Technical Director of Bristol Engine Div., Rolls-Royce Ltd., and Perry W. Pratt, Vice President and Chief Scientist of United Aircraft Corp., had been selected to share $10,000 Goddard Award for their separate work in developing gas turbine engines. Goddard Award, named for late rocket pioneer Robert H. Goddard, was awarded annually to "a person who has made a brilliant discovery or a series of outstanding contributions over a period of time, in the engineering science of propulsion or energy conversion."

    Prof. Rene H. Miller, head of MIT Dept. of Aeronautics and Astronautics, would receive Sylvanus Albert Reed Award for "outstanding contributions" to rotary-wing aircraft. Dr. Robert D. Fletcher, USAF Air Weather Service's Deputy Chief of Staff for Aerospace Sciences, would receive Robert M. Losey Award for "outstanding and dedicated leadership and service" for 30 yr to aeronautical meteorology. (AIAA Release; NASA Biog, 11/29/68; *NYT*, 12/31/68, 52)

December 19: NASA Aerobee 150 MI sounding rocket launched from WSMR carried Cornell Univ. experiment to 100.3-mi (161.3-km) altitude to examine sky in far infrared (5 micron to 1,600 microns) using copper-doped-germanium, two gallium-doped-germanium, and indium-antimonide detectors in conjunction with telescope. Rocket and instrumentation performed satisfactorily. Preliminary results indicated detectors functioned properly. (NASA Rpt SRL)

- NASA issued Management Instruction establishing Committee on Extra-Vehicular Activities (EVA), activities performed in space by astronaut outside space vehicle. Seven-member committee, serving for two-year period, would provide Deputy Associate Administrator for Manned Space Flight with recommendations on overall NASA EVA planning and development by: identifying EVA capabilities which must be defined and developed to support manned space flight activities; identifying ground-based, orbital, and lunar surface experiments to establish required EVA capabilities; providing recommendations on short- and long-term EVA program plans; reviewing proposed EVA flight experiments and making recommendations; and maintaining awareness of EVA-related activities of organizations other than NASA. (NMI 1152.36)

- NASA announced renewal through August 1970 of 10-yr contract with NAS which provided one-year appointments for post-doctoral and senior post-doctoral scientists and engineers specializing in space-related work to conduct research at NASA field centers. NASA would pay NAS $2,390,-500 to carry out program. (NASA Release 68–218)
- AEC announced it had conducted underground nuclear test of about one-megaton yield at Nevada Test Site. It was similar to April 26 test. Tests had been described in press as largest continental explosions announced by AEC. Newsmen were permitted to witness test for first time in 10 yr. (AEC PIO; AEC Release L–288; Wilson, *W Post*, 12/20/68, A3; Hill, *NYT*, 12/20/68, 1; 12/25/68; AP, *NYT*, 12/18/68, 35)
- U.N. General Assembly, by vote of 96–0, approved plans for permanent body of 42 members to study means of reserving seabed for peaceful purposes and of exploring resources beyond national jurisdiction. (Estabrook, *W Post*, 12/20/68, A12)

December 20: Cosmos CCLXI was launched by U.S.S.R. into orbit with initial apogee of 670 km (416 mi), perigee of 207 km (128.5 mi), 93-min period, and 71° inclination. Satellite carried scientific equipment for studies of upper layers of earth's atmosphere and nature of Northern Lights, with participation of research institutions and observatories of Bulgaria, Hungary, East Germany, Poland, Romania, U.S.S.R., and Czechoslovakia. Satellite reentered Feb. 12, 1969. (*SF*, 5/69, 165; GSFC *SSR*, 12/31/68; 2/15/69)

- NASA announced completion of X–15 flight research program [see Jan. 21]. On final flight Oct. 24—199th in series which began June 8, 1969—NASA test pilot William H. Dana flew rocket-powered aircraft to 255,000-ft altitude. Attempt at 200th flight Dec. 20 was canceled because of adverse weather conditions.

    In NASA–USAF–USN program initiated in 1954, flights by three X–15 aircraft manufactured by North American Rockwell Corp. had included more than 82 min of flight at speeds exceeding mach 5 and total flight time of more than 30 hr. Peak altitude reached was 354,200 ft (67.04 mi) and top speed was 4,520 mph (mach 6.7)—speeds and altitudes never before attained by any vehicle fully controlled by pilot from launch to landing. It had set two official world altitude records of 246,740 ft and 314,750 ft previously. X–15 flight program provided knowledge applicable to design and development of future spacecraft and commercial supersonic aircraft and data on aerodynamic heating in high-speed flight, which could cause deterioration of aircraft structural integrity. X–15 remained only aircraft capable of studying phenomena at hypersonic speeds, space-equivalent flight, and reentry flight.

    It also had served as test bed for new components and subsystems, subjecting them to hypersonic flight environment. In 1962 four X–15 pilots received Robert J. Collier Trophy from President Kennedy for "the greatest achievement of aeronautics or astronautics in America, with respect to improving the performance, safety, or efficiency of air and space vehicles." (NASA Release 68–221; NASA SP–60; NASA EP–9; AP, *NYT*, 12/21/68, 73; *SBD*, 12/23/68, 236)

- Workmen loading super-cold oxygen into Apollo 8 service module discovered gas had been contaminated, apparently by nitrogen used to flush tanks. Oxygen would be changed and tanks refushed and launch was expected to take place on schedule (Lannan, W *Star*, 12/20/68, A5)
- Dr. Robert H. Guest, professor of organizational behavior in Amos Tuck School of Business Administration, Dartmouth College, was sworn in as consultant by Dr. Thomas O. Paine, Acting NASA Administrator. Dr. Guest would serve on Management Advisory Panel. (NASA Release 68–220)

December 21–27: NASA's *Apollo 8* (AS–503), second manned mission in Apollo lunar landing program and first manned mission to orbit moon, was successfully launched from KSC Launch Complex 39 at 7:51 am EST by Saturn V booster. Primary objectives were to demonstrate crew, space vehicle, and mission support performance during manned Saturn V mission with command and service module (CSM) and to demonstrate performance of nominal and selected backup lunar orbit rendezvous (LOR) mission activities—including translunar injection; CSM navigation, communications, and midcourse corrections; and CSM consumables assessment and passive thermal control. All launch events occurred as planned and spacecraft—carrying Astronauts Frank Borman (commander), James A. Lovell, Jr. (CM pilot), and Willam A. Anders (LM pilot)—entered initial orbit with 118.4-mi (190.6-km) apogee, 113.8-mi (183.2-km) perigee, 32.51° inclination, and 88.2-min period.

At 10:42 am EST 3rd stage burned for second time, injecting spacecraft into lunar trajectory, and astronauts began journey to become first men to leave earth's gravitational field. Stage and instrument unit separated as planned and service module propulsion system was fired to increase separation distance from 3rd stage, which was trailing 500–1,000 ft behind spacecraft, spewing unused propellants. Crew fired service propulsion system (SPS) engine for 2.4 sec, correcting trajectory and increasing velocity by 25 fps.

Second midcourse maneuver, scheduled for second day, was canceled because trajectory was already so accurate that burn would have required velocity change of only 0.7 fps. Borman reported illness, apparently from 24-hr intestinal virus or from reaction to sleeping pills being used during space flight for first time, and Lovell and Anders reported nausea. Crew took navigation sightings and conducted first TV transmission, showing spacecraft interior and earth from 138,690-mi altitude and demonstrating food preparation and movements in weightlessness. Signals were received at ground stations and transmitted to NASA Mission Control Center in Houston before release live to commercial networks.

Second TV transmission, on third day, showed excellent pictures of earth from 201,365-mi altitude, including view of Western Hemisphere in sunlight. Crew pointed out North Pole, South America, Cape

Horn, and Baja California and noted that U.S. East Coast was very cloudy. Earth, they said, was beautiful; water looked royal blue, land areas brown, and clouds bright white. Reflection off earth was much greater than off moon. SM reaction control system's four rockets were fired for 12 sec to reduce velocity by 2 fps and to make approach to moon closer to 60 mi at nearest point.

On fourth day, Christmas Eve, communications were interrupted as *Apollo 8* passed behind moon and astronauts became first men to see moon's far side. SPS engine was fired for 4 min 2 sec, reducing speed by 2,994 fps and placing spacecraft in lunar orbit with 193.6-mi (310.6-km) apolune and 69.1-mi (111.2-km) perilune. In third telecast Anders described lunar surface as "whitish gray, like dirty beach sand with lots of footprints on it. Some of these craters look like pick-axes striking concrete, creating a lot of fine dust." After spacecraft passed behind moon at end of second revolution, SPS engine burned for 10 sec to reduce speed by 135 fps and to circularize orbit with 70.0-mi (112.6-km) apolune and 69.6-mi (112.0-km) perilune. Lovell said astronauts had "a grand view" of the lunar surface and confirmed that prospective landing sites were satisfactory. He reported that at about two minutes before sunrise a fan-shaped white haze appeared just behind moon's limb. Crew continued landmark sightings and named numerous unnamed lunar features after other astronauts, NASA officials, and friends. They conducted communications experiment which showed that radio signal from earth to *Apollo 8* and back to earth took three seconds to make 460,000-mi round trip.

Third TV transmission during ninth revolution showed heavily impacted mountains described by Anders as "a vastness of black and white, absolutely no color. The sky up here is also rather forbidding, foreboding extents of blackness with no stars visible when we're flying over the moon in daylight. You can see by the numerous craters that this planet has been bombarded through the eons with numerous small asteroids and meteoroids, pock-marking the surface of every square inch. And one of the amazing features of the surface is . . . that most of the craters . . . have a round mound type of appearance instead of sharp jagged rocks. All, only the newest of features have any sharp definitions to them, and eventually they get eroded down by the constant bombardment of small meteoroids." The moon is "a very dark and unappetizing place. . . ."

Crew read verses from first chapter of Genesis and wished viewers "good night, good luck, a Merry Christmas and God bless all of you—all of you on the good earth."

On fifth day, Christmas Day, while spacecraft was behind moon completing its 10th revolution, SPS engine was fired for 3 min 23 sec, increasing spacecraft velocity by 3,523 fps and propelling *Apollo 8* back toward earth. Spacecraft left lunar gravity at 201,807 mi above earth. At 104 hours mission elapsed time 14-sec reaction-control-system burn increased velocity by 5 fps. Fifth TV transmission showed spacecraft interior, controls, and food preparation. Data analysis revealed sixth midcourse correction would not be necessary because of accuracy of course. Astronauts reported they had slept well and were in "very good shape." Sixth TV transmission showed earth from 112,125-mi altitude.

December 21–27: NASA's Apollo 8 *carried Astronauts James A. Lovell, Jr., William A. Anders, and Frank Borman (left to right above) out of earth's field of gravity and into man's first orbit of moon. In photo at right, Saturn V lifts off from* KSC *launch complex 39 to put command and service module into initial orbit.*

On sixth day crew prepared for reentry and SM separated from CM on schedule. Parachute deployment and other reentry events were nominal and *Apollo 8* splashed, apex down, in Pacific about 5,100 yd from recovery ship U.S.S. *Yorktown* at 10:51 am EST Dec. 27, 147 hr after launch and precisely on time. According to prior planning, helicopters and aircraft hovered over spacecraft and pararescue personnnel were not deployed until local sunrise, 50 min after splashdown. Crew was then picked up and reached recovery ship at 12:20 pm EST.

All primary *Apollo 8* mission objectives and detailed test objectives were achieved and, in addition, five not originally planned. All launch vehicle and spacecraft systems performed according to plan. Engineering accomplishments included use of ground network with onboard navigational techniques to sharpen accuracy of lunar orbit determination

and successful use of Apollo high-gain antenna, four-dish unified S-band antenna that deployed from SM after separation from 3rd stage. Mission proved capability of Apollo CSM and crew, as well as ground support and control systems, to operate out to lunar distances and return through the earth's atmosphere at lunar velocity.

Apollo 8 was fifth Apollo mission to date, second manned Apollo mission, first manned mission on Saturn V launch vehicle, and first manned operation of Apollo system under conditions for which it was designed. Earlier unmanned Apollo flights had yielded all spacecraft information possible without crew on board. *Apollo 4* (launched Nov. 9, 1967) and *Apollo 5* (launched Jan. 22, 1968) had both been highly successful, completing inflight tests of all major pieces of Apollo hardware. *Apollo 6* (launched April 4), despite launch vehicle problems, had attained four of five primary objectives with the spacecraft recovered in excellent condition. First manned Apollo mission, *Apollo 7* Oct. 11–22, had achieved all primary objectives and had verified opera-

December 21–27: *Rising earth, 240,000 miles away, greets* Apollo 8 *astronauts as they come from behind moon after lunar orbit insertion burn.*

tion of spacecraft for lunar-mission duration. Apollo program was directed by NASA Office of Manned Space Flight; MSC was responsible for Apollo spacecraft development, MSFC for Saturn V launch vehicle, and KSC for launch operations. Tracking and data acquisition was managed by GSFC under overall direction of NASA Office of Tracking and Data Acquisition. (NASA Proj Off; NASA Release 68–208; NASA Special Releases; W *Star*, 12/21/68, A1; W *Post*, 12/21–28/68, A1; *NYT*, 12/21–28/68, 1; B *Sun*, 12/21–28/68, A1; MSC *Roundup*, 1/10/69, 3)

December 21: President Johnson sent congratulatory message to *Apollo 8* astronauts: "I am confident that the world's finest equipment will strive to match the courage of our astronauts. If it does that, a successful mission is assured." (PD, 12/30/68, 1738)

• Apollo Program Director, B/G Samuel C. Phillips, told post-launch press conference at KSC early portion of *Apollo 8* mission had been perfect: ". . . Apollo 8 is now on its way to the Moon. And . . . I certainly envy the crew the magnificent views that they might have at this point as they were describing it to us. . . . To be able to see the entire earth. I think by this time in their flight, they are something over 10,000 miles high and moving at about 25,000 miles an hour. . . . The countdown . . . was—during most of the night actually a hit, and even into the late minutes of this morning, the count was proceeding with comfortable margins of time. The flight of the Saturn V was flawless in all of the maneuvers it was expected to make. The performance of the spacecraft . . . has similarly been flawless." (Transcript)

December 22: *New York Times* commented on *Apollo 8* flight: "Space contains more than enough opportunity for fruitful application of the energies that all mankind can devote to its exploration, development and eventual settlement. There is no need here for wasteful rivalry deriving from earthbound nationalistic and political ambitions. In the face of the most breathtaking challenge humanity has ever faced, the only rational response is cooperation to make space an arena of unity and international brotherhood. Man's hopes and prayers ride with the pre-Christmas voyagers. After them must come ships bearing the United Nations flag, each carrying men of different citizenship, language, political and religious convictions and color." (*NYT*, 12/22/68, E10)

- New York Academy of Sciences investigative committee of 22 members directed "preliminary report" to President Johnson, President-elect Richard M. Nixon, and Congress on intensifying crisis in U.S. science and education which they said stemmed from cuts in Federal Government support. Investigation covering 84 academic institutions and work of 193 research scientists had shown "potential solutions of such problems as poverty, racial discrimination, population control, air and water pollution, cancer and cardiovascular disease, mental illness, mass transportation, housing and education are not being pursued because of lack of continuing support." Committee recommended diverting Federal science funds earmarked for buildings and other capital equipment to use on research programs and in training scientific manpower to "preclude a serious shortage in the near future, a crippling one within five years." (UPI, *W Post*, 12/23/68, A20)

December 24: *Intelsat–III F–2* began carrying segment of coverage of *Apollo 8* to Europe. It transmitted moon pictures from capsule and relayed TV coverage of Pacific splashdown to Europe and Puerto Rico. Intelsat II satellites over Atlantic and Pacific each reserved about 100 voice circuits for NASA support communications with *Apollo 8*. Ats II and *III* augmented commercial communications coverage and transmitted limited number of weather photos. (OSSA; NASA Release 69–6)

- One out of every four persons on earth—nearly one billion people in 64 countries—heard Christmas Eve reading of Genesis by moon orbiting *Apollo 8* crew either on radio or TV, according to *TV Guide*. Delayed broadcasts same day reached 30 additional countries. "The fantastic success of TV on that flight echoed around the world." (*TV Guide*, 5/10–16/69, 9)

- *New York Times* said of *Apollo 8* telecast of earth: ". . . the drama and interest of yesterday's view of earth from space transcended any prosaic considerations of practical utility. Rather the excitement these pictures aroused among millions of stay-at-homes flowed from the visual evidence they provided of man's successful entrance into a completely new realm, one which poses challenges, opportunities and dangers such as the human species has never before faced. And yesterday's pictures provided a sobering perspective on man's puny earthly works

and rivalries, reminding all humanity that nature is the basic antagonist, not other men." (*NYT*, 12/24/68, 22)
- FAA announced report *Friction Effects of Runway Grooves, Runway 18–36 Washington National Airport* [AD 678 645 (DS 68–21)], result of nine-month test series at Washington National Airport, indicated runway grooving, designed to prevent hydroplaning by increasing drainage of water, might also enhance braking effectiveness of aircraft on wet runways [see Nov. 17]. (FAA Release T 68–48)
- U.S.S.R. announced through Tass successful completion of experiment in which three researchers spent from Nov. 5, 1967, to Nov. 5, 1968, in isolated chamber consisting of living compartment and greenhouse linked to outside world by videotelephone. Purpose of experiment was to test man's ability to live in isolation for year, using water and oxygen regenerated from waste products and dehydrated food supplemented by greenhouse-grown vegetables; study effects of various factors on human organism and establish optimum conditions for long isolation; and evaluate effectiveness of self-contained life-support systems based on regeneration of waste products. Daily requirements of astronauts on long voyage included 700 g of food, 2.4 kg of drinking water plus 5.5 kg of water for other purposes, and 800 g of oxygen, amounting to 11 tons of supplies for one-year space voyage. Report said there were no significant changes in body weight and temperature of experimenters except one lost eight to nine pounds before his weight stabilized in five to six months. Electrocardiogram remained unchanged and no dehydration was observed. Pulse and respiration rates had fluctuated before assuming lower level than at start of experiment. Researchers Gherman A. Manovtsev, Andrey N. Bozhko, and Boris N. Ulybshev were reported in good health at conclusion of experiment. (Shabad, *NYT*, 12/25/68, 38)

December 25: The space age, no doubt, will produce many future heroes and many other historic accomplishments, *New York Times* said, "but even now it is plain that yesterday Astronauts Borman, Lovell and Anders assured themselves of immortality as the first men literally to break the shackles of earth and travel successfully to another destination in this suddenly shrunken solar system." (*NYT*, 12/25/68, 30)

December 26: U.S.S.R. launched *Cosmos CCLXII* from Kapustin Yar into orbit with 791-km (491.5-mi) apogee, 264-km (164-mi) perigee, 95.2-min period, and 48.4° inclination. Equipment was functioning normally. (*SBD*, 12/30/68, 257; GSFC *SSR*, 12/31/68)
- *Apollo 8* lunar flight was voted top news story of 1968 in Dec. 24 repolling of editors of Associated Press member newspapers, radio, and TV stations. Previous poll, completed before Dec. 21–27 mission, had selected assassinations of Sen. Robert F. Kennedy and Rev. Martin Luther King as No. 1 and No. 2 stories of year. (W *Star*, 12/26/68, A8)
- *Washington Post* said of *Apollo 8:* "Above all, perhaps, this Christmas Eve at the moon and Christmas Day on the way home have told us more about our earth than about the moon. One of the astronauts had wondered on his way to the moon whether, if he were a traveller from another planet, he would think that intelligent life existed on Earth. The answer, from Captain Lovell at least, is that our planet is 'a grand oasis in the great vastness of space.' That is an awing insight and it reminds us that man has far to go here at home to fulfill the Christmas

promise of Peace on Earth, Good Will toward Men." (*W Post*, 12/26/68, A20)

- *Times of India* termed *Apollo 8*, "the most magnificent achievement in space to date." It was "by any reckoning a tremendous achievement of science and technology" and "decidedly the most daring adventure man has ever undertaken.' (*Times of India*, 12/26/68, 8)
- In *Washington Post*, columnist Joseph Kraft said post-Apollo programs had been sharply cut in Congress and space spending had been compared unfavorably with money for pressing internal needs. "In this situation, it makes sense for this country to disengage while it is ahead. There is no need for the United States to race Russia to every new milestone in space. On the contrary, what the United States wants is a program closely connected to explicit American requirements—a program of exploration for its own sake, not for the sake of beating the Russians. In that way, this country can continue to develop a capability in space, without having to respond in a panic to the ups and downs that are necessary part of the space business." (*W Post*, 12/26/68)

December 26–31: During 135th meeting of AAAS in Dallas, Tex., Catholic Univ. of America assistant professor of international relations John M. Logsdon gave paper on lunar landing decision at session of Society for the History of Technology. Prof. Logsdon said President Eisenhower had planned to end manned space flights after Project Mercury but his successor, President Kennedy, had "calculated the costs, weighed the needs, and finally decided that 'whatever mankind must undertake, free men must fully share.'"

Outgoing Presidential Science Adviser, Dr. Donald F. Hornig, recommended reexamination of concept of Federal Dept. of Science since science "has now assumed such importance to the nation that its position would be stronger if it had a voice in the Cabinet." He advocated annual report on state of science similar to annual economic report and said Office of Science and Technology "could eventually evolve in an office of planning and analysis, looking broadly at national problems with some scientific or technological component, but extending well beyond the purely technical areas." Under questioning, Dr. Hornig said he agreed with Dr. J. Herbert Hollomon, President of Univ. of Oklahoma and former Assistant Secretary of Commerce, that Federal Dept. of Science, with NSF as its core, might encompass oceanographic agencies, high-energy-physics research currently funded by AEC, ESSA, Bureaus of Census and Labor Statistics, geophysics branches of Geological Survey, and some NIH programs. Dr. Hollomon also suggested NASA be added when it could be included "without having it become the tail that wags the dog." Dr. Hornig insisted no massive "science agency" should be created to usurp supervision of Nation's science effort.

Bacteria might have caused gastrointestinal upset suffered by astronaut Frank Borman on *Apollo 8* mission, Dr. Rudolf H. T. Mattoni, head of Biological Systems Div. of Nuclear Utility Services, Hawthorne, Calif., told meeting. Effects of weightlessness on bacteria on *Biosatellite II* flight (Sept. 7–9, 1967) suggested that lack of gravity might have allowed common, normally benign, intestinal bacteria to cause illness like Borman's.

Drs. Bouilin Browning of St. Thomas Univ. and Irwin Oster of Bowling Green State Univ. reported first experimental evidence that weightlessness can cause chromosomal damage of genetic significance, based on sperm cells flown 45 hr aboard Biosatellite.

Cornell Univ. map expert, Prof. Arthur J. McNair, said sophisticated photographic mapping by satellites at 140-mi altitude would provide faster, cheaper, broader, and more detailed coverage than now possible by airplane mapping. Single map-making satellite photo, he said, would be equivalent to 1,000 photos taken by aircraft. U.S. could be fully mapped in one year from photos from satellite in near-polar orbit for four weeks. Another 11 mo would be needed for data processing.

Dr. Robert H. Hardie of Vanderbilt Univ. said planet Pluto had appeared to be dimming for past 10 yr. It was moving in 248-yr orbital period to point where it reflected little sunlight and its surface temperature had dropped two degrees. He speculated that planet froze into mass of stone and solid nitrogen as temperatures reached $-250°$ C when facing away from the sun. When bathed in sunlight, planet warmed to $-200°$ C and formed reflecting puddles which astronomers saw as variations in light intensity. (Text; UPI, *W Post*, 12/28/68, A9; Lannan, *W Star*, 12/30/68, A3; AP, *W Post*, 12/30/68, A6)

December 27: At White House press conference President Johnson discussed *Apollo 8*'s effect on U.S. position in space race: "We are very pleased with the progress we have made. . . . Each side has different examples of its achievements. But in the 10 to 11 years since Sputnik I . . . when we didn't even have a space committee in the Congress, when we were talking about the basketball up there in the air, when we have weathered the storms that have brewed—everyone who wanted to cut anything, the first thing they wanted to cut was the space program —when we have seen the editorial professors inform us that there was really no value in doing all of this anyway, it gives me great pleasure now to see the thrill that even they are getting out of it."

President described anxiety about complex mission and said he had repeatedly asked himself whether U.S. was ready, whether date selected was best one, whether every possible precaution had been taken, and whether every man had performed his requirements. "About all you can do . . . is to pick men that you have confidence in, that you trust, give them the support they need, and then hold on." Remembering he had "recommended this goal for this decade" to President Kennedy, he said: "There have been many pitfalls every step of the way. I don't know how many folks have just wanted to abandon it, clip it, cut it, take the money for the cities or the war or just anything else. Space has been a whipping boy.

"So when you see the day approaching when visions, and dreams, and what we said to the Congress when we created the Space Administration back in 1958 are becoming reality, you naturally are hopeful."

Later, in telephone message to *Apollo 8* astronauts, President congratulated crew and said: "You have made us very proud to be alive at this particular moment in history. You have made us feel akin to those Europeans nearly five centuries ago who heard stories of the New World for the first time. . . . My thoughts this morning went back to more than 10 years ago . . . when we saw Sputnik racing through the skies, and we realized that America had a big job ahead of it.

"It gave me so much pleasure to know that you men have done a large part of that job." (*PD*, 12/30/68, 1744–50)

December 27

- International comment on *Apollo 8* mission:
  In statement distributed by Soviet Embassy in Washington, D.C., Boris Petrov, Chairman of Council for International Cooperation in Investigation and Utilization of Outer Space, U.S.S.R. Academy of Sciences, hailed "hardware" and "courage" of *Apollo 8* astronauts but called attention to "considerable accent" placed on "automatic devices" in Soviet space program. He said, "The Soviet Union is engaged in a large-scale program of planned studies and exploration of outer space, which provides for the investigation of the earth's upper atmosphere and of near-earth outer space and studies of sun-earth relationships and of our closest space neighbors—the moon, venus and mars, and later, on more remote planets."

  Cuban National Radio called mission "a total success."

  Tass: "Due tribute should be paid to the courage and mastery of Frank Borman, William Anders and James Lovell who have accomplished this outstanding scientific and technical experiment. The successful flight of Apollo 8 ushers in a new stage in the history of space exploration." Ten Soviet cosmonauts telegraphed Apollo crew congratulations for "another milestone in scientific and technical progress."

  Pope Paul VI, in message to President Johnson: "Giving thanks to God for the successful completion of the magnificent enterprise of the Apollo 8 mission, we congratulate you and the people of the United States of America and particularly the intrepid space travelers, and invoke divine blessing upon all contributing to this noble achievement."

  U.N. Secretary General U Thant: "The powerful thrust of the rocket engines, the awesome speeds attained in flight, the vast distances traveled, the precision of navigation and the fidelity of communications, aural and visual, all stagger the imagination and defy the comprehension of man."

  Emperor Haile Selassie of Ethiopia: "a great milestone in man's continued search of the unknown."

  Gov. Gen. Roland Michener of Canada, in message to President Johnson: Canadians had followed astronauts' flight with "admiration for their courage and the technical skill of all who have a part in the enterprise."

  Prime Minister Eisaku Sato of Japan to President Johnson: feat showed America's "courage and the high level of scientific technology."

  In Kaduna, Nigeria, *New Nigerian* said *Apollo 8*, "Apart from being the supreme technological achievement . . . of all time," was "another exciting chapter in man's eternal quest to triumph over his own natural environment." Same skills and resources which were helping to conquer space "could easily be diverted to giving man a better life on earth itself—if only all of us would allow the good that is inherent in all mankind to assert itself." (AP, B *Sun*, 12/28/68, 1; *New Nigerian*, 12/27/68, 1)

- San Francisco *Chronicle* said: "In their meteoric flitting about the skies the crew of the Apollo 8 did more than thrill their fellow Americans and pump up the national ego. They drew commendations from directors of the Soviet space project whose considerable achievement they had outstripped." They made believers of "informed skeptics such as Sir Bernard Lovell, director of Britain's Jodrell Bank Observatory, who had more than ever scoffed at the value of the moon venture. (They

December 27

also drove the stubborn adherents of the flat-earth theory into confusion.) All this they accomplished in faultless style." (SF *Chornicle*, 12/27/68, 30)

- Associated Press quoted Mrs. Madalyn Murray O'Hair, who was instrumental in having prayer removed from U.S. public schools, as saying she would register complaints with NASA against reading by *Apollo 8* astronauts of portions of Genesis during lunar orbit and would organize mail campaign to bar prayer from space. (AP, *W Post*, 12/28/68, A9)
- *Washington Post* said Harvard Univ. scientist Dr. George B. Kistiakowsky, who in 1959 was chief science adviser to President Eisenhower, had said in taped interview for broadcast on CBS radio network Dec. 29 that putting man on moon would not compare to great scientific achievements such as breaking genetic code. "This [moon flight] is an adventure.... it's different from Darwin's travels," which led to his discovery of genetic evolution. (Aarons, *W Post*, 12/28/68, A8)
- *Apollo 7* mission (Oct. 11–22) was adjudged successful by NASA. All launch vehicle systems had performed satisfactorily throughout expected lifetime and spacecraft systems functioned with few minor anomalies, which were countered, preventing loss of systems support. Splashdown occurred within one mile of guidance system target point and recovery of flight crew and CM was successful. All test objectives had been successfully accomplished. (NASA Proj Off)
- *Apollo 6* (launched April 4) was adjudged unsuccessful by NASA. Although three of five primary mission objectives had been fully accomplished and two partially accomplished, overall mission was not a success. *Apollo 6* had entered elliptical parking orbit instead of planned circular orbit when 2nd stage engines shut down prematurely and 3rd stage failed to reignite on command. (NASA Proj Off)
- NASA announced it would convert to civil service operation during next 18 mo work performed in 810 contractor positions at GSFC, to bring Center's operations into accord with Civil Service Commission guidelines prohibiting Government supervision of contractor employees. Many incumbent contractor employees would be offered Government positions and only small portion of contractors' activities would be affected in many cases. Conversion would be completed by June 1970 and NASA would continue to rely on industry to considerable extent for support services. (NASA Release 68–223)
- MSFC announced it had issued Boeing Co. $8,429,047 supplemental agreement extending from October 1968 to March 1970 maintenance and operation of Saturn V development facility at MSFC and providing for mechanical ground support equipment and logistics work. (MSFC Release 68–283)
- NASA announced award by LeRC of $3,448,762 cost-plus-award-fee contract to Honeywell, Inc., for Centaur launch vehicle guidance system including management, engineering, repair, and modification support during 1969. (NASA Release 68–222)
- In *Science* Dr. J. C. G. Walker, Yale Univ. geologist and geophysicist, and N. W. Spencer, Chief of GSFC's Laboratory for Atmospheric and Biological Sciences, said thermosphere probe experiments had provided largest body of rocket mass-spectrometer data obtained. Tests had been conducted jointly by scientists at GSFC and Space Physics Research Lab.

at Univ. of Michigan to determine temperatures of earth's upper atmosphere. Since 1962, concentration and temperature of molecular nitrogen in upper atmosphere had been measured in 22 successful flights under varying conditions of solar activity, from launch sites at Fort Churchill, Manitoba, Wallops Island, Va., and Vega Baja, Puerto Rico. Scientists had concluded that upper atmosphere consisted of mixture of gases—electrons, ions, and neutral particles—each of which had distinct temperature. Multiplicity of temperatures could be expected in upper atmosphere of planets, though differences between planetary atmospheres must be more striking than similarities because of differences in atmospheric composition and in distance from sun. It was likely that absence of permanent magnetic field on either Mars or Venus caused further substantial differences between upper atmospheres of these planets and upper atmosphere of earth. (*Science*, 12/27/68, 1437–41)

- National Science Foundation announced median annual salary of U.S. scientists in 1968 was $13,000, up $1,200 (10%) from 1966. Self-employed scientists earned highest median, $18,000, in 1968, with those employed by industry, business, and nonprofit organizations second at $14,700. Federal Government scientists reported same median salary—$13,500—as those employed on yearly basis by educational institutions. Single largest users of scientific talent were educational institutions (40%). Industry and business employed 32%, and 10% were civilians in Federal Government. Information was based on responses by 298,000 U.S. scientists to NSF's biennial National Register survey. (NSF Release 68–181)

- *New York Times* said article in December issue of *Australian Quarterly* by Australian National Univ. lecturer Robert Cooksey had suggested U.S. "space research facility" at Pine Gap near geographical center of Australia might be station designed to guide orbital missiles fired from U.S. to targets in Communist China. Article had caused flurry of questions about base in Australian press. Later *New York Times* story said U.S. officials in Washington had denied suggestion and said Pine Gap installation was joint U.S.-Australian space research facility established by agreement in 1966. (Trumbull, *NYT*, 12/29/68, 30)

- DOD announced appointment to Defense Science Board of Dr. Arthur T. Biehl, Associate Director for Advanced Study, Lawrence Radiation Laboratory; Dr. Lewis M. Branscomb, Chairman of Joint Institute for Laboratory Astrophysics, Univ. of Colorado; Daniel J. Fink, General Manager for Space Systems, General Electric Corp.; and Dr. Charles M. Herzfeld, Technical Director of Defense Space Group (R&D), International Telephone and Telegraph Corp. (DOD Release 1139–68)

December 28: *Intelsat-III F–2* (launched Dec. 19) was used between Etam, W. Va., and Raisting, Germany, under emergency authorization to back up interruption of service on TAT–4 cable. (NASA Proj Off)

- Finnish professor Arrno Niini said photos of earth brought back by *Apollo 8* astronauts might show tiny ring of small dust particles 200 to 350 mi above earth. It would be discernible only in pictures taken with sun behind camera and with sufficiently sensitive film. (UPI, *NYT*, 12/30/68, 2)

- In Moscow interview with Turin, Italy, newspaper *La Stampa*, Soviet space scientist Prof. Leonid I. Sedov said U.S.S.R. was concentrating on perfecting unmanned spaceships for exploration of celestial bodies deeper in space than moon. "There does not exist at present a similar project [to Apollo 8] in our program. In the near future we will not send a man around the moon. We start from the principle that certain problems can be resolved with the use of automatic soundings." Sedov hailed *Apollo 8* mission as "a great scientific conquest." (UPI, *W Post*, 12/29/68, A4)

- President Johnson announced promotion of *Apollo 8* Astronaut William A. Anders (Maj., USAF) to lieutenant colonel under his policy of granting one-grade promotion to military astronaut after his first successful space mission. (Maynard, *W Post*, 12/29/68, A4; *PD*, 1/6/69, 5)
- Polish noncommunist party newspaper *Zycie Wrszawy* said of *Apollo 8* crew: "We were all with them during those five days. . . . We congratulate them heartily. At such a time we do not think about politics and we forget about the country from which they come." Paper also said, "Only a few changes would be enough for the monstrous rocket to carry a nuclear warhead into orbit instead of astronauts." (UPI, *C Trib*, 12/29/68)
- In *Paris Match*, Raymond Cartier said, "With Apollo 8 the summit is human daring is attained." (*Paris Match*, 12/28/68, 37)
- *Chicago Tribune* said, "Now that Apollo 8 and its three astronauts are home from their historic trip around the moon, we can safely call it one of the most memorable Christmas gifts ever given to the American people and mankind." (*C Trib*, 12/28/68, 12)
- *The Economist:* "What did they think, those three men of Apollo 8 who risked their lives and their sanity to fly to the moon only to report that it looked like grey plaster of paris? What should we earthbound ones think? In New York City, there are at least 2,000 people who would rather have watched a football game and were sufficiently incensed to telephone the television networks and tell them so. The blame is not the astronauts'. A whole series of photographs, some taken from instruments actually on the moon's surface, some in black-and-white, some in colour, had already warned them what to expect. The buck lies on the desks of the men who for the past 13 years have directed the United States' $32 billion space programme, and spent 70% of it on getting men into space without planning anything constructive for them to do when they got there. . . . The cost of a manned moonshot is put at around $1 billion, and for that sum you could get a whole programme of unmanned moon launches. . . .

    "But man does not live by science alone. . . . the greatest achievements of men in space have so far been in the realm of the human spirit. . . . Apollo 8 is part of the unceasing restlessness, invention and ambition of our kind. Have we really any reason to believe that man's evolution has come to a stop after a bare half million years on earth? It requires arrogance, a closed mind and absolutely no sense of history . . . to say that sending men into space is an utter waste of time." (*The Economist*, 12/28/68, 112)
- Neither Peking Radio nor New China News Agency covered *Apollo 8* Mission. (*N Va Sun*, 12/28/68, 1)

December 29: Yomiuri of Tokyo said of *Apollo 8:* "This splendid voyage is hailed around the world as 'mankind's greatest feat' and 'a symbol of hope for the future'. . . . For the U.S. and the Soviet Union, it may be a matter of serious concern as to which puts men on the moon first. . . . This aspect is not important for mankind in general. The two superpowers should regard space exploration as a project of the human race and not a narrow issue of national prestige." (*Yomiuri*, 12/29/68, 2)

- International Flat Earth Society said in London that earth was definitely flat despite pictures from *Apollo 8*. Society Secretary Samuel Shenton said moon was circular but there was no proof it was a globe. (Reuters, B *Sun*, 12/30/68)
- In Washington *Sunday Star*, William Hines reported results of query of eight eminent scientists on most important single decision related to science and technology which faced President-elect Richard M. Nixon early in his administration. Consensus was: shaping of long-term science policy. Also cited were need for reexamination of priorities and goals, with firm decision on extent to which science and higher education should receive Government support; creation of Dept. of Science in cabinet or stronger science-Government communication lines; reassessment of space funding and other "big science" projects, including proton accelerators; greater emphasis on social goals; vigorous antipollution efforts; and more imaginative use of science and technology as instruments of national policy. None of eight scientists advocated abandonment of space program after culmination of Apollo project. Alvin M. Weinberg, Director of Oak Ridge National Lab., said of space program, "This thing takes so very much money that it's hard to get excited about any other decision in science until this one has been made. What should be its level in the 1970s? We have become accustomed to something like four or five billion dollars a year, but I doubt very much that this is a level the new President is likely to concur in." (W *Star*, 12/29/68, C3)
- Noting what he called "curious conspiracy of silence on Russia's capabilities and intention" for almost eight years, William Hines in Washington *Sunday Star* said: "One of the many small acts of positive statesmanship which Richard M. Nixon could profitably perform early in his administration would be to tell the American people fully and frankly just what the United States government knows about the Soviet space program. . . . It would lend credence to Mr. Nixon's professed policy of openness. It would compromise no significant secrets. And it would enable the American public to make judgments about the future of the U.S. space program at a time when vital decisions along this line would be coming due." (W *Star*, 12/29/68, C4)

December 30: Defense Secretary designate, Rep. Melvin R. Laird (R-Wis.), named David Packard, chairman of Hewlett-Packard Co., California electronics firm, to be Deputy Secretary of Defense in Nixon Administration. (Beecher, *NYT*, 12/31/68, 1; *WSJ*, 12/31/68; *Aero Daily*, 12/31/68)

- Cost-plus-fixed-fee contract for Cal Tech's operation of Jet Propulsion Laboratory as a major NASA installation was renewed by NASA and Cal Tech through Dec. 31, 1971. Cal Tech staffed and operated JPL; prop-

December 30

erty, facilities, and equipment were owned by Government. (NASA Procurement Off; NASA Release 69–2)

- In *Pravda*, Prof. Boris Petrov, Soviet guidance mechanisms specialist, said U.S.S.R. preferred not to send men to moon at this point, though unmanned *Zond V* and *Zond VI* "were adapted for piloted flight." (Reuters, *W Post*, 12/31/68, A3)
- Cleveland *Plain Dealer* editorial: "Those who argue that the country should be spending more money on important domestic programs are correct. But these increases should not have to come at the expense of American space exploration and newly-won world prestige. NASA's needs deserve high priority. There should be no lost opportunities. When the first American sets foot on the moon next year, his accomplishment should symbolize a beginning, not an ending." (C *Plain Dealer*, 12/30/68)
- In Washington *Evening Star*, David Lawrence asked, "What was really the big 'miracle' in the voyage of the American astronauts to the moon and back?" It could have happened, "and the rest of the world would not have witnessed the dramatic arrival of the astronauts aboard an aircraft carrier in the middle of the Pacific Ocean or the pictures sent from outer space for several days if it had not been for another great feat of science—transmission of television and radio from artificial satellites direct to every continent of the world." (W *Star*, 12/30/68, A9)

December 31: White House announced from Johnson City, Tex., that President Johnson would present NASA Distinguished Service Medals to *Apollo 8* Astronauts Frank Borman, James A. Lovell, Jr., and William A. Anders during "Astronaut Day" ceremonies in Washington, D.C., Jan. 9, 1969. (AP, W *Star*, 12/31/68, A5; UPI, *NYT*, 1/1/69, 10; *W Post*, 1/1/69, 2)

- Soviet test pilot Eudard V. Yelyan test-flew Tu-144, U.S.S.R.'s delta-wing supersonic transport, on successful 38-min maiden flight from airport near Moscow. Tass reported "the equipment on board the airliner operated normally." Aircraft was designed to carry 120–135 passengers at speeds to 1,600 mph over 4,000-mi range. It was equipped with four Kuznetsov N.K.-144 turbofan engines. Tass said sonic boom "is almost not felt" on earth because Tu-144 reached maximum speed at high altitudes.

Later, *New York Times* editorial commented: "The long international debate about supersonic airliners is sure to be stepped up in the wake of the news that the Soviet entry, the TU–144, has made its first flight. Since it has not been claimed that this initial trial reached supersonic speeds, the likelihood is that the TU–144 was kept subsonic on its first outing. But Soviet airplane designers and builders have manufactured enough supersonic military planes to suggest that on a subsequent flight the TU–144 will fly faster than the speed of sound. . . . In this country, the Soviet first will undoubtedly spur those who want to pour still more Government funds into the Boeing project to build an advanced supersonic plane that is faster and larger than the TU–144. Fortunately, this country's technological prestige is now higher than ever in the wake of the Apollo 8 moon journey. That fact should help the incoming Nixon Administration to see that there are better uses for the nation's scarce resources than to engage in a supersonic plane race whose economics are dubious and whose product's contribution to noise pollution is all too loud." (Shabad, *NYT*, 1/1/69, 1; UPI, W *Star*, 1/1/69, A7; Winters, B *Sun*, 1/1/69, A1; *NYT*, 1/2/69, 30)

- *Earth Resources Satellite System*, report by Subcommittee on Space Science and Applications for Subcommittee on NASA Oversight, House Committee on Science and Astronautics, was transmitted to committee and subcommittee. ERS system "unquestionably presents NASA with . . . perhaps the best possible opportunity to achieve tangible economic returns from the substantial investment already made by the American taxpayer in the U.S. space program." Pace of program to date had been "much too leisurely" and financial support, "inadequate." Report recommended "NASA concentrate a much larger portion of its efforts and resources on this project" and "give the most serious consideration to the recommendation of the National Academy of Sciences' summer study to the effect that funding for the applications satellite program should be at least doubled, perhaps tripled." (Text)

- National Science Foundation released *Federal Funds for Research, Development, and Other Scientific Activities: Fiscal Years 1967, 1968, and 1969*. Federal obligations for basic research, applied research, and development (plant excluded) totaled $16.5 billion in FY 1967 and were expected to total $16.2 billion in FY 1968, first drop since 1955. Obligations had been expected to total $17.3 billion in FY 1969, but, because of appropriations and apportionment actions, probably would be even less than in 1968.

  Basic research obligations totaled $2.0 billion in FY 1967 and were expected to be $2.1 billion in 1968. Applied research obligations totaled $3.3 billion in FY 1967, with $3.3 billion estimated for 1968. Originally expected increases in FY 1969 obligations, to $2.4 billion for basic research and $3.6 billion for applied research, probably would not occur because of cutbacks in FY 1969 budget. Development obligations were $11.3 billion in 1967, highest ever reported, but were expected to drop to $10.8 billion in 1968. Originally expected rise to $11.3 billion in 1969 probably would not be achieved. In FY 1967, obligations for R&D plant totaled $620 million, with estimated $517 million for 1968 and $777 million for 1969. DOD, NASA, and AEC provided bulk of R&D funds, with 90% average share during 1960-66 and 85% during 1967-69. DOD, NASA, and AEC together supported 60% of Federal research total and more than 95% of development total in 1968.

  During 1967-69, 80% of Federal R&D funds were scheduled for extramural performers, chiefly U.S. industrial firms. In 1968 they received 60%. (Text)

- New York Gov. Nelson B. Rockefeller announced *Apollo 8* astronauts had accepted invitation to come to New York for special day of observance Jan. 10. (Fox, *NYT*, 1/1/69, 1)

- Senator-elect Barry Goldwater (R-Ariz.) test-rode F-111A and took controls during part of 90-min flight to and from Nellis AFB, Nev. Later he said aircraft had been victim of politics, not by party, but by "bungling in the Defense Department." Goldwater said he had not opposed aircraft but objected to way DOD had awarded contract to General Dynamics Corp. He said USAF had done good job in "taking the bugs out of this airplane." (AP, *W Post*, 1/1/69, A9)

- USAF awarded contracts totaling $28.8 million to Fairchild Hiller Corp., McDonnell Douglas Corp., and North American Rockwell Corp. for contract definition for ZF-15A advanced air superiority fighter aircraft. (DOD Release 1147-68)

During December: NASA issued *Objectives and Goals in Space Science and Applications—1968.* NASA Office of Space Science and Applications had participated in 1968 agency-wide planning to detail program objectives and options from which program could be built. Many tools required for future space program had already been developed and many future ventures would require only modest improvements. Spacecraft pointing accuracies and stability would improve, and their lifetimes would increase. More powerful transmitters would communicate data across ever-widening expanses. Spacecraft weight would increase and man would have increasing capability to work and navigate in space. Advances in chemical propulsion, introduction of nuclear and electric propulsion, and new combinations of existing stages, would permit growth of launch vehicle capability to meet mission demands.

FY 1969 support of program recognized need for austerity and provided for continuance of existing programs at economical level and initiation of only "projects of great merit, including those where a unique opportunity might be lost." Future emphasis would be on applications of space and space technology for benefit of man: surveying earth's resources, TV broadcast from space, and weather forecasting. Knowledge of Mars, Venus, Mercury, Jupiter, and other outer planets would be expanded. Introduction of larger, more accurate telescopes would provide man perhaps with "his greatest step in understanding the nature of his universe." (Text)

- Senate Committee on Aeronautical and Space Sciences issued *Tenth Anniversary, 1958–1968* to meet requests for information concerning its historical background, activities, jurisdiction and procedures, legislative record, membership, and staff assistance. Report contained National Aeronautics and Space Act of 1958, as amended, related legislation—including NASA's funding history—and Communications Satellite Act of 1962. (Text)
- Global military expenditures ranked first in world public expenditures. They had risen from $132 billion to 1964 to $138 billion in 1965, $159 billion in 1966, and estimated $182 billion in 1967—record high level. Preliminary data, said U.S. Arms Control and Disarmament Agency Report, *World Military Expenditures 1966–67*, said current military spending exceeded by 40% world's expenditures on education by all levels of government and was more than three times worldwide public health expenditure. (Text; Shackford, *W Post*, 1/24/69, A21)

During 1968: In 1968 U.S. orbited 64 spacecraft and U.S.S.R., 74. U.S. total included 43 launched by DOD. NASA's 21 included satellite orbited as secondary payload.

Highlight of NASA's 10th anniversary year was success in carefully planned series of Apollo missions—including first two manned flights in Apollo lunar landing program. Unmanned *Apollo 5* and *Apollo 6* completed inflight tests of all major pieces of Apollo hardware. First manned mission, *Apollo 7*, carried three astronauts around earth for 11 days, verifying spacecraft operation before splashing down precisely on target. Mission included live TV transmission from space, largest number of inflight restarts of SPS to date, and new record of 781 manhours in space. Apollo program climaxed Dec. 21–27 with highly successful six-day *Apollo 8* mission on which three-man crew demonstrated operation of spacecraft systems in lunar environment while world watched live TV coverage via satellite. Spacecraft orbited moon 10 times, providing man with his first trip out of earth's gravitational field and first look at moon's far side and proving capability of crew, spacecraft, and support and control systems to operate out to lunar distances and return through earth's atmosphere at lunar-return velocity.

Unmanned *Surveyor VII*, NASA's first launch in 1968 and last spacecraft in Surveyor series, softlanded on moon, conducted on-site analyses of lunar soil, and took part in laser-detection communications experiment.

Applications satellites launched included *Intelsat–III F–2* comsat for ComSatCorp and *Essa VII* and *VIII* meteorological satellites for ESSA. Scientific achievements included orbiting of NASA's *Ogo V, Aao II, Explorer XXXVI, XXXVIII, XXXIX,* and *XL* satellites and Naval Research Laboratory's *Explorer XXXVII. Oao II*—heaviest and most complicated U.S. automated spacecraft ever launched—took first UV photos of stars, returning previously unobtainable data. *Explorer XXXVI* was successfully used by GSFC as target in first daylight operational laser tracking. *Pioneer IX*, launched into solar orbit, carried *Tetr II* pickaback and ejected it into earth orbit where it served as tracking target for NASA's Manned Space Flight Network.

Some 157 meteorological sounding rockets and 100 scientific sounding rockets were launched, and RAM C–II spacecraft was launched on suborbital mission to obtain data on radio attenuation during reentry.

X–15 rocket research aircraft made 13 flights, ending its flight program with 199 missions, including 154 at mach 4 or greater, 109 at mach 5 or above, and 4 at greater than mach 6. NASA–USAF flight research continued with 13 flights of XB–70 supersonic aircraft. USAF's C–5A, world's largest military cargo aircraft, made its maiden flight and three of the aircraft completed 31 flights for 88-hr total flying time. Design of SST was reevaluated and program was delayed until new fixed-wing design was selected, with prototype construction expected to begin in 1969. NASA's HL–10 lifting-body vehicle completed 12 successful glide flights and first powered flight.

DOD space program included orbiting of 8 IDCSP comsats, 1 Lincoln Experimental Satellite, 5 Orbiting Vehicle research satellites, and 2 environmental research satellites; static test-firings of Titan III–M 1st stage; and near completion of MOL launch complex at Vandenberg AFB.

In joint NASA–AEC effort, propulsion technology highlights included power tests of Phoebus 2A nuclear reactor, first tests of cold-flow test engine in flight configuration in new test stand, and first test-firings of Pewee 1 fuel element test-bed reactor.

U.S.S.R. launched 74 payloads, including 64 Cosmos satellites, 1 Luna, 1 Proton, 3 Zond, and 2 Soyuz spacecraft, and 3 Molniya I comsats. Manned *Soyuz III* rendezvoused but did not dock with unmanned *Soyuz II* launched one day earlier. Unmanned *Zond V* and *VI* completed circumlunar flights and were recovered. *Zond IV* circled moon and apparently crashlanded. U.S.S.R.'s Tu-144 delta-wing supersonic transport successfully made 38-min maiden flight. (*Pres Rpt 68;* NASA Release 68–219; B *Sun,* 1/18/69, 28; Shabad, *NYT,* 1/1/69, 1; UPI, W *Star,* 1/1/69 A7)

- Major administrative events affecting NASA and its role in space were resignation of Administrator James E. Webb in October, after 7½ yr as NASA's head, and budget cuts necessitating program and personnel reductions.

  Press marked Webb retirement with praise for many accomplishments of U.S. space program during his leadership. Webb, planning to devote time to interests in education and urban and foreign affairs, continued to serve as consultant to Acting Administrator Dr. Thomas O. Paine.

  NASA FY 1969 budget request of $4.37 billion, already smallest since 1963 and $700 million below FY 1968 request, was further trimmed by Congress because of urgent national needs in other areas, particularly Vietnam war and urban problems. Authorization of $4.013 billion was lowered to appropriation of $3.995 billion—$375.12 million below original request. Revenue and Expenditure Control Act of 1968 again reduced NASA funds, as well as funds for DOD and other agencies.

  Dr. Paine in October announced NASA interim operating plan limited to $3.85 billion for FY 1969. Cutbacks included "sharply limited and deferred" Apollo Applications program, end to Saturn IB and V booster production, 50% cut in advanced mission studies, reduced lunar and planetary exploration program, one-year delay in NERVA development, slips in Biosatellite program, and reductions in astronomy, physics, and basic research. In November Dr. Paine said that below $4-billion level NASA could no longer "hold together our hard-won capabilities and utilize them effectively in critical programs; some of them would have to be dropped entirely." (EH)

- In its international cooperation program, NASA successfully launched three satellites for European Space Research Organization: International Radiation Investigation Satellite *Iris I* (*Esro II–B*), *Aurorae* (*Esro I*) to study aurora borealis and related polar ionosphere phenomena, and Highly Eccentric Orbiting Satellite *Heos A* for interplanetary physics research. From eight nations, 35 investigators were selected to carry out experiments with first lunar surface material to be retrieved by NASA. Four nations contributed experiments flown on NASA spacecraft and 122 sounding rockets were launched in scientific programs with eight countries. Geodetic satellite observations were carried out with 34 countries; aeronautical research was conducted with 4 countries; and 52 countries had APT facilities to receive cloud cover pictures from U.S. satellites. (NASA Releases 68–219, 68–204; *Pres Rpt 68*)

- *Uspekhi SSSR v Issledovanii Kosmicheskogo Prostranstva. Pervoye Kosmicheskoye Desyatiletiye 1957–1967* [U.S.S.R. Achievements in Space Research (First Decade in Space, 1957–1967)] was published as Sputnik anniversary edition by Nauka Publishing House, Moscow. Dr. Anatoly A. Blagonravov, Academician and chief editor, wrote introduction. Text included results of scientific studies, experiments, and exploration. Appendix listed satellites, spacecraft, and their characteristics. Joint Publications Research Service of U.S. Dept. of Commerce published translation, JPRS 47,311, on Jan. 24, 1969. (Texts)

Appendix A

SATELLITES, SPACE PROBES, AND MANNED SPACE FLIGHTS

A CHRONICLE FOR 1968

The following tabulation was compiled from open sources by Leonard C. Bruno of the Science and Technology Division of the Library of Congress. Sources included the United Nations Public Registry; the *Satellite Situation Report* issued by the Operations Control Center at Goddard Space Flight Center; public information releases of the Department of Defense, NASA, ESSA, and other agencies, as well as those of the Communications Satellite Corporation. Russian data are from the U.N. Public Registry, the *Satellite Situation Report,* translations from Tass News Agency, statements in the Soviet press, and international news services' reports. Data on satellites of other foreign nations are from the U.N. Public Registry, the *Satellite Situation Report,* government announcements, and international news services' reports.

It might be well to call attention to the terms of reference stated or implied in the title of this tabulation. This is a listing of payloads that have (*a*) orbited; (*b*) as probes, ascended to at least the 4,000-mile altitude that traditionally has distinguished probes from sounding rockets, etc.; or (*c*) conveyed one or more humans into space, whether orbit was attained or not. Furthermore, only flights that have succeeded—or at least can be shown by tracking data to have fulfilled our definition of satellite or probe or manned flight—are listed. Date of launch is referenced to local time at the launch site. An asterisk by the date marks those dates that are one day earlier in this tabulation than in listings which reference to Greenwich Mean Time. A double asterisk by the date marks those dates of Soviet launches which are a day later in this compilation than in listings which reference to Greenwich Mean Time.

World space activity in 1968 continued at about the same pace as 1967. There was a slight decline in total successful launches—119 against 127 in 1967—and a decline in total payloads orbited—138 against 159 in 1967, a good deal under the record 160 of 1965. The difference between launches and payloads is of course accounted for by the multiple-payload launches

(DOD is the principal user of this system, with 9 multiple-payload launches orbiting a total of 26 payloads and with as many as 8 payloads on one Titan III–C vehicle; NASA had 2 multiple launches of 4 payloads against one multiple launch and 2 payloads in 1967; the U.S.S.R. again had none).

Of the 1968 world total, the United States launched 45 boosters carrying 64 payloads (compared with 55 and 87 in 1967), the U.S.S.R. launched 74 (compared with 67 in 1967); 6 of NASA's total were non-NASA missions— *Esro II, Aurorae, Heos–A, Intelsat-III F–2, Essa VIII.*

If 1967 was characterized by the large number of U.S. lunar flights (seven compared with none for the U.S.S.R.), 1968 was distinguished by the reappearance of manned flight. The U.S. moved closer to its lunar landing goal with two manned Apollo successes, and the Soviet Union demonstrated its manned capability in the *Soyuz III* flight. In 1968 both the United States and the Soviet Union successfully circled the moon. The Soviet's *Zond V* first accomplished the feat with an unmanned craft, while the United States was first in putting men around the moon in *Apollo 8.*

As we have cautioned in previous years, the "Remarks" column of these appendixes is never complete because of the inescapable lag behind each flight of the analysis and interpretation of results.

ASTRONAUTICS AND AERONAUTICS, 1968

Launch Date	Name, Country, International Designation, Vehicle	Payload Data	Apogee (st mi)	Perigee (st mi)	Period (min)	Inclination (degrees)	Remarks
Jan. 7	Surveyor VII (United States) 1968-1A Atlas-Centaur	Total weight: 2,293 lbs (weight at launch, including 1,446-lb retromotor, propellants, etc; weight of Surveyor lander at moon, 689 lbs). Objective: Softland on moon; obtain postlanding TV pictures of lunar surface; determine relative abundance of chemicals in lunar soil by alpha-scattering experiment; manipulate lunar material with surface sampler in view of TV camera. Payload: 10′-high × 14′-dia (around 3 extended landing gear) spacecraft; triangular frame with mast supporting rotatable planar-array antenna and solar panel, 2 folding booms deploying conical omnidirectional antennas, alpha-scattering experiment, thermal compartment housing decoder and signal processing equipment, 2 altitude radar antennas, 2 convex beryllium mirrors, surface-sampler experiment, survey TV camera, retromotor, and small bar magnet attached to footpad.	Soft landed on moon				Last of Surveyor series. Soft landed successfully 1/9/68 near designated target area, crater Tycho. Transmitted over 21,000 excellent TV pictures, some stereo. Performed alpha-scatter and moon-digging experiments; detected small amount of iron group elements, indicating highlands less dense than seas. Cameras detected laser beams directed from earth to spacecraft. Photographed earth and Jupiter.
Jan. 11	Explorer XXXVI (Geos II) (United States) 1968-2A Thor-Delta	Total weight: 460 lbs. Objective: To establish one world datum and to adjust all local datums to common center-of-mass of earth, providing geodetic control to ±10 meters relative to center-of-mass coordinate system. Payload: 48″-wide octahedron topped by truncated pyramid, for total spacecraft height of 32″; below protrudes 24″-dia	978	671	112.2	106.8	Thrust-Augmented Thor-Delta vehicle put satellite in orbit. Second of GEOS series. Carries experiments to study earth's gravitational field, to improve accuracy of satellite tracking systems and worldwide geodetic datum. Also studies effects of earth's atmosphere on modulated laser beam as part of long-range laser communications system.

ASTRONAUTICS AND AERONAUTICS, 1968

Launch Date	Name, Country, International Designation, Vehicle	Payload Data	Apogee (st mi)	Perigee (st mi)	Period (min)	Inclination (degrees)	Remarks
		hemisphere antenna; inside spacecraft are clock, memory corputer system, telemetry, 3 power systems, and geodetic instruments which include 2 C-band transponders and 1 C-band passive reflector; face of the satellite surfaced with solar panels, 4 flashing-light beacons, 4 silvered quartz corner reflectors for laser experiments; 60' boom deploys for gravity-gradient stabilization; antennas.					Argon laser successfully triggered satellite radio 3/27/68. Still in orbit, transmitting on command.
Jan. 16	Cosmos CXCIX (U.S.S.R.) 1968-3A Not available	Total weight: Not available. Objective: Continuation of Cosmos scientific satellite series. Payload: Not available.	137	99	88.2	65.2	Reentered 2/1/68.
Jan. 17	DOD spacecraft (United States) 1968-4A Thor-Agena D	Total weight: Not available. Objective: Develop space-flight techniques and technology. Payload: Not available.	335	285	94.5	75.1	Still in orbit.
Jan. 18	DOD spacecraft (United States) 1968-5A Titan III-B-Agena D	Total weight: Not available. Objective: Develop space-flight techniques and technology. Payload: Not available.	254	77	89.8	111.4	Reentered 2/4/68.
Jan. 19	Cosmos CC (U.S.S.R.) 1968-6A Not available	Total weight: Not available. Objective: Continuation of Cosmos scientific satellite series. Payload: Not available.	334	322	95.1	74	Still in orbit.
Jan. 22	Apollo 5 (United States) 1968-7A Saturn IB	Total weight: 70,200 lbs (weight in earth orbit, including S-IVB stage, lunar module, and nose cone). Objective: Verify operation of LM descent and ascent propulsion systems, including restart, and LM structure; evaluate staging and launch ve-	188	101	88.3	31.6	First earth orbital test of Apollo spacecraft lunar module; verified ascent and descent stage propulsion systems, including restart and throttle operations. CSM was replaced by a dummy nose cone. LM ascent stage reentered 1/24/68; descent stage

ASTRONAUTICS AND AERONAUTICS, 1968

Date	Spacecraft	Objective/Payload					Remarks
Jan. 24	DOD spacecraft (United States) 1968-8A Thor-Agena D	hicle performance. Payload: 58.4'-long S-IVB/IU/spacecraft lunar module adaptor/lunar module/nose cone replacing CSM; cameras; telemetry.					reentered 2/12/68. No attempts at recovery planned or made.
	and	Total weight: Not available. Objective: Develop space-flight techniques and technology. Payload: Not available.	269	112	90.6	81.5	Two payloads launched with single Long-Tank Thrust-Augmented Thor-Agena D booster. 8A reentered 2/27/68.
	DOD spacecraft 1968-8B	Total weight: Not available. Objective: Develop space-flight techniques and technology. Payload: Not available.	338	294	94.7	81.6	8B still in orbit.
Feb. 6	Cosmos CCI (U.S.S.R.) 1968-9A Not available	Total weight: Not available. Objective: Continuation of Cosmos scientific satellite series. Payload: Not available.	203	126	89.7	64.9	Reentered 2/14/68.
Feb. 20	Cosmos CCII (U.S.S.R.) 1968-10A Not available	Total weight: Not available. Objective: Continuation of Cosmos scientific satellite series. Payload: Not available.	283	131	91.2	48.4	Reentered 3/24/68.
Feb. 20	Cosmos CCIII (U.S.S.R.) 1968-11A Not available	Total weight: Not available. Objective: Continuation of Cosmos scientific satellite series. Payload: Not available.	748	737	109.2	74	Still in orbit.
Mar. 1*	DOD spacecraft (United States) 1968-12A Scout	Total weight: Not available. Objective: Provide navigation satellite capability. Payload: Not available.	711	640	106.9	89.9	Still in orbit.
Mar. 2	Zond IV (U.S.S.R.) 1968-13A Not available	Total weight: Not available. Objective: Study remote areas of near-earth space, and check new onboard systems and apparatus. Payload: Not available.	180 Parking orbit	131	89.5	51.6	Launched from parking orbit and acquired an apogee comparable to lunar altitude (240,000 mi). Speculation was that reentry attempt failed, heat shield burned up.

343

ASTRONAUTICS AND AERONAUTICS, 1968

Launch Date	Name, Country, International Designation, Vehicle	Payload Data	Apogee (st mi)	Perigee (st mi)	Period (min)	Inclination (degrees)	Remarks
Mar. 4	Ogo V (United States) 1968-14A Atlas-Agena D	Total weight: 1,347 lbs. Objective: Acquire correlative scientific data at high information rates in magnetic fields, energetic particles, and plasma disciplines; demonstrate technological capability for extended operations, 135 days or more, of OGO 3-axis stabilization system in highly elliptical orbit; conduct electric-field and radio and gamma-ray astronomy observations. Payload: 67″ x 32″ x 31″ rectangular spacecraft containing 9 plasma experiments, 4 field measurement experiments, 7 solar and galactic cosmic ray experiments, and 4 optical and radio emission experiments; deployed from ends of spacecraft are 2 22′ booms and 4 4′ booms, 2 6′ x 7.5′ solar panels, 2 orbital plane experimental packages, and 2 30′ dipole experiment antennas; active 3-axis and spin stabilization; solar supply to 2 28-v nickel cadmium batteries.	92,078	168	3,795.5	31.3	Provided first evidence of electric fields in low shock, and first observations of hydrogen cloud surrounding earth (geocorona). Also observed new particle and field phenomena.; noted presence of electric field discontinuities in solar wind. Still in orbit, transmitting on command.
Mar. 5	Cosmos CCIV (U.S.S.R.) 1968-15A Not available	Total weight: Not available. Objective: Continuation of Cosmos scientific satellite series. Payload: Not available.	524	168	95.7	70.9	Reentered 3/2/69.
Mar. 5	Cosmos CCV (U.S.S.R.) 1968-16A Not available	Total weight: Not available. Objective: Continuation of Cosmos scientific satellite series. Payload: Not available.	181	122	89.3	65.6	Reentered 3/13/68.
Mar. 5	Explorer XXXVII (United States) 1968-17A Scout	Total weight: 198 lbs. Objective: To continue and augment overall solar x-ray monitoring program; to per-	545	324	98.8	59.4	Second joint Naval Research Laboratory-NASA SOLRAD satellite. Provides real-time solar data through COSPAR to

ASTRONAUTICS AND AERONAUTICS, 1968

						scientific community. Still in orbit, transmitting on command.
Mar. 13	DOD spacecraft (United States) 1968-18A Titan III-B-Agena D	form temporal measurements of x-ray emission intensity and spectral quality of solar flare emission; to provide real-time solar-monitoring information to all interested participants. Payload: 30" (dia) x 27" 12-sided spacecraft; central band contains x-ray photomultipliers, Geiger tubes, photomultipliers, solar aspect systems, and attitude control and spin nozzles; 2 folding turnstile antenna systems project at 45° angles; 24 7" x 10" solar cell panels cover vertical surfaces; 8 telemetry systems; 2 attitude and primary spin replenishment systems; nickel-cadmium batteries.				
		Total weight: Not available. Objective: Develop space-flight techniques and technology. Payload: Not available.	260	82	89.9	99.9	Reentered 3/24/68.
Mar. 14	Cosmos CCVI (U.S.S.R.) 1968-19A Not available	Total weight: Not available. Objective: Continuation of Cosmos scientific satellite series. Payload: Not available.	398	372	97.1	81.2	Weather satellite carries TV and infrared cameras. Still in orbit.
Mar. 14	DOD spacecraft (United States) 1968-20A Thor-Agena D and	Total weight: Not available. Objective: Develop space-flight techniques and technology. Payload: Not available.	242	114	90.2	83	Two payloads launched with single Long-Tank Thrust-Augmented Thor-Agena D booster. 20A reentered 4/10/68.
	DOD spacecraft 1968-20B	Total weight: Not available. Objective: Develop space-flight techniques and technology. Payload: Not available.	326	299	94.6	83.1	20B still in orbit.
Mar. 16	Cosmos CCVII (U.S.S.R.) 1968-21A Not available	Total weight: Not available. Objective: Continuation of Cosmos scientific satellite series. Payload: Not available.	213	130	89.8	65.6	Reentered 3/24/68.

ASTRONAUTICS AND AERONAUTICS, 1968

Launch Date	Name, Country, International Designation, Vehicle	Payload Data	Apogee (st mi)	Perigee (st mi)	Period (min)	Inclination (degrees)	Remarks
Mar. 21	*Cosmos CCVIII* (U.S.S.R.) 1968–22A Not available	Total weight: Not available. Objective: Continuation of Cosmos scientific satellite series. Payload: Not available.	173	126	89.2	64.9	Reentered 4/2/68.
Mar. 22	*Cosmos CCIX* (U.S.S.R.) 1968–23A Not available	Total weight: Not available. Objective: Continuation of Cosmos scientific satellite series. Payload: Not available.	587	541	103.1	65.8	Still in orbit.
Apr. 3	*Cosmos CCX* (U.S.S.R.) 1968–24A Not available	Total weight: Not available. Objective: Continuation of Cosmos scientific satellite series. Payload: Not available.	232	123	90.2	81.3	Reentered 4/11/68.
Apr. 4	*Apollo 6* (United States) 1968–25A Saturn V	Total weight: 265,050 lbs (weight in earth orbit, including S-IVB stage, command and service module, and lunar module test article). Objective: Demonstrate structural and thermal integrity and compatibility of launch vehicle and spacecraft; demonstrate separation of 1st, 2nd, and 3rd stages; verify operation of propulsion, guidance and control, and electrical systems; evaluate performance of emergency detection system in closed-loop configuration; demonstrate mission support facilities and operations required for launch, mission, and CM recovery. Payload: 114.3′-long S-IVB/IU/Apollo adapter/command and service module/lunar module test article; cameras; telemetry.	223	107	89.8	31.6	Launch vehicle performance near nominal, but 2 of 5 2nd-stage J-2 engines shut down prematurely, causing remaining 2nd-stage engines and 3rd-stage engine to burn longer than planned. When 3rd stage failed to reignite on command, alternate mission was assumed and spacecraft placed into high trajectory with 13,823-mi apogee. Since insufficient propellant remained after extended burn, 2nd SPS burn not attempted; CM reentered 9 hr 50 min after launch. Despite S-IVB launch vehicle and S-IVB anomalies, 4 of 5 primary mission objectives were achieved.
Apr. 6	*OV 1–13* (United States) 1968–26A Atlas-F	Total weight: 290 lbs. Objective: Determine flexibility of cadmium sulfide solar cells in space environment; meas-	5,789	346	199.5	100.5	Two spacecraft launched with single Atlas booster. *OV 1–13* equipped with 9 experiments to study effects of

ASTRONAUTICS AND AERONAUTICS, 1968

	OV $I-14$ and 1968–26B	ure proton/electron energy spectra and angular distribution of electrons. Payload: 30″ x 27″ (dia) cylinder. Total weight: 300 lbs. Objective: Obtain high-resolution measurements of proton/electron flux, spectra, decay, and time variations. Payload: 30″ x 27″ (dia) cylinder.	6,173	348	207.8	100	cosmic radiation and nature of friction and wear on specific materials in space. Includes 8 experiments to measure electromagnetic interference at specific orbital attitudes. Still in orbit.
Apr. 7	$Luna$ XIV (U.S.S.R.) 1968–27A Not available	Total weight: Not available. Objective: Conduct studies of near-lunar space. Payload: Not available.	(Lunar orbital data) 541				$Luna$ XIV entered lunar orbit 4/10; transmitted telemetry but not photographic signals. Still in orbit.
Apr. 9	$Cosmos$ $CCXI$ (U.S.S.R.) 1968–28A Not available	Total weight: Not available. Objective: Continuation of Cosmos scientific satellite series. Payload: Not available.	960	124	102.1	81	Reentered 11/10/68.
Apr. 14	$Cosmos$ $CCXII$ (U.S.S.R.) 1968–29A Not available	Total weight: Not available. Objective: Achieve automatic docking in space. Payload: Not available.	124	112	88.3	51.6	On 4/15/68 $Cosmos$ $CCXII$ docked automatically with $Cosmos$ $CCXIII$. Second automatic docking in space, photographed by TV cameras in both spacecraft. $Cosmos$ $CCXII$ reentered 4/19/68.
Apr. 15	$Cosmos$ $CCXIII$ (U.S.S.R.) 1968–30A Not available	Total weight: Not available. Objective: Achieve automatic docking in space. Payload: Not available.	158	116	89.1	51.6	On 4/15/68 $Cosmos$ $CCXIII$ docked automatically with $Cosmos$ $CCXII$. Second automatic docking in space, photographed by TV cameras in both spacecraft. $Cosmos$ $CCXIII$ reentered 4/20/68.
Apr. 17	DOD spacecraft (United States) 1968–31A Titan III–B– Agena D	Total weight: Not available. Objective: Develop space-flight techniques and technology. Payload: Not available.	262	79	89.9	111.4	Reentered 4/29/68.
Apr. 18	$Cosmos$ $CCXIV$ (U.S.S.R.) 1968–32A Not available	Total weight: Not available. Objective: Continuation of Cosmos scientific satellite series. Payload: Not available.	230	124	90.1	81.3	Reentered 4/26/68.

Launch Date	Name, Country, International Designation, Vehicle	Payload Data	Apogee (st mi)	Perigee (st mi)	Period (min)	Inclination (degrees)	Remarks
Apr. 19**	Cosmos CCXV (U.S.S.R.) 1968-33A Not available	Total weight: Not available. Objective: Continuation of Cosmos scientific satellite series. Payload: Not available.	250	158	91.1	48.4	Carried 8 telescopes for radiation studies. Reentered 6/30/68.
Apr. 20	Cosmos CCXVI (U.S.S.R.) 1968-34A Not available	Total weight: Not available. Objective: Continuation of Cosmos scientific satellite series. Payload: Not available.	165	121	89.1	51.8	Reentered 4/28/68.
Apr. 21	Molniya I-8 (U.S.S.R.) 1968-35A Not available	Total weight: Not available. Objective: Develop and further improve satellite radio and TV communications system. Payload: Satellite with transmitter, command system, orientation system, orbit correction device, power supply.	24,680	257	718	65	Still in orbit.
Apr. 24	Cosmos CCXVII (U.S.S.R.) 1968-36A Not available	Total weight: Not available. Objective: Continuation of Cosmos scientific satellite series. Payload: Not available.	113	93	87.6	62.2	Reentered 4/26/68.
Apr. 25	Cosmos CCXVIII (U.S.S.R.) 1968-37A Not available	Total weight: Not available. Objective: Continuation of Cosmos scientific satellite series. Payload: Not available.	Less than 1 orbit				Possible test of fractional orbital bombardment system. Reentered 4/25/68.
Apr. 26	Cosmos CCXIX (U.S.S.R.) 1968-38A Not available	Total weight: Not available. Objective: Continuation of Cosmos scientific satellite series. Payload: Not available.	1,085	140	104.7	48.4	Space record set with 3rd Cosmos launched in 3 days and 9th spacecraft in 12 days. Reentered 8/2/69.
May 1	DOD spacecraft (United States) 1968-39A Thor-Agena D	Total weight: Not available. Objective: Develop space-flight techniques and technology. Payload: Not available.	176	115	88.6	83.1	Reentered 5/15/68.
May 7	Cosmos CCXX (U.S.S.R.) 1968-40A Not available	Total weight: Not available. Objective: Continuation of Cosmos scientific satellite series. Payload: Not available.	469	421	99	74	Still in orbit.

ASTRONAUTICS AND AERONAUTICS, 1968

Date	Name	Description	Col1	Col2	Col3	Col4	Remarks
May 16*	Iris I (U.S.–ESRO) 1968–41A Scout	Total weight: 164 lbs. Objective: Investigate x-ray and corpuscular radiation of sun and electron component of cosmic radiation. Payload: 33.5″ × 30″ (dia) 12-sided satellite, with each side carrying 2 areas of solar cells; cylindrical thrust tube provides longitudinal structural integrity; experiment instrumentation located at top, bottom, and around satellite midband; telemetry and data-storage system, real-time telemetry; tape recorder; 2 antennas deployed parallel to satellite's longitudinal axis; nickel-cadmium battery; thermal control system; spin stabilized.	673	204	98.9	97.2	First satellite designed, developed, and constructed by ESRO. NASA provided launch vehicle, conducted launch operations, and supplied data and tracking acquisition. Satellite (also called Esro II–B) performs solar astronomy and cosmic ray studies; 6 of 7 experiments functioning satisfactorily. Still in orbit, transmitting on command.
May 23	DOD spacecraft (United States) 1968–42A Thor-Burner II	Total weight: Not available. Objective: Develop space-flight techniques and technology. Payload: Not available.	561	509	102.1	98.9	Still in orbit.
May 24	Cosmos CCXXI (U.S.S.R.) 1968–43A Not available	Total weight: Not available. Objective: Continuation of Cosmos scientific satellite series. Payload: Not available.	1,294	133	108.3	48.4	Reentered 8/31/69.
May 31**	Cosmos CCXXII (U.S.S.R.) 1968–44A Not available	Total weight: Not available. Objective: Continuation of Cosmos scientific satellite series. Payload: Not available.	323	175	91.3	70.9	Speculation was that satellite might be solar flare monitoring replacement for Cosmos CCIV, launched 3/5/68. Reentered 10/11/68.
June 1	Cosmos CCXXIII (U.S.S.R.) 1968–45A Not available	Total weight: Not available. Objective: Continuation of Cosmos scientific satellite series. Payload: Not available.	608	137	89.9	72.9	Reentered 6/9/68.
June 4	Cosmos CCXXIV (U.S.S.R.) 1968–46A Not available	Total weight: Not available. Objective: Continuation of Cosmos scientific satellite series. Payload: Not available.	193	104	89.1	51.8	Reentered 6/12/68.
June 5	DOD spacecraft (United States) 1968–47A Titan III–B	Total weight: Not available. Objective: Develop space-flight techniques and technology. Payload: Not available.	262	85	89.9	110.5	Reentered 6/17/68.

Launch Date	Name, Country, International Designation. Vehicle	Payload Data	Apogee (st mi)	Perigee (st mi)	Period (min)	Inclination (degrees)	Remarks
June 11	Cosmos CCXV (U.S.S.R.) 1968-48A Not available	Total weight: Not available. Objective: Continuation of Cosmos scientific satellite series. Payload: Not available.	322	154	92.2	48.4	Reentered 11/2/68.
June 12	Cosmos CCXXVI (U.S.S.R.) 1968-49A Not available	Total weight: Not available. Objective: Continuation of Cosmos scientific satellite series. Payload: Not available.	399	371	96.8	81.2	Weather satellite carries TV and infrared cameras. Still in orbit.
June 13	IDCSP XIX, XX, XXI, XXII, XXIII, XXIV, XXV, XXVI (United States) 1968-50A B, C, D, E, F, G, H Titan III-C	Total weight: 800 lbs (100 lbs for each satellite). Objective: Establish an interim defense communication satellite system. Payload: 36" x 32" symmetrical 24-face polyhedron satellite, spin stabilized; 40-watt solar cell array; x-band transmitter.	21,027–21,401	20,954–20,976	1,355.5–1,350.6	00.1	All 8 spacecraft went into near-synchronous orbits, joining 18 other IDCSP satellites launched previously. Still in orbit.
June 18	Cosmos CCXXVII (U.S.S.R.) 1968-51A Not available	Total weight: Not available. Objective: Continuation of Cosmos scientific satellite series. Payload: Not available.	168	125	89.2	51.8	Reentered 6/26/68.
June 20	DOD spacecraft (United States) 1968-52A Thor-Agena D and DOD spacecraft 1968-52B	Total weight: Not available. Objective: Develop space-flight techniques and technology. Payload: Not available. Total weight: Not available. Objective: Develop space-flight techniques and technology. Payload: Not available.	251 322	113 273	90.3 94.1	85 85.1	Two payloads launched with single booster. 52A reentered 7/16/68. 52B still in orbit.
June 21	Cosmos CCXXVIII (U.S.S.R.) 1968-53A Not available	Total weight: Not available. Objective: Continuation of Cosmos scientific satellite series. Payload: Not available.	150	126	88.9	51.6	Reentered 7/3/68.
June 26	Cosmos CCXXIX (U.S.S.R.) 1968-54A Not available	Total weight: Not available. Objective: Continuation of Cosmos scientific satellite series. Payload: Not available.	204	188	89.8	72.9	Reentered 7/4/68.
July 4	Explorer XXXVIII (United States)	Total weight: 607 lbs (including apogee kick motor).	3,656 After	898	157	59.4	Satellite launched initially into elliptical transfer orbit.

ASTRONAUTICS AND AERONAUTICS, 1968

Date	Name/Designation	Description	Weight (lb)	Perigee (mi)	Apogee (mi)	Incl. (°)	Remarks
July 5	1968-55A Thrust-Augmented Improved Thor-Delta	Objective: Measure intensity and direction of radio signals from cosmic sources in 0.5-10-mhz range; obtain data for study of dynamic spectra and decay rates of sporadic radio bursts. Payload: 31" x 36" (dia) cylindrical spacecraft covered with lightweight skin of aluminum honeycomb; cylinder capped at each end by truncated cones; affixed are 4 2' fixed solar paddles, 1 630' long libration damper boom, 4 extendable 750' X-shaped booms, 2 elements forming dipole antenna measuring 120' tip to tip; 4 9-step frequency receivers; apogee-kick motor; turnstile telemetry and command antenna; spin stabilized by 2 8' x ¾" (dia) fiberglass booms; nickel-cadmium batteries.	apogee motor firing, 3,654	3,641	224.4	59.2	On 7/7/69 apogee kick motor fired, placing satellite into circular orbit. On 7/22, successfully gravity-gradient stabilized and all antennas partially deployed. On 10/8, 4 V antennas deployed full and final length of 750' (1,500' tip-to-tip) and damper boom to 315' (630' tip to tip). Discovered that earth, like Jupiter, sporadically emits low-frequency radio signals, suggesting earth's ionosphere and magnetic field are similar to Jupiter's. Still in orbit, transmitting on command.
July 5	Cosmos CCXXX (U.S.S.R.) 1968-56A Not available	Total weight: Not available. Objective: Continuation of Cosmos scientific satellite series. Payload: Not available.	338	176	92.8	48.4	Reentered 11/2/68.
July 6	Molniya I-9 (U.S.S.R.) 1968-57A Not available	Total weight: Not available. Objective: Develop and further improve satellite radio and TV communication system. Payload: Satellite with transmitter, command system, orientation system, orbit correction device, power supply.	24,734	246	714.6	65	Still in orbit, transmitting.
July 10	Cosmos CCXXXI (U.S.S.R.) 1968-58A Not available	Total weight: Not available. Objective: Continuation of Cosmos scientific satellite series. Payload: Not available.	243	128	89.6	64.9	Reentered 7/18/68.
July 11	OV I-15 (United States) 1968-59A Atlas F and	Total weight: 300 lbs. Objective: Measure density of atmosphere from 100 to 300 mi and its variations with position and time. Payload: 30" x 27" (dia) cylinder.	1,032	94	103	89.9	Two payloads launched with single booster. OV I-15 reentered 11/6/68.

351

Launch Date	Name, Country, International Designation, Vehicle	Payload Data	Apogee (st mi)	Perigee (st mi)	Period (min)	Inclination (degrees)	Remarks
	OV 1-16 1968-59B	Total weight: 600 lbs. Objective: Measure density of atmosphere from 90 to 110 mi. Payload: 23"-dia brass ball, omnidirectional antenna.	286	82	90.4	89.8	Densest (162 lb-per-cu-ft) satellite orbited by U.S. Could orbit closer to earth than any previous spacecraft. Reentered 8/19/68.
July 16	Cosmos CCXXXII (U.S.S.R.) 1968-60A Not available	Total weight: Not available. Objective: Continuation of Cosmos scientific satellite series. Payload: Not available.	221	124	89.4	65.3	Reentered 7/24/68.
July 18	Cosmos CCXXXIII (U.S.S.R.) 1968-61A Not available	Total weight: Not available. Objective: Continuation of Cosmos scientific satellite series. Payload: Not available.	935	124	101.9	81.9	Reentered 2/7/69.
July 30	Cosmos CCXXXIV (U.S.S.R.) 1968-62A Not available	Total weight: Not available. Objective: Continuation of Cosmos scientific satellite series. Payload: Not available.	183	129	89.5	51.8	Reentered 8/5/68.
Aug. 6	DOD spacecraft (United States) 1968-63A Atlas-Agena D	Total weight: Not available. Objective: Develop space-flight techniques and technology. Payload: Not available.	24,769	19,685	1,436	9.9	First unannounced launch from ETR in 5 yrs; long-duration orbit desired. Still in orbit.
	DOD spacecraft (United States) 1968-64A Titan III-B-Agena D	Total weight: Not available. Objective: Develop space-flight techniques and technology. Payload: Not available.	266	88	90.5	109.5	Reentered 8/16/68.
Aug. 7	DOD spacecraft (United States) 1968-65A Thor-Agena	Total weight: Not available. Objective: Develop space-flight techniques and technology. Payload: Not available.	161	100	88.7	82.1	Reentered 8/27/68.
Aug. 8	Explorer XXXIX (United States)	Total weight: 20.5 lbs. Objective: Extend measurements	1,548	425	117.9	80.6	Two payloads launched with single booster. Air Density

Date	Name/Designation	Payload/Objective	Weight	Apogee	Perigee	Inclination	Remarks
	1968-66A Scout and	of latitudinal, seasonal, and solar cycle variations in the upper air density; compare measurements obtained with other satellites to gain insight into sources of atmospheric heating. Payload: 12′ (dia) polka-dotted inflatable sphere, constructed of 4-ply laminate of alternating layers of 1/2-mil aluminum foil and polyethylene-terephtholate plastic film; beacon; 4 groups of solar cells; telemetry; nickel-cadmium batteries.					balloon was ejected, inflated fully, and separated from Injun Explorer. Beacon modulating satisfactorily. Still in orbit, still transmitting.
	Explorer XL 1968-66B	Total weight: 153 lbs. Objective: Make direct measurement of downflux of charged particles into atmosphere; study geomagnetically trapped charged particles with emphasis on their spectra, spatial distribution, and time variations; study very low frequency radio emissions in the ionosphere. Payload: 29″ x 30″ (dia) hexagonal cylinder with 2 5′ AFCRL experiment booms, 2 electrostatic boom antennas, telemetry and command antenna, magnetic loop antenna; 4 experiments; magnetic stabilization; clock subsystem; tape recorder; solar cells; nickel-cadmium batteries.	1,573	424	118.3	80.6	*Injun V* Explorer separated from 4th stage rocket motor and was despun. Experiment booms extended and experiments turned on. Anomaly occurred in power subsystem and spacecraft operating in power budgeting mode. Quantity of data obtainable reduced, but quality of data unaffected. Still in orbit, still transmitting.
Aug. 9	*Cosmos CCXXXV* (U.S.S.R.) 1968-67A Not available	Total weight: Not available. Objective: Continuation of Cosmos scientific satellite series. Payload: Not available.	176	126	89.3	51.8	Reentered 8/17/68.

ASTRONAUTICS AND AERONAUTICS, 1968

Launch Date	Name, Country, International Designation, Vehicle	Payload Data	Apogee (st mi)	Perigee (st mi)	Period (min)	Inclination (degrees)	Remarks
Aug. 10	Ats IV (United States) 1968–68A Atlas-Centaur	Total weight: 860 lbs. Objective: Conduct carefully instrumented gravity gradient orientation experiment to provide basic design information and determine feasibility and usefulness of gravity gradient as a means of stabilizing synchronous orbit applications satellites in presence of perturbing effects generated within the satellite or caused by the environment. Payload: 72″ x 56″ (dia) cylinder with 2 360° circular cylindrical solar panel sections separated by 25″ cylindrical midsection compartment, 22,000 solar cells, 4 123′ booms which deploy in X configuration and stabilize spacecraft, 2 45′ damper booms, telemetry and command antennas, payload and control sensors, TV camera, Image Orthicon Camera (day/night), 4 experiments, omnidirectional antenna, gravity-gradient and active stabilization.	476	115	94.5	29.1	Failure of Centaur stage to ignite for a second burn prevented achievement of transfer orbit before injection into synchronous orbit. Highly elliptical orbit precluded return of meaningful data. Reentered 10/17/68.
Aug. 16	Esss VII (United States) 1968–69A Thor-Delta	Total weight: 325 lbs. Objective: Place and operate Advanced Vidicon Camera System in sun-synchronous orbit with local equator crossing time between 2:35 pm and 2:55 pm, so that daily AVCS pictures of entire earth can be obtained regularly and dependably. Payload: 22″ x 42″ (dia) 18-sided hatbox-shaped polygon, with 18″ receiving antenna and 4 22″ transmitting whip	918	895	114.9	101.7	Esss VII was put into near-polar orbit by Thrust-Augmented Long-Tank Thor-Delta booster. All systems functioned normally; after extensive postlaunch evaluation, satellite was turned over to ESSA 9/3/68 to operate. Still in orbit, transmitting on command.

ASTRONAUTICS AND AERONAUTICS, 1968

Date	Name	Description					Remarks
		antennas, containing 2 AVCS cameras, FM transmitters, 2 spin control systems (magnetic coil; solid-propellant rockets); 4 solar radiation sensors; 63 nickel-cadmium batteries; 10,020 n-on-p solar cells.					
Aug. 27	*Cosmos CCXXXVI* (U.S.S.R.) 1968-70A Not available	Total weight: Not available. Objective: Continuation of Cosmos scientific satellite series. Payload: Not available.	390	367	96.8	56	Still in orbit.
Aug. 27	*Cosmos CCXXXVII* (U.S.S.R.) 1968-71A Not available	Total weight: Not available. Objective: Continuation of Cosmos scientific satellite series. Payload: Not available.	201	124	89.7	65.4	Reentered 9/4/68.
Aug. 28	*Cosmos CCXXXVIII* (U.S.S.R.) 1968-72A Not available	Total weight: Not available. Objective: Continuation of Cosmos scientific satellite series. Payload: Not available.	130	126	88.6	51.7	Reentered 9/1/68.
Sept. 5	*Cosmos CCXXXIX* (U.S.S.R.) 1968-73A Not available	Total weight: Not available. Objective: Continuation of Cosmos scientific satellite series. Payload: Not available.	163	123	89.1	51.8	Reentered 9/13/68.
Sept. 10	DOD spacecraft (United States) 1968-74A Titan III-B	Total weight: Not available. Objective: Develop space-flight techniques and technology. Payload: Not available.	200	89	89.1	106	Reentered 9/25/68.
Sept. 14	*Cosmos CCXL* (U.S.S.R.) 1968-75A Not available	Total weight: Not available. Objective: Continuation of Cosmos scientific satellite series. Payload: Not available.	176	126	89.3	51.8	Reentered 9/21/68.
Sept. 15	*Zond V* (U.S.S.R.) 1968-76A Not available	Total weight: Not available. Objective: Study physical characteristics of outer space in area of the moon. Payload: Not available.	Circumlunar flight				First successful circumlunar flight, reentry, and recovery. Carried biological experiments and took pictures of moon and earth. Reentered 9/21/68.
Sept. 16	*Cosmos CCXLI* (U.S.S.R.) 1968-77A Not available	Total weight: Not available. Objective: Continuation of Cosmos scientific satellite series. Payload: Not available.	203	123	89.7	65.4	Reentered 9/24/68.

Launch Date	Name, Country, International Designation, Vehicle	Payload Data	Apogee (st mi)	Perigee (st mi)	Period (min)	Inclination (degrees)	Remarks
Sept. 18	DOD spacecraft (United States) 1968-78A Thor-Agena D and	Total weight: Not available. Objective: Develop space-flight techniques and technology. Payload: Not available.	243	111	90.1	88	Two payloads launched with single booster. 78A reentered 10/8/68.
	DOD spacecraft 1968-78B	Total weight: Not available. Objective: Develop space-flight techniques and technology. Payload: Not available.	318	312	94.7	88.2	Reentered 9/28/69.
Sept. 20	Cosmos CCXLII (U.S.S.R.) 1968-79A Not available	Total weight: Not available. Objective: Continuation of Cosmos scientific satellite series. Payload: Not available.	251	167	91.2	70.9	Reentered 11/13/68.
Sept. 23	Cosmos CCXLIII (U.S.S.R.) 1968-80A Not available	Total weight: Not available. Objective: Continuation of Cosmos scientific satellite series. Payload: Not available.	185	128	89.5	71.8	Reentered 10/4/68.
Sept. 26	OV II-5 (United States) 1968-81A Titan III-C and	Total weight: 250 lbs. Objective: Investigate radiation at synchronous altitudes. Payload: 23" x 23" x 24" near-cubic spacecraft.	22,232	21,827	1,418.1	3	Four spacecraft launched with single booster. Still in orbit.
	Ers XXVIII 1968-81B and	Total weight: 21.5 lbs. Objective: Investigate radiation in Van Allen belt. Payload: Octahedron.	22,236	115	630.8	26.4	Still in orbit, transmitting when in sunlight.
	Ers XXI 1968-81C and	Total weight: 28 lbs. Objective: Measure transfer of heat in liquid Freon in a zero-G environment. Payload: Not available.	22,225	22,220	1,435.8	3	Still in orbit.
	Les VI 1968-81D	Total weight: 360.4 lbs. Objective: Provide test data on an expanded UHF tactical communications satellite operating with aircraft, ship, and ground terminals.	22,236	22,233	1,435.9	2.9	Still in orbit.

ASTRONAUTICS AND AERONAUTICS, 1968

Date		Name	Payload/Objective				Remarks
Oct.	2	Cosmos CCXLIV (U.S.S.R.) 1968-82A Not available	Payload: 5½' x 4' (dia) cylinder; spin stabilized; geostationary.	98	87	87.4	49.6	Possible test of fractional orbital bombardment system. Reentered 10/2/68.
Oct.	3	Cosmos CCXLV (U.S.S.R.) 1968-83A Not available	Total weight: Not available. Objective: Continuation of Cosmos scientific satellite series. Payload: Not available.	299	169	92	70.9	Reentered 1/15/69.
Oct.	3	Aurorae (U.S.–ESRO) 1968-84A Scout	Total weight: 189 lbs. Objective: Perform integrated study of high-latitude ionosphere, including particle experiments, auroral photometry, and ionospheric experiments. Payload: 39" x 30" (dia) cylinder; sensor booms extending tip to tip, 95.6"; 60" height; carries 8 experiments; 7,120 n-on-p solar cells; battery; high- and low-speed telemetry and command system; passive magnetic stabilization.	952	159	102.8	98.8	Satellite (also called Esro 1) placed into near-polar orbit. ESRO responsible for experiment instrumentation, delivery, equipment and personnel necessary for mating, and spacecraft testing. NASA provided launch vehicle, conducted launch operations, and supplied tracking and data acquisition. All experiments operating, and returning good data. Still in orbit, transmitting on command.
Oct.	5	Molniya I-10 (U.S.S.R.) 1968-85A Not available	Total weight: Not available. Objective: Develop and further improve satellite radio and TV communications system. Payload: Satellite with transmitter, command system, orientation system, orbit correction device, power supply.	24,630	267	711.9	64.8	Satellite to relay telephone and telegraph communications and TV programs to far nothern and far eastern U.S.S.R. and to Central Asia. Still in orbit.
Oct.	5	DOD spacecraft (United States) 1968-86A Thor-Agena D	Total weight: Not available. Objective: Develop space-flight techniques and technology. Payload: Not available.	316	300	94.5	74.9	Still in orbit.
Oct.	7	Cosmos CCXLVI (U.S.S.R.) 1968-87A Not available	Total weight: Not available. Objective: Continuation of Cosmos scientific satellite series. Payload: Not available.	197	90	89.1	65.3	Reentered 10/12/68.

Launch Date	Name, Country, International Designation, Vehicle	Payload Data	Apogee (st mi)	Perigee (st mi)	Period (min)	Inclination (degrees)	Remarks
Oct. 11	*Cosmos CCXLVII* (U.S.S.R.) 1968-88A Not available	Total weight: Not available. Objective: Continuation of Cosmos scientific satellite series. Payload: Not available.	213	184	89.9	65.4	Reentered 10/19/68.
Oct. 11	*Apollo 7* (United States) 1968-89A Saturn IB	Total weight: 67,669 lbs (weight in earth orbit, including S-IVB stage, instrument unit, spacecraft LM adapter, and command and service modules). Objective: Demonstrate CSM crew performance, CSM rendezvous capability, and crew/vehicle/support facilities performance; qualify heat shield; determine environmental-system-radiator heat rejection and thermal degradation; qualify J-2 engine augmented spark ignition (ASI). Payload: 111'-long S-IVB/IU/ Apollo adapter/Block II command and service modules; cameras; telemetry.	178	138	89.6	31.6	*Apollo 7* spacecraft—carrying Astronauts Walter M. Schirra, Jr., Donn F. Eisele, and R. Walter Cunningham—launched successfully from Launch Complex 34, KSC. First manned Apollo flight, first U.S. manned flight since 11/66. Confirmed operation of all major Apollo inflight systems except LM. Crew in CSM simulated docking with S-IVB stage to within 4'-5', later rendezvoused to within 70 ft of stage. SPS fired successfully 8 times. First live commercial TV from U.S. manned space flight, first flight test of Apollo spacesuits. CM separated from SM, landed in Atlantic 10/22/68 after 163 revolutions and 260 hrs 9 min of flight.
Oct. 19	*Cosmos CCXLVIII* (U.S.S.R.) 1968-90A Not available	Total weight: Not available. Objective: Continuation of Cosmos scientific satellite series. Payload: Not available.	337	294	94.7	62.2	Still in orbit.
Oct. 20	*Cosmos CCXLIX* (U.S.S.R.) 1968-91A Not available	Total weight: Not available. Objective: Continuation of Cosmos scientific satellite series. Payload: Not available.	1,341	305	112.1	62.3	Still in orbit.
Oct. 23	DOD spacecraft (United States) 1968-92A Thor-Burner II	Total weight: Not available. Objective: Develop space-flight techniques and technology. Payload: Not available.	529	497	101.3	99	Still in orbit.

ASTRONAUTICS AND AERONAUTICS, 1968

Date	Spacecraft	Description					
Oct. 25	Soyuz II (U.S.S.R.) 1968-93A Not available	Total weight: Not available. Objective: Serve as target for joint experiments with manned spacecraft. Payload: Not available.	142	119	88.6	51.7	Flew dual mission with Soyuz III, launched 10/26/68. Reentered 10/28/68.
Oct. 26	Soyuz III (U.S.S.R.) 1968-94A Not available	Total weight: Not available. Objective: Perfect rendezvous techniques in orbit and perform joint experiments with target vehicle. Payload: 3-unit spacecraft with 2 crew cabins, 4 TV cameras, 2 large wing-like solar panels.	127	114	88.3	51.7	Flew dual mission with unmanned Soyuz II, launched 10/25/68. Automatic maneuvering brought spacecraft within 656' of Soyuz II on first orbit. Cosmonaut Georgy T. Beregovoy later performed manual maneuvers; no docking accomplished. Live TV transmissions showed inflight activity. Soyuz III softlanded 10/30/68, completing 94 hrs 51 min and 64 orbits.
Oct. 31	Cosmos CCL (U.S.S.R.) 1968-95A Not available	Total weight: Not available. Objective: Continuation of Cosmos scientific satellite series. Payload: Not available.	525	468	100.6	74	Still in orbit.
Oct. 31	Cosmos CCLI (U.S.S.R.) 1968-96A Not available	Total weight: Not available. Objective: Continuation of Cosmos scientific satellite series. Payload: Not available.	140	106	88.3	64.7	Reentered 11/18/68.
Nov. 1	Cosmos CCLII (U.S.S.R.) 1968-97A Not available	Total weight: Not available. Objective: Continuation of Cosmos scientific satellite series. Payload: Not available.	1,335	330	112.4	62.8	Still in orbit.
Nov. 3	DOD spacecraft (United States) 1968-98A Thor-Agena D	Total weight: Not available. Objective: Develop space-flight techniques and technology. Payload: Not available.	177	108	88.8	82.1	Reentered 11/23/68.
Nov. 6	DOD spacecraft (United States) 1968-99A Titan III-B	Total weight: Not available. Objective: Develop space-flight techniques and technology. Payload: Not available.	249	90	89.8	106	Reentered 11/20/68.

Launch Date	Name, Country, International Designation, Vehicle	Payload Data	Apogee (st mi)	Perigee (st mi)	Period (min)	Inclination (degrees)	Remarks
Nov. 8	Pioneer IX (United States) 1968-100A Thor-Delta and Tetr II 1968-100B	Total weight: 148 lbs. Objective: Collect scientific data on electromagnetic and plasma properties of interplanetary medium for period covering 6 or more passages of solar activity centers; acquire data when solar event of high scientific value occurs; provide synoptic study of solar-interplanetary relations; provide target for MSFN checkout of equipment and training of personnel by launching Test and Training Satellite (Tetr II) as secondary payload. Payload: 35" x 37" (dia) cylindrical spacecraft, with boom protruding from the top containing communications antennas; sides are covered with solar cells except for narrow band carrying experiments and 3 6' booms, 2 for orientation jets and 1 for the magnetometer; data storage; transmitter, batteries; total of 8 experiments. Total weight: 40 lbs. Objective: See Pioneer IX. Payload: 11" on-a-side octahedron, with top apex supporting S-band antenna and sides covered with solar cells; S-band transmitter, receiver; batteries.	0.99 au 582	0.75 au 231	297.55 (days) 97.9	0.09 33	Pioneer IX was launched by Thrust-Augmented Improved Thor-Delta. As Delta 3rd stage burned out, Tetr II pickaback satellite was spring-ejected into earth orbit; Pioneer IX continued on into heliocentric orbit. All experiments functioned and returned good data. Still in orbit. See Pioneer IX. Still in orbit.
Nov. 10	Zond VI (U.S.S.R.) 1968-101A Not available	Total weight: Not available. Objective: Conduct scientific explorations along flight route and in near-lunar space. Payload: Not available.	Circumlunar flight				Second successful Soviet circumlunar flight, reentry and recovery. Zond VI circled moon at minimum distance of 1,504 mi. Carried biological specimens for ra-

ASTRONAUTICS AND AERONAUTICS, 1968

Date	Name	Details				Remarks
Nov. 13	Cosmos CCLIII (U.S.S.R.) 1968-102A Not available	Total weight: Not available. Objective: Continuation of Cosmos scientific satellite series. Payload: Not available.	209	134	90	65.4	Reentered 11/18/68.
Nov. 16	Proton IV (U.S.S.R.) 1968-103A Not available	Total weight: 37,478 lbs. Objective: Study the nature of cosmic rays of high and superhigh energies and their interaction with atomic nuclei. Payload: Not available.	293	154	91.7	51.5	Proton IV, largest unmanned scientific satellite to date. Reentered 7/24/69.
Nov. 21	Cosmos CCLIV (U.S.S.R.) 1968-104A Not available	Total weight: Not available. Objective: Continuation of Cosmos scientific satellite series. Payload: Not available.	206	130	89.8	65.4	Reentered 11/29/68.
Nov. 29	Cosmos CCLV (U.S.S.R.) 1968-105A Not available	Total weight: Not available. Objective: Continuation of Cosmos scientific satellite series. Payload: Not available.	197	131	89.6	65.4	Reentered 12/7/68.
Nov. 30	Cosmos CCLVI (U.S.S.R.) 1968-106A Not available	Total weight: Not available. Objective: Continuation of Cosmos scientific satellite series. Payload: Not available.	760	743	109.4	74	Still in orbit.
Dec. 3	Cosmos CCLVII (U.S.S.R.) 1968-107A Not available	Total weight: Not available. Objective: Continuation of Cosmos scientific satellite series. Payload: Not available.	272	167	91.6	70.9	Reentered 3/5/69.
Dec. 4	DOD spacecraft (United States) 1968-108A Titan III-B	Total weight: Not available. Objective: Develop space-flight techniques and technology. Payload: Not available.	454	94	89.3	106.2	Reentered 12/12/68.

The Nov. 13 remarks also include: "radiation experiments, automatic camera for lunar photography. Landed 11/17/68 after skip-lob reentry and aerodynamic lift for deceleration."

Launch Date	Name, Country, International Designation, Vehicle	Payload Data	Apogee (st mi)	Perigee (st mi)	Period (min)	Inclination (degrees)	Remarks
Dec. 5	Heos-A (U.S.-ESRO) 1968-109A Thor-Delta	Total weight: 240 lbs. Objective: Investigate interplanetary magnetic fields and study solar and cosmic-ray particles outside magnetosphere during period of maximum solar activity. Payload: 27.3″ x 48.6″ (dia) octagonal cylinder carrying 8 experiments, spin-stabilized with telemetry and command system and omnidirectional antenna, 8,576 solar cells around periphery of cylinder, optical sensors on base, and magnetometer booms extending from top; overall height 94.3″.	138,881	268	6,749	28.3	Heos-A launched into elliptical orbit by Thrust-Augmented Improved Thor-Delta; 3rd ESRO mission orbited by NASA, first reimbursable mission. All experiments operating satisfactorily and returning good data. Barium vapor cloud experiment successfully accomplished 3/18/69, with cloud visible to naked eye. Still in orbit, transmitting on command.
Dec. 7	Oao II (United States) 1968-110A Atlas-Centaur	Total weight: 4,400 lbs (including 1,000 lbs of scientific instrumentation). Objective: Obtain precise astronomical observations from above earth's atmosphere of stars, planets, nebulae, galaxies, and interplanetary and interstellar matter in the regions of their electromagnetic radiation spectrum that do not penetrate earth's atmosphere; i.e., ultraviolet, x-ray, and gamma-ray regions. Payload: 10′ x 7′ octagonal cylinder, from which extend 8 solar panels for total width of 21′; attitude control system, 6 star trackers; vertical trusses and horizontal shelves form equipment bays around central tube to give spacecraft extreme rigidity and minimize thermal distortions;	486	479	100.4	35	Oao II, developed by University of Wisconsin and Smithsonian Astrophysical Laboratory, is heaviest and most complex automated spacecraft developed by U.S. All instruments functioning normally and making precision observations of relatively unexplored ultraviolet region of spectrum. Still in orbit, transmitting on command.

ASTRONAUTICS AND AERONAUTICS, 1968

Date	Name	Description	Col1	Col2	Col3	Notes
Dec. 10	*Cosmos CCLVIII* (U.S.S.R.) 1968–111A Not available	inertial balance booms, 2 transmitters, 4 receivers; 2 experiments (WEP and Smithsonian celescope) carry 11 telescopes; 3 nickel-cadmium batteries; over 109,000 solar cells. Total weight: Not available. Objective: Continuation of Cosmos scientific satellite series. Payload: Not available.	188	128	89.5	64.9	Reentered 12/18/68.
Dec. 12	DOD spacecraft (United States) 1968–112A Thor-Agena D and	Total weight: Not available. Objective: Develop space-flight techniques and technology. Payload: Not available.	148	109	88.6	81	Two payloads launched with single booster. 112A reentered 12/28/68.
	DOD spacecraft 1968–112B	Total weight: Not available. Objective: Develop space-flight techniques and technology. Payload: Not available.	916	862	114.4	80.3	112B still in orbit.
Dec. 14	*Cosmos CCLIX* (U.S.S.R.) 1968–113A Not available	Total weight: Not available. Objective: Continuation of Cosmos scientific satellite series. Payload: Not available.	827	132	100.3	48.4	Reentered 5/6/69.
Dec. 15	*Essa VIII* (United States) 1968–114A Thor-Delta	Total weight: 300 lbs. Objective: Operate 2 Advanced Picture Transmission (APT) camera systems in a sun-synchronous orbit crossing equator between 8:40 and 9:00 am daily; provide regular daily global cloud coverage. Payload: 22" x 42" 18-sided hatbox-shaped polygon, with 18" receiving antenna and 4 22" transmitting whip antennas; containing 2 APT camera systems, FM transmitters, 2 spin-control systems (magnetic coil; small solid-propellant rockets); 8 solar radiation sensors; 63 nickel-cadmium batteries; 10,020 n-on-p solar cells.	903	875	114.6	101.9	*Essa VIII* was put into nearly polar, sun-synchronous orbit by Thrust-Augmented, Long-Tank Thor-Delta. All systems functioned normally; after extensive post-launch evaluation, satellite was turned over to ESSA on 12/24/68 to operate. Still in orbit, operating on command.

Launch Date	Name, Country, International Designation, Vehicle	Payload Data	Apogee (st mi)	Perigee (st mi)	Period (min)	Inclination (degrees)	Remarks
Dec. 16	*Cosmos CCLX* (U.S.S.R.) 1968-115A Not available	Total weight: Not available. Objective: Continuation of Cosmos scientific satellite series. Payload: Not available.	24,591	317	712.3	64.9	Still in orbit.
Dec. 18*	*Intelsat-III F-2* (United States) 1968-116A Thor-Delta	Total weight: 322 lbs (in synchronous orbit; 632 lbs at lift-off, including 310 lbs of apogee motor fuel). Objective: Place satellite and apogee motor into proper transfer orbit, provide tracking and telemetry and back-up calculations through the transfer orbit so satellite can be injected into synchronous orbit for commercial communications by ComSat Corp. Payload: 78″ x 56″ (dia) cylindrical satellite, including mechanically despun receive/transmit and omnidirectional antennas on top of spacecraft; outer cylindrical sleeve covered with 10,720 solar cells; 2 frequency translation mode repeaters with design capability for 1,200 high-quality, 2-way voice or 4 TV channels, or a mix; attitude and control system; 1 battery.	22,590 After apogee motor firing, 22,328	162 21,833	642.9 1,436	28.8 .79	*Intelsat-III F-2* was put into good transfer orbit by Thrust-Augmented, Long-Tank Thor-Delta. Apogee motor fired 12/20, putting satellite into synchronous orbit over Atlantic. Commercial operations began 12/28. On 6/29/69, despun antenna ceased functioning; service reestablished on later repositioning of antenna. Still in orbit, still transmitting.
Dec. 20**	*Cosmos CCLXI* (U.S.S.R.) 1968-117A Not available	Total weight: Not available. Objective: Study upper atmospheric layers of the earth and obtain data on the nature of the Northern Lights. Payload: Not available.	416	129	93	71	Reentered 2/12/69.
Dec. 21	*Apollo 8* (United States) 1968-118A Saturn V	Total weight: 284,390 lbs (weight in earth orbit, including S-IVB stage, instrument unit, spacecraft LM adapter, LM test article, and CSM).	118 Lunar orbit, 194	114 69	32.5	88.2	*Apollo 8* launched successfully; during 2nd earth orbit, Astronauts Frank Borman, James A. Lovell, Jr., and William A. Anders fired

	Objective: Demonstrate crew, space vehicle, mission support facilities performance during manned Saturn V mission with CSM; demonstrate performance of nominal and selected backup lunar orbit rendezvous mission activities, including translunar injection and CSM navigation, communications, midcourse corrections, consumables assessment, and passive thermal control. Payload: 114-ft-long S-IVB/IU Apollo adapter/Block II command and service modules/lunar module test article B; cameras; telemetry.	491	164	95.2	48.4	S-IVB stage engine, boosting spacecraft into translunar coast. On 12/24, spacecraft entered lunar orbit for 20 hrs (10 revolutions). Astronauts surveyed potential landing sites, took photos, and transmitted live TV of lunar surface. On 12/25, spacecraft motor fired to lunar escape speed to begin transearth coast. CM separated from SM on 12/27/68, reentered earth's atmosphere, and was picked up by U.S.S. *Yorktown* after total 147-hr flight. First men to orbit moon, first manned Saturn V flight, first manned launch from KSC Complex 39, Pad A. Fastest and farthest man had traveled to date.
Dec. 26 *Cosmos CCLXII* (U.S.S.R.) 1968-119A Not available	Total weight: Not available. Objective: Continuation of Cosmos scientific satellite series. Payload: Not available.				Reentered 7/18/69.

*Local time at site; 1 day later by Greenwich time.
**Local time at site; 1 day earlier by Greenwich time.

Appendix B

CHRONOLOGY OF MAJOR NASA LAUNCHES

JANUARY 1, 1968, THROUGH DECEMBER 31, 1968

This chronology of major NASA launchings in 1968 is intended to provide an accurate and ready historical reference, one compiling and verifying information previously scattered over several sources. It includes launchings of all rocket vehicles larger than sounding rockets launched either by NASA or under "NASA direction (e.g., NASA provided vehicles, launch facilities, and performed the launches for ComSatCorp's *Intelsat-III F–2*, for ESSA's *Essa VII* and *VIII*, and for ESRO's *Iris I*, and *Aurorae* and *Heos–A* in 1968). NASA sounding rocket launches are published annually by the Goddard Space Flight Center Historian in *Goddard Projects Summary: Satellites and Sounding Rockets.*

An attempt has been made to classify the performance of both the launch vehicle and the payload and to summarize total results in terms of primary mission. Three categories have been used for evaluating vehicle performance and mission results—successful (S), partially successful (P), and unsuccessful (U). A fourth category, unknown (Unk), has been provided for payloads where vehicle malfunctions did not give the payload a chance to exercise its main experiments. These divisions are necessarily arbitrary, since many of the results cannot be neatly categorized. Also they ignore the fact that a great deal is learned from missions that may have been classified as unsuccessful.

Date of launch is referenced to local time at the launch site. Sources used were all open ones, verified where in doubt from the project offices in NASA Headquarters and from NASA Centers. For further information on each item, see Appendix A of this volume and the entries in the main chronology as referenced in the index. Prepared May 1969 by Dr. Frank W. Anderson, Jr. Deputy NASA Historian (EH).

Date	Name (NASA Code)	General Mission	Launch vehicle (site)	Performance Vehicle	Performance Payload	Performance Mission	Remarks
Jan. 7	Surveyor VII (Surveyor G)	Scientific lunar landing probe	Atlas-Centaur (ETR)	S	S	S	Surveyor VII softlanded 1/9 on ejecta blanket of crater Tycho; during 1st lunar day transmitted 21,274 TV photos, operated its 3 scientific experiments. On 2/12 was turned on again, operated until 2/20. Last of Surveyor series.
Jan. 11	Geos II or Explorer XXXVI (GEOS-B)	Applications geodetic satellite	Thrust-Augmented Thor-Delta (WTR)	S	S	S	Geos II provided more than required 90 days of precise spacecraft positioning using gravity gradient system; also checked geodetic positioning of 40 control points on earth.
Jan. 22	Apollo 5 (AS-204)	Launch vehicle and spacecraft development	Saturn IB (ETR)	S	S	S	Apollo 5 was launched into earth orbit; lunar module, in 1st flight test, sparated, fired its ascent and descent propulsion systems several times.
Mar. 4	Ogo V (OGO-E)	Scientific satellite, geophysical	Atlas-Agena D (ETR)	S	S	S	Ogo IV was launched into highly elliptical earth orbit (92,078/168), studied particle wave processes through magnetosphere into geomagnetic tail.
Mar. 5	Explorer XXXVII (Solar Explorer B)	Scientific satellite, solar	Scout (WS)	S	S	S	Explorer XXXVII's orbit was not circular as planned but sufficed for measuring solar x-rays in ascending solar cycle.
Apr. 4	Apollo 6 (AS-502)	Launch vehicle and spacecraft development	Saturn V (ETR)	U	Unk	U	Apollo 6 was launched into earth orbit though 2 of 5 2nd-stage engines stopped prematurely. The 3rd-stage engine failed to restart but command module achieved intended 12,000-nm apogee, reentered, and was recovered by U.S.S. Okinawa. Three of 5 primary objectives were met but mission was not a success.
Apr. 27	Reentry F	20,000 fps reentry test	Scout (WS)	S	S	S	Reentry F reentered at 19,750 fps, impacted within 5 mi of aiming point near Bermuda; telemetry good.
May 16*	Iris I (Esro II-B)	Scientific satellite, radiation	Scout (WTR)	S	S	S	Iris I (International Radiation Investigation Satellite) operated 6 of 7 experiments.
May 18	Nimbus B	Meteorological earth satellite	Thrust-Augmented Thor-Agena D (WTR)	U	Unk	U	Nimbus B was destroyed by Range Safety Officer when booster malfunctioned; debris fell into Pacific off California coast. Nimbus payload included 2 25-w

ASTRONAUTICS AND AERONAUTICS, 1968

Date	Name	Purpose	Launch Vehicle			Remarks
July 4	Explorer XXXVIII (RAE-A)	Scientific satellite, radio astronomy	Thrust-Augmented Thor-Delta (WTR)	S	S	SNAP-19 radioisotope electric power generators. First use of Thorad-Agena D.
Aug. 8	Explorer XXXIX (Air Density) and Explorer XL (Injun V)	Scientific satellites, ionospheric	Scout (WTR)	S	S	Explorer XXXVIII demonstrated required 30 days of gravity gradient stabilization with 920' antenna, returned significant data on low-frequency radio emissions from celestial sources.
Aug. 10	Ats IV (ATS-D)	Applications technology satellite	Atlas-Centaur (ETR)	U	Unk	Explorer XXXIX was launched into polar orbit as companion satellite to Explorer XL and on same Scout booster. It successfully inflated to its 12' dia. Explorer XL successfully deployed its booms and operated its experiments.
Aug. 16	Essa VII (TOS-E)	Operational weather satellite	Thor-Delta (WTR)	S	S	Centaur 2nd stage failed to restart and Ats IV failed to separate from Centaur; failure made satellite uncontrollable and prevented exercise of gravity gradient stabilization experiment.
Aug. 22	RAM C-II	Reentry probe	Scout (WS)	S	S	Essa VII was launched by NASA for ESSA; 1st Long-Tank Thor used by NASA with Delta, 1st operational 2-burn booster for a TOS launch, 1st operational 2-burn mission for Delta.
Sep. 18	Intelsat-III F-1	Operational communications satellite	Thor-Delta (ETR)	U	Unk	RAM C-II reentered at 17,000 mph, measured amount of electrons and ions that built up around spacecraft.
Oct. 3	Aurorae (Esro I)	Scientific satellite, ionospheric	Scout (WTR)	S	S	Launch vehicle blew up at 9-mi altitude, destroying satellite.
Oct. 11	Apollo 7 (AS-205)	Orbital manned Apollo flight	Saturn IB (ETR)	S	S	Launched by NASA for ESRO, Aurorae went into good orbit; all 8 experiments operated in integrated study of high-latitude ionosphere.
						First manned Apollo flight confirmed operation of all major Apollo inflight systems except LM, including 8 burns of critical SPS motor; 1st live commercial TV from space; Astronauts Walter M. Schirra, Jr., Donn F. Eisele, and R. Walter Cunningham landed in Atlantic on 10/22 on 164th revolution, were picked up by helicopter from U.S.S. Essex, after 260-hour flight.

Date	Name (NASA Code)	General Mission	Launch vehicle (site)	Performance Vehicle	Performance Payload	Performance Mission	Remarks
Nov. 8	*Pioneer IX* (Pioneer D) and	Scientific probe, radiation	Thor-Delta (ETR)	S	S	S	Launched into heliocentric orbit; all 8 experiments returned good data on electromagnetic and plasma properties of space between earth and sun.
	Tetr II	Test and training satellite					Training satellite went into earth orbit to provide training for MSFN.
Dec. 5	*Heos-A*	Scientific satellite, radiation	Thor-Delta (ETR)	S	S	S	Launched by NASA for ESRO, 1st launching of foreign satellite on cost-reimbursable basis.
Dec. 7	*Oao II*	Scientific satellite, radiation	Atlas-Centaur (ETR)	S	S	S	Heaviest and most complex automated satellite launched by U.S. to date; all experiments operated.
Dec. 15	*Essa VIII* (TOS-F)	Operational meteorological satellite	Thor-Delta (WTR)	S	S	S	Launched by NASA for ESSA into near-polar orbit; turned over to ESSA 12/24; 18th straight success in Tiros program.
Dec. 18*	*Intelsat-III F-2* (Intelsat III-B)	Operational communications satellite	Thor-Delta (ETR)	S	S	S	Launched into elliptical orbit by NASA for ComSat-Corp; on 12/20 ComSatCorp fired apogee motor to circularize synchronous orbit and put satellite on station over South Atlantic; began commercial operation on 12/28.
Dec. 21	*Apollo 8* (AS-503)	Manned lunar orbital Apollo flight	Saturn V (ETR)	S	S	S	After 2 revolutions in earth orbit, *Apollo 8*'s 3rd-stage motor was fired to escape velocity; on 12/24 Astronauts Frank Borman, James A. Lovell, Jr., and William A. Anders went into orbit around the moon, stayed there for 20 hrs (10 revolutions), transmitted live TV of lunar surface; on 12/25 they fired spacecraft motor to lunar escape speed, reentered earth's atmosphere 12/27, landed in the Pacific, and were picked up by helicopter from U.S.S. *Yorktown* after 147 hr 11 sec flight. First manned Saturn V flight, 1st men to escape earth's gravity, 1st men to orbit moon.

*Date of launch per time at launch site.

Appendix C

CHRONOLOGY OF MANNED SPACE FLIGHT, 1967–1968

This chronology contains basic information on all manned space flights during 1967 and 1968 and, taken with Appendix C to the 1965 and 1966 editions of this publication, provides a summary record of manned exploration of the space environment through 1968. The information was compiled by William D. Putnam, Assistant NASA Historian for Manned Space Flight.

The year 1967 saw only one flight by man into space and that mission of the Soviet spaceship *Soyuz I* ended in the tragic death of Cosmonaut Vladimir M. Komarov. Although there were no U.S. flights in 1967, tragedy also struck the American program with the death of astronauts Virgil I. Grissom, Edward H. White II, and Roger B. Chaffee inside their spacecraft on the launch pad at Cape Kennedy while they were conducting tests preparatory to the first manned flight in the Apollo program.

It was October of 1968 before each nation resumed manned space flight, with *Apollo 7* the initial manned test of America's spacecraft for the lunar landing program and the *Soyuz III* the first Soviet manned flight since the crash of *Soyuz I* 18 months earlier. The epochal flight of *Apollo 8* was man's first trip outside the earth's environs to lunar orbit. This landmark in human exploration demonstrated the ability of the Apollo-Saturn V system to perform the mission for which it was designed and built—man's landing on the lunar surface and safe return to earth.

By the end of 1968, the United States had conducted a total of 18 manned space flights, 15 in earth orbit and one circumlunar voyage, with a total of 22 different crewmen. Six of the 22 American astronauts had participated in two flights each, and two had flown three times. The Soviet Union had conducted a total of 10 manned flights, all in earth orbit, with 12 different crewmen. Cosmonaut Komarov's fatal flight in 1967 was his second. Cumulative totals for manned spacecraft hours on flight had reached 1,430 hours 53 minutes for the United States and 554 hours 4 minutes for the Soviet Union. Cumulative total man-hours in space were 3,215 hours and 628 hours 40 minutes, respectively.

Data on U.S. flights are the latest available to date within NASA. Although minor details are subject to modification as data are refined, major aspects of all U.S. manned flights remain subject to direct observation by interested citizens of the world, with a significant portion of recent missions seen live on worldwide television.

Date Launched	Date Recovered	Designation	Crew	Weight (lb)	Revolutions	Maximum Distance from Earth (st mi)	Duration	Remarks
1967								
Apr. 23	Apr. 24	*Soyuz I*	Vladimir M. Komarov	Not available	16	139	26 hrs 40 min	First Soviet manned flight since 3/65. Cosmonaut Komarov was killed in crash while landing near Orenburg in Ural Mountains. Tumbling of spacecraft caused premature reentry and apparently caused straps on main parachute to become twisted on opening at altitude of 7 km. Fouled parachute failed to slow descent sufficiently and Cosmonaut Komarov became 1st man to die in space flight. Problems were also apparently experienced with both communications and control equipment throughout flight.
1968								
Oct. 11	Oct. 22	*Apollo 7*	Walter M. Schirra, Jr. Donn F. Eisele R. Walter Cunningham	67,669	163	282	260 hrs 9 min	First manned flight of Apollo spacecraft, first U.S. manned flight since 11/66. Liftoff from KSC Launch Complex 34 on Saturn IB booster injected S-IVB 2nd stage, instrument unit (IU), spacecraft lunar module adapter (SLA), and command and service module (CSM) into initial orbit. Crew successfully completed manual control test, then separated CSM from S-IVB, transposed CSM, and simulated docking with spent stage, approaching to within 4-5 ft. Later CSM rendezvoused with S-IVB in different orbit. All spacecraft systems were extensively checked out, with 8 service propulsion system (SPS) firings and several orbital path changes. Final SPS burn deorbited CSM for reentry. CM successfully separated from SM and landed in Atlantic at 7:11 am EDT 10/22/68. First flight test of Apollo spacesuits, first live commercial TV from U.S. manned space flight (7 broadcasts). Hot meals and increased spacecraft volume added to crew comfort. Astronauts developed colds in flight but had no aftereffects. All mission objectives were achieved, as well as 3 added tasks.
Oct. 26	Oct. 30	*Soyuz III*	Georgy T. Beregovoy	Not available	60	157	94 hrs 51 min	First manned Soviet mission since Cosmonaut V. M. Komarov was killed in *Soyuz I* mission. "Two-compartment" spacecraft with Cosmonaut Beregovoy

Date	Mission	Crew			Duration	Description
						sleeping in separate compartment from "pilot's station." Flew dual mission with unmanned *Soyuz II*, launched 10/25/68. During first revolution *Soyuz III* approached to within 200 m (656 ft) of *Soyuz II*, initially by automatic piloting; subsequent operations were performed by Beregovoy. On 10/27, *Soyuz III* changed orbital path to "conduct scientific, technical, medical, and biological experiments and research." Beregovoy, at 47 oldest man to fly in space, transmitted live TV of cabin interior. *Soyuz III* approached *Soyuz II* second time before *Soyuz II* reentry 10/28, remained in orbit until 10/30/68, softlanded "with the use of aerodynamics" in preset area of U.S.S.R.
Dec. 21	*Apollo 8*	Frank Borman James A. Lovell, Jr. William A. Anders	284,390	10 (of moon)	147 hrs	Launch from KSC Launch Complex 39, Pad A, at 7:51 am EST on first manned Saturn V inserted spacecraft and S-IVB combination into parking orbit to check spacecraft and ground systems. Translunar injection was initiated at 10:41 am EST by reigniting S-IVB for 5 min 9 secs. CSM separated from S-IVB and made 2 midcourse corrections during translunar coast, which was devoted to navigation sightings, 2 TV transmissions, and systems checks. At 4:59 am EST on 12/24, 246.9-sec SPS maneuver inserted CSM into 194- by 69-mi lunar orbit, later circularized at about 70 mi. During 20 hrs 11 min in lunar orbit, astronauts made 10 revolutions of moon, made 2 TV transmissions to earth, surveyed landing sites, and pinpointed landmarks. At 1:10 am EST on 12/25, engine was ignited behind moon to accelerate spacecraft out of lunar orbit and initiate transearth coast. One midcourse correction was made. At 10:22 am EST on 12/27, CM separated from SM and reentered earth's atmosphere, splashing down in Pacific less than 1 nm from recovery ship, U.S.S. *Yorktown*. Man's first voyage to lunar distance. Highest speed attained by man, 24,625 mph. All objectives achieved, plus 4 tasks added during flight. All equipment and procedures for lunar landing proved out except LM and actual landing sequence.

Appendix D

ABBREVIATIONS OF REFERENCES

Listed are abbreviations for sources cited in the text. This list does not include all sources provided in the chronology, for some of the references cited are not abbreviated. Only those references which appear in abbreviated form are listed below. Abbreviations used in the chronology entries themselves are cross-referenced in the Index.

A&A	AIAA's magazine, *Astronautics & Aeronautics*
A&A 68	NASA,: *Astronautics and Aeronautics 1968* [this publication]
ABC	American Broadcasting Company
AEC Release	Atomic Energy Commission News Release
Aero Daily	*Aerospace Daily* newsletter
Aero Tech	*Aerospace Technology* magazine (now *American Aviation*)
AFJ	*Armed Forces Journal* magazine
AFFTC Release	Air Force Flight Test Center News Release
AFHF Newsletter	*Air Force Historical Foundation Newsletter*
AFNS Release	Air Force News Service Release
AFOSR Release	Air Force Office of Scientific Research News Release
AFRPL Release	Air Force Rocket Propulsion Laboratory News Release
AFSC *Newsreview*	Air Force Systems Command's *Newsreview*
AFSC Release	Air Force Systems Command News Release
AF/SD	*Air Force and Space Digest* magazine
AFSSD Release	Air Force Space Systems Division News Release
AIA Release	Aerospace Industries Association News Release
AIAA *Facts*	American Institute of Aeronautics and Astronautics' *Facts*
AIAA *News*	American Institute of Aeronautics and Astronautics' *News*
AIAA Release	American Institute of Aeronautics and Astronautics News Release
Amer Av	*American Aviation* magazine (formerly *Aerospace Technology*)
AP	Associated Press news service
ARC *Astrogram*	NASA Ames Research Center's *Astrogram*
ARC Release	NASA Ames Research Center News Release
Av Daily	*Aviation Daily* newsletter
Av Wk	*Aviation Week and Space Technology* magazine
B News	*Birmingham News* newspaper
B Sun	*Baltimore Sun* newspaper
Bus Wk	*Business Week* magazine
Can Press	Canadian Press news service
CBS	Columbia Broadcasting System
C Daily News	*Chicago Daily News* newspaper
ComSatCorp Release	Communications Satellite Corporation News Release
CQ	*Congressional Quarterly*
CR	*Congressional Record*

CSM	*Christian Science Monitor* newspaper
CTNS	Chicago Tribune News Service
C Trib	*Chicago Tribune* newspaper
DJ	Dow Jones news service
DOD Release	Department of Defense News Release
DOT Release	Department of Transportation News Release
EH	NASA Historical Staff (Code EH)
ERC Release	NASA Electronics Research Center News Release
ESSA Release	Environmental Science Services Administration News Release
FAA Release	Federal Aviation Administration News Release
FonF	*Facts on File*
FRC Release	NASA Flight Research Center News Release
FRC X-Press	NASA Flight Research Center's *FRC X-Press*
GE Forum	*General Electric Forum* magazine
Goddard News	NASA Goddard Space Flight Center's *Goddard News*
GSFC Release	NASA Goddard Space Flight Center News Release
GSFC *SSR*	NASA Goddard Space Flight Center's *Satellite Situation Report*
H Chron	*Houston Chronicle* newspaper
H Post	*Houston Post* newspaper
JPL *Lab-Oratory*	Jet Propulsion Laboratory's *Lab-Oratory*
JPL Release	Jet Propulsion Laboratory News Release
KC Star	*Kansas City Star* newspaper
KC Times	*Kansas City Times* newspaper
KSC Release	John F. Kennedy Space Center, NASA, News Release
LA *Her-Exam*	Los Angeles *Herald-Examiner* newspaper
Langley Researcher	NASA Langley Research Center's *Langley Researcher*
LaRC Release	NASA Langley Research Center News Release
LA Times	*Los Angeles Times* newspaper
LATNS	Los Angeles Times News Service
LC *Info Bull*	Library of Congress *Information Bulletin*
LeRC Release	NASA Lewis Research Center News Release
Lewis News	NASA Lewis Research Center's *Lewis News*
Marshall Star	NASA George C. Marshall Space Flight Center's *Marshall Star*
M Her	*Miami Herald* newspaper
MJ	*Milwaukee Journal* newspaper
M News	*Miami News* newspaper
MSC Release	NASA Manned Spacecraft Center News Release
MSC *Roundup*	NASA Manned Spacecraft Center's *Space News Roundup*
MSFC Release	NASA George C. Marshall Space Flight Center News Release
M Trib	*Minneapolis Tribune* newspaper
NAA *News*	National Aeronautic Association *News*
NAC Release	National Aviation Club News Release
NAE Release	National Academy of Engineering News Release
NANA	North American Newspaper Alliance
NAR Release	North American Rockwell Corp. News Release
NAR *Skywriter*	North American Rockwell Corp. *Skywriter*
NAS Release	National Academy of Sciences News Release
NASA Ann	NASA Announcement
NASA Auth Hearings	NASA Authorization [1969] Hearings
NASA Hq PB	NASA Headquarters Personnel Bulletin
NASA Hq *WB*	NASA Headquarters *Weekly Bulletin*
NASA Int Aff	NASA Office of International Affairs
NASA *LAR* VII/8	NASA Legislative Activities Report, Vol. VII, No. 8
NASA Proj Off	NASA Project Office
NASA Release	NASA Headquarters News Release
NASA Rpt SRL	NASA Report of Sounding Rocket Launching
NASA SP–4006	NASA Special Publication #4006

NASC Release	National Aeronautics and Space Council News Release
NAS–NRC Release	National Academy of Sciences-National Research Council News Release
NAS–NRC–NAE *News Rpt*	National Academy of Sciences-National Research Council-National Academy of Engineering *News Report*
Natl Obs	*National Observer* magazine
NBC	National Broadcasting Company
NGS Release	National Geographic Society News Release
NMI–	NASA Management Instruction-
NN	NASA Notice
N News	*Newark News* newspaper
NSC Release	National Space Club News Release
NSF Release	National Science Foundation News Release
N Va Sun	*Northern Virginia Sun* newspaper
NY News	*New York News* newspaper
NYT	*New York Times* newspaper
NYTNS	New York Times News Service
Oakland Trib	*Oakland Tribune* newspaper
Omaha W–H	*Omaha World-Herald* newspaper
O Sen	*Orlando Sentinel* newspaper
PAO	Public Affairs Office
PD	National Archives and Records Service's *Weekly Compilation of Presidential Documents*
P *EB*	Philadelphia *Evening Bulletin* newspaper
P *Inq*	Philadelphia *Inquirer* newspaper
PIO	Public Information Office
PMR Release	USN Pacific Missile Range News Release
Pres Rpt 68	*United States Aeronautics & Space Activities 1968*, Report to the Congress from the President of the United States
P *SB*	Philadelphia *Sunday Bulletin* newspaper
SA	*Space Aeronautics* magazine
SBD	*Space Business Daily* newsletter
Sci Amer	*Scientific American* magazine
SciServ	Science Service news service
SD	*Space Digest* magazine
SD Union	*San Diego Union* newspaper
SF	*Space Flight* magazine
SF Chron	*San Francisco Chronicle* newspaper
SP	*Space Propulsion* newsletter
SR	*Saturday Review* magazine
SSN	*Soviet Sciences in the News*, published by Electro-Optical Systems, Inc.
St. Louis G–D	*St. Louis Globe-Democrat* newspaper
St Louis P–D	*St. Louis Post-Dispatch* newspaper
Testimony	Congressional testimony, prepared statement
Text	Prepared report or speech text
Transcript	Official transcript of news conference or Congressional hearing
UPI	United Press International news service
USGS Release	U.S. Geological Survey News Release
US News	*U.S. News and World Report* magazine
WH Release	White House News Release
WJT	*World Journal Tribune* newspaper
W News	*Washington Daily News* newspaper
W Post	*Washington Post* newspaper
WSJ	*Wall Street Journal* newspaper
WS Release	NASA Wallops Station News Release
W *Star*	Washington *Evening Star/Sunday Star* newspaper

INDEX AND LIST OF ABBREVIATIONS AND ACRONYMS

A

AA. See Apollo Applications program.
AAS. See American Astronautical Society.
AAAS. See American Assn. for the Advancement of Science.
ABM. See Antiballistic missile system.
ABMA. See Army Ballistic Missile Agency.
Abramson, Rudy, 229
ABRES. See Advanced Ballistic Reentry System.
Acceleration, effects of, 86, 190, 200–201
Accelerator, 91, 93, 154, 316
Accident, 95, 96, 157
  aircraft
    F–111A, 2, 72, 75, 80, 84, 87, 97, 102, 108, 109, 117, 119, 134, 160, 195, 207, 224, 258
    F–111B, 2
    Mig–15, 70
    SR–71, 8
    T–38, 128
    X–15, 12, 164–165
  Lunar Landing Research Vehicle, 257
  Lunar Landing Training Vehicle, 305
  spacecraft
    Apollo AS–204, 25, 51, 161, 191, 215, 221–222, 222–223, 229, 237, 245, 249, 264, 308
    Soyuz I, 70, 264
    Sputnik IV, 16
ACS. See Attitude Control System.
Adams, Dr. Mac C., 57, 118–119, 150, 283
Adams, Maj. Michael J. (USAF), 164–165
Adapticom, 143
Adelaide, Australia, 123
Advanced Ballistic Missile Defense Agency, 43
Advanced Ballistic Reentry System (ABRES), 292
Advanced Manned Strategic Aircraft (AMSA), 81
Advanced Research Projects Agency (ARPA) 40, 42, 43, 145, 190
Advanced Videcon Camera System (AVCS), 192–193
AEC. See Atomic Energy Commission.
Aegerter, Dr. Irene, 302
Aerobee (sounding rocket), 62, 87, 150 auroral experiment, 36
  instrumentation test, 17, 65, 175
  launch test, 211, 214
  solar astronomy, 221, 224
  stellar data, 67
  ultraviolet astronomy, 115
  upper atmosphere data, 206
  x-ray astronomy, 30, 40–41, 128
150 A, 141, 289–290, 291
150 MI
  infrared astronomy, 49–50, 316
  solar astronomy, 97, 118, 232, 256–257
  stellar data, 247–248, 282
  upper atmosphere data, 90, 127, 238, 289–290
  x-ray astronomy, 169, 256–257, 303
Aeroflot, 163
Aerojet-General Corp., 59, 75, 172
  Space Div., 86
Aeronautics, 172, 249, 249–250, 291
  anniversary, 11, 303, 313–314
  award, 15, 70–71, 84, 209, 259, 263, 286, 299, 300, 303, 316
  cooperation, 118, 238, 336
  employment, 289
  exposition, 233
  funds for, 18–19, 33–34, 71, 126, 161–162, 237–238, 239
  general aviation, 71, 137–138, 179, 187–188, 189, 194. 196, 278
  military, 239–240, 313, 317–318
  NASA program, 19, 25, 33–34, 47, 48, 118, 125, 161–162, 166, 179, 187, 225–226, 236–237, 238, 248, 255–256, 288–289, 313, 317–318
  noise abatement. See Noise, aircraft.
  research (see also X–15 and XB–70), 19, 25, 38–39, 56, 69–70, 118, 124–125, 166, 168. 197, 237, 247, 296
Aeroplane Structures, 91
Aerospace Communications Award, 77
Aerospace Corp., 156
Aerospace Defense Command, 9
Aerospace Industries Assn. (AIA), 6, 124–125, 134, 288–289, 315–316
Aerospace industry, 6, 134, 272, 278–279, 289, 297, 315–316
Aerospace Medical Assn., 109
Aerospace Safety Panel, (NASA), 122–123
AFCRL. See Air Force Cambridge Research Laboratories.

AFETR. See Air Force Eastern Test Range.
AFHRL. See Air Force Human Resources Laboratory.
Africa, 31, 311
AFSC. See Air Force Systems Command.
AFSCF. See Air Force Satellite Control Facility.
Agena (booster) (see also Atlas-Agena and Thor-Agena), 11, 26, 242, 293
Agnew, Dr. Harold, 122-123
Agreement,
  astronaut, rescue, 161, 299
  international, 27, 165, 298
  -NASA-Brazil, 177-178
    -ESRO, 113, 240
    Norway, 177-178
    -Spain, 177-178
    -Sweden, 177-178
    -USAF, 83
    U.K.-U.S., 59-60
  U.S.-Australia, 257, 329
    -Romania, 292
    -South Vietnam, 88
    -U.S.S.R., 127, 149-150, 157, 159, 164, 166
Agriculture, 16, 62, 85, 106, 205, 257
Agriculture, Dept. of 299
AIA. See Aerospace Industries Assn.
AIAA. See American Institute of Aeronautics and Astronautics.
Air Defense Command, 9
Air Force Academy, 104-105
Air Force, Assn., 217
Air Force Cambridge Research Laboratories (AFCRL), 10, 170, 209-210, 247, 250, 302
Air Force Eastern Test Range (AFETR), 148
Air Force Human Resources Laboratory (AFHRL), 150
Air Force Satellite Control Facility (AFSCF), 189
Air Force Systems Command (AFSC), 5, 39-40, 54, 81, 237-238, 272
  Aeronautical Systems Div., 61
  Aerospace Medical Div., 141
  Air Force Human Resources Laboratory, 150
  associateships, 8
  contract, 2
  rocket engine, reusable, 2
  Space and Missiles Systems Organization, 244, 297
  test, 70
Air Holdings, Ltd., 72
Air pollution, 169, 251, 323
Air Products and Chemicals, Inc., 79
Air traffic control, 104, 203, 238
  contract, 82
  FAA regulation, 1-2, 194, 196, 199-200, 206, 248
  NASA role in, 187, 225-226
  press comment, 188, 196
  satellite use in, 217
Air Transport Assn. of America, 286
Air transportation, 99, 104, 187-188, 194, 206, 217, 232
  NAE report, 187
  R&D, 47, 224, 225-226
Airborne Auroral Expedition (1968), 10-11, 17-18, 72
Aircraft (see also individual aircraft, such as C-5A, F-111, X-15, XB-70), 288-289, 291
  accident, 2, 8, 12, 72, 75, 80, 97, 102, 108, 109, 117, 119, 128, 134, 157, 160, 164-165, 195, 207, 224, 258
  air show, 105, 117
  antisubmarine, 187
  award, 76
  bomber, 76, 112, 240, 280
  cargo, 16, 19, 54, 81, 146-147, 150, 155-156, 208, 231, 282, 288, 312, 335
  collision study, 179
  communications test, 111, 205, 229
  cost, 146, 202-203, 282, 288
  delta-wing, 332-333
  fighter, 76, 130, 165, 187, 195, 201, 217, 232-233, 272, 281, 314
  flying boat, 160
  foreign, 40, 49, 52, 105, 112, 115, 126, 130, 139, 151, 156, 165, 194, 195, 198, 217, 233, 259, 294, 332-333
  general-aviation, 71, 137-138
  helicopter, 18, 219, 226, 251
  Hovercraft, 175
  hypersonic, 226, 237, 300, 317
  interceptor, 112, 156, 171-172
  navigation, 187, 217
  noise. See Noise, aircraft.
  personal, 207
  reconnaissance, 8, 40
  record, 11-12, 207, 299, 317-318
  regulations, 1-2, 116, 130, 194, 196, 206, 300
  research (see also X-15 and XB-70), 8, 18, 34, 39, 72, 81, 184, 276-277, 286, 313, 324, 335
  safety, 187
  sonic boom. See Sonic boom.
  statistics, 71, 120, 137-138
  STOL, 16, 139, 217, 219, 224, 226, 231
  tracking, 12
  transport (see also Supersonic transport)
    air-bus, 72
    jet, 134, 157, 197
    military, 293
  STOL, 16, 139, 179, 217, 219, 224, 226, 231
  variable-sweep-wing, 206, 220-221
  Vietnam War use, 195
  V/STOL, 33, 34, 47, 104, 134, 166, 168, 177, 219, 224, 238, 288
  VTOL, 16, 33, 40, 47, 105, 135, 219, 224, 230-231, 276

wind-tunnel testing, 40, 59, 219
Aircraft carrier, 238
Airglow, 177, 256
Airlines (see also Air traffic control and Supersonic transport), 107–108, 135, 196, 203, 206, 286
  aircraft, 45, 71–72, 139, 217, 224–225, 232–233
  forecast, 137–138, 278
  highjacking, 189
Airlock module, 218–219
Airmail, 113
Airports (see also Air traffic control; Noise, aircraft; and individual airports, such as Washington National Airport), 150–151, 179, 199–200
  FAA plans, 82, 137–138, 248, 278, 296
  facilities, 122, 187, 286
  funds for, 95, 168
  meeting, 187–188
  NASA role, 187, 225–226
  regulation, 189, 194, 203, 206, 300, 302
  runway research, 215, 286, 324
  statistics, 137–138
Alabama, 73–74, 272
Alabama State Employment Service, 272
Alaska, 10–11, 303
Alaska, Univ. of, 14, 49
Alberstadt, Milton, 296
Aldrin, L/C Edwin E., Jr. (USAF), 109, 183, 196
Alexander, George, 161
Alexandria, Va., 152
Algeria, 138
Algranti, Joseph S., 305
All-weather landing system (AWLS), 248
Allen, M/G Brooke E. (USAF, ret.), 49–50
Allen, H. Julian, 70, 260–261, 263, 298
Allen, William M., 45
Allenby, Dr. Richard J., Jr., 2
Allis-Chalmers, 294
Allison Div., General Motors Corp., 247
Allott, Sen. Gordon L., 162, 258
Alouette II (Canadian satellite), 65, 106
Alpha-scattering instrument, 3, 6
ALSEP. See Apollo Lunar Surface Experiment Package.
Alter, Dr. Dinsmore, 221
Altman, Dr. David, 5
Aluminum Co. of America, 152
Alvarez, Prof. Luis W., 266–267
Amazon River, 62
American Academy of Arts and Sciences, 309
American Airlines, 40
American Assembly, 105
American Assn. for the Advancement of Science (AAAS), 325
American Astronautical Society (AAS), 79, 111, 161–163, 262–263
  Space Flight Award, 89
American Bar Assn., 202
American Federation of Television and Radio Artists, 251
American Helicopter Society, 116
American Independent Party, 271
American Institute of Aeronautics and Astronautics (AIAA), 48, 268, 293
  award, 15, 66, 71, 133, 145, 260–261, 316
  meetings, 38–39, 55, 65, 105, 131, 157, 202, 247, 259–261, 301–302, 316
  National Capital Section, 2
American Institute of Physics, 312
American National Balloon Race, 188
American Nuclear Society, 274
  International Meeting, 280–281
American Oceanic Organization, 284
American Ordnance Assn., 39–40
American Physical Society, 49, 91
American Science and Engineering, Inc., 30, 62, 128, 240 303
American Society for Public Administration, 66
American Society of Mechanical Engineers, 313
American Telephone and Telegraph Co. (AT&T), 82, 154, 276, 307
American Univ., 1
Ames Research Center (ARC), 42, 138, 181, 274
  award, 70–71, 260–261, 263, 283
  Biomedical Research Branch, 273
  experiment, 65, 175
  meeting, 135–136
  personnel, 263, 298
  research, 2, 16–17, 21–22, 78, 166, 285–286, 286–287
    aircraft, 179, 184
  Space Science Div., 316
  VTOL Facility, 135
Ammonia, 312
AMSA. See Advanced Manned Strategic Aircraft.
Analytical Services Inc., 26–27
Anchorage, Alaska, 303
Anders, L/C William A. (USAF)
  Apollo 8 flight, 318–323, 327
    preparations for, 59, 183, 196, 222–223, 261–262, 278
  award, 332
  Humphrey, Vice President Hubert H., visit with, 287
  press conference, 285, 289, 304–305
  promotion, 330
  White House visit, 306
Anderson, Sen. Clinton P., 30, 170, 235
Anderson, John D., 275–276
Anderson, Prof. Kinsey A., 15
Anderson, William, 177
Andover, Me., 315
Andoya, Norway, 219–220, 223
Andrew, G. M., 11
Angaran, Jack G., 44
Animal experiments, space, 46, 47, 86, 95, 105, 151, 181, 218, 283–284
Ann Arbor, Mich., 85

Anniversary, 221, 298, 313–314
  Government, 12, 113
  industry, 26
  MSFC, 149
  NASA, 64–65, 135, 235, 239, 253, 334–335
  satellite, 15, 22–24, 30, 49–50, 63–65, 80, 145
  U.S.S.R., 45–46, 273, 336–337
Antarctic Treaty, 298
Antarctica, 46, 72, 89, 127, 293
Antenna, 5, 49, 133, 167, 171, 219, 238–239, 241–242, 290, 293
  spacecraft, 34–35, 153, 184, 224, 246
Antiballistic missile (ABM) system, 28
  agreement, 127, 159
  contract, 72
  Europe (NATO), 88
  funds for, 122, 133, 138, 171, 242
  limitation of, 145
  U.S.S.R., 145, 159
APL. See Applied Physics Laboratory.
Apollo (program) (see also Apollo Applications program), 41, 45, 102–103, 108–109, 131, 145, 148, 215, 218, 237, 281, 311
  astronaut. See Astronaut.
  communications, 205, 289, 312, 315
  criticism, 25, 191, 289, 308
  facilities, 34–35, 256
  fire. See Accident, spacecraft, Apollo AS–204.
  funds for, 19, 35, 104, 110, 135–136, 175, 185, 241–242, 311
  House staff study, 51
  landing site, 35–36, 66, 265, 266, 319
  launch
    Apollo 5 (AS–204), 13, 334
    Apollo 6 (AS–502), 77–78, 334
    Apollo 7 (AS–205), 150–253, 334–335
    Apollo 8 (AS–503), 318–322, 335
  management, 2, 8–9, 20, 25, 157, 218–219, 224, 303, 315
  plans for, 33, 35, 63, 66, 92–93, 97, 158, 196, 223, 227, 257, 278, 279–280
  press comment, 80–81, 81–82, 161, 222–223, 253, 261–262, 265, 271, 323–324, 324–325, 327–328
  progress, 92–93, 122, 136, 141, 144, 158, 173, 181, 186, 191, 197, 201, 209
  test, 63, 78, 158, 173, 181, 197, 201, 209
  tracking, 205, 289
  training, 2, 158, 181, 223, 227, 266
Apollo (spacecraft), 13, 77–78, 92–93, 251–253, 319–322
  ascent propulsion system, 13
  command and service module, 11, 201, 206, 208, 250, 251, 318, 320
  command module, 13, 42, 77, 122, 144, 181, 186, 197, 251–253
  control, 77–78, 136, 251–252, 319–322
  descent propulsion system, 13
  emergency detection system, 77
  environmental control system, 61–62, 134, 186, 251–252
  escape device, 77–78, 210
  equipment, 167, 282–283, 295
  fire prevention, 4–5, 13–14, 61, 163, 237, 252
  heat shield, 77
  landing system, 59, 144, 158
  launch. See Apollo (program).
  launch vehicle. See Saturn.
  life support system, 61–62, 144, 251–253
  lunar module, 1, 13, 17, 37, 63, 92, 144, 173, 181, 191, 197, 205–206, 227
  materials, 4–5
  recovery, 70, 77, 320
  service module, 13, 42, 63, 122, 144, 181, 186, 197, 251–252, 318, 321
  test, 1, 17, 25, 33, 37, 42, 63, 77, 81, 136, 143, 158, 181, 186, 201, 206, 209, 227
Apollo 4 (AS–501) (spacecraft), 13, 78, 92, 253, 321–322
Apollo 5 (AS–204), 17, 78, 253, 321–322, 334–335
  launch, 13
Apollo 6 (AS–502), 42, 63, 70, 83, 92–93, 135, 166, 205–206, 253, 266, 321–322, 328, 334–335
  launch, 77–78
  press comment, 80–81, 81–82
Apollo 7 (flight), 259, 265–266, 278, 312, 321–322
  launch, 250–253
  preparations for, 136, 144, 158, 180, 181, 197, 206, 209, 222–223, 227
  press comment, 208, 222–223, 253, 261–262, 264–265, 265, 266
  press conference, 219
  results, 251–253, 334–335
Apollo 7 (AS–205) (spacecraft), 204, 261
  command and service module, 201, 206, 223, 250–251
  command module, 122, 136, 144, 181, 251
  lunar module, 144, 181, 205
  service module, 122, 136, 144, 181, 251
  test, 136, 144, 158, 181, 186, 197, 201, 205, 208
Apollo 8 (flight), 261, 266, 315, 330, 330–331
  criticism, 308, 311
  Johnson, President Lyndon B., message, 322
  launch, 318, 320
  preparations for, 196, 197, 201, 208, 227, 265–266, 278, 311–312, 313
  press comment, 323, 323–324, 324–325, 327–328, 329–330, 330–331
  press conference, 223, 278, 285, 289, 304–305, 322–323
  results, 321–322, 326, 330, 334–335
Apollo 8 (AS–503) (spacecraft), 315
  command and service module, 196, 201, 266, 318
  command module, 266
  lunar module, 196

service module, 319, 320, 321
test, 197, 201, 265–266, 318
Apollo 9 (flight), 311
Apollo 10 (flight), 279–280
Apollo Applications (AA) program, 51–52, 108–109
  contract, 79, 202, 294–295
  funds for, 19, 35, 41, 101, 102–103, 110, 135–136, 182, 240–241, 255, 256, 311, 325, 336
  House staff study, 51
  management, 218–219, 293, 303, 315
  plans for, 86, 223, 243, 336
Apollo Lunar Surface Experiment Package (ALSEP), 51–52, 68, 282–283, 316
Apollo, Pa., 2
Apollo 204 Review Board, 4–5, 25, 245
Apollo Telescope Mount (ATM), 33, 109, 110, 128, 275
  contract, 62, 80, 85–86, 186, 202, 240
  experiment, 80
  funds for, 182, 241
  management, 218–219
  use of, 202, 243
Applications Technology Satellite (ATS), 8
Applied Physics Laboratory (APL) (Johns Hopkins Univ.), 8, 86
Apra Heights, Guam, 290
APS. See Ascent propulsion system.
APT. See Automatic Picture Transmission.
Aquanaut, 41–42
ARC. See Ames Research Center.
Arcas (sounding rocket), 67, 86, 214
Arcas I, 60, 175
Arches of Science Award, 217–218
Arctic, 127, 302
Arecibo (Puerto Rico) Ionospheric Observatory, 37–38, 54, 64–65, 238, 275–276, 284
Argentina, 46, 160–161, 293
Ariel (program), 152
Ariel III (U.K. satellite), 105
Arizona, 154–155
Arizona, Univ. of, 133
  Planetary Laboratory, 12–13
Arkansas, 249
Arlington, Va., 293
Armed Forces Management Assn., 299
ARMS (Application of Remote Manipulators in Space), 186
Armstrong, Neil A., 105, 116, 196, 211, 257
Army Ballistic Missile Agency (ABMA), 22
Army Electronics Command, 170–171
Army Map Service, 266
Arnold Engineering Development Center, 135, 163
Arons, Daniel M., 127
ARPA. See Advanced Research Projects Agency.
ARTS. See Automated radar tracking system.
Ascension Island, 5

Ascent propulsion system (APS) (Apollo), 13
Asher, Dr. Harold, 160
Asia, 31
Association for the Advancement of Medical Instrumentation, 165
Association of Man in Space, 276
Association of Universities for Research in Astronomy, Inc. (AURA), 292
Asteroid, 133, 291
Astheimer, Robert W., 215
Astin, Dr. Allen V., 5
Astris (West German rocket), 296
Astrobee 1500 (sounding rocket), 88
Astronaut (see also Cosmonaut; Extravehicular activity), 187, 218, 223, 261–262, 295, 296, 305
  accident, 105, 128, 218, 308
  achievements, 22–24, 66, 236–237, 253, 326, 334–335
  Apollo mission, 66, 81–82, 85, 97, 116–117, 136, 183, 196, 223, 227, 250–253, 278, 282–283, 311–312, 313, 318–322
  crew assignment, 59, 92, 97, 183, 196, 279–280
  former, 203–204, 286
  hazards, 61–62, 131, 176–177, 220, 278, 283–284, 296, 300
  honors, 11–12, 28, 103, 109, 110, 259, 269, 286, 303, 306, 314, 330, 332 333
  physiology 61–62, 113, 168, 181, 205, 211, 218, 283–284, 286–287, 291
  press conference, 220, 276
  scientist-astronaut, 36, 51–52, 200
  space rescue. See Space rescue treaty.
  spacesuit, 57, 176–177, 236–237, 283–284
  training, 22, 52, 59, 92, 105, 134, 144, 158, 173, 181, 186, 200, 208, 211, 223, 227, 266, 309
Astronaut Day, 332
Astronautics Engineer Award, 151
Astronomy (see also individual observatories, planets, and satellites such as *Oao I, Oao II*; Pulsar; Radioastronomy; Star), 142, 151, 155, 221, 290, 292, 294, 312
  NASA program, 44, 241, 243, 280–281, 336
  solar, 56–57, 153, 219, 224, 266, 269, 277, 296, 300–301
  stellar, 128–129, 208–209, 231, 238, 247, 272, 284, 292, 304, 310
  ultraviolet, 310
  x-ray, 30, 62, 104, 256–257, 303
AT&T. See American Telephone & Telegraph Co.
Atlantic II (communications satellite). See *Intelsat-II F–3*.
Atlantic Ocean, 130–131, 193, 216
  Apollo 7 splashdown, 251
  satellite communications over, 5, 39, 80, 139, 314–315, 323
  sounding rocket, 141, 144, 172

Atlas (booster), 80, 148, 159
Atlas-Agena D (booster), 1, 54–55, 179, 185
Atlas-Burner II (booster), 193
Atlas-Centaur (booster), 3, 8, 39, 55, 118, 184–185, 281–282, 304
ATM. See Apollo Telescope Mount.
Atmosphere, 29, 56, 62, 63–64, 105, 168, 177–178, 182–183, 217, 328–329
 National Atmospheric Sciences Program, 101–102
Atom, man-made, 6
Atomic bomb, 265
Atomic Energy Commission (AEC) (see also NASA–AEC Space Nuclear Propulsion Office; NERVA, Rover, SNAP, and Vela programs), 82, 117, 181, 242, 325
 Annual Report to Congress . . . for 1967, 25–26
 artificial elements, production of, 6
 award, 198, 217–218, 221
 Brookhaven National Laboratory, 154
 budget, 19, 170
 contract, 75
 cooperation, 65, 70, 91, 131, 156, 170, 335
 food irradiation test, 2
 High Energy Physics Advisory Panel, 154
 nuclear power, peaceful use of, 205, 228
 nuclear reactor, 131, 143, 156, 205, 335
 nuclear rocket engine, 65, 170, 288–289, 335
 nuclear test, 108, 335
 nuclear test detection, 91
 personnel, 143
 Plowshare, Project, 110
Atoms for Peace Award, 198
ATS. See Applications Technology Satellite.
Ats I (Applications Technology Satellite), 26–27, 185
Ats II, 119, 185, 323
Ats III, 37, 62, 144, 155, 185, 229, 232, 251–252, 323
Ats IV, 184–185, 225, 257, 261, 303
ATS–D, 1
ATS–E, 303
ATS–F, 120, 209
ATS–G, 120, 209
Attitude control system (ACS), 30, 121, 247–248, 282
Au, Benjamin, 184
Aubiniere, Gen. R. (France), 60
Ault, G. Mervin, 283
AURA. See Assn. of Universities for Research in Astronomy, Inc.
Auriga (constellation), 155
Aurora borealis, 240, 302, 336
Aurorae (Esro I) (satellite), 240, 336
Auroras, 10–11, 17–18, 31–32, 49, 53–54, 219–220, 223
Australia, 35, 46, 118, 120, 137, 170–171, 206, 224, 248, 253–254, 257, 290, 329
Australian Air Force, 118
Australian National Univ., 290, 329
Automated radar tracking system (ARTS), 248
Automatic Picture Transmission (APT), 50, 97, 155, 295, 311, 336
Avco Corp., 150
AVCS. See Advanced Vidicon Camera System.
Aviation Day (U.S.S.R.), 195
Aviation/Space Writers Assn., 315–316
Awards
 civic, 281, 290, 300
 Government, 2, 7, 11–12, 29, 38, 39, 51, 91, 117, 164, 194, 221, 263, 269, 283, 298, 306, 314, 332
 institutions, 110, 194, 217–218
 military, 11–12, 104–105, 314
 society
  aeronautics, 15, 66, 70–71, 84, 116, 209, 259, 263, 286, 303, 314, 316
  astronautics, 15, 57, 89, 106, 133, 145, 151, 194, 235, 260–261, 276, 300, 316
  military, 199, 263
  science, 122, 198, 309
AWLS. See All-weather landing system.
Axford, Dr. W. Ian, 216–217

B

B–17 (Flying Fortress), 146, 296
B–29 (Superfortress), 146
B–52 (Stratofortress)
 accident, 154
 HL–10 flights, 63, 131, 139, 224–225, 240, 261, 279, 306, 335
Babcock, Harold D., 82
Baehr, Edward F., 104
Baikonur Cosmodrome, 126, 135, 158–159, 172, 183, 264, 269
Baja California, 318–319
Baker, Dr. Robert, L., Jr., 172
Bal Harbour, Fla., 109
Balcarce, Argentina, 293
Ball Brothers Research Corp., 85–86
Balloon, 151–200
 drag, 151–152
 Explorer XXIV, 182
 Explorer XXXIX, 182, 310
 race, 188
 record, 209–210, 240
 research, use, 68, 209–210, 275
 Stratoscope II, 116
 tracking, 144
 Venus probe, 163
Bally Clayton comet, 201
Bally-Urban, John, 201
Baltimore, Md., 119, 179
Banks, Peter M., 216–217
Barbados Oceanographic Meteorological Experiment (BOMEX), 262
Barger, Allan R., 163, 199
Barium, 157, 160, 161, 223, 308
Barnes Engineering Co., 215

Barreria do Inferno Range, Brazil, 130–131
Barrett, Dr. Alan H., 200
Barrios, Joseph, 285–286
Battelle Memorial Institute, 312–313
Battery, 78–79
Baum, Dr. Werner A., 131
BBC. See British Broadcasting Co.
Beams, Jesse W., 38
Bear (U.S.S.R. turboprop aircraft), 40
Becker, John V., 300
Beggs, James M., 26, 118–119, 164, 274–275, 288–289
Beheim, Milton A., 104
Belgium, 5, 60, 89, 292, 300–301
Bell Aerosystems Co., 107, 127, 145, 227–228
Bell, Rep. Alphonzo, 262–263
Bell Telephone Laboratories, 26–27
Bellingshausen station, Antarctica, 46
Bells, Tex., 120
Beloit College, 116–117
Bendix Corp., 59, 61, 202, 226
Bendix, Field Engineering Corp., 12, 97–98, 146
Benedict, Howard, 178
Benn, Minister of Technology Anthony W. (U.K.), 49, 85
Bennett, Dr. Ivan L., Jr., 126
Bennett, Ralph Kinney, 173–174
Beregovoy, Georgy T., 264–265, 266, 272
Bergaust, Erik, 161, 222–223
Berkey, Donald C., 15
Bermuda, 97
Berne, Univ. of, 68
Berry, Dr. Charles A., 109, 164, 291, 313
Beryllium, 312–313
Betts, L/G Austin W. (USA), 29–30
Beverly Hills, Calif., 303
Biehl, Dr. Arthur T., 329
Biennial Guidance Test Symposium, Fourth, 272
Big Dipper (constellation), 310
Big Three Industrial Gas and Equipment Co., 45
Bing, R.H., 115
Bioastronautics. See Space biology.
Biology, 60, 126
Biosatellite (program), 161, 241, 313, 336
Biosatellite II, 46, 325–326
Biosatellite C, 47
Biosatellite D, 1, 47, 95
Biosatellite E, 47
Biosatellite F, 47
Biosensor, 246
Birch, Prof. A. Francis, 38
Birlandeanu, Alexandru, 140, 156
Bisplinghoff, Dr. Raymond L., 1, 5
Bisson, Edmond E., 283
Black Brandt IV (Canadian sounding rocket), 106, 130–131
Blagonravov, Dr. Anatoly A., 336–337
Blondstein, J.L., 190–191
Blue Streak (U.K. rocket), 296
BMWF. See Germany, West Ministry of Scientific Research (BMWF).
Bochum Observatory (W. Germany), 54, 211–212, 222
Boeing Co., 5
 award, 15
 booster, 171
  Saturn V, 35, 39, 104, 127, 266, 275
 contract, 35, 39, 104, 127, 242–243, 266, 275, 328
 employment, 229–230, 250, 272
 Lunar Orbiter, 12–13, 24
 Minuteman, 312–313
 noise suppression, 47
 Space Div., 229–230
 space station, 242–243
 supersonic transport, 11, 45, 66–67, 150, 202–203, 217, 220–221, 259–260, 270, 285, 332–333
 Vertol Div., 219, 224
Boeing 707 (jet transport), 163, 248
Boeing 727 (jet transport), 15
Boeing 747 (jet transport), 104, 232
BOMEX See Barbados Oceanographic Meteorological Experiment.
Bonn, W. Germany, 201, 279, 300
Boone, Adm. Walter F. (USN, Ret.), 283
Boosted-Arcas II (sounding rocket), 178, 219, 262
Boosted-Dart (sounding rocket), 178
Borman, Col. Frank (USAF), 332
 Apollo 8 flight, 325, 327
 flight, 318, 324
 plans for, 183, 196, 223, 261, 278
 award 332
 Humphrey, Vice President Hubert H., visit with, 287
 Press conference, 285, 304–305
 White House visit, 306
Born, Gordon S., 281
Boston, Mass., 49, 217, 231, 309
Boulogne, France, 175
Bourdon, Allen, 49
Bourguiba, President Habib (Tunisia), 118
Bowie, Tex., 97
Boyd, Secretary of Transportation Alan S., 137–138, 189, 200
Bozhko, Andrey N., 324
Brackett, Ernest W., 83
Bradbury, Dr. Norris E., 94
Brand, Vance D., 134, 144
Branscomb, Dr. Lewis M., 329
Brazil, 32, 39, 130–131, 138, 177–178
Breguet III (STOL aircraft), 217
Breguet 941 (French aircraft), 139
Breit, Prof. Gregory, 38
Brewer, Frank G., Trophy, 71
Brewster Flat, Wash., 82–83, 103
Brezhnev, Leonid, I., 72
Bridges, Thomas J., 39
Brillante, Gabe, 227
Bristol Siddely Engines, Ltd., 198
British Aircraft Corp., 70, 217
British Broadcasting Co. (BBC), 289

British Interplanetary Society, 52, 73, 185
British Railways, 175
Bronk, Dr. Detlev W., 76–77
Brooke, Sen. Edward W., 25, 210
Brookhaven National Laboratory, 154
Brooks AFB, Tex., 11, 150, 233
Brooks, Harvey, 115
Brown, Duane, 156
Brown Engineering Co., 79–80, 110
Brown, Secretary of the Air Force, Dr. Harold, 112, 130, 206, 310
Brown, Col. Henry W. (USAF), 2
Browning, Dr. Bouilin, 325–326
Bryce National Park, 154–155
Bubble chamber, 154
Buckley, Edmond C., 17, 283
Buckley, James, 231
Budget, Bureau of, 286
Buffalo, N.Y., 228
Bulgaria, 189–190, 317
Bull, L/ Cdr. John S. (USN), 113, 164
Bunker Hill AFB, Ind., 39, 110
Burcham, Dr. Donald P., 152–153
Bureau of Labor Statistics, 325
Bureau of the Census, 325
Burke, Dr. Bernard, 238–239
Burnham, Frank A., 126
Butz, J.S., Jr., 297
Bykovsky, Valery F., 46
Byrd, Adm. Richard E. (USN), 293
Byrd Station, Antarctica, 36

C

C–5A (Galaxy) (military cargo transport), 15, 18–19, 81, 104, 231, 312–313
  cost, 146, 282, 288
  Johnson, President Lyndon B., statement, 54
  press comment, 150–151, 155–156
  test flights, 146–147, 150–151, 208, 335
C–54 (turboprop aircraft), 148
C–130 (transport aircraft), 9, 293
C–131 (research aircraft), 81
C–46 (Caravan), 296
CAAS. See Computer-aided approach spacing system.
CAB. See Civil Aeronautics Board.
Cable, underwater, 5
Cactus test, 97
Cahn, M.S., 11
Cain, Dan L., 199
Cajun-Dart (sounding rocket), 143
California, 8, 137, 204, 219
California Div. of Highways, 286
California Institute of Technology (Cal Tech), 5, 43, 79, 248–249, 287 299
  award, 38, 309
  contract, 331–332
  Guggenheim Aeronautical Laboratory, 267
  pulsar signal research, 114–115
  symposium on space research, 29–30
California, Univ. of, 1, 5, 38, 51–52, 65, 68, 128, 216–217
  Berkeley, 172, 266–267, 299. 312
  Davis Chronic Acceleration Lab., 218
  Irvine, 115
  Los Angeles, 17, 29–30, 88, 181, 247, 267, 302
  San Diego, 31–32, 90, 98, 103
  School of Medicine, 5
Californium, 6, 181–182
Callaghan, Richard L., 283
Callahan, M/G Daniel F. (USAF, Ret.), 150
Cal Tech. See California Institute of Technology.
Cambridge, U.K., 265
Camera, 49–50, 145
  Apollo 6, 78
  Apollo 7, 227, 250–252
  Apollo 8, 312
  Ats III, 29
  Ats IV, 185
  Essa VII, 192
  Essa VIII, 311
  reconnaissance, 200
Cameron, Dr. Roy E., 89
Campbell, John P., 116
Canada, 46, 54, 65, 113, 170–171, 285
  cooperation, 1, 10–11, 64–65
  satellite, 1, 64–65
  sounding rocket, 106, 130–131
Canadian Aeronautics and Space Institute (CASI), 157
Canberra, Australia, 312
Cannon, Sen. Howard W., 98
Canyon de Chelly National Monument, 154–155
Cape Horn, 318–319
Cape Kennedy, Fla. (see also Eastern Test Range and Kennedy Space Center), 62, 136, 148, 227, 272
Carbon dioxide, 302
Cardion Electronics, 143
Carlson, Sen. Frank, 62
Carlson, Dr. Loren D., 5
Carnegie Institute of Washington, 166
Carnegie-Mellon Univ., 110, 299
Carpenter, Cdr. M. Scott (USN), 116–117, 143
Carswell AFB, Tex., 160, 313
Cartier, Raymond, 330
Case Western Reserve Univ., 1
CASI. See Canadian Aeronautics and Space Institute.
CAT. See Clear air turbulence.
Cate, Dr. James L., 1
Catholic Univ. of America, 325
Cattaneo, Dr. Alfred G., 36–37
Centaur (booster) (see also Atlas-Centaur), 1, 82, 144, 148, 261, 275, 329
Centaure (French sounding rocket), 31, 108, 118
Centaurus (constellation), 273
Central Bay (moon), 35
Central Intelligence Agency (CIA), 208
Cernan, L/Cdr. Eugene A. (USN), 173, 279

Cerro Tololo, Chile, 292
cets. See Conférence Européenne sur les Télécommunications par Satellites.
Chaffee, L/Cdr. Roger B. (usn), 103
Chanute, Octave, 134
Chanute, Octave, Award, 66
Charlottesville, Va., 151
Cheprakov, Dr. Viktor A., 195
Chernyi, Dr. Gorimir G., 268
Chew, Peter T., 245
Chicago, Ill., 15, 112, 188, 200, 203, 300, 302
Chicago Sanitary District, 281
Chicago, Univ. of, 1, 90
Chile, 46, 160, 219, 292
Chimpanzee experiment, 283
China, Communist, 208, 298, 329
 missile threat, 18–19, 32, 88, 138, 143
 nuclear test, 200, 330
 weapons, 159
Christchurch, New Zealand, 240
Christmas Island, 116
Chrysler Corp., 61, 91, 150, 250, 272, 296
 Space Div., 84, 239
Churchill Research Range, Canada (see also Ft. Churchill, Canada), 287, 295
 Airborne Auroral Expedition, 10–11, 17–18
 launch
  Aerobee 150, 36, 175
  Javelin, 54, 113, 308
  Nike-Apache, 90, 129
  Nike-Cajun, 216, 256, 309–310
  Nike-Tomahawk, 14, 31–32, 43–44, 45, 49, 65, 67, 90, 94, 98, 103
cia. See Central Intelligence Agency.
Civil Aeronautics Board (cab), 40, 187–188
Civil Service Commission, 168, 328
Clark, Evert, 10, 43, 93–94, 148
Clark, R/A John E. (usn, Ret.), 17, 42
Clarke, Arthur C., 72, 119, 142, 147
Clayton, Pat, 201
Clear air turbulence (cat), 215
Clearinghouse for Federal Scientific and Technical Information, 97, 233
Cleveland Clinic, 302
Clifford, Secretary of Defense Clark M., 98, 138, 160, 261, 263, 284, 296, 308
 appointment, 12, 45
Cloud, 29, 50, 62, 85, 251, 311, 336
cm. See Command module.
cnae. See Comissão Nacional de Atividades Espaciais.
Cocoa Beach, Fla., 220
Cohen, Prof. Paul J., 38
Cohen, Secretary of Health, Education, and Welfare Wilbur J., 169
Cold Flare, Project, 168
Cold Regions Research and Engineering Laboratory, 36
Cole, Dandridge M., Memorial Lecture, 131
Cole, Kenneth S., 38

College Park, Md., 111, 154, 245
Collier, Robert J., Trophy, 84, 106, 317–318
Collins, l/c Michael (usaf), 168, 183
Collision-avoidance system (cas), 129, 177, 238
Colloquium on the law of Outer Space, XIth, 258
Colorado, 137
Colorado, Univ. of, 90, 129, 168, 169, 197, 238, 254, 290, 329
 Joint Institute for Laboratory Astrophysics, 329
Columbia Radiation Laboratory, 169
Columbia Univ., 38, 117
 Graduate School of Business, 103
Comet, 155, 201
Comisión Nacional de Investigación del Espacio (conie), 178
Comissão Nacional de Atividades Espaciais (cnae), 131, 178
Command module (cm), 11, 14, 183
 Apollo/Saturn
  AS–202, 13
  AS–203, 13
  AS–501, 13
  AS–502 (Apollo 6), 77
 Apollo 7, 144, 186, 197, 201, 251
 Apollo 8, 265
Command and service module (csm), 11, 15, 181, 208, 277
 Apollo 6, 77
 Apollo 7, 197, 201, 223, 250, 251
 Apollo 8, 206, 265, 321
 Apollo 10, 281
Commerce, Dept. of, 89, 195, 299, 325, 336–337
Committee for Environmental Information, 108
Communications, 307–308
 deep space, 5, 6, 10, 16, 61, 118, 203, 237, 246
 global, 76, 111, 127, 237
 international, 10, 69, 73, 80, 82–83, 95, 137, 196–197, 230, 293
 laser use in, 6, 183
 military use, 60, 79, 111, 132, 134, 137, 139, 166, 180, 217, 228
Communications satellite (see also individual satellites: *Intelsat-III F–2*, *Molniya I–9*, etc.)
 anniversary, 80
 award, 77
 contract, 9, 38, 149, 183, 184, 204, 209, 244, 259
 cooperation, 128
  international, 10, 17, 52, 82–83, 85, 95, 111, 136–137, 139, 279, 292
 ground station, 39, 69, 80, 88, 103, 108, 111, 137, 230, 290, 302, 303, 306–307
 launch, 1, 178, 213–214, 335
  failure, 216, 273
  Initial Defense (idcsp), 121, 132, 137, 166, 335

387

Intelsat-III F-2, 314–315, 335
Les VI, 228
Molniya I-8, 88
Molniya I-9, 155
Molniya I-10, 244
plans for, 1, 38, 39, 88, 137, 189
military, 9, 59, 79, 111, 121, 132, 137, 139, 166, 180, 229
use of, 1, 39, 60, 80, 82–83, 93, 109, 209
Communications Satellite Act, 334
Communications Satellite Corp. (ComSatCorp), 128
Atlantic II. See *Intelsat-II F-13*
contract, 38, 149, 183, 184, 204, 244, 259
Early Bird. See *Intelsat I*.
FCC regulation, 111, 204
ground station, 39, 69, 80, 88, 103, 108, 111, 204, 230, 290, 302, 303, 307
Intelsat I (Early Bird), 27, 39, 80, 111, 145, 315
Intelsat II, 189, 204, 323
Intelsat-II F-2 (Intelsat II-B; Pacific I), 82–83, 103, 111
Intelsat-II F-3 (Intelsat II-C; Atlantic II), 5, 39, 111, 315
Intelsat-II F-4 (Intelsat II-D; Pacific II), 111, 302
Intelsat III, 111, 149, 216, 225, 314–315
Intelsat-III F-1, 216, 225, 273, 315
Intelsat-III F-2, 314–315, 323, 329, 335
Intelsat-III F-4, 315
Intelsat III ½, 38
Intelsat IV, 38, 111, 137, 244, 259
Pacific I. See *Intelsat-II F-2*.
Pacific II. See *Intelsat-II F-4*.
revenues, 69, 93, 111–112, 169, 258
satellite program, 69, 88, 93, 111, 183, 184, 216, 243, 259, 307–308
services, 111, 145, 196–197, 230, 276, 293, 302, 307–308
Computer-aided approach spacing system (CAAS), 248
Computer Applications, Inc., 123
Computer Sciences Corp., 123, 172
Computer Softwear Management and Information Center (COSMIC), 310
Computers, 172, 228, 238, 267–268, 276, 293
NASA, 120–121, 164, 205, 246, 310
contract, 3, 75, 79, 154, 239
Concorde (U.K.-France) supersonic transport, 49, 115, 198, 217, 259–260, 294
Condon, Dr. Edward U., 99, 169
Conférence Européene sur les Télécommunications par Satellites (CETS), 52, 85, 139, 279
Conference on Basic Environmental Problems of Man in Space, 286–287
Conference on Exploration and Peaceful Uses of Outer Space, 189–191, 204, 211
Conference on Pavement Grooving and Traction Studies, 286
Congress, 43, 60, 75, 101, 115, 131–132, 159, 160, 168, 175–176, 180, 225–226, 235, 301, 323
McNamara, Secretary of Defense Robert S., report to, 29
President's budget message, 18–20
President's report to Congress, *United States Aeronautics and Space Activities, 1967*, 20–21, 148
R&D, 122, 156–157, 258–259, 299
space program, 24, 65, 87, 110, 139, 147, 199, 210, 213, 218, 221–222, 233, 242, 287, 325, 326, 336
Joint Committee on Atomic Energy, 159, 169, 170, 211
Joint Economic Committee, Subcommittee on Economy in Government, 282
Congress, House of Representatives, 15–16, 93, 127, 135, 140, 197, 218, 229
bills introduced, 32, 227
bills passed, 102–103, 108, 130, 135–136, 171, 218, 233
Committee on Appropriations, 144, 146, 166, 170, 284
Subcommittee on Department of Defense, 130
Subcommittee on Independent Offices and HUD, 32
Committee on Government Operations, 137
Subcommittee on Military Operations, 137
Subcommittee on Special Studies, 163
Committee on Interstate and Foreign Commerce, 116
Committee on Science and Astronautics, 16, 33–35, 47, 53, 57, 58, 99, 102–103, 172, 239
Subcommittee on Advanced Research and Technology, 41, 44, 47, 50, 53, 225–226, 238
Subcommittee on Manned Space Flight, 41, 58, 103, 222
Subcommittee on NASA Oversight, 4–5, 51, 68, 99, 333
Subcommittee on Research and Technical Programs, 70
Subcommittee on Science, Research, and Development, 10, 60, 135, 194–195
Subcommittee on Space Science and Applications, 41, 43, 44, 49, 333
Committee on Ways and Means, 134
Congress, Senate, 131, 135, 152, 239–240, 348, 292
bills introduced, 30
bills passed, 129, 165–166, 167, 168, 175, 210, 227, 242
nuclear nonproliferation treaty, 158, 249

resolution, 127
space rescue treaty, 161
Committee on Aeronautical and Space Sciences, 210, 246–247
  Apollo AS-204 accident, 4–5, 25
  DOD, 69–70
  NASA budget, 30, 48–49, 50, 57, 65, 90–91, 93, 94, 98, 110 240–242
  NASA contracts, 86
  NERVA program, 57, 93, 94, 98, 170
  Tenth Anniversary, 1958–1968, report, 334
Committee on Appropriations, 170
  Subcommittee on Dept. of Defense, 29
  Subcommittee on Independent Offices, 140
Committee on Armed Services, 20, 29, 71, 76, 138
  Preparedness Investigating Subcommittee, 155, 224, 231, 245–246
Committee on Commerce, 26
  Subcommittee on Aviation, 137–138
Committee on Foreign Relations, 56, 159, 165, 210, 215–216, 225, 230
Committee on Government Operations, 71
Committee on Interior and Insular Affairs, 68
Committee on Nuclear Safeguards, 228
Committee on Post Office and Civil Service, 62
CONI. See Comisión Nacional de Investigación del Espacio.
Connecticut, 224
Conrad, Max, 207
Conte, Rep. Silvio O., 127
Contract (see also under agencies, such as NASA, USAF)
  cost-plus-award-fee, 59, 122, 143, 145, 146, 239, 245, 289, 308, 328
  cost-plus-fixed-fee, 79, 84, 104, 173, 239, 264, 331–332
  cost-plus-incentive-award-fee, 146
  cost-plus-incentive-fee, 9, 39, 123, 184, 194, 202, 298
  fixed price, 59, 86, 120, 130, 209
  fixed-price-incentive-fee, 59
  incentive-fee, 144
  study, 62, 187, 275
Convair 990 (*Galileo*) (jet aircraft), 10–11 17, 72, 184
Cook, Don, 298
Cook, Gen. Orval (USAF, Ret.), 123
Cook, William H., 15
Cooksey, Robert, 329
Coon, Grant W., 2
Cooper, L/C L. Gordon (USAF), 279, 280, 306
Cooper, Sen. John S., 242
Copenhagen, Denmark, 163
Coralie (French rocket), 296
Corddry, Charles W., 40, 156
Corliss, William R., 82

Cornell Aeronautical Laboratory, Inc., 31, 81, 228, 279
Cornell, Univ., 31, 49,–50, 54, 171, 172, 228, 238, 284, 292, 316, 326
  Center for Radiophysics and Space Research, 37–38
  Graduate School of Aerospace Engineering, 146
Corpus Christi, Tex., 227
Cortright, Edgar M., 65, 101, 122
COSMIC. See Computer Softwear Management and Information Center.
Cosmic ray, 93, 208–209 284–285, 300–301
Cosmonaut, 28, 83, 116–117, 190, 212, 230, 253, 254
  accident, 264
  commemorative stamp, 85
  death, 70, 72, 264
  interview, 46
  medal, gold, 309
  Soyuz 3 flight, 264–265, 266, 272, 277
  space flight, effects of, 211
  space rescue treaty, 299
  Vostok I flight, 17, 309
  woman, 148
Cosmodrome, 126, 130, 132, 135, 158–159, 165, 172, 183, 264–265, 269, 279, 289, 295
Comos CX (U.S.S.R. satellite), 287
Cosmos CLXXXIV, 2
Cosmos CXCIX, 9
Cosmos CC, 11
Cosmos CCI, 32
Cosmos CCII, 42
Cosmos CCIII, 42
Cosmos CCIV, 57
Cosmos CCV, 57
Cosmos CCVI, 61
Cosmos CCVII, 66
Cosmos CCVIII, 63, 180
Cosmos CCIX, 67
Cosmos CCX, 76
Cosmos CCXI, 82
Cosmos CCXII, 84, 87, 88–89, 119, 178
Cosmos CCXIII, 84, 87, 88–89, 119, 178
Cosmos CCXIV, 87
Cosmos CCXV, 87, 128, 130
Cosmos CCXVI, 88
Cosmos CCXVII, 92
Cosmos CCXVIII, 93, 178, 180
Cosmos CCXIX, 95
Cosmos CCXX, 106
Cosmos CCXXI, 120
Cosmos CCXXII, 123
Cosmos CCXXIII, 126
Cosmos CCXXIV, 126
Cosmos CCXXV, 130
Cosmos CCXXVI, 132
Cosmos CCXXVII, 135
Cosmos CCXXVIII, 139
Cosmos CCXXIX, 143
Cosmos CCXXX, 154
Cosmos CCXXXI, 158, 180

Cosmos CCXXXII, 163–164
Cosmos CCXXXIII, 165
Cosmos CCXXXIV, 172, 180
Cosmos CCXXXV, 183
Cosmos CCXXXVI, 201
Cosmos CCXXXVII, 201
Cosmos CCXXXVIII, 201
Cosmos CCXXXIX, 206
Cosmos CCXL 211
Cosmos CCXLI, 213
Cosmos CCXLII, 219
Cosmos CCXLIII, 223
Cosmos CCXLIV, 239
Cosmos CCXLV, 240
Cosmos CCXLVI, 245
Cosmos CCXLVII, 248
Cosmos CCXLVIII, 259
Cosmos CCXLIX, 259
Cosmos CCL, 267
Cosmos CCLI, 267
Cosmos CCLII, 269
Cosmos CCLIII, 279
Cosmos CCLIV, 289
Cosmos CCLV, 295
Cosmos CCLVI, 296
Cosmos CCLVII, 298
Cosmos CCLVIII, 308
Cosmos CCLIX, 310
Cosmos CCLX, 313
Cosmos CCLXI, 317
Cosmos CCLXII, 324
Cotton, Col. Joseph F. (USAF), 67, 145
Council for the Advancement of Science Writing, 284
Couve de Murville, Prime Minister Maurice (France), 294
Cowen, Robert C., 229–230
CP–1919 (pulsar), 114–115
Crab Nebula, 52, 62, 272, 284, 292
Crane, Robert M., 283
Cromley, Ray, 257
Crooker, John H., 40
Cryogenics, 101, 122
CSM. See Command and service module.
Cuba, 42, 189, 247
Cudaback, Dr. David, 119, 128
Cunningham, R. Walter, 276
   Apollo 7 flight, 250, 252, 276
      preparations for, 97, 158, 173, 220, 227
   award, 269
   press conference, 220, 276
   White House visit, 306
Curtin, M/G Robert H. (USA, Ret.), 102, 120
Curtis, Sen. Carl T., 129
Curtiss, Prof. W. David, 31
Cushman, Ralph E., 102
Cygnus (constellation), 52, 200
Czechoslovak Academy of Sciences, 269
   Astronomical Institute, 254
Czechoslovakia, 161, 189–190, 205, 215–216, 223, 247, 309, 317

D

Dacca, East Pakistan, 18
Daddario, Rep. Emilio Q., 10, 127, 194–195
Dallas, Tex., 112, 325
Dana, William H., 53, 78, 132, 198, 262, 317
Darcey, Robert J., 283
Darmstadt, W. Germany, 300–301
Dartmouth College, 117
   Amos Tuck School of Business Administration, 318
Davies, Merton E., 248–249
Davis, Frank W., 155
DC–8 (jet airliner), 293
Debré, Foreign Minister Michel (France), 250
Debus, Dr. Kurt H., 89, 256
Deep Space Network (DSN), 5, 6, 10, 118, 203, 209
Deep Submergence Search Vehicle, 227
Defense Communications Agency, 166
Defense, Dept. of (DOD) (see also U.S. Air Force, U.S. Army, and U.S. Navy), 40, 42, 156, 301
   Advanced Research Projects Agency, 40, 42, 145
   aircraft, 71, 75, 146, 155, 160, 245–246, 282, 284, 288, 333
   annual report to Congress, 29
   award, 7, 51
   budget, 18–20, 143, 146, 166, 233, 242, 243, 261, 336
   communications satellite system, 9, 137, 166, 180–181
   contract, 9, 61, 278–279, 298, 334
   cooperation, 9, 65, 70, 137
      FAA, 238
      NASA, 20, 33, 70, 84, 156, 225, 237, 310
   criticism, 83–84, 333
   Defense Science Board, 7, 76–77, 329
   facilities, 231
   missile program, 61, 133, 137, 143, 155, 193, 242
   nuclear submarine, 169, 211, 261
   personnel, 12, 40, 51, 331
   R&D, 69–70, 101–102, 136–137, 186, 239–240, 290, 302
   reentry vehicle program, 8
   space program, 32, 69–70, 126, 137, 149, 180–181, 190, 208, 255, 335
   underground atomic tests study, 303–304
Defense Production Act, 83
Defense Projects Support Office (DPSO), 84
Defense Satellite Communications System (DSCS), 137
De Gaulle, President Charles (France), 233
Deimos (Mars moon), 281
DeLong, Dr. Earl H.,

Delta (booster) (see also Thor-Delta), 59, 273, 300
Dembling, Dr. Paul G., 12, 127, 258, 283
Denmark, 5, 154, 240, 292
Denver, Colo., 112, 161
Des Moines, Iowa, 207
Descent propulsion system (DPS) (Apollo), 13
Detroit, Mich., 199
DFC. See Distinguished Flying Cross.
Diamonds, industrial, 158
Diebold, John, Lecture, 232
Diehl, William, 49
Dietlein, Dr. Lawrence F., 291
Di Luzio, Frank C., 123
Disarmament, 11, 27, 32, 59, 63, 149, 160, 164
Distinguished Civilian Service Medal (DOD), 7
Distinguished Flying Cross (DFC), 11–12
Distinguished Service Medal, 314
Distinguished Service Medal (NASA), 263, 269, 283, 332
Distinguished Service Medal (USN), 29
Ditoro, Dr. Michael J., 143
Dmitriyev, Prof. A., 249
DO 31 (VTOL transport aircraft), 105
Dobbins AFB, Ga., 146, 230–231
Docking, 236
  automatic, 24, 84, 85, 87, 88, 93, 119
  manned, 24, 84, 250, 280
DOD. See Defense, Dept. of.
Doeker, Robert, 158
Dominion Observatory (Canada), 134
Donely, Philip, 283
Dong Hoi, North Vietnam, 69
Donlan, Charles J., 51, 65, 122
Dorman, Bernhardt L., 172
Dornier-Werke, GmbH, 105, 151
DOT. See Transportation, Dept. of.
Doty, Robert C., 88
Douglas Aircraft Co., Inc. See McDonnell Douglas Corp.
Douglas United Nuclear, Inc., 87–88, 122
Dover, Del., 176–177
Dover, U.K., 175
Downey, Calif., 122, 167
DPS. See Descent propulsion system.
DPSO. See Defense Projects Support Office.
Draco (constellation), 310
Drake, Dr. Frank D., 54, 238, 284
Dryden, Dr. Hugh L., 175
Dryden, Hugh L., Memorial Fund, 175
Dryden Research Lecture Award, 15
DSCS. See Defense Satellite Communications System.
DSN. See Deep Space Network.
Dubinin, Prof. Nikolay P., 200
DuBridge, Dr. Lee A., 267, 299, 301, 310, 314
Dudley Observatory (Albany, N.Y.), 129, 185, 197, 300, 310
Duff, Brian M., 117
Duffy, Robert T., 30–31

Duke, Capt. Charles M., Jr. (USAF), 59, 280
Duke Univ., 255, 299
  Medical Center, 40, 115
Dulles International Airport, 179, 224
Dunbar, William R., 104
Duncan, Robert C., 283
Dunn, M/G Carroll H. (USA), 32
Dunn, Dr. Richard, 296
Dupree, Dr. A. Hunter, 1
Dvorscak, Bernie J., 230–231

E

Eaker, L/G Ira C. (USAF, Ret.), 125
Early Bird (communications satellite). See *Intelsat I.*
Earth
  gravitational field, 318, 335
  horizon, 132
  light sources, 177
  magnetic field, 72, 308
  magnetic tail, 138
  mapping, 325
  photographs of, 251, 318, 323, 324, 329
  resources measurement, 33–34, 174
  shape, 10–11, 331
  size, 10–11
Earth Resources Observation Satellite (EROS) program, 7–8, 19, 39, 62, 333
Earth Resources Satellite System (Congressional report), 333
Earth Resources Survey Program, 39, 51, 85, 173, 215, 279
Earthquake, 191
Easter Island, 112
Eastern Airlines, 72, 139, 217
Eastern Test Range (ETR) (see also Cape Kennedy and Kennedy Space Center)
  contract, 83, 123
  launch, 1, 9, 59, 105, 244, 278
    Apollo/Saturn, AS-204, 13
    Atlas-Agena D, 54–55, 179
    Atlas-Centaur, 184, 304
      AC-15, 3
    failure, 216, 225
    Long-Tank Delta, 216
    Minuteman III, 193
    Poseidon, 193
    satellite, 3, 54–55, 111, 132, 179, 184, 216, 228, 273, 300, 314–315
    Thrust-Augmented Improved Thor-Delta, 273
    Thrust-Augmented Long-Tank Delta, 314–315
    Thrust-Augmented Thor-Delta, 244, 300
    Titan III–C, 132, 228
Eastman Kodak Co., 292
EC–135N (Apollo Range Instrumentation aircraft), 148
Echo program, 152
Echo I (communications satellite), 60, 120
Eclipse, lunar, 250

Ecology, 126, 135
Economic Club of Detroit, 37
Ecuador, 37
Eddystone, Pa., 219
Editorial comment. See Press comment.
EDP Technology, Inc., 228, 279
EDS. See Emergency detection system.
Education, 235, 249–250, 271, 314, 323, 331, 334
Edwards AFB, Calif. (see also Air Force Flight Test Center)
  flight
    accident, 2
    F–111A, 2
    HL–10 (lifting body vehicle), 63, 131, 139, 224–225, 240
    X–15, 53, 78, 164, 198, 211, 262
    XB–70, 8, 38, 67, 269, 313
    XB–70A, 145, 167, 208, 258, 298
EG&G, Inc., 123
Eggers, Dr. Alfred J., Jr., 55
Eglin AFB, Fla., 4
Ehricke, Dr. Kraft A., 163
Eielson AFB, Alaska, 168
Eisele, Maj. Donn F. (USAF)
  Apollo 7 flight, 250, 251
    preparations for, 97, 158, 173, 220, 227
  Apollo 10 flight, 279–280
  award, 269
  press conference, 220
  White House visit, 306
Eisenhower, President Dwight D., 325, 328
Ekers, Dr. Ronald D., 114
Eklund, Sigvard, 198
El Centro, Calif., 8
ELDO. See European Launcher Development Organization.
ELDO F–7 (ELDO satellite), 296
Electric power, 275
Electromagnetic pulse (emp), 228
Electro-Mechanical Research Co., 111
Electron, 31, 60, 62–63, 67, 173, 175, 198, 256, 329
Electronics Research Center (ERC) (NASA), 39, 123, 145, 164, 179, 283
Electro-Optical Systems, Inc., 268
Ellington AFB, Tex., 105, 116, 257
Elms, James C., 165
Emergency detection system (EDS) (Apollo), 77
Emme, Dr. Eugene M., 1
Emmons, Dr. Howard W., 5
Employment of Scientists and Engineers in the United States, 1950–1966 (NSF report), 234
Engine (see also individual engines, such as F–1, H–1)
  aircraft, 313
    gas turbine, 316
    jet, 130, 197, 198, 226, 232
      Quiet Research Engine Program, 47, 197, 248

supersonic transport, 146, 233, 332
tubofan, 145, 158
turboprop, 148, 217
electric, 25–26, 303
flying platform, 107
hypersonic, 47, 299
lunar module, 127, 227–228
nuclear (see also NERVA), 44, 53, 57, 93, 94, 156, 241–242, 302
reusable, 2
rocket, 2, 62–63, 68–69, 261, 275, 279, 308, 317
  hybrid, 3–4, 9
  test, 13, 17, 36, 42, 77–78, 92, 143, 184
space station, 141–142
vernier, 6
Engineers, 234, 289
Engle, Capt. Joseph H. (USAF), 134, 144, 280
English Channel, 175
ENTEL, 293
Enthoven, Dr. Alain C., 259
Environment, 135
Environmental Science Services Administration (ESSA)
  budget, 19
  cooperation 66, 262, 335
  Man's Geographical Environment: Its Study from Space, 73
  personnel, 131
  research, 101–102, 110, 266, 325
  satellite, 1, 49, 111, 149, 232, 311
    launch, 193, 335
  weather modification, 194–195
ERC. See Electronics Research Center (NASA).
EROS. See Earth Resources Operation Satellite.
Ers XXI (OV V–4) (Environmental Research Satellite), 228, 335
Ers XXVII (OV V–2), 228, 335
Escape system, 78, 210
ESO. See European Southern Observatory.
ESRO. See European Space Research Organization.
Esro I (ESRO satellite). See *Aurorae (Esro I)*.
ESRO II–A (International Radiation Investigation Satellite), 113
Esro II–B. See *Iris I (Esro II–B)*.
ESSA. See Environmental Science Services Administration.
Essa I (meteorological satellite), 193
Essa II, 50, 193
Essa III, 193
Essa IV, 193, 232
Essa V, 193
Essa VI, 193, 295
Essa VII (TOS–E), 193, 311, 335
Essa VIII (TOS–F), 311, 335
Etam, W. Va., 230, 315, 329
Ethiopia, 137, 327
ETR. See Eastern Test Range.

Etzel, Prof. James E., 281
EURATOM. See European Atomic Energy Community.
Euromart, 233
Europe (see also European Launcher Development Organization; European Space Research Organization), 160
  communications, 38, 73, 139, 230, 312, 323
  cooperation, 136–137, 139
  space funding, 136
  tracking station, 137
European Atomic Energy Community (EURATOM), 11, 92
European Conference on Satellite Communications. See Conférence Européene sur les Télécommunications par Satellites.
European Launcher Development Organization (ELDO)
  launch
    ELDO F-7, 296
  merger proposal, 136–137
  policy, 279
  president elected, 60
  U.K. membership, 85, 95, 233
European Nuclear Research Center, 139
European Southern Observatory (ESO), 292
European Space Conference, Third, 279
European Space Authority (proposed), 279
European Space Research Organization (ESRO), 60
  launch
    satellite, 1
      Aurorae (Esro I), 240
      Heos–A, 301, 336
      Iris I (Esro II–B), 113, 336
    sounding rocket, 108, 118
  membership, 5, 85
  merger proposed, 136
  satellite, 95, 149, 168–169, 214, 240, 336
Eurospace Conference, Third U.S.-European, 136–137
EUV: extreme ultraviolet.
EVA. See Extravehicular activity.
Evans, Albert J., 47
Evanston, Ill., 284
Everett, Wash., 232
Evvard, John C., 283
Exceptional Scientific Achievement Medal (NASA), 263, 283
Exceptional Service Medal (NASA), 194, 269, 283
Exhibit, 190
Exosphere, 82
Explorer (program), 44
Explorer I (satellite), 22, 29–30, 37
Explorer XVI, 177
Explorer XXII, 10, 183
Explorer XXIII, 177
Explorer XXIV, 182
Explorer XXV (Injun IV), 182

Explorer XXVII, 183
Explorer XXVIII, 153–154
Explorer XXIX, 8, 183
Explorer XXX (IQSY Solar Explorer), 56–57
Explorer XXXV (IMP–E), 167
Explorer XXXVI (Geos II; GEOS–B), 1, 7, 8, 16, 42–43, 71, 183, 259, 335
Explorer XXXVII (Solar Explorer B), 56–57, 335
Explorer XXXVIII (Radio Astronomy Explorer RAE-A), 153, 167, 210, 224, 246, 335
Explorer XXXIX (Air Density Explorer C), 182, 310, 335
Explorer XL (Injun V), 182, 310, 335
Extraterrestrial life, 54, 142, 301, 302
Extravehicular activity (EVA), 24, 168, 177, 190, 236, 316–317

F

F-1 (rocket engine), 30, 257, 275
F-4 (Phantom) (fighter aircraft), 3, 14, 217
F-4J, 76
F-5 (supersonic fighter aircraft), 281
F-12 (interceptor aircraft), 112, 171–172
F-14A (USN interceptor aircraft), 76, 98–99, 167, 171, 230–231, 314
F-15A (aircraft). See ZF-15A.
F-104 (Starfighter) (aircraft), 165
F-106 (interceptor aircraft), 171–172
F-106B, 233
F-106X, 171–172
F-111 (supersonic fighter), 130, 206, 231, 245–246, 259, 270
  accident, 2, 195
  cost, 130, 146
F-111A, 117, 160, 201, 316, 333
  accident, 2, 72, 80, 84, 87, 97, 102, 108, 109, 119, 134, 160, 195, 224, 258
  defense of, 75, 333
  test 155, 207, 258
  Vietnam War use, 69, 71, 72, 84, 104, 109, 155
F-111B, 76, 130, 167
  accident, 2
  funds for, 14, 19, 71, 98, 159
  weight, 160
F-111C, 206, 248
FAA. See Federal Aviation Administration.
FAI. See Fédération Aéronautique Internationale.
Fairbank, Dr. William M., 91
Fairbanks, Alaska, 35, 143, 192
Fairchild-Hiller Corp., 120, 209, 289, 334
Fairhall, Defence Minister Allen (Australia), 206
Fangataufa Atoll, 200
FB-111 (supersonic bomber), 201, 259, 284, 286
FB-111A, 160
FCC. See Federal Communications Commission.

FDA. See Food and Drug Administration.
FDVU. See Flight Design Verification Unit.
Federal-Aid Airport Program (FAAP), 95
Federal Air Quality Advisory Board, 169
Federal Aviation Act of 1958, 130
Federal Aviation Administration (FAA), 26, 96, 108, 150, 175, 207
  air traffic control, 1–2, 82, 137–138, 168, 187, 196, 206, 238, 248, 300, 302
  airports, 95, 137–138, 231, 278, 296, 302, 324
  appropriations, 19, 168
  contract, 82, 231
  cooperation, 168, 238
  forecast, 71, 137–138
  landing system, 248
  noise, aircraft, 116, 130, 135
  regulations, 1–2, 116, 130, 194, 206, 300
  transport, supersonic (see also Supersonic transport), 19, 66–67, 135, 150, 158
    design and development, 45, 202–203, 260
Federal Communications Commission (FCC), 111, 154, 204, 276, 308
Federal Contract Research Center, 122
Federal Council for Science and Technology, 271
  Interdepartmental Committee for Atmospheric Sciences, 101–102
Federal Department of Science (proposed), 325
Fédération Aéronautique Internationale (FAI), 111, 309
Felberg, Fred H., 283
Fels Planetarium, 27
Ferguson, Gen. James (USAF), 5, 141, 237, 272
Fermi, Enrico, Award, 221, 298
Fermi, Enrico, Institute, 90
Findlay, Dr. John W., 30
Finger, Harold B., 35, 48, 133
Fink, Daniel J., 329
Finney, John W., 92
The First Four Stages, 17
Fitzgerald, A. E., 282
Fjeldbo, Gunnar, 276
Flammability tests, 13–14, 57, 61, 265
Flax, Dr. Alexander H., 283
Flemming, Arthur S., Award, 39
Fletcher, Dr. Robert D., 316
Flickinger, B/G Don, Jr. (USAF), 28
Flight design verification unit (FDVU), 128
Flight Research Center (FRC) (NASA), 12, 42, 48, 68, 186, 313
Flight Safety Foundation, 313
Flight Test, Simulation and Support Conference, 65, 66
Flock, Dr. Warren L., 129
Flood control, 62
Florida, 109

FOBS. See Fractional orbital bombardment system.
Food and Drug Administration (FDA), 2
Food irradiation, 2
Fordham Univ., 287
Ft. Churchill, Canada (see also Churchill Research Range, Canada), 29, 72, 329
Fort Myers, Fla., 146
Fort Worth, Tex., 206
Foster, Dr. John S., Jr., 69–70, 155
Foster, William C., 295
Founders Medal (NAE), 43
Fowler, Prof. William A., 115
Foxbat (U.S.S.R. fighter aircraft), 130
Fractional orbital bombardment system (FOBS), 55–56, 93, 94, 102, 178, 239, 255
France, 46, 60
  aircraft, 217
  balloon, 166
  Concorde (France-U.K. supersonic transport), 115, 198, 260, 294
  cooperation, 5, 31, 55, 113, 134, 214, 233, 292, 300–301
    U.S.S.R., 10
  hydrogen bomb, 153, 156, 200, 207–208
  missile program, 153, 200
  nuclear attack, 138
  nuclear test, 156, 180, 294
  sounding rocket, 31, 296
Frantz, Dr. Joe B., 1
FRC. See Flight Research Center.
Freibaum, Jerome, 128
French Guiana, 172
Frenkel, Dr. Lothar, 39
Freon TF, 157
Friction Effects of Runway Grooves, Runway 18–36, Washington National Airport, 324
Friel, Dr. Patrick J., 43
Friendship Airport, 179, 224
Fruehaf, Capt. David E. (USAF), 8
Frutkin, Arnold W., 30, 239, 283
Fucino, Italy, 80
Fuel, 94, 303
  hydrogen, 1
  leak, 92
  liquid, 3–4, 9, 274
  liquid hydrogen, 79, 101, 266
  nitrogen, 45
  oxygen-hydrogen, 2
  plutonium 238, 116
  safety, 95
  solid, 3–4, 42, 68, 193, 274
Fuel cell, 246, 251–252, 294–295
Fuhrmeister, Paul F., 283
Fullbright, Crooker, Freeman & Bates, 40
Fulton, Fitzhugh L., Jr., 8, 38, 50, 67, 131, 167, 208, 269, 298
FX (aircraft). See ZF–15A.

G

Gagarin, Col. Yuri A. (U.S.S.R.), 17, 70, 72, 190, 309
Galapagos Islands, 189
Gamma (star), 67
Gamma ray, 52, 54–55, 181, 281
Gamow, Dr. George, 197
Ganczak, Edward G., 107
Gandhi, Prime Minister, Mrs. Indira, 30–31
GAO. See General Accounting Office.
Garbuzov, Finance Minister Vasily (U.S.S.R.), 309
Gardner, Lester D., Lecture, 104
Gas turbine, 316
GCA Corp., 45, 46–47, 168
Geiss, Dr. Johannes, 68
Geller, Justice Abraham N., 279
Gemini (program), 66, 80, 103, 161 211, 215, 221–222, 236–237, 269, 281
Gemini (spacecraft), 148
Gemini X (flight), 168
Gemini XI (flight), 168
General Accounting Office (GAO), 86, 226
General Aircraft Corp., 224
General aviation, 47
  air traffic, 187–188, 189, 194, 196
  aircraft, 71, 137
  airports, 278
  collision avoidance system, 179
  flying hours, 71, 137
General Dynamics Corp.
  contract, 167, 187, 278, 333
  Convair Div., 8, 131, 207
  Electric Boat Div., 302–303
  F–111A, 117, 155, 224, 333
  F–111B, 159, 167
  F–111C, 206
  VFX aircraft, 187
General Electric Co., 215, 329
  Aircraft Engine Group, 296
  Apollo Systems Div., 79
  ARMS (Application of Remote Manipulators in Space), 186
  award, 15
  contract, 40, 54, 120, 149, 172, 173, 186, 201, 202–203, 209, 262, 278
  laboratory, undersea, 101
  Mississippi Test Facility, 173
  "Quiet jet engine," 197
  Re-Entry Systems Div., 97
  reentry vehicle, 40
  spacecraft, 120, 172, 181, 209
  supersonic transport engine, 150, 202–203
  Technical Military Planning Operation (TEMPO), 26
  turbofan engine, 54, 201, 276, 282
  XB–70 engine, 149
General Motors Corp., Allison Div., 248
General Precision Systems, Inc., 22
Geneva Disarmament Conference, 59, 62, 132
Geneva, Switzerland, 32, 59, 127, 149, 164
Gentry, Maj. Jerauld R. (USAF), 63, 139, 224–225, 306
Geocorona, 55
Geodetic satellite, 1, 7–8, 16, 42–43, 71, 183, 335
Geology, 51–52
Geophysics Corp. of America, 198
Georgia, Univ. of, 310
Geos II (GEOS-B; *Explorer XXXVI*) (geodetic satellite), 1, 7–8, 16, 42–43, 71, 183, 259, 335
Germany, East, 317
Germany, West, 46
  aircraft, 105, 151, 165
  Bochum Observatory, 54, 211–212, 222
  cooperation, 5, 137, 160, 165, 201, 292, 298, 300–301
  Ministry of Scientific Research (BMWF), 75–76, 160–161, 308
  nuclear nonproliferation treaty, 11
  space program, 137, 151, 160–161, 201, 300
GFT: ground elapsed time.
GHOST. See Global Horizontal Sounding Technique.
Gilruth, Dr. Robert R., 14, 28, 75, 142
Giroux, Dennis, 44
Glassboro (N.J.) State College, 127
Glenn, Col. John H., Jr. (USMC, Ret.), 203–204, 215
Global Atmospherics Research Program (GARP), 101–102
Global Horizontal Sounding Technique (GHOST), 240
Globe Exploration Corp., 29
Goddard, Dr. Robert H., 28, 57, 316
Goddard, Robert H., Award, 15, 316
Goddard, Dr. Robert H., Lecture, 55–56
Goddard, Dr. Robert H., Memorial Trophy, 57
Goddard Memorial Dinner, 57–58
Goddard Memorial Symposium, 58–59.
Goddard Space Flight Center (GSFC), 216, 225, 294
  Apollo 6, 78
  Apollo 7, 253
  Apollo 8, 322
  award, 283
  contract, 45, 47, 61, 123, 289
  facilities, 123
  Laboratory for Atmospheric and Biological Sciences, 328–329
  Laboratory for Theoretical Studies, 153–154
  laser experiment, 71, 183, 335
  personnel, 35, 131, 152, 168, 328
  Radio Astronomy Explorer, 153
  satellite monitoring
    Ats II, 119
    Ats III, 29, 37, 144
    Ats IV, 185
    Essa VII, 191–193

Essa VIII, 311
Explorer XXXVI, 71, 183, 259, 335
Explorer XXXVIII, 153
Molniya I-8, 110
Oao II (OAO-A), 304, 310
Ogo V (OGO-E), 55
Satellite Tracking Center, 12, 143, 146, 205
sounding rocket experiments, 43-44
  astronomical, 62-63, 67, 115, 118, 121, 129, 175, 198, 232
  atmospheric data, 13, 60, 62, 64-65, 76, 82, 130-131, 168, 175, 213-214, 216, 219-220, 223, 273, 287, 291, 309-310
  instrumentation test, 17, 31, 70
  launch test, 211, 214
  parachute recovery system test, 126-127
  vehicle performance test, 31
Gold, Thomas, 37-38
Goldberg, Ambassador Arthur J., 130
Goldberg, Leo, 243-244
Goldberger, Dr. Marvin L., 93
Goldstein, Dr. Richard M., 114-115, 171, 283, 291
Goldstone Tracking Station, 114-115, 122, 171, 209, 291, 312
Goldwater, Barry M., 333
Gomez Palacios, Mex., 295
Goodell, Sen. Charles E., 210
Goodyear Tire and Rubber Co., 194
GOR-16 (general ocean research ship), 310
Gordon, James, Trophy, 188
Grants, 44-45, 46, 140-141, 151, 314
Gravity, 141, 238-239
  artificial, 268, 291
Gray, Dr. Clinton W., 273
Gray, Harold E., 107-108
Gray, Robert H., 216
Greater Lameshur Bay, Virgin Islands, 101
Grechko, Defense Minister Marshal Andrey A. (U.S.S.R.), 45-46
Green Bank, W. Va. 161, 238, 272
Green State Univ., 325-326
Greenbelt, Md., 205
Greenland, 11
Greenwood, A. H. C., 217
Grew, Gary W., 177
Griffith Planetarium, 221
Grissom AFB, Ind., 39, 110
Grissom, L/C Virgil I. (USAF), 103, 110
Gromyko, Foreign Minister Andrey A. (U.S.S.R.), 145
Gross National Product, 141
Group Achievement Award (NASA), 164, 283
Grumman Aircraft Engineering Corp., 14, 86, 167, 187, 191, 227-228, 314
GSFC. See Goddard Space Flight Center.
Guam, 137
Guest, Dr. Robert H., 318

Guggenheim Aeronautical Laboratory (Cal Tech), 267
Gulf of Mexico, 59, 73-74, 181, 251
Guppy (undersea research vehicle), 156
Gurovsky, Dr. N. N., 287
Gurshteyn, A. A., 61
Gurtler, C. A., 177

H

H-1 (rocket engine), 9, 207
Hackerman, Norman, 115
Haeussermann, Dr. Walter, 111
Hage, George H., 315
Haggerty, James J., Jr., 95, 117, 253
Haglund, Howard H., 151, 194
Hahn, Dr. Otto, 171, 265
Haile Selassie, Emperor of Ethiopia, 327
Hailstone model test, 247
Haise, Fred W., Jr., 183, 196
Halaby, Najeeb E., 107-108
Hall, Donald A., 104
Hall, Ernest, 49
Hall, R. Cargill, 129
Halle, Louis J., 81
Halpern, Rep. Seymour, 227
Ham (chimpanzee), 273
Hamburger Flugzeugbau, 151
Hamilton, Harry H., 283
Hammett, Prof. Louis P., 38
Handler, Dr. Philip, 40, 115, 255, 299, 314
Haney, Paul, 36
Hanover Air Show, 105
Hanover, N. H., 36
HAPPE. See High Altitude Particle Experiment program.
Harder, Dr. James A., 172
Hardie, Dr. Robert H., 326
Hardy, Dr. Harriet L., 312-313
Hardy, Rep. Porter, Jr., 163
Harlow, Prof. Harry F., 38
Harmon International Aviator's Trophy, 259, 299
Harper, Charles W., 26, 38-39, 47, 225-226, 247
Harr, Dr. Karl G., Jr., 124-125, 134, 315-316
Harriman, N. Y., 105
Harrington, Dr. Charles D., 87, 122-123
Harris, R. Emerson, 128
Harris, S. T., 88
Harvard College Observatory, 161, 224
  Solar Satellite Project, 243
Harvard Univ., 5, 38, 115, 133, 134, 159, 303-304, 328
  Graduate School of Business Administration, 232
Hatfield, Sen. Mark O., 119, 210
Haueter, Paul E., Award, 116
Haughney, Louis C., 11, 72
Hauser, John A., 91
Hawaii, 116, 132, 137

Hawker Siddeley Aviation Co., 277
Hawker Siddeley Group, Ltd., 278
Hawthorne, Calif., 325
Hayes International Corp., 80
Haymes, Dr. R. C., 52
Haynes, Charles G., 207
Haystack, Mass., 49
Hazelhurst Field, N.Y., 11
Health, Education, and Welfare, Dept. of, 169, 271
Heart, 291
  artificial, 194, 302
Heat shield, 212
Helgeson, Bob P., 128
Helicopter, 219
  astronaut pickup, use for, 251
  commercial use, 225–226
  military, 18–19, 146
Helium, 82, 257, 308
Henry, Dr. Richard C., 87
Hensleigh, Walter E., 160
HEOS (Highly Eccentric Orbiting Satellite), 1, 149, 214
Heos–A, 300–301, 336
Herzfeld, Dr. Charles M., 329
Hess, Dr. Harry H., 90
Hess, John L., 95
Hess, Dr. Wilmot N., 36
Hevelius, Johnannes, 28
Hewlett-Packard Co., 331
High Altitude Particle Experiment (HAPPE) program, 68
Hill, Louis W., Space Transportation Award, 260–261
Himmel, Dr. Seymour C., 104
Hines William, 129, 244–245, 312, 331
HL–10 (lifting-body vehicle)
  test flight, 63, 131, 139, 224–225, 240, 335
    powered, 279, 306
    unsuccessful, 261
  test results, 67–68
Hoberg, Otto A., 283
Hodge, John D., 39
Hodgson, Alfred S., 207, 283
Hodgson, Dr. Gordon W., 22
Hoff, Dr. Nicholas J., 5
Holloman AFB, N. Mex., 117, 272, 283–284
Hollomon, Dr. J. Herbert, 325
Honda (comet), 201
  No. 6, 155
Honda, Minori, 155
Honeywell, Inc., 8, 328
Hooker, Dr. Stanley G., 316
Hornbeck, Dr. J. A., 32
Horne, Walter B., 118
Hornig, Dr. Donald F., 139–140, 156, 292, 301, 304, 325
Horton, Bob, 293
Hosenthien, Hans H., 283
Hotz, Robert, 271, 293
Housing and Urban Development, Dept. of (HUD), 108, 165–166, 218, 227, 243, 271
Houston, Tex., 53, 82–83, 118, 188, 220, 246, 267
Hovercraft, 175
Howick, George J., 230
HP 1506 (pulsar), 161
HRE. See Hypersonic Research Engine program
Hubbard, Samuel H., 8–9
Huber, William C., 145
HUD. See Housing and Urban Development, Dept. of.
Hughes Aircraft Co., 3, 5, 84, 106, 111, 160, 204, 244, 259, 272
  Satellite Systems Laboratories, 76
  Space Systems Div., 77
Hughes, Howard R., 303–304
Hughes Nevada Operations, 303–304
Huguenin, Dr. George R., 161
Hulburt, E. O., Center for Space Research, 87
Humphrey, Vice President Hubert H., 229
  astronauts, visit with, 287
  Earth Resources Observation Satellite, 7
  election results, 271–272
  NASA visit, 287
  nuclear nonproliferation treaty, 215–216, 230
  ocean exploration, 138
  space program, 106–107, 178–179, 235–236, 254
Hungary, 189–190, 317
Hunt, Graham R., 170, 250
Huntsville, Ala., 220, 272
Hurricane Gladys, 251
Hurricane Naomi, 295
Hydrogen bomb, 152, 156, 200, 207–208
Hyland, Lawrence A., 84, 106–107
Hynek, Dr. J. Allen, 169–170, 172, 254
Hypersonic aircraft, 226, 237, 300, 317–318
Hypersonic Research Engine (HRE) program, 47

I

IAEA. See International Atomic Energy Agency.
IAF. See International Astronautical Federation.
IAU. See International Astronomical Union.
IBM. See International Business Machines Corp.
Icarus (asteroid), 133, 291
ICBM. See Intercontinental ballistic missile.
ICL Industry, Inc., 176–177
Iconoscope, 43
IDSCP. See Initial Defense Communications Satellite Program.
IEEE. See Institute of Electrical and Electronics Engineers.

IES. See Institute of Environmental Sciences.
Ignatius, Secretary of the Navy Paul R., 14
IGY. See International Geophysical Year.
Illinois Institute of Technology, 15
Illinois, Univ. of, 63, 168
Ilyushin-62 (U.S.S.R. airliner), 126, 163
Image Orthicon Camera (IOC), 185
IMP-E (Interplanetary Monitoring Platform). See *Explorer XXXV* (IMP-E).
IMP-G, 1
India
  cooperation, 27, 30–31, 75–76, 196–197, 273
  Dept. of Atomic Energy, 31
  ground station, 27
  nuclear nonproliferation treaty, 32, 138
  nuclear power, 205
  satellite, 27
  Thumba Equatorial Rocket Launching Station, 30–31, 60, 75–76, 273
  weapons, 130–131
Indian National Commission for Space Research (INCOSPAR), 30–31, 75–76
Indian Ocean, 196–197, 212, 250–251
Industrial Research, Man of the Year award, 15
Informatics, Inc., 169
Information retrieval, 48, 154
Infrared sensor, 206–207, 208
Initial Defense Communications Satellite Program (IDCSP), 121, 132, 137, 166, 335
Injun IV. See *Explorer XXV*.
Injun V. See *Explorer XL*.
Institute for Strategic Studies, 210
Institute of Electrical and Electronics Engineers (IEEE), 82, 119
Institute of Environmental Sciences (IES), 79
Institute of High Energy Physics, 316
Institute of Strategic Studies, 309
INTELSAT. See International Telecommunications Satellite Consortium.
Intelsat (communications satellite), 1, 39
Intelsat I (Early Bird), 27, 39–40, 80, 111, 145, 314–315
Intelsat II, 189, 204, 323
Intelsat-II F-2 (Intelsat II-B; Pacific I), 82–83, 103, 111
Intelsat-II F-3 (Intelsat II-C; Atlantic II), 5, 39, 111, 315
Intelsat-II F-4 (Intelsat D-II; Pacific II), 111, 302
Intelsat III, 111, 149, 216, 225, 315
Intelsat-III F-1, 216, 225, 273, 315
Intelsat-III F-2, 314–315, 323, 329, 335
Intelsat-III F-4, 315
Intelsat III ½, 38
Intelsat IV, 38, 111, 137, 244, 259
Interagency Aircraft Noise Abatement Program, 154–155
Intercontinental ballistic missile (ICBM), 3, 18–19, 29, 42, 138, 191, 195, 210, 312
"Intermediate 20" (booster), 275
Interior, Dept. of, 101, 290–291, 299
International Academy of Astronautics, 254
International Aeronautical Exposition (proposed), 233
International Aerospace Exposition, 295
International Air and Space Salon, Third, 126
International Astronautical Federation (IAF), 258, 286–287
  Congress, 246–247
International Astronomical Union (IAU), 27–28, 298
  Commission on Solar Activity, 254
International Atomic Energy Agency (IAEA), 92, 132, 198, 211, 228
International Center for Theoretical Physics, 198
International Balloon Race, 188
International Bank for Reconstruction and Development, 16
The International Biological Problem —Its Meaning and Needs House study), 60
International Biological Program, 60, 126
International Business Machines Corp. (IBM), 79, 169
  Federal Systems Div., 82, 173
  Space Guidance Center, 77
International Congress for Noise Abatement, 117
International Congress of Genetics, 200–201
International cooperation (see also Nuclear nonproliferation treaty), 127
  aircraft, 9, 49, 105–106, 117, 118, 126, 134, 157–158, 163, 165, 198
  astronomy, 72
  law, 12, 127
  meteorology, 127
  military, 9, 127, 145, 146, 149–150, 160, 164, 166, 170–171, 191
  nuclear power, 11, 154
  oceanography, 56, 72, 138
  science and technology, 39, 60, 126, 140, 141, 156, 158, 159, 247, 255, 257
International cooperation, space (see also European Launcher Development Organization; European Space Research Organization; International Telecommunications Satellite Consortium; Space rescue treaty), 2, 20–21, 30
  law, 12, 127, 258, 281

398

satellite, 19, 52, 134, 137, 189, 233, 236
  Europe, 10, 85
  NASA–ESRO, 113, 214, 240
    -Canada, 1, 11, 72
    -India, 196–197
    -U.K., 1
    -U.S.S.R., 2
  U.S.
    -Japan, 82
    -South Vietnam, 88
    -U.K., 9
    -U.S.S.R., 66, 99, 281
  U.S.S.R.-France, 10
sounding rocket, 29, 154
  India-France, 31
    -U.S.S.R., 31
  NASA-Brazil, 131, 177
    -Canada, 64–65, 106
    -Germany, West, 75–76, 160
    -India, 30–31, 75–76, 273
    -Spain, 177–178
    -Sweden, 177–178
space research, 141, 204, 210, 254
  Europe, 136–137, 279
  U.K.-U.S.S.R., 52
  U.S.-Australia, 329
    -Germany, West, 201
    -Switzerland, 68
    -U.S.S.R., 202
tracking, 137
  U.S.-Australia, 34–35
    -Spain, 34–35
International Council on Human Environment, 127
International Decade of Ocean Exploration, 177
International Flat Earth Society, 331
International Geophysical Year (IGY), 29–30, 73, 166–167
International Institute for Peace and Conflict Research, 285
International Institute of Space Law, 258
International Polar Year, 166–167
International Radiation Investigation Satellite (IRIS), 113
International Research and Marketing Corp., 230
International Satellite for Ionospheric Studies (ISIS), 1, 106, 149
International Space Research Committee, 136–137
International Symposium on Bioastronautics and the Exploration of Space, Fourth, 141–142
International Symposium on Genetic Effects of Space Environment, 200–201
International Telecommunications Satellite Consortium (INTELSAT), 27, 60, 80, 239, 293
  communications satellite, 39, 80, 183, 184, 189–190, 216, 243–244, 259
  Interim Communications Satellite Committee, 35, 80
  membership, 26, 46
International Telephone and Telegraph Corp. (IT&T), 329
Interplanetary Monitoring Platform (IMP). See *Explorer XXXV* (IMP-E) and IMP-G.
Intersputnik, 189–190
Invention, 91
IOC. See Image Orthicon Camera.
Ion propulsion, 25–26, 303
Ionosphere, 63, 73
  composition, 46–47, 60, 130, 132, 168, 175, 198, 273
  diurnal changes, 64–65
  probe, 106, 149, 336
  properties of, 31, 168, 240, 262
Iowa, Univ. of, 29–30, 113, 182
Ireland, 299
Iris I (*Esro II–B*) (International Radiation Investigation Satellite), 113, 168–169, 240, 336
Irwin, Maj. James B. (USAF), 279–280
ISIS–A (International Satellite for Ionospheric Studies), 1, 106, 149
Isolation experiment, 324
Italcable, 293
Italy
  cooperation, 5, 46, 95, 108, 165, 233, 293, 300–301
  ground station, 69
  nuclear nonproliferation treaty, 32
  satellite, 296
ITT World Communications, Inc., 82–83, 154, 293, 307–308

J

J–2 (rocket engine)
  Apollo 6 and *7*, 77–78, 92–93, 251–252
  contract, 257, 275
  failure, 92, 119–120, 135, 252–253
  test, 220
J85–19 (VTOL engine), 230–231
Jackass Flats, Nev., 131, 143
Jackson, Sen. Henry M., 228
Jackson and Moreland, 231
Jaffe, Leonard, 30–31
James, Lee B., 20, 112
Jamesburg, Calif., 302
Japan, 46, 82–83, 146, 155, 166, 312
Japanese Defense Agency, 166
Jastrow, Dr. Robert, 147–148
Javelin (sounding rocket), 54, 82, 113, 308
JC–130 (Hercules) (turboprop aircraft), 148
Jennings, Mark, 247
Jet Propulsion Laboratory (JPL) (Cal Tech), 22–23, 79, 148, 331–332
  award, 15, 89, 106, 151, 194, 260–261, 283
  Deep Space Network, 6, 209
  Goldstone Tracking Station, 114–115, 122, 171, 209, 291, 312

Mariner Project, 118, 122, 199
  personnel, 15, 17, 42, 129, 152–153, 171, 229–230, 267, 275–276
Pioneer VI, 209
Surveyor Project, 3, 106, 151
symposium on space research, 29–30
Jodrell Bank Experimental Station, 42, 54, 83, 211–212, 222, 253–254, 289
Johns Hopkins Univ., 8, 36, 86, 173
Johnsen, Edwin G., 82
Johnson City, Tex., 1, 269, 332
Johnson, President Lyndon B., 301, 304, 323, 327
  aircraft, 54
  Apollo 8 flight, 322, 326
  appointments and nominations by, 12, 45, 115, 131, 143, 169, 175, 211
  astronauts, 322, 326, 330, 332
    flight pay, 203
  awards by, 38, 51, 57, 269, 298, 332
  awards to (National Space Club), 235
  balance of payments deficit, 1
  bills signed, 135–136, 152, 197, 243
  budget, 18–19, 60, 233
    NASA, 135–136, 152, 294
  communications satellite system, 76
  disarmament, 164
  international cooperation, 20–21, 127
  meteorology, 127
  metric system, 197
  noise abatement, 271
  nuclear nonproliferation treaty, 132, 149, 158, 230, 249
  oceanography, 60
  press conference, 1
  resignations accepted by, 40
  space program, 18–19, 20–21, 53, 179, 202, 208, 235, 248, 267, 292
  space rescue treaty, 89, 127, 161, 299
  supersonic transport, 144, 207
  Webb, James E., 235, 269
  Wright Brothers Day, 313–314
Johnson, Mrs. Lyndon B., 292
Johnson, Lyndon B., Presidential Library, 292
Joint Oceanographic Institutions for Deep Earth Sampling (JOIDES), 311
Joint Publications Research Service, 336–337
Jolly Green Giant (helicopter), 146
Jonash, Edmund R., 104
Jones, John W., 43
Jones, Dr. Walton W., 286–287
Jordon, L/C Joe B. (USAF), 2
Journal of the Royal Aeronautical Society, 91
JPL. See Jet Propulsion Laboratory.
Judi-Dart (sounding rocket), 31
Jupiter (planet), 120–121
  atmosphere, 16–17, 142
  exploration of, 19, 30, 44, 51, 79, 148, 153–154, 188–189, 275, 280–281, 334
  life on, 51, 142
  photographs of, 6
Jupiter C (booster), 22
Jurgens, Raymond F., 37–38

K

Kaminski, Heinz, 54, 222
Kansas City, Mo., 207
Kaplan, Dr. Joseph, 29–30
Kaplan, Dr. Lewis D., 283
Kapustin Yar, U.S.S.R., 130, 154, 310
Karth, Rep. Joseph E., 58–59, 131–132
KC–135 (flying ionospheric laboratory), 302
Keldysh, Mstislav V., 272
Kennedy, President John F., 147–148, 212–213, 317–318, 325, 326
Kennedy, John F., International Airport, 163, 187–188, 189, 194, 206
Kennedy, Sen. Robert F., 326
Kennedy Space Center (KSC)
  Apollo/Saturn, 77, 78, 136, 144, 186–187, 250, 252, 318, 321–322
  astronauts at, 172, 208
  award, 89
  contract, 45, 172, 239
  facilities, 210, 256
  high-altitude research, 143
  personnel, 150
  spacecraft delivery and shipments to, 22, 30, 101, 122, 181, 186
  spaceport, 1
  visits to
    Humphrey, Vice President Hubert H., 287
    Johnson, Mr. Lyndon B., 292
    Robb, Mrs. Lynda, 292
    U.S.S.R. scientists, 247
Kentucky, Univ. of, 291
Kerwin, Joseph P., 134, 144
Kiefer, Paul J., 145
Killian, Dr. James R., 303–304
King George Island, 46
King, Guy H., 44
King, Rev. Martin Luther, 324
Kirkman, Don, 215
Kiruna Range, Sweden, 29, 70, 118, 130, 133, 262
Kistiakowsky, Dr. George B., 38, 303–304, 328
Kitt Peak National Observatory, 6, 119, 128
Kittikachorn, Prime Minister Thanom (Thailand), 108
Kitty Hawk Memorial Trophy, 286, 303
Klein, Milton, 44, 65, 94
Kling Report, 231
Kling, Vincent, and Associates, 231
Kliore, Dr. Arvydas J., 199, 275–276
Kluttz, Jerry, 168
Knight, Maj. William J. (USAF), 11–12, 66, 95, 164, 211, 259, 299
Knoppers, Dr. Anthonie T., 16
Knowles, Stephen, 185
Kobzarez, Alexander, 126

Kocjan, Barbara E., 68
Kokusai Denshin Denwa Co., Ltd., 82–83
Komarov, Col. Vladimir M. (U.S.S.R.), 70, 264
Kometsky, Dr. George, 206
Kordes, Dr. Eldon E., 313
Korea, 109, 137
Korolev, Sergei P., 17
Kosygin, Premier Alexey N. (U.S.S.R.), 72, 149, 189–190
Kraft, Joseph, 325
Kranzberg, Dr. Melvin, 1
Kraushaar, Dr. William, 93
Krivsky, Dr. L., 269
KSC. See Kennedy Space Center.
Kubat, Jerald K., 9
Kubis, Dr. Joseph F., 287
Kuchel, Sen. Thomas H., 121
Kuers, Werner R., 266, 272–273
Kuiper, Dr. Gerard P., 12–13
Kurashiki Astronomical Observatory, 155
Kuznetsov N.K.–144 (U.S.S.R. turbofan engine), 332
Kwajalein Test Site, 72

L

L–1011 (jet airbus), 72, 72
Laboratory, 12–13, 17–18, 59, 91, 101, 151, 266
LaGow, Herman E., 283
La Guardia Airport, 179, 189, 194, 206
Laika (dog, U.S.S.R.), 46
Laird, Rep. Melvin R., 309, 331
Lambert Field, Mo., 207
Lancaster, Calif., 279
Land, Dr. Edwin H., 38
Langley Aeronautical Laboratory, 281
Langley Research Center (LaRC) (NASA), 144, 177, 300
  award, 116, 283
  collision-avoidance system, 179
  contract, 87, 123
  Digital Computer Complex Group, 164
  Explorer program, 182
  Flight Control Research Facility, 164
  Lunar Orbiter program, 24
  noise alleviation conference, 247
  paraglider, 187
  personnel, 15, 65, 101, 134–135, 173
  runway grooving research, 125, 286
  RAM program, 198–199
  VTOL aircraft, 276–277
Langmuir probe, 45, 46–47
Lankard, J. P., 169
Lannan, John, 205, 215, 230, 237, 311
Lapp, Dr. Ralph E., 308, 311
Larsen, Dr. Finn J., 238
Las Vegas, Nev., 108
Laser, 312
  experiments, 6, 10, 71, 183, 262, 282–283
  measurement of, 39
  production of, 169
  use of, 6, 183, 259

Latin America, 230
Launch Complex 13, 55
Launch Complex 34
  Saturn IB test, 136, 144, 173, 186
  Saturn/Apollo
    launch, 250
    mating, 186, 197, 204, 206
  side-wire escape way test, 210
Launch Complex 37, 256
Launch Complex 39, 318
Launch Escape System (LES) (Apollo), 78
Launch vehicle (see also individual launch vehicles such as Atlas-Centaur, Saturn, etc.), 36, 37, 185
  U.S., 235–236, 246
  U.S.S.R., 33–34, 36, 93, 186–187
Lawrence, David, 332
Lawrence Radiation Laboratory (Univ. of Calif.), 329
Learjet, 179
Leasco Systems and Research Corp., 169, 245
Leavitt, William, 28
Lebedev, Vladimir, 81
Lederer, Jerome F., 315
Lefkowitz, Louis J., 279
Leonov, L/C Aleksey A. (U.S.S.R.), 81, 85, 190
LES. See Launch Escape System.
Les V (Lincoln Laboratory Experimental Satellite), 79, 137, 228
Les VI, 228, 335
Lesher, Dr. Richard L., 35
Levitt, Dr. I. M., 27–28
Lewellyn, John A., 200
Lewis Research Center (LeRC) (NASA)
  Aerospace Safety Research and Data Institute, 95–96
  Atlas-Agena, 54–55
  Atlas-Centaur, 3, 8
  award, 281, 283, 300
  budget, 41
  contract, 59, 82, 144, 329
  noise abatement, 248
  organization, 104
  personnel, 26, 32, 101, 104, 118, 134–135
  Propulsion Systems Laboratory, 214–215
  research, 302
  SERT II, 26
  supersonic transport engine, 233
Lewis, Dr. W. Deming, 36
LEX (French sounding rocket), 9
Ley, Willy, 28
LFU. See Lunar flying unit.
Libby, Dr. Willard F., 17, 302
Library of Congress, 202, 203–204
Lick Observatory, 119
Liepmann, Dr. Hans W., 5, 15
Lifting body vehicle, 65
  HL–10, 63, 131, 139, 224–225, 240, 261, 279, 306, 335
  M2–F2, 67–68
  X–24, 42

Lilly, William E., 35, 48–49
Lima, Peru, 146
Lindbergh, Anne Morrow, 290
Lindbergh, Charles A., 104, 290, 306
Ling-Temco-Vought A–7 (fighter aircraft), 259
Ling-Temco-Vought, Inc., 276–277
Ling-Temco-Vought (LTV) Aerospace Corp., 123, 167, 187
  Range Systems Div., 3
Linville, Dr. John R., 5
Liquid hydrogen, 79–80
Little Dipper (constellation), 161, 310
LM. See Lunar module.
LLRV. See Lunar Research Vehicle.
LLTV. See Lunar Landing Training Vehicle.
LMS. See Lunar module simulator.
Local scientific survey module (LSSM), 51–52
Lockheed Aircraft Corp., 5, 43, 128, 293
  C–5A (cargo transport), 146–147, 285, 312–313
  contract, 62, 173, 187, 225, 278
  Deep Submergence Search Vehicle, 227
  F–104 (Starfighter), 165
  L–1011 airbus, 72
  Marietta, Ga., plant purchase, 231
Lockheed-California Co., 48
Lockheed Electronics Co., 123, 308
Lockheed-Georgia Co., 54, 146–147
Lockheed Missiles and Space Co., 148
  contract, 11, 26, 61, 120, 123, 151, 298
  missile, 61
  pollution study, 158
  sounding rocket experiment, 40–41, 256–257
  Space Systems Div., 130
Logsdon, Prof. John M., 325
Logunov, Dr. Anatoli A., 316
London Imperial College, 70
London, U.K., 89–90, 117, 185, 210, 278
Long-Tank Delta (booster), 216
Long-Tank Thrust-Augmented Thor (LTTAT)-Agena D (booster), 16, 61
Look, 99
LOR. See Lunar orbit rendezvous.
Los Alamos (N. Mex.) Scientific Laboratory, 91, 94, 123
Los Altos Morning Forum, 119
Los Angeles, Calif., 112, 186, 188, 221, 265, 312
  meeting in, 65, 66, 262–263
Los Angeles International Airport, 197–198
Los Angeles Junior Chamber of Commerce, 121
Losey, Robert M., Award, 316
Louisiana, 73–74
Lovelace, Dr. W. Randolph, II, 28
Lovell, Sir Bernard, 42, 54, 211–212, 222, 289, 327–328
Lovell, Capt. James A., Jr. (USN)
  Apollo 8 flight, 318, 322–323, 327
  preparations for, 59, 183, 196, 223, 261–262, 278
  award, 332
  Humphrey, Vice President Hubert H., visit with, 287
  press conference, 285, 304–305
  White House visit, 306
Low, Robert J., 169–170
Lowell Observatory, 219
Lowery, Mrs. Barbara, 153–154
LRV. See Lunar roving vehicle.
LSSM. See Local scientific survey module.
LTA. See Lunar Test Article.
LTV. See Ling-Temco-Vought, Inc.
LTV Service Technology Corp., 145
Lucas, Dr. W. R., 303
Luedecke, M/G Alvin R. (USAF, Ret.), 283
Lugo, Fernando, 41–42
Lukasik, Dr. Stephen J., 40
Lukens, Matthias E., 5
Luna IX (U.S.S.R. lunar probe), 61
Luna XIV, 81, 83, 87, 335–336
Lunar flying unit (LFU), 51–52
Lunar Landing Research Vehicle (LLRV), 105, 116, 257
Lunar Landing Training Vehicle, (LLTV) 305
Lunar module (LM), 19–20, 116, 183, 218–219
  contract, 86
  flight test, 1, 191
    No. 1, 13, 17, 37, 63
    No. 2, 78, 92
  ground test, 143, 173, 181
  Lunar Module 3, 158, 173, 181, 196
  manned, 127, 191, 279–280
Lunar module simulator (LMS), 227–228, 266
Lunar orbit rendezvous (LOR), 318
Lunar Orbiter (program), 41, 281, 301–302
Lunar Orbiter I, 24
Lunar Orbiter II, 24
Lunar Orbiter III, 24
Lunar Orbiter IV, 24, 266
Lunar Orbiter V, 12–13, 24, 193–194, 266
Lunar roving vehicle (LRV), 295
Lunar Science Institute, 53, 267
Lunar Test Article (LTA), 205–206
Lundin, Bruce T., 32, 118–119
Luskin, Harold T., 15, 48, 65, 293, 315
Lysenko, Trofim D., 298

M

M2–F2 (lifting-body vehicle), 67–68
McCaffery, Robert J., 283
McCarthy, Sen. Eugene J., 159
McCartin, Matthew J., 83
Macomber, Frank, 156
McConnell, Gen. John P. (USAF), 39, 122, 130, 152
McCormack, James, 111
McCormack, Rep. John W., 16
McCormick, Leander, 151

McDermott, Sgt. Robert (USA), 112
McDivitt, L/C James A. (USAF), 97, 167, 196, 306
MacDonald, Dr. James E., 169
McDonald Observatory, 294
McDonnell Douglas Corp., 22, 44, 47, 260–261
McDonnell, James S., 260–261
McIver, Dr. R. G., 283–284
McKee, Gen. William F. (USAF, Ret.), 150, 233
Mackin, Prof. J. Hoover, 188
McKinsey Foundation lecture, 103, 109, 115
McNair, Prof. Arthur J., 326
McNamara, Secretary of Defense Robert S., 12, 20, 29, 51, 53, 75, 93–94, 112, 229, 270
Madrid, Spain, 312
Magnetometer, 130–136
Magnetosphere, 88, 138–139, 153–154, 210–211
Magnuson, Sen. Warren G., 26, 209, 314
Mallick, Donald L., 8, 131, 167, 208, 298–299
Management Services, Inc., 122
Managing the Environment (House report), 135
Manhattan Project, 311
Manila, Philippines, 103
Manke, John A., 131, 240, 279
Mann, Dr. David E., 43
Manned Orbiting Laboratory (MOL), 32, 105, 126, 208, 255, 293
  appropriations, 19, 20, 122, 166, 171, 308–309
  cooperation, 20, 70, 102
  launch complex, 335
  test, 20
Manned space flight (see also Apollo program, Gemini program, and flights; Astronaut; Cosmonaut; Manned Orbiting Laboratory; *Soyuz 3* flight; Space biology; and Space station), 73–74, 80–81, 281, 304–305
  achievements, 24, 66, 165, 190, 215, 236–237, 253, 269, 324–325, 326, 327–328, 334–335
  criticism, 25, 51, 257
  EVA. See Extravehicular activity.
  funding, 19–20, 33–34, 35, 53, 98, 102–103, 140, 218, 311, 336
  hazards, 25, 200–201, 230, 278, 311
  long-duration, 35, 219
  lunar landing, manned. See Moon landing, manned.
  military potential, 205
  policy and plans
    U.S., 33–34, 35, 41, 47, 51–52, 55, 61–62, 63, 86, 92–93, 94–95, 97, 108–109, 112–113, 141, 173, 180, 188–189, 191, 194, 196, 215, 223, 243, 245, 265–266, 276, 282–283, 311, 312, 314, 328

    U.S.S.R., 34, 40, 116–117, 178, 222, 223, 254, 264–265, 267–268, 277–278
  safety, 4–5, 25, 51, 92–93, 246, 264, 265–266, 273
Manned Space Flight Network (MSFN), 5, 12, 97–98, 258, 273–274, 312, 335
Manned Spacecraft Center (MSC) (NASA), 39, 222–223, 224, 227, 246, 291
  Apollo spacecraft, 78, 253, 322
  astronauts at, 164, 168, 208, 220, 285, 304–305
  award, 75
  contract, 14, 22, 146
  Lunar Landing Test Vehicle (LLTV), 305
  Lunar Receiving Laboratory, 267
  management, 218–219
  patent, 145
  Space Physics Div., 130–131
  Space Environment Simulation Laboratory, 144
  spacecraft test. See Apollo (spacecraft).
  visits to
    Bourguiba, President Habib (Tunisia), 118
    Johnson, President Lyndon B., 53
    Nixon, President-elect Richard M., 207
    Webb, James E., 53
Manovtsev, Gherman A., 324
Man's Geophysical Environment: Its Study from Space (ESSA study), 73
Mansfield, Sen. Mike, 249
Manson, M/C Hugh B. (USAF), 11–12
March, James G., 115
Marder, Murrey, 161
Marietta, Ga., 231
Marine Resources Council, 284
Marine Science Affairs—A Year of Plans and Progress (President's report), 60
Mariner (program), 8, 33–34, 240–241, 281–282
Mariner F, 118
Mariner G, 118
Mariner II (Venus probe), 199
Mariner III (Mars probe), 118
Mariner V (Venus probe), 10, 162, 199, 241, 275–276
Mark 11 (reentry vehicle), 8
Mark 17, 8
Marks, Leonard H., 239
Mars (planet) (see also *Mariner III* and Voyager program), 162
  atmosphere, 328–329
  exploration of, 219, 280–281, 327, 334
  funding, 18, 162, 188–189, 241, 281–282
  manned, 259
  plans for, 118, 162, 182, 188–189, 248–249, 281–282, 301–302
  spacecraft, 18, 43, 44, 78–79, 118,

212, 248–249, 281–282, 287–288
  unmanned, 51, 66, 78–79, 212, 215, 248–249
 life on, 97, 188–189
 magnetic field, 329
 photographs, 215, 249, 301–302
 surface, 170, 249, 301–302
Marshall Islands, 72
Marshall Space Flight Center (MSFC) (NASA), 57–58, 66, 68, 91, 98, 143, 275
 anniversary, 149, 235
 Apollo Telescope Mount, 62, 85–86, 218–219, 240
 Astrionics Laboratory, 59
 award, 235, 283
 contract, 62, 79, 85–86, 110, 120, 122, 240, 257, 266, 295
  Saturn, 75, 77, 79, 127, 207, 239, 257, 262, 275, 328
 employment, 149, 168
 launch vehicle. See Saturn.
 Lunar roving vehicle (LRV), 295
 management, 219
 Manufacturing Engineering Laboratory, 272–273
 meeting, 88
 marine transportation, 110
 organization, 303
 Pegasus program, 203
 personnel, 20, 45, 47–48
 Safety Office, 121
 Saturn I Workshop, 21–22, 202, 218–219
 Space Sciences Laboratory, 110
 Test Laboratory, 144
Martin Marietta Corp., 42, 163, 199
Maryland, Univ. of, 67, 128, 173, 198
Maser, 312
Mason, Roy, 59–60
Massachusetts General Hospital, 312–313
Massachusetts Institute of Technology (MIT), 1, 5, 57, 139–140, 148, 224, 268, 303–304, 312
 Department of Aeronautics and Astronautics, 53, 316
 Div. of Sponsored Research, 14
 experiment, 170
 research, 49, 200, 238–239
 School of Engineering, 99
Masursky, Dr. Harold, 254–255
Materials technology, 39, 53, 144, 150–151, 157, 162–163, 177, 234, 251, 253, 265, 267–268
The Mathematical Sciences: A Report, 291–292
Mathematics, 291–292, 293
Mathews, Charles W., 65, 86, 108
Mathews, L/C Wayne (USAF, Ret.), 128
Matthews, N. Whitney, 152
Mattoni, Dr. Rudolph H. T., 325
Max Planck Institute, 130, 133, 160, 214
Maxwell, M/G Jewell C. (USAF), 66–67, 202–203, 207, 217

May, Tom R., 146–147, 155–156
Mazur, Daniel G., 225
Mechling, A. L., Inc., 110
Meeker, Leonard C., 281
"Meet the Press" (TV program), 276
Meghreblian, Dr. Robert V., 152–153
Meitner, Dr. Lise, 265
Melbourne, Australia, 306–307
Melbourne, Fla., 156
Mellon Institute Award, 110
Melnikov, N., 226
Memorandum of Understanding, 137
Merck & Company, Inc., 16
Mercury (planet), 30, 184, 188–189, 280–281, 334
Mercury (program), 66, 102–103, 161, 220, 221–222, 236–237, 281, 325
Mercury (spacecraft), 148
Mesa Verde National Park, 155
Messerschmitt-Bölkow, 151
Messmer, Defense Minister Pierre (France), 200
Meteor, 5, 185–186, 287
Meteorite, 5, 120
Meteoroid, 177
Meteorological satellite (see also individual satellites, such as *Ats III, Ats IV, Essa VI, Essa VII, Essa VIII*), 185, 229, 232, 237, 306–307
 cooperation,
  U.S.–U.S.S.R., 2
 ground station, 97
 Nimbus program, 1, 33–34, 49, 60–61, 116, 145, 241–242, 243
 photographs by, 97, 229, 311
 Tiros program, 49, 191, 311
 U.S.S.R., 2
Meteorology, 101–102, 235, 236–237, 251
 award, 316
 cooperation, 306–307
  NASA–ESSA, 111, 262, 335
  U.S.–U.S.S.R., 65–66
 forecasting, 85, 102, 257–258, 295, 334–335
 hailstone research, 247
 satellite. See Meteorological satellite.
 sounding rocket experiments, 10, 168
 U.S.S.R., 2, 61
 World Weather Watch, 19, 127, 155, 306–307
Metric system, 197
Mexico, 39, 59, 295
Mexico City, Mexico, 295
Mi-6 (U.S.S.R. helicopter), 126
Mi-10, 126
Miami Beach, Fla., 178–179
Miami, Univ. of, 91–92
Michel, F. Curtis, 92
Michener, Gov. Gen. Roland, 327
Michigan, Univ. of, 39, 64–65, 85, 115, 168, 182–183, 183–184, 256, 287
 Space Physics Research Laboratory, 328–329
Michoud Assembly Facility (MSFC), 3,

91, 171, 173, 227, 296
Micrometeoroid, 61, 118, 129–130, 130, 177, 177–178, 185–186, 197, 310–311,
Middle East, 205
MiG–15 (U.S.S.R. fighter aircraft), 70
MiG–23 (Foxbat) (U.S.S.R. fighter aircraft), 171–172, 195
"The Military Balance" (report), 210
Military technology, 29, 43, 102, 334
Milky Way (galaxy), 114–115, 208–209, 224, 247
Miller, Frederic H., 150
Miller, Rep. George P., 15–16, 32, 235
Miller, Dr. Rene H., 224, 316
Minashin, Vladimir, 189–190
Mineola, N. Y., 11
Miniaturization, 186, 237, 267–268
Minnesota, Univ. of, 88, 177
Minott, Peter O., 183
Minuteman (ICBM), 3, 148, 237, 312–313
Minuteman I, 193
Minuteman II, 18–19, 193
Minuteman III, 159, 193, 195
MIRV. See Multiple Independently Targetable Reentry Vehicle.
Missile, 126, 130, 272, 276, 284
  air-to-air, 98, 160
  antimissile, 72, 88, 133, 138, 143, 159, 180, 242
  ballistic intercontinental (ICBM), 3, 18, 29, 42, 138, 191, 195, 210, 289, 312
  contract, 61
  detection, 208
  foreign
    Communist China, 29
    France, 153
    U.S.S.R., 29, 42, 55, 159, 210, 272, 289
  medium-range, 29
  nuclear, 29, 228
  orbital, 55, 93, 94, 178, 255, 329
  underwater-to-surface, 19, 61, 153, 159, 193, 195
  limitation of, 149–150, 159, 166, 188, 191, 195, 209
Mississippi, 73–74
Mississippi Test Facility (MTF), 73–74
  contract, 173
  test, 30, 36, 73–74, 122, 186
MIT. See Massachusetts Institute of Technology.
Mitchell, L/Cdr. Edgar D. (USN), 279–280
Mitchell, Ind., 39
Moffet, Dr. Alan, 114–115
Mojave Desert Ground Station, Calif., 78–79, 146, 155, 291
MOL. See Manned Orbiting Laboratory.
Molniya (U.S.S.R. communications satellite), 10
Molniya I–8, 88, 110, 335–336
Molniya I–9, 155, 335–336
Molniya I–10, 244, 335–336
Mondale, Sen. Walter F., 25

Mondane, France, 134
Mongolia, 189–190
Monkey experiment, 47, 95, 151, 181, 273, 283–284
Montreal, Canada, 157, 163, 295
Moon, 120–121, 136, 142, 153–154, 158, 167, 177, 190, 221, 294, 330–331
  crater, 50, 78, 250, 254–255, 298, 319
  eclipse, 250
  exploration of, 22–24, 34, 41, 51–52, 54, 66, 68, 72–73, 79, 211–212, 222, 232, 242, 245, 248–249, 259, 267, 277, 278, 281, 285, 287–288, 289, 308, 311, 327, 329–330, 336
  equipment, 51–52, 105, 106
  laboratory, 79
  landing, 33–34, 35–36, 213, 215, 218, 220, 222, 233
    manned, 19, 33, 46, 63, 79, 80–81, 81–82, 87, 97, 105, 112, 141, 178, 186, 191, 194, 215, 222, 237, 242, 245, 246, 248–249, 250, 254, 255–256, 259, 265, 267, 277, 282, 287–288, 289, 295, 312, 314, 325, 328, 332
    simulated, 105, 292
    soft, 3, 6, 35–36, 335
    unmanned, 112–113, 289
  landing simulator, 257
  landing site, 35–36, 51–52, 66, 265, 266, 319
  Lunar Science Institute, 53, 267
  nomenclature, 28
  photographs, 6, 23, 35–36, 54, 61, 78, 215, 266, 298, 326
  probe, 3, 6–7, 42, 81, 83, 278, 282–283
  surface, 3, 6–7, 37, 50, 52, 61, 68, 78, 181–182, 188, 193–194, 250, 254–255, 316–317, 319, 330, 334–335, 336
Moore, Wendell F., 107
Moorer, Adm. Thomas H. (USN), 14, 314
Moree, Australia, 120
Morris, Mildred V., 283
Morton, Dr. Louis, 1
Moscow (U.S.S.R.), 10, 211–212, 222, 316
  accident, 70
  airline service, 163, 194
  airport, 332
  anniversary ceremony, 45–46
  nuclear nonproliferation treaty signing, 149
  press conference, 119, 272
  weather data exchange, 306–307
Moscow Univ., 268
Mt. Hamilton, Calif., 119
Mount Hopkins, Ariz., 262
Mount Locke, Tex., 294
Mt. Palomar Observatory, 82, 300
Mt. Wilson Observatory, 82, 300
Mrazek, Dr. William A., 122–123
MSC. See Manned Spacecraft Center.
MSFC. See Marshall Space Flight Center.

MSFN. See Manned Space Flight Network
MTF. See Mississippi Test Facility.
Mueller, Dr. George E., 4–5
   Apollo, 157, 191, 265–266, 289
   Apollo Applications, 41
   award, 199
   launch vehicle, 185
   space program, 37, 51, 103, 157, 210
   systems engineering, 157, 246
   technology utilization, 246, 293–294
   U.S.S.R. space program, 186–187
Mulholland, Dr. J. Derral, 120–121
Muller, Paul M., 193–194
Multiple Independently Targetable Reentry Vehicle (MIRV), 188, 191, 193, 195, 205
Mundt, Sen. Karl E., 71
Murder on Pad 34, 161
Murphy, Charles, 40
Murphy, James T., 121
Murray, Dr. Bruce C., 43, 248–249
Murray, Grover, 115
Mururoa Atoll, 153, 156, 207–208
Myers, Boyd C., II, 163, 283
Myskowski, Edwin T., 152

N

NAA. See National Aeronautics Assn.
Nabrit, Dr. Samuel M., 143
NACA. See National Advisory Committee for Aeronautics.
NAE. See National Academy of Engineering.
NAR. See North American Rockwell Corp.
NAS. See National Academy of Sciences.
NASA–AEC Space Nuclear Propulsion Office, 65, 75, 94
NASA Apollo Applications Program Office, 79
NASA Apollo Site Selection Board, 35–36
NASA Board of Contract Appeals, 83
NASA Contract Adjustment Board, 83
NASA Committee on Extra-Vehicular Activities, 316–317
NASA Communications Network (NASCOM), 5, 61, 246
NASA Historical Advisory Committee, 1
NASA Lunar and Planetary Missions Advisory Board, 29–30
NASA Management Advisory Panel, 91–92
NASA Manned Space Flight Safety Office, 51, 63, 315
NASA Office of Advanced Research and Technology (OART), 34, 44, 57, 67–68, 97, 121, 163, 203, 262
NASA Office of Aerospace Safety Research Programs, 121
NASA Office of Facilities, 120
NASA Office of Manned Space Flight (OMSF), 2, 32, 35, 78, 84, 112, 122, 253, 293, 322
NASA Office of Space Science and Applications (OSSA) 2, 24, 34, 50, 55, 56–57, 152–153, 182, 185, 304, 334
NASA Office of Technology Utilization, 82, 310
NASA Office of Tracking and Data Acquisition (OTDA), 17, 78, 253, 262, 321–322
NASA Pasadena (Calif.) Office, 243
NASA Post Apollo Advisory Group, 17
NASA Scientific and Technical Information Facility, 154, 169
NASC. See National Aeronautics and Space Council.
NASCOM. See NASA Communications Network.
Natal, Brazil, 68, 130–131
Nathan, Dr. Robert, 154
National Academy of Engineering (NAE), 31, 75, 175, 177
   Aeronautics and Space Engineering Board, 187
   Founders Medal, 43
National Academy of Sciences (NAS), 255, 312
   annual meeting, 93
   applications satellite study, 27, 36, 73, 76–77, 117, 124–125, 219, 241–242
   Committee on SST-Sonic Boom, 56
      Subcommittee on Human Response, 134
      Subcommittee on Research, 8
   Computer Science and Engineering Board, 134
   contract, 317
   cooperation, 31, 177, 216–217
   Lunar Science Institute, 53, 267
   report, 7, 36, 152, 291–292, 293
   space program, 36, 90
   Space Science Board, 188, 219, 275, 276
   symposia, 46, 216, 217
   UFO study review, 99
National Advisory Committee for Aeronautics (NACA), 102, 134–135, 175, 281
National Aeronautics and Space Act, 237, 248, 306, 334
National Aeronautics and Space Administration (NASA) (see also NASA centers, programs, satellites, and related headings, such as Ames Research Center, Apollo program, *Essa VII*), 12, 38, 103, 109, 115, 127, 128, 133, 134, 143, 144, 145, 146, 198, 218, 225, 247, 275, 280, 287, 291, 294, 295, 301, 325
   accomplishments, 10–11, 22–24, 36, 48, 66, 162–163, 215, 232, 235–238, 239, 248, 249, 253
   Aerospace Safety Advisory Panel, 32, 87, 122
   agreement. See Agreement.
   Airborne Auroral Expedition, 10–11, 17, 72
   anniversary, 63, 80, 149, 235, 237, 239, 253, 334
   Apollo 204 Review Board. See Apollo 204 Review Board.

astronaut. See Astronaut.
awards and honors, 2, 39, 70–71, 89, 91, 194, 199, 235, 260, 276, 281, 286, 300, 303, 306, 316
  Distinguished Service Medal, 263, 269, 283, 332
  Exceptional Scientific Achievement Medal, 263, 283
  Exceptional Service Medal, 194, 269, 283
  Group Achievement Award, 164, 283
budget, FY 1969, 7, 18–20, 24, 56, 59, 75, 99, 119, 126, 129, 135–136, 139–140, 168, 170, 175–176, 182, 188, 190, 213, 233, 250, 262, 272, 276, 287, 288, 334, 336
  bills signed, 152, 243
  House consideration
    appropriations, 32, 108, 140, 218, 243
    authorization, 15–16, 32, 33–35, 41, 43, 44, 47, 48, 50, 53, 99, 101, 102, 127, 135–136, 152, 170
  interim operating plan, 182, 240–242
  press comment, 85, 90, 129, 133, 139, 141, 182, 194, 215, 229, 255–256, 324
  Senate consideration
    appropriations, 140, 165–166, 227, 240–242, 243
    authorization, 30, 48–50, 57, 90–91, 93, 94, 98, 110, 129, 170
  Vietnam war, effect of, 15, 56, 99, 131, 146, 215, 245, 336
budget, FY 1970, 255, 274, 294
conference, 109, 247, 286
contract, 86, 187, 317, 331
  aircraft, 149
  balloon, 68
  computer services, 3, 75, 79, 82, 123, 144, 145, 239
  data management, 262
  engine, 9, 26, 35, 113, 149, 207, 227, 248, 257, 262, 264, 275
  facilities, 59, 78–80, 97–98, 122, 146, 169, 214, 226, 245, 326, 331
  fuel, 79, 266
  fuel cell, 294
  guidance and navigation, 8, 14, 77, 82, 328
  instrumentation, 80
  launch services, 8, 39, 59, 75
  launch vehicle, 8, 35, 39, 84, 99, 104, 123, 127, 144, 149, 226, 275
  nuclear propulsion, 59, 75
  sounding rocket, 143
  space equipment, 22, 85–86, 87, 130, 202, 295
  space station, 62, 86–87, 202, 242–243
  spacecraft, 60–61, 86–87, 120, 146, 151, 163, 172, 209, 281–282
  study, 61, 87, 275

support services, 9, 12, 59, 61, 75, 79–80, 104, 110, 120, 123, 172, 173, 239, 289, 308
telescope, 184, 202, 240
tracking, 12, 97–98, 143, 146
cooperation, 235–236
  AEC, 65, 82, 156, 170, 335
  Agriculture, Dept. of, 299
  Commerce, Dept. of, 299
  ComSatCorp, 335
  DOD, 20, 33, 69–70, 84, 156, 238, 310
  ESSA, 110, 262, 335
  FAA, 168
  Geological Survey, 118
  Interior, Dept. of, 101, 299
  Naval Research Laboratory, 56–57
  USAF, 38, 42, 63, 83, 156, 168, 261, 335
  USN, 101, 299
cooperation, international. See International cooperation, space; and Sounding rocket, international programs.
criticism, 86, 133
  Apollo AS–204 accident, 25, 223
  Apollo 8 flight, 289, 328
employment, 24, 35, 73–74, 85, 115–116, 141, 149, 168, 182, 213, 237, 250, 256, 272, 276, 328, 336
facilities, 33–35, 164, 242, 256, 296
Historical Advisory Committee, 1
launch, 1
  Apollo 5 (AS–204), 13, 334
  Apollo 6 (AS–502), 77, 334
  Apollo 7 (AS–205), 250–253, 334–335
  Apollo 8 (AS–503), 318–322, 335
  balloon (Explorer XXXIX), 182
  cost, 37
  failure, 116, 184–185, 216, 225, 243, 261
  postponed, 13
  probe, 1
    Pioneer IX (Pioneer D), 273–274
    South Atlantic Anomaly Probe, 130–131
    Surveyor VII (Surveyor G), 1, 3, 335
  reentry test
    RAM C–II, 198–199, 258, 335
    reentry F experiment, 97, 193
  satellite, 335–336
    Ats IV (ATS–D), 184–185
    Aurorae (Esro I), 240
    Esro II–B, 113
    Essa VII (TOS–E), VIII (TOS–F), 191–193, 311
    Explorer XXXVI, XXXVII, XXXVIII, XXXIX, XL, 7–8, 56–57, 153–154, 182
    Heos-A, 300–301
    Intelsat-III F–2, 314–315
    Oao II (OAO–A2), 304
    Ogo V (OGO–E), 54–55
  sounding rocket, 1, 62–63

Aerobee 150, 17, 30, 36, 40, 65, 67, 104, 115, 128, 175, 206, 211, 214, 221, 224
Aerobee 150A, 141, 289–290, 291
Aerobee 150 MI, 49–50, 90, 97, 118, 121, 169, 170, 219, 232, 238, 247–248, 256–257, 282, 290, 303, 316
Arcas, 86, 214
Arcas I, 60, 175
Astrobee 1500, 88
Black Brandt IV, 106, 130–131
Boosted Arcas II, 219, 262
Javelin, 54, 82, 113, 308
Nike-Apache, 45, 46–47, 64–65, 70, 76, 90, 126–127, 129–130, 133, 168, 181, 182–183, 185–186, 197, 198, 273, 287, 300, 310–311
Nike-Cajun, 9, 29, 32–33, 68, 168, 213–214, 216, 255, 256, 287, 288, 291, 309–310
Nike-Tomahawk, 13, 14, 31–32, 43–44, 45, 49, 53–54, 60, 62–63, 64–65, 67, 90, 94, 98, 173, 183, 198, 219–220, 221, 223, 256
Pacemaker, 144
management, 24, 115, 157, 218–219, 222–223, 284
Management Advisory Panel, 91–92, 206
manpower. See Employment.
organization, 33, 49, 84, 95–96, 104, 116, 120, 121, 303, 316–317
patents, 145, 187
personnel, 2, 8–9, 15, 26, 29, 32, 45, 51, 53, 65, 68, 75, 83, 87, 89, 91–92, 101, 104, 109, 121, 128, 152–153, 156, 160, 163, 207, 212–213, 215, 218, 221–222, 224, 229, 230, 233, 239, 245, 246–247, 268, 272–273, 298, 315, 328
 appointment, 5–6, 17, 26, 29, 32, 48, 51, 87–88, 92, 117, 118, 122–123, 150, 160, 206, 209, 233, 264, 318
 death, 152, 173, 293, 298, 315.
 resignation, 13, 17, 117, 118, 122–123, 128, 150, 172, 230, 244, 336
 retirement, 112, 243, 266, 272–273, 298
procurement, 35, 41
programs, 1, 14, 22–24, 162, 182, 215, 241–242
 aeronautics, 19, 25, 33, 34, 47, 48, 57, 104, 109–110, 118, 125, 162, 166, 179, 187, 197, 225, 237, 238, 247, 248, 256, 270, 288, 313, 317
 astronomy, 18, 19, 24, 30, 33, 34, 41, 43, 44, 51, 56, 148, 162, 188, 194, 201, 243, 280, 281, 301, 312
 earth resources, 7, 19, 50, 73, 85, 125, 257, 334
 international, 2, 20–21, 177, 202, 210, 254, 336
 manned space flight, 4, 19, 25, 34, 35, 51, 53, 55, 63, 73, 86, 92, 97, 103, 108–109, 112–113, 140, 141, 180, 188–189, 191, 194, 196, 223, 243, 245, 265, 276, 278, 282, 311, 314, 327–328
 meteorology, 11, 34, 49, 102, 137, 145
 nuclear propulsion, 18, 19, 33, 34, 44, 53, 57, 65, 93, 94, 98, 170, 288–289
 sounding rocket, 1, 10, 336
 space medicine, 151, 165, 181, 211, 218, 273, 313
 space science, 19, 47, 52, 165, 334
 technology utilization, 20, 36, 41, 48
 tracking and data acquisition, 34–35, 50, 65, 137
Research and Technology Advisory council, 5
Semiannual Report, 249
supersonic transport, 26, 47, 131, 233, 237
test, 37, 135, 210, 247
 aircraft, 166, 168, 276
 collision-avoidance system, 179
 ion engine, 303
 launch vehicle, 21–22, 36, 91, 101, 166, 206
 lifting-body vehicle, 42, 63, 65, 67–68, 131, 139, 224–225, 237, 240, 261, 279, 306, 335
 materials, 144
 nuclear, 131, 143, 156
 parachute, 8, 158, 261, 309
 spacecraft, 13, 17, 42–43, 61–62, 63–64, 78–79, 136, 144, 158, 172, 173, 181, 193, 197, 201, 206, 208, 227
 tracking, 144
 VTOL aircraft, 276
translations, 233
universities, 35, 39, 48, 90–91, 109–110, 115, 139–140, 151, 161, 249–250, 275, 280–281
 grants, 129, 151
X–15. See X–15.
National Aeronautics and Space Council (NASC), 7, 31, 79, 105–106, 112, 141, 202, 274–275, 292, 308
National Aeronautics Assn. (NAA), 40, 84, 209, 314
National Air and Space Museum, 106, 113
National Airport Plan, 278
National Airport System Plan, 137–138, 278
National Airspace System, 82
National Armed Forces Museum (proposed), 160
National Atmospheric Sciences Program— Fiscal Year 1969 (Office of Science and Technology report), 101–102
National Bureau of Standards, 5, 197
National Center for Atmospheric Research and Information, 52, 206, 240, 257–258
National Conference of United Press In-

ternational Editors and Publishers, 246
National Conference on Industrial Research, 15
National Council for Social Studies, 294
National Council on Marine Resources and Development, 138, 300
National Environmental Satellite Center, 233–234
National Geodetic Satellite Program (NGSP), 7–8
National Geographic Society, 5, 122
National Institute of Social Sciences, 290
National Institutes of Health (NIH), 38, 325
National Medal of Science, 38
National Park Service, 154–155
National Pilots Assn., 203
National Press Club, 22–23
National Radio Astronomy Observatory, 161, 200, 208–209, 238, 272
National Research Council (NRC), 8, 31, 117, 175, 190, 275
    Committee on Polar Research, 73
    Space Science Board, 188, 190, 194, 201, 275
National Research Council of Canada, 72
Naval Research Laboratory (NRL), 184
National Science Board, 115, 255
National Science Foundation (NSF), 36, 89, 190, 194–195, 255, 312, 325
    Employment of Scientists and Engineers in the United States, 1950–1966, 234
    Federal Funds for Research, Employment, and Other Scientific Activities: Fiscal Years 1967, 1968, and 1969, 333
    funds for, 101–102, 127, 227
    grants, 44–45
    National Register survey, 329
    Research and Development in Industry, 1966, 290
    Reviews of Data on Science Resources, 99–100
National Sea Grant College Program Act, 44–45
National security, 55–56, 57–58, 99, 162, 147–148, 194–195, 207, 210, 235–236
National Space Club, 55–56, 86, 112–113, 235, 293–294
    Astronautics Engineer Award, 194
    Goddard Memorial Trophy, 57
National Sporting Aviation Council, 111
National Zoo, 273
NATO. See North Atlantic Treaty Organization.
Natural Environment Panel, 154–155
Naugle, Dr. John E., 34, 41, 44, 49, 50, 89, 225, 243, 280
Naval Air Facility, Calif., 144, 158
Naval Air Missile Test Center, 42
Naval Missile Center, 59
Naval Ordnance Test Station, China Lake, Calif., 214
Naval Research Laboratory (NRL), 56, 87, 99, 128, 247–248, 335
    Rocketsonde Branch, 131
Naval Ship Systems Command, 310
Navigation satellite, 53, 187, 215, 217, 255
Navy Yard, Washington, D.C., 41–42
Needles, Calif., 209–210
Nellis AFB, Nev., 224, 258, 316, 336
Nelson, Sen. Gaylord, **68**
Nepal, 299
Neptune (planet), 79, 148
Nerry, D. P., 66
NERVA. See Nuclear Engine for Rocket Vehicle Application.
NET. See Nuclear Emergency Team.
Netherlands, 5, 55, 113, 155, 165, 233, 292, 298
Neubert, E.W., 121
Neumann, Gerhard, 296
Nevada, 25–26, 196
Nevada Test Site, 317
New Hampshire, Univ. of, 31–32, 90, 98, 103
New Jersey, 137, 224
New Mexico, 249
New Mexico State Univ., 201
New Orleans, La., 31, 55, 272
New Tanay, Philippines, 103
New York Academy of Sciences, 139–140, 325
New York, N.Y., 88, 107, 117, 163, 179, 187–188, 200, 203, 217, 224, 300, 302
    meetings, 14, 66–67, 128, 247, 258, 290, 315
New York Univ., 38, 71
Newark Airport, 206
Newark, N.J., 194
Newby, Dave, 303
Newell, Dr. Homer E., 14, 28, 85, 294
Newfoundland, 40
Newport News (Va.) Ship Building Co., 167
News conference. See Press conference.
Ney, E. P., 177
NGSP. See National Geodetic Satellite Program.
Nichols, Mark R., 283
Nicks, Oran W., 301–302
NIH. See National Institutes of Health.
Niini, Prof. Arrno, 329
Nike-Apache (sounding rocket), 30, 43–44
    electron measurement, 45, 46–47, 181, 198, 287
    instrumentation test, 70
    ionosphere experiments, 64–65, 130, 198, 273
    magnetic field measurement, 181
    micrometeoroid sampling, 129–130, 130, 133, 185–186, 197, 198, 310–311
    parachute test, 126–127
    upper atmosphere data, 45, 46–47, 63–64, 76, 89, 128–129, 168, 182–183,

183–184, 287, 300
Nike-Cajun (sounding rocket)
  electron measurement, 9
  parachute test, 32–33, 213–214
  upper atmosphere data, 29, 32–33, 68, 168, 214, 216, 255, 256, 287, 288, 291, 309–310
Nike-Tomahawk (sounding rocket), 43–44
  electron measurement, 31–32, 60, 62–63, 67, 90, 94, 98, 103, 173, 198, 256
  instrumentation test, 31–32
  ionosphere experiments, 14, 31–32, 62–63, 64–65, 90, 94, 98, 102
  upper atmosphere data, 13, 45, 49, 53–54, 64–65, 183, 219–220, 221, 223
Nike–X (antiballistic missile system), 43
Nikolayeva-Tereshkova, Maj. Valentina (U.S.S.R.), 148
Nimbus I (meteorological satellite), 145
Nimbus II, 145
Nimbus B, 1, 49, 116, 145, 225, 241–242
Nimbus, B2, 145, 241–242
Nimbus D, 33, 49, 60–61
Nimbus E, 49
Nimbus F, 49
Nitric acid, 168
Nitrogen, 45, 61
Nitze, Paul H., 159, 259
Nixon, President-elect Richard M., 309, 332–333
  election, 271–272
  nuclear nonproliferation treaty, 210, 216
  oceanography, 300
  research and development, 244, 314, 324
  science and technology, 244, 299, 305, 310, 314, 323, 331
  space program, 207, 235, 249, 311, 312, 331
Nobel Prize, 267, 271
Noise, aircraft (see also Sonic boom), 36–37, 38, 104, 142, 197–198, 271, 290–291
  International Conference for Noise Abatement, 117
  NASA program, 34, 47, 57, 104, 109–110, 197, 225–226, 236–237, 238, 247, 248
  regulation, 116, 135
Noise and the Sonic Boom in Relation to Man (report), 290–291
Noise—Sound Without Value (report), 271
North American Rockwell Corp. (NAR), 151–152, 163
  accident, 157
  aircraft, 149, 167, 187, 317, 333
  Apollo spacecraft, 4–5, 167
  contract, 59, 149, 167, 187, 188, 333
  rocket engine test, 101
  Rocketdyne Div., 10, 126, 207, 227–228, 257, 275

  Space Div., 11
North Atlantic Treaty Organization (NATO), 99, 137, 139, 243
  Nuclear Planning Group, 88
North Korea, 247
North Pole, 72, 168, 318–319
Northrop Corp., 11, 281
  Norair Div., 63, 67–68, 87, 135, 184
Northrop Systems Laboratories, 151
Northwestern Univ., 172, 254
Norton Research Corp., 172
Norway, 89, 177–178, 240
Notre Dame Univ., 112
Noyes, Robert W., 243–244
NRC. See National Research Council.
NRL. See Naval Research Laboratory.
NSF. See National Science Foundation.
Nuclear accelerator, 139
Nuclear clock, 170–171
Nuclear Engine for Rocket Vehicle Application (NERVA), 44, 94, 143, 289
  funds for, 18, 57, 98, 110, 135–136, 140, 167, 170, 182, 241–242, 289, 336
  NERVA I, 33
  NERVA II, 33
  NERVA XE–1, 156
Nuclear fallout, 110
Nuclear fission, 220, 265, 298
Nuclear Materials Equipment Corp. (NUMEC), 2
Nuclear nonproliferation treaty, 228
  Geneva Disarmament Conference consideration, 11, 32, 59, 62
  ratification, 295
  Senate hearings, 158, 159, 165, 210, 215–216, 236, 249
  signing, 149
  U.N. consideration, 11, 59, 62, 130, 132, 133, 138, 158
  U.S. draft proposal and consideration, 11, 32, 92
  U.S.S.R. draft proposal and consideration, 11, 32, 92, 145
Nuclear power, 92, 171, 205
Nuclear propulsion, 72–73, 75, 93, 94, 97, 98, 132, 274
Nuclear reactor, 131, 143, 156, 221, 298, 335
Nuclear Rocket Development Station, 25–26
Nuclear submarine, 42, 153, 160, 169, 211, 224, 261, 302–303
Nuclear test, 108, 110, 180, 200, 208, 228, 317
Nuclear test detection satellite, 19, 25–26, 91
Nuclear Utility Services, Biological Systems Div., 325–326
Nuclear weapons (see also Disarmament and Nuclear nonproliferation treaty), 285, 330
  Communist China, 200
  France, 153, 156, 180, 200, 266, 294
  Japan, 166

NATO consideration, 88
U.S., 160, 191, 211, 224
U.S.S.R., 29, 42, 45–46, 210, 255
Nuclear Week, 117
NUMEC. See Nuclear Materials Equipment Corp.

O

Oak Ridge National Laboratory, 28, 331
Oakland, Calif., 221
OAO. See Orbiting Astronomical Observatory.
Oao I (Orbiting Astronomical Observatory), 304
Oao II (OAO–A2), 304, 310, 335
OART. See NASA Office of Advanced Research and Technology.
Oberbeck, Verne R., 78
Objectives and Goals in Space Science and Applications, 1968 (NASA Report), 334
O'Bryant, Capt. William T. (USN, Ret.), 2
Observatory satellite. See Orbiting Astronomical Observatory; Orbiting Geophysical Observatory; Orbiting Solar Observatory.
Ocean of Gagarin (proposed name for Ocean of Storms), 190
Ocean of Storms (moon), 35–36, 190
Ocean Science and Engineering, Inc., 152
Oceanography
 grants, 44–45
 international aspects, 56, 117–118, 138, 177, 317
 research, 101, 152, 227, 262, 295, 310, 311
 satellite use in, 73, 85, 117, 262
 U.S. program, 27, 105, 167, 284, 300, 325
O'Connor, B/G Edmund F. (USAF), 20, 45, 303
Oettinger, Dr. Anthony G., 134
Office National d'Etudes et de Recherches Aérospatiales (ONERA), 9, 134
Office of Naval Research (ONR), 312
Office of Science and Technology (President's), 101–102, 325
OGO. See Orbiting Geophysical Observatory.
Ogo I (Orbiting Geophysical Observatory), 55
Ogo II, 54
Ogo III, 55
Ogo IV, 55, 72
Ogo V (OGO–E), 1, 54–55, 132–133, 335
OGO–F, 1, 19
O'Hagan, Michael, 278
O'Hair, Mrs. Madalyn Murray, 328
O'Hare International Airport, 206
Ohio, 227
Ohio State Univ., 257–258
Okinawa, 137
Oklahoma, 249

Oklahoma, Univ. of, 325
O'Leary, Dr. Brian T., 92, 200
Omega Position Location Equipment (OPLE), 144
OMSF. See NASA Office of Manned Space Flight.
ONERA. See Office National d'Etudes et de Recherches Aérospatiales.
ONR. See Office of Naval Research.
Onsager, Prof. Lars, 266
Operation Tektite, 101
OPLE. See Omega Position Location Equipment.
Orbiting Astronomical Observatory (OAO), 1, 8, 99, 243
Orbiting Geophysical Observatory (OGO), 19, 44
Orbiting Primate Experiment, 151
Orbiting Solar Observatory (OSO), 1, 243
Orbiting Vehicle (research satellite), 335
Ordahl, Douglas D., 3–4
Ornithopter, 282
Ortona, Ambassador Egidio (Italy), 276
OSO. See Orbiting Solar Observatory.
Oso II (Orbiting Solar Observatory), 177
Oso III (Orbiting Solar Observatory), 93
Oso IV, 243–244
OSO–G, 244
OSSA. See NASA Office of Space Science and Applications.
Oster, Dr. Irwin, 325–326
OTDA. See NASA Office of Tracking and Data Acquisition.
Otolith experiment, 86
O'Toole, Thomas, 225, 303, 312–313
Outstanding Young Men of America, 129
OV I–15 (orbiting vehicle research satellite), 159
OV I–16, 159
OV II–5, 228
OV V–2. See *Ers XXVIII*.
OV V–4. See *Ers XXI*.
Ovshinsky, Stanford R., 276
Owens-Illinois Co., 184
 Fecker Div., 184
Oxygen, 61–62, 318

P

P–1127 (U.K. VTOL aircraft), 276–277
Pacemaker (booster), 144
Pacemaker, cardiac, 26
Pacific I (communications satellite). See *Intelsat-II F–2*.
Pacific II (communications satellite). See *Intelsat-II F–4*.
Pacific Missile Range (PMR), 42
 Aero-Mechanical Branch, 67
Pacific Ocean, 254, 258, 268
 Apollo 6, 77–78
 Apollo 8, 320, 323
 Ats III, 37
 communications satellite, 69, 103
 French nuclear test, 153, 156, 180
 missile launch, 3, 121–122
 Sealab III experiment, 295

U.S.S.R. rocket test, 116, 121–122
Pacific Science Center, 217–218
Packard, David, 331
Packard, Robert F., 2–3
Page Communications Engineers, Inc., 88
Page, William A., 283
Paine Field, Wash., 201
Paine, Dr. Thomas O., 257, 318, 366
   aeronautics, 225
   Apollo 8 flight, 266, 278
   awards by, 283
   budget, 240–242, 255, 286, 288, 294
   nomination as Deputy Administrator, 26, 32, 68, 212
   space program, 223, 288
Pakistan, 138, 198
Pakistan Space and Upper Atmosphere Research Committee, 18
Palestine, Tex., 116, 257
Pan American World Airways, 107–108, 163, 232
Parachute
   Apollo, 8, 144, 158
   funds for, 126
   steerable, 57, 112
   test, 8, 32–33, 57, 126, 144, 158, 213, 227, 261, 309
   Voyager, 227
Parafoil (steerable parachute), 112
Paraglider, 187
Parawing (steerable parachute), 57
Pardoe, K. C. C., 278
Parin, Vasili, 105
Parker, Dr. John A., 265, 283
Parker, P. J., 234
Parker, Robert A. R., 113
Parkinson, William H., 243
Parsons, John F., 263, 298
Particles, charged, 31, 53, 65, 68, 182, 216–217
Patent, 43, 107, 131, 143, 145, 151–152, 156, 172, 187
Paternotte de la Vailee, A., 60
Paul VI, Pope, 327
Paumalu, Hawaii, 302
PCA. See Polar Cap Absorption.
Pecora, W. T., 191
Peddie, Norman W., 89
Pegasus I (meteoroid detection satellite), 203
Pegasus II, 203
Pegasus III, 203
Pell, Sen. Claiborne, 56, 105
Pennsylvania, Univ. of, 160
Perception of Space and Time in the Cosmos, 81
Percy, Sen. Charles H., 25, 210
Perdasdefogu, Sardinia, 108
Perkinson, William J., 205
Perrin AFB, Tex., 158
Perry, Robert L., 1
Perseid meteor shower, 185–186
Peru, 160
Petrov, Prof. Boris, 327–332

Petrov, Prof. Georgy I., 287–288
Petrovich, Prof. Georgy V., 36
Pewee 1 (nuclear reactor), 335
Pezdirtz, Dr. George F., 39
Philadelphia, Pa., 189, 231, 247
Philco-Ford Corp., 61, 194
   Space and Re-Entry Systems Div., 9, 231, 244
Philippines, 137
Philips, Ronald J., 230
"Phillips Report," 25
Phillips, L/G Samuel C. (USAF), 92–93, 97, 191, 196, 266, 278, 322–323
Phobos (Mars moon), 281–282
Phoebus 2A (nuclear reactor), 25–26, 131, 143, 156, 335
Phoenix, Ariz., 288
Phoenix (missile), 14, 98–99, 160
Photography zenith tube (PZT), 134
Photography. See Advanced Vidicon Camera System; Automatic Picture Transmission; Earth, photographs; Moon, photographs.
Photometer, 67, 82, 121
Physical Research Laboratory, Ahmedabad, India, 273
Physics, 93, 102, 154, 275, 325
Physics of the Earth in Space—A Program of Research: 1968–1975, 275
Pickering, Dr. William H., 15, 22–23, 29, 79, 89, 260
Pine Gap, Australia, 329
Pinkel, I. Irving, 96
Pioneer (program), 26, 44, 273–274
Pioneer VI (interplanetary probe), 209, 274
Pioneer VII, 138, 274
Pioneer VIII, 112, 138–139, 274
Pioneer IX (Pioneer D), 1, 273, 335
Piper Aztec (light aircraft), 207
Piper Twin Comanche (light aircraft), 179
Pippard, Alfred J. S., 91
Pittsburgh Des Moines Steel Co., 214–215
Pittsburgh, Pa., 210
Pittsburgh, Univ. of, 36
Planetary Exploration, 1968–1975 (report), 188–189
Planets, life on. See Extraterrestrial life.
Plant experiments, space, 46
Plastics, 39, 265
Plesetsk Cosmodrome, U.S.S.R., 126, 132, 165, 279, 289, 295
Plohr, H. Warren, 104
Plotkin, Dr. Henry H., 71, 183
Plowshare, Project, 110
Pluto (planet), 326
PMR. See Pacific Missile Range.
Podgorny, President Nikolay V. (U.S.S.R.), 72
Point Barrow, Alaska, 29, 216, 255, 256, 287, 288, 291, 310
Pokrovsky, Prof. Georgi, 95, 119

Poland, 189
Polar Cap Absorption (PCA), 129
Polar wind, 216
Polaris (missile), 9, 148, 193
Polaroid Corp., 38
Pollack, Martin A., 39
Pollution control, 10, 66, 323
Ponnamperuma, Dr. Cyril A., 16, 22
Port of New York Authority, 5, 26, 188
Porter, Dr. Richard W., 136–137
Portland, Ore., 119
Poseidon (missile), 18–19, 61, 159, 193, 195, 312
Pratt & Whitney Div., United Aircraft Corp. 248
Pratt, Perry, W., 316
Presidential Medal of Freedom, 51, 306
Presidential Task Force on Communications Policy, 27
President's Advisory Committee on Supersonic Transport, 207, 270
President's Commission and Council on Marine Sciences, 27
President's Foreign Intelligence Advisory Board, 45
President's Science Advisory Committee, 40, 76–77, 270
Press comment
  air traffic congestion, 188, 196, 199–200
  airports, 302
  antiballistic missile (ABM) system, 143
  Apollo AS–204, accident, 161, 222–223, 264
  Apollo 6 flight, 80–81, 81–82
  Apollo 7 flight, 253, 261–262, 264, 265
  Apollo 8 flight, 323–32, 324–325, 327–328, 329–331
  C–5A cargo aircraft, 150, 155
  disarmament, 149–150, 161
  European Launcher Development Organization (ELDO), 233
  F–111, 270
  F–111A, 75
  fractional orbital bombardment system, 102, 255
  International Biological Program, 126
  lunar landing, 261, 271
  mathematics, 293
  moon, exploration of, 255, 281
  Multiple Independently Targetable Reentry Vehicle, 188, 191, 195, 205
  National Academy of Sciences, 152
  NASA, 223, 284
  NRC Space Science Board, 194, 201
  nuclear nonproliferation treaty, 128, 133
  oceanography, 117
  Paine, Dr. Thomas O., 215
  research and development, 152
  science and technology, 271, 301
  sonic boom, 115, 135, 152
  space program, national, 15, 90, 101, 185, 194, 229, 234, 253, 255, 256, 271, 284, 293, 332
    budget, 85, 90, 129, 133, 139, 141, 182, 194, 215, 229, 255, 256, 325
  space race, 102, 266, 271, 325
  supersonic transport, 220, 263, 270, 332
  U.S.S.R. space program, 87, 255, 267, 271
  Webb, James E., 215, 218, 221, 229
  Zond V mission, 222, 223, 226
Press conference, 40
  airports, 194
  Apollo 5 flight, 63
  Apollo 6 flight, 63, 92
  Apollo 7 flight, 220, 277
  Apollo 8 flight, 196, 278, 285, 304, 322, 326, 330
  ARMS (Application of Remote Manipulators in Space), 186
  astronaut, 164
  balance of payments deficit, 1
  electronics, 276
  hydrogen bomb, 200
  manned space flight, 63, 286
  science and technology, 314
  sonic boom, 11, 146
  Soyuz III, 272
  space program, national, 15, 212–213, 254, 314, 328
  space suit, 176–177
  Surveyor VII, 50
  U.S.S.R. space program, 15, 226, 254, 267–268, 286–287, 329–330
  Webb, James E., 212–213
  Zond V, 222, 226–227,, 267
Princess Ragnhild Coast, 89
Princeton Univ., 90, 93, 116, 221, 298
Princeton Univ. Observatory, 104, 282
Pritchard, Capt. J. Laurence (RAF), 91
Probe (see also individual probes, such as *Mariner V, Pioneer VIII, Pioneer IX, Venus IV*)
  interplanetary, 1, 18, 19, 41, 44, 84, 140, 148, 178, 188–189, 201, 241, 273–274, 280–281, 336
  lunar. See *Luna XIV, Lunar Orbiter V, Zond V, Zond VI*.
  Mars, 118, 188–189, 248–249, 301–302
  Venus, 17, 163, 188–189, 248–249, 275–276
The Promise of Space, 147
Propeller Club, 119
"Prospects in Aeronautics Research and Development" (Wright Brothers Lecture), 38–39
Proton IV (U.S.S.R. space station), 284–285, 335
Proton accelerator, 316
Prugh, Peter H., 272
Puckett, Dr. Allan E., 5
Puerto Rico, 37, 54, 64–65, 230, 238, 323
Pulsar, 114, 119, 128, 161, 171, 231, 238, 253, 272, 284, 292
Puppis (constellation), 67
Purdue Univ., 103, 281
PZT. See Photographic zenith tube.

Q

Quaide, Dr. William L., 78

Quark (theoretical elementary particle of matter), 316
Quasar (quasi-stellar object), 239, 309
Queen Maud Land, 89
Quiet Research Engine Program, 47, 197, 248
Quito, Ecuador, 146

R

Radar, 7–8, 49, 80, 129, 136, 145, 179, 181, 238, 248, 272, 291
Radiation, 277
  cosmic, 46, 87, 115, 200, 228, 272
  effects, 46, 62, 168
  gamma, 52, 55, 181–182, 281,
  measurement, 115, 131, 168, 257, 282
  nuclear, 110
  sensor, 257
  solar, 56, 121, 168, 269, 296, 300
Radio Astronomy Explorer-A, 1
Radio Corporation of America (RCA), 43, 75, 86, 239, 293
Radio signal, 54, 136, 153, 184, 199, 210, 229, 238, 249, 262, 319
Radioactivity, 56
Radioastronomy, 54, 189, 200, 231, 238, 272, 312
Radiotelescope, 65, 219, 238
Raffensperger, M. J., 84
Raisting, W. Germany, 329
RAM (Radio Attenuation Measurement) Project, 198
RAM C–I (spacecraft), 199
RAM C–II, 198–199, 258, 335
Ramey, James E., 143
RAND Corp., 1, 104, 166, 248–249
Rat experiment, 286–287, 289–290, 291
Raymond Loewy/William Snaith, Inc., 86–87
Razdow, Dr. Adolph, 121
RCA. See Radio Corporation of America.
RCA Communications, Inc., 154, 307
RCA Laboratories, 43
RCA Services Co., 120, 143
Reconnaissance satellite
  U.S., 22–24, 109, 185–186, 206–207, 208, 228, 255, 262–263, 278
  U.S.S.R., 22–24, 178, 180, 202, 262–263
Record
  aircraft, 11–12, 207, 299, 317–318
  balloon, 209–210, 240
  spacecraft, 334–335
Redstone Arsenal, Ala., 72
Reed, Robert D., 65
Reed, Sylvanus Albert, Award, 15, 263, 316
Reentry, 17
  Apollo 7, 251
  control, 277–278
  heating, 97, 163, 181, 193, 226–227, 230
  radio attenuation, 193, 198–199, 258, 335
  spacecraft debris, 164
  test, 193, 198–199
  vehicle 163, 194, 292

Reentry Heating Project, 97, 193
Rees, Dr. Eberhard F. M., 32, 122–123
Reese AFB, Tex., 200
Reeves, Edmond M., 243–244
Reid, Dr. Henry J., 173
Reining, Dr. Henry, Jr., 32
Relativity theory, 49, 238–239
Rembaum, Dr. Alan, 283
Rendezvous, 145, 186, 191, 295
  U.S., 22–24, 136, 190, 220, 223, 236–237, 250
  U.S.S.R., 40, 263, 264–265, 335–336
Renzetti, Dr. N. A., 209
Report to Congress from President of United States, *United States Aeronautics and Space Activities, 1967*, 20–21, 148
Republican Coordinating Committee, 99, 105
Republican National Convention, 178–179
Research and development (R&D), 46, 105, 150
  aeronautics, 38–39, 187, 225–226, 248, 296
  benefits, 127, 244, 271, 275, 294
  employment, 234, 289, 290
  Federal support, 27, 101–102, 127, 131–132, 139–140, 140–141, 152, 156–157, 180, 208, 244, 258, 271, 291–292, 294, 301, 323
  funds for, 27, 101–102, 127, 131–132, 136–137, 139–140, 141, 258–259 296, 333
  DOD, 20, 131–132, 239–240, 290, 299
  NASA, 19, 165–166, 290, 299
  U.S.S.R., 309
Research and Development in Industry, 1966 (NSF report), 290
Research and Technology Advisory Council (NASA), 5
Resler, Prof. Edwin L., 146
Resolute Bay, Canada, 175
Retriever (NASA motor vessel), 59
Retromotor, 6, 17
Reuss, Rep. Henry S., 70
Revenue and Expenditure Control Act of 1968, 140, 145–146, 336
Reynolds Electrical and Engineering Co., 122–123
Rhode Island, Univ. of, 131
Riccitiello, Salvatore, 265
Rice Univ., 52, 53, 92, 181
Richey, B. J., 135
Rickover, V/A Hyman G. (USN), 83–84. 160, 169, 261
Riegel, Dr. Kurt, 247
Risk, Don C., 41–42
Robb, Mrs. Lynda, 292
Roberts, Charles S., 26–27
Rochester, Univ. of, 292
Rockefeller, Gov. Nelson B., 180, 333
Rockefeller Public Service Award, 281, 300
Rockefeller Univ., 76–77
Rogallo, Francis M., 187

Rogallo, Gertrude, 187
Rogallo wing (paraglider), 187
Rogers, Rep. Paul G., 247
Rogers, William P., 309
Rolls-Royce, Ltd., 72
    Bristol Engine Div., 316
Romania, 32, 156, 189–190, 292, 317
Romanian National Council of Scientific Research, 140
Roosa, Capt. Stuart A. (USAF), 59
Rooster experiment, 218
Rosamond Dry Lake, 261
Rosen, Dr. Harold A., 77
Rosman, N.C., 119, 143
Roush, Rep. J. Edward, 99
Rover (program), 19, 94, 123, 156, 170
Rowan, Carl T., 263–264
Royal Aeronautical Society, 70–71
Royal Greenwich Observatory (U.K.), 134
Royal Norwegian Council for Scientific and Industrial Research, 177–178
Rubey, Prof. William W., 267
Rudolph, Dr. Arthur, 112, 283
Rudolph, Capt. Gerald T. (USAF), 297
Rumford, Sir Benjamin Thompson, Count, 309
Rumford Premium, 309
Rusk, Secretary of State Dean, 159, 247
Russell, Sen. Richard B., 138
Rutgers Univ., 12, 224
Ryan Aeronautical Co., 166, 177, 276–277
Ryan, Rep. William F., 25

S

SAAP. See South Atlantic Anomaly Probe.
Sacramento Peak Observatory, 296
Sadeh, Dror, 184
Sänger, Dr. Eugen, 300
Safety, 32, 95–96, 122–123, 157, 163, 220, 237, 251–253, 264, 273
Sagan, Dr. Carl, 172
Sage (Semi-Automatic Ground Environment) system, 238
Saigon, S. Vietnam, 104
St. Clair, Wade, 117
St. Helena island, 257
St. Louis, Mo., 108, 207
St. Thomas Univ., Tex., 325–326
Sakharov, Prof. Andrey D., 159, 195–196
Salam, Abdus, 198
Salisbury, Dr. John W., 170, 250
Samfield, Edwin, 145
San Antonio, Tex., 141
San Clemente Island, Calif., 143, 295
San Diego, Calif., 104, 207
San Francisco, Calif., 42
San Jose, Calif., 285–286
San Miguel Island, 116, 242
Sanders Associates, Inc., 75
Sanders, Newell D., 104
Sandia Corp., 32
Sandpiper (missile), 3–4
Santa Cruz, Calif., 51–52
Santa Monica, Calif., 166–167

Santa Susana, Calif., 113
Santiago, Chile, 146
Satellite Tracking and Data Acquisition Network (STADAN), 5, 12, 104, 109
Sato, Prime Minister Eisaku, 327
Saturn (planet), 79, 148, 275
Saturn I (booster), 21–22, 149
Saturn I Workshop (spacecraft), 162, 218–219
    contract, 62
    design review meeting, 22, 66
    funds for, 102–103, 182
    plans for, 33, 108–109, 186, 202, 241, 275
Saturn IB (booster), 140, 148, 149, 210, 218–219, 275
    contract, 9, 61, 75, 77, 175–176, 207, 239
    engine
        H–1, 9, 207
        J–2, 119–120
    launch
        AS–204, 13
        AS–205, 173, 250–253
    program, 11, 20, 24, 33, 92–93, 175–176, 212–213, 241, 336
    stage, S–IVB test, 10, 13
    test, 13, 21–22, 91, 136, 144, 173, 186, 242
Saturn V (booster), 44, 78, 93, 148, 149, 190, 191, 274
    capability, 52, 80–81, 81–82, 98, 186, 236, 255–256
    contract, 35, 75, 77, 104, 127, 163, 175–176, 226, 264, 275
    engine
        F–1, 30, 257, 275
        J–2, 77, 92, 119–120, 135, 220, 252
    launch
        AS–502, 63, 77–78, 166, 328
        AS–503, 144, 173, 318–322
        AS–504 (preparations for), 197
    program, 11, 18, 24, 33–34, 42, 63, 92–93, 97, 98, 140, 144, 162, 175–176, 181, 212–213, 232, 242, 321–322, 336
    stage
        1st (S–IC), 166, 173
            test, 77
        2nd (S–II), 30, 36, 47–48, 53, 173, 197
            test, 77, 122
        3rd (S–IVB), 10, 63, 173, 186
            test, 77
Saturn V Workshop (spacecraft), 10, 20, 33, 35, 109, 142, 182
Schafer, Col. George E. (USAF), 11
Scherer, Capt. Lee R. (USN, Ret.), 2
Schindlar, William R., 283
Schirra, Capt. Walter M., Jr., (USN), 261–262, 305
    Apollo 7 flight, 250–251, 252
    preparations for, 97, 158, 173, 220, 227
    honors, 269, 286, 303, 314

press conference, 220, 276
White House visit, 306
Schmidt, Edward J., 264
Schmidt, Dr. Maarten, 309
Schneider, William C., 223, 315
Schriever and McKee Associates, 233
Schriever, Gen. Bernard A. (USAF, Ret.), 55–56, 99
Schwartz, I. R., 109–110
Schweickart, Russell L., 97, 167, 196
Science (see also National Academy of Sciences), 188–189
  award, 38, 217–218, 266–267, 271
  benefits, 110, 195–196, 301
  Government support of, 90–91, 93, 110, 131–132, 139–140, 270, 271, 301, 314, 323
  human needs, 16, 110, 301
  national policy and goals, 10, 38–39, 90–91, 98, 135, 178–179, 180–181, 207, 244, 299, 301–302, 305, 314, 331
  President's Science Advisory Committee, 40, 76–77, 270
  U.S.S.R., 63, 93, 195–196, 244
Science, Dept. of (proposed), 325, 331
Science Advisory Committee (President's). See President's Science Advisory Committee.
Science and Public Policy Studies Group, 284
Scientific Balloon Flight Station, 116
Scientist-astronaut, 36, 51–52, 200
Scientists, 99–100, 152, 156–157, 234, 247, 254–255, 257, 289, 290, 301, 323, 329
  women, 99–100
Scorpius (constellation), 247–248, 273
Scott, L/C David R. (USAF), 97, 167, 211
Scout (booster), 53, 56, 97, 113, 123, 182, 193, 198, 240, 255–256
Scout (sounding rocket), 160–161, 308
Scrag (U.S.S.R. weapon), 255
Scrimp, Project, 171
Sea of Tranquility (moon), 35–36
Seaborg, Dr. Glenn T., 117, 143, 205, 211, 217–218, 228
Seal Beach, Calif., 30
Sealab I (underwater laboratory), 295
Sealab II, 143, 295
Sealab III, 41–42, 143, 295
Seamans, Dr. Robert C., Jr., 3, 5–6, 26, 53, 57, 156, 268
Seattle, Wash., 217–218
Seckbach, Dr. Joseph, 302
SECOR (geodetic satellite), 7–8
The Security of Japan and Prospects for 1970. (study), 166
Security Research Council, 166
Sedov, Prof. Leonid I., 212, 254, 267–268, 329–330
Seifert, Dr. William W., 99
Seismometer, 282–283
Seitz, Dr. Frederick, 7, 76–77, 90, 255, 267
Selenographica, 27–28

Sensor, 2, 39, 50, 11
Sentinel (antiballistic missile system), 18–19, 72, 88, 122, 133, 138, 159, 171, 175, 242
Sentinel System Command, 72
Serpukhov, U.S.S.R., 154, 316
SERT II (Space Electric Rocket Test), 26
Service module (SM), 13, 14, 63, 136, 144, 186, 197, 201
Service propulsion system (SPS), 77–78, 136
Service Technology Corp. (LTV), 123
Seryogin, Col. Vladimir S. (U.S.S.R.), 70, 72
Sewage sludge, 281
Shafer, Edward M., 83
SHAPE (Supersonic High Altitude Parachute Experiment) Project, 261, 309
Shapiro, Dr. Irwin I., 49
Sheldon, Dr. Charles S., II, 202
Shenton, Samuel, 331
Sherman, Harvey, 26
Shillito, Thomas B., 26
Shipley, Rep. George E., 284
Siding Spring Mountain Observatory, 290
Siebel, Dr. Mathias P., 272–273
Sikorsky Aircraft Div., United Aircraft Corp., 38
Sikorsky, Igor I., 38
Silver, Brent W., 148
Silverstein, Dr. Abe, 41, 281, 283, 300
Simpson, Ernest C., 15
Simpson, Dr. John A., 90–91
Simpson, Robert W., 224
Siple, Dr. Paul Allman, 293
Sjogren, William L., 193–194
Skolnikoff, Prof. Eugene B., 284
Skylark (U.K. sounding rocket), 70, 123
Skynet (U.K. communications satellite), 1, 60, 121, 137
SLAC. See Standford Linear Accelerator.
Sloan Foundation, 284
Sloan School of Management, 53
SM. See Service module.
Smart, Gen. Jacob E. (USAF, Ret.), 237, 299
Smelt, Dr. Ronald, 5
Smith, Dr. Arthur H., 218
Smith, Secretary of Commerce C. R., 40, 127
Smith, Francis B., 249–250
Smith, Prof. Frederick E., 115
Smith, Dr. Graham, 253–254
Smith, Dr. Harlan J., 294
Smith, Dr. Henry J., 89
Smith, Sen. Margaret C., 30, 86
Smithsonian Astrophysical Observatory (Cambridge, Mass.), 201, 262, 304
Smithsonian Institution, 106, 120, 160
Smyth, Henry DeWolf, 198
SNAP–8 (nuclear reactor), 25–26, 57, 59
SNAP–19 (radioisotope generator), 242
Snyder, Conway W., 283
Society for the History of Technology,

325
Society of Automotive Engineers Space Technology Conference, 108–109
Solar corona, 97
Solar flare, 56–57, 110, 121, 128, 157–158, 168, 198, 210–211, 243–244, 266, 296
Solar Particle Alert Network (SPAN), 121
Solar Pointing Aerobee Rocket Control System (SPARCS), 65
Solar wind, 68, 91, 282–283
Solar Wind Composition Experiment, 68
Solid propellant, 3–4, 43, 68–69, 274
Sonic boom
  damage, 56, 117, 142–143, 154–155, 227, 290–291
  regulation of, 115, 116, 130, 271
  research 8, 11, 109–110, 142–143, 146, 158, 313
  National Academy of Sciences report, 8, 56, 142–143, 152
  supersonic transport, 8, 11, 56, 66–67, 109–110, 115, 117, 125, 135, 142–143, 146, 290–291, 332–333
Sonnett, Dr. Charles P., 316
Sorokin, Dr. Peter P., 169
Sounding rocket (see also individual sounding rockets: Aerobee 150A, Aerobee 150 MI, Arcas, Astrobee 1500, Boosted Arcas, Boosted Dart, Javelin, Nike-Apache, Nike-Cajun, Nike-Tomahawk, Skylark)
  foreign
    U.K., 123
  international programs, 29, 30, 64, 75, 123, 336
    ESRO, 108, 118
    NASA-Brazil, 68, 130, 177
      -Canada, 11, 14, 29, 36, 43, 49, 53, 54, 67, 90, 94, 98, 103, 106, 113, 128, 175, 329
      -Germany, West, 75, 130, 133
      -India, 30, 60, 75, 273
      -Norway, 177–178
      -Puerto Rico, 64–65
      -Spain, 177–178
      -Sweden, 29, 130, 177–178
South America, 184–185, 318–319
South Atlantic Anomaly Probe (SAAP), 130–131, 173
South Georgia, 37
South Pole, 89
Southeast Asia, 195
Southern California, Univ. of, 82
Southwest Astronomical Conference, 201
Soviet Academy of Sciences, 63, 272, 298
  Council for International Cooperation in Investigation and Utilization of Outer Space, 327
Soviet Corps of Cosmonauts, 70
Soviet Women's Committee, 148
Sowers, L/C Robert G. (USAF), 8
Soyuz (U.S.S.R. spacecraft), 40, 222, 268
Soyuz I, 70

Soyuz II, 263, 264–265, 335–336
Soyuz III, 264–265, 269, 272, 277, 335–336
SP–5B (Martin Marlin) (flying boat), 160
Spaatz, Gen. Carl A. (USAF, Ret.), 104–105
Space Age Law Conference, 62
Space Applications Summer Study, 1967, Interim Report, 36
Space biology, 232, 241–242
  animal experiments, 46, 47, 86, 95, 105, 151, 181, 218, 272, 273
  atmosphere, artificial, 61, 173, 251, 283, 287, 289–290, 291
  contract, 86, 151
  drugs, use of 211
  environment, effects, 44, 47, 88, 101, 143, 158, 165, 324
  life support system, 44, 86, 144, 176–177, 236, 283
  medical benefits, 2, 57, 165, 186, 246, 302
  psychology, 287
  radiation, effects, 46, 62, 121, 168, 200, 257, 272, 277
  symposium, 141–142, 200, 291
  weightlessness, 86, 190, 286
    effects 81, 46, 200–201, 268, 273
    prolonged, 151, 181, 218
Space Disturbance Forecast Center, 157–158
Space law treaty, 12, 127, 258, 281
Space, military use of (see also Manned Orbiting Laboratory), 329
  communications, 190–191, 217, 228
  reconnaissance, 178, 179, 202, 206–207, 208, 262–263
  space station, 69–70, 208
  U.S., 69–70, 105, 178, 190–191, 206–207, 208, 262–263
  U.S.S.R., 93, 102, 178, 202, 205, 229, 255, 262–263
Space, peaceful use of, 178–179, 245, 249, 257, 263–264, 309, 334–335
Space program, national (see also individual programs, such as Apollo program), 94, 112–113, 178
  achievements, 33, 37, 48, 53, 66, 79, 212–213, 215, 218, 221–222, 232, 235–238, 248, 249, 253, 306, 324–328, 334–335
  international, 336
  management, 66, 109, 157, 296–297
  manned space flight, 22–24, 66, 165, 190, 215, 236–237, 253, 269, 334–335
  benefits. See Space results.
  budget, 33, 34–35, 36, 44, 48, 50, 51, 57, 65, 66, 85, 90, 94, 98, 99, 102, 129, 131–132, 138, 140, 161–162, 165–166, 173, 178, 182, 202, 209, 212–213, 218, 229, 239, 240–242, 255, 276, 287, 288, 299, 332
  cost, 37, 62, 72–73, 85, 109, 237, 253,

255–256, 276
criticism, 25, 51, 86, 101, 103, 133, 257, 265, 289, 325
education, benefits to, 20, 124–125, 235, 249–250
employment, 73–74, 85, 115, 168, 182, 213, 229–230, 256, 272, 289
House staff study, 51
Humphrey, Vice President Hubert H., 106–107, 235, 254
international aspects (see also International cooperation), 2–3, 20–21, 99, 124–125, 127, 141, 204, 232, 254
Johnson, President Lyndon B., 18–19, 20–21, 53, 178–179, 202, 208, 212 235–236, 248, 267, 292
lunar landing. See Moon, landing.
manned space flight. See Manned space flight.
military, 55–56, 85, 102, 105, 122, 147–148, 179, 190–191, 205, 208, 255, 262–263
Nixon, President-elect Richard M., 207, 235, 249, 293, 311, 331
objectives, 86, 106, 140, 147–148, 235–237
policy, 41, 44, 48, 55, 57–58, 62, 92, 97, 98, 99, 105, 112–113, 124–125, 188–189, 212–213, 215, 245, 253, 263–264, 293, 299, 308, 314, 325, 331
post-Apollo, 29–30, 41, 44, 47, 51, 66, 79, 85, 124–125, 162, 180, 182, 218, 222, 243, 257, 259, 274–275, 301–302, 303, 311
    budget, 101, 182, 188–189, 212–213, 233, 281–282, 325
    suggested programs, 51–52, 94, 148, 188–189, 194, 201, 275, 280–281, 334
significance, 43, 90–91, 98, 103, 106–107, 115, 141, 161–162, 199, 235–237, 249, 275
    international, 20–21, 199, 210
U.S. vs. U.S.S.R. See Space race.
Vietnam War, effect of, 15, 56, 99, 131–132, 146, 215, 245, 336
Space race, 15–16, 41, 90–91, 147–148, 180, 212–213, 249, 253, 326
booster, 33, 34, 186–187, 209, 213
criticism, 257, 325
funds, 37, 174, 213, 229–230
manned space flight, 22–24, 34, 40, 41, 51, 180, 222, 226–227, 253, 269, 326
military, 42, 87, 93–94, 102, 174, 180, 188, 223, 255
moon, 22–24, 41, 87, 112–113, 119, 173–174, 186–187, 222, 229–230, 249, 261, 266, 308, 311
payload, 36, 93, 112, 209
planetary flights, 23–24, 34, 36, 41, 43, 112–113, 188–189, 222
press comment, 102, 266, 324–325
Space rescue treaty, 89–90, 127, 161, 215, 225, 246, 258, 299
Space results (see also Earth; Moon; Mars; Venus; individual probes, satellites, and rockets), 2, 24, 31–32, 34, 53, 56, 66, 72–73, 79, 90, 94, 106–107, 119, 124–125, 141, 190, 232, 255–256, 333, 334
agriculture, 62, 106, 117, 125, 257
aircraft, 20, 57
astronomy, 180, 231, 276
communications, 20, 27, 33, 34, 36, 82–83, 106, 233–234, 235, 246, 254, 257, 267, 334
earth sciences, 20, 180, 191, 215, 254, 257, 275, 276, 334
economic benefits, 33, 36, 49, 58, 73–74, 103, 109, 117, 202, 234, 235, 257
education, 20, 235, 249–250
engineering, 66, 106, 234, 246, 294, 310
geology, 117, 231, 235, 257
medicine, 20, 57, 82–83, 165, 234, 235, 246, 285–286, 302
meteorology, 20, 27, 32–33, 36, 49, 106, 117, 232, 236–237, 254, 257–258, 334
military, 2–3, 20, 22–23, 57, 185, 208, 217, 237–238, 275
navigation, 20, 37, 106, 217, 257
oceanography, 27, 60, 106, 231
photography, 106, 185, 231
political, 2–3, 202, 204, 232, 254
social science, 58, 202
technology, 20, 91, 103, 124–125, 190, 204, 234, 235–238, 249–250, 267–268, 274–275, 297
Space station (see also Manned Orbiting Laboratory; Saturn I Workshop; Saturn V Workshop), 73, 141–142, 163, 185, 259, 275, 293–294, 303
contract, 62, 86–87, 202, 242–243
U.S.S.R., 54, 119, 211–212, 222, 226, 230, 68, 277–278, 283
Spacecraft (see also individual spacecraft, such as Apollo, Lunar Orbiter, Luna, Mariner, Surveyor)
braking, 17, 277
communications, 319
control, 24, 185, 218–219, 226–227, 264–265
design, 10, 48, 66, 86, 111, 202–203, 218–219, 252, 311
development testing, 3
electrical systems, 4–5, 251
environment control system, 47, 61–62, 134, 173, 186
equipment, 8, 153, 167, 179, 191–192, 218–219, 282–283, 303
escape system, 4–5, 78, 210, 252
exhibit, 190
extravehicular equipment, 316–317
heating, 193, 212
instrumentation, 55, 78, 83, 172, 210–211
landing system, 3, 59, 78–79, 116, 144, 158
launch system, 88, 251–252

life support system, 4–5, 13–14, 61–62, 144, 173, 176–177, 212, 236–237, 242–243, 273
  propulsion. See Engine and individual launch vehicles, such as Saturn.
  recovery, 65, 268
  reentry control system (see also reentry), 198–199, 230, 277
  reusable, 72–73, 259
Spacecraft debris, 42, 164, 205–206
Spacecraft, model, 65
Spacesuit, 176–177, 283–284
SPACO, Inc., 79–80
Spain, 5, 35, 69, 111, 177–178
SPAN. See Solar Particule Alert Network.
SPARCS. See Solar Pointing Aerobee Rocket Control System.
Sparkman, Sen. John J., 248
Sparrow, J. G., 177
Spartan (missile), 72
Spaulding, Dr. Roland H., 71
Special Report on Underground Testing, 303–304
Spectrograph, 67
Spectroheliograph, 118, 128, 240
Spectrometer, 60, 62–63, 80, 130–131, 257–258, 273, 328–329
Spence, Roderick W., 123
Spencer, N. W., 328–329
Sperry Rand Corp., 59, 123
Spirit of St. Louis (aircraft), 104
Sport aviation, 40
Springfield, Va., 233
SPS. See Service propulsion system.
Sputnik I (U.S.S.R. satellite), 139–140
Sputnik IV, 17
SR–71 (reconnaissance aircraft), 8
Sri Racha, Southeast Asia, 108
SRN4 (hovercraft), 175
SST. See Supersonic transport.
STADAN. See Satellite Tracking and Data Acquisition Network.
Stafford, Maj. Thomas P. (USAF), 73, 279–280
Standard Telephone & Cables, Ltd., 278
Stanford Linear Accelerator (SLAC), 93, 154
Stanford, Neal, 237
Stanford Univ., 5, 38, 91, 93, 154, 275–276
  Sloan Fellow, 151, 194
Stanley, Hubert Ray, 283
Star, 272
  formation, 247
  neutron, 238, 284
  photographs, 310, 335
  radiation, 30
  radio signals from, 54, 208–209, 231, 238
  study of, 128–129, 208–209, 231, 238, 247, 272, 284, 304, 310
Star Tracking Rocket Attitude Positioning (STRAP) system, 30
Starr, Dr. Chauncey V., 5
State, Dept. of, 2–3, 89–90, 109, 189–190, 247, 255, 281
Status of Actions Taken on the Apollo 204 (NASA report) 4–5
Stein, Jerome, 281
Stennis, Sen. John C., 246–247
Stevens, Col. Robert L. (USAF), 207
Stever, Dr. H. Guyford, 299
Stewart, I. A., 295
Stockholm, Sweden, 266–267, 285
STOL (short takeoff and landing) aircraft, 139, 179, 217, 219, 224, 225–226, 231
STOLport, 179, 278
Stoltenberg Science Minister, Dr. Gerhard (W. Germany), 136–137, 201
Stoney, William E., Jr., 224
Stranraer, Scotland, 111
STRAP. See Star Tracking Rocket Attitude Positioning.
Strass, H. Kurt, 121
Stratoscope II (balloon-borne telescope), 116
Strughold, Dr. Hubertus, 233
A Study of NASA University Programs, 249–250
Stuhlinger, Dr. Ernst, 303
Sturmthal, L/C Emil (USAF), 38, 50, 145, 269
Sturtevant, Prof. Alfred H., 38
Styles, Paul L., 92
Submarine, nuclear, 42, 152, 160, 169, 211, 224, 261, 302–303
Sud-Aviation (France), 198
Suitland, Md., 155
Sullivan, Leo J., 146–147
Sullivan, Thomas E., 39
Sullivan, Walter, 231, 277, 292
Sun (see also Solar corona; Solar flare, Radiation, solar; etc.), 56–57, 65, 68, 102, 109, 118, 128, 153–154, 180, 203, 243–244
Sun Shipbuilding and Dry Dock Co., 156
Sunblazer (probe), 19, 35
Sunspot, N. Mex., 296
Sunspots, 154, 277
Suojanen, Dr. Waino W., 91–92
Suomi, Verner E., 229
Super Jolly (helicopter), 146
Super Loki Dart (sounding rocket), 143
Superior Engineering Co., 29
Supersonic High Altitude Parachute Experiment. See SHAPE Project.
Supersonic transport (SST) (see also Concorde and Tu–144), 104
  benefits, 66–67
  cost, 150, 202–203
  criticism, 115, 135, 263, 270, 332–333
  design and development, 45, 67, 158, 217, 220–221, 226, 259, 263, 285, 335
  flight plans, 45, 198, 217
  foreign, 49, 115, 126, 194, 198, 217, 259–260, 294, 332, 336
  funds for, 19, 43, 115, 144, 285
  hazards, 121, 164–165
  NASA participation in research, 26, 47,

131, 233, 237
President's Advisory Committee on Supersonic Transport, 207, 270
press comment, 115, 220–221, 263, 270, 332–333
reservations, 285
sonic boom, 8, 11, 56, 67, 109–110, 112, 115, 117, 125, 135, 142–143, 146, 271, 290–291, 332–333
Survey of Views of Leading Industrial Executives on the National Space Program (House report), 99
Survey Satellite (SURSAT), 156
Surveyor (program), 1, 3, 41, 51–52, 85, 106, 151, 281, 301, 314, 335
Surveyor I (lunar probe), 7
Surveyor II, 7
Surveyor III, 7
Surveyor IV, 7
Surveyor V, 7
Surveyor VI, 7, 232
Surveyor VII (Surveyor G), 1, 3, 6–7, 42, 50, 335
Svestaka, Dr. Zdanek, 254
Sweden, 5, 32, 46, 118, 177–178, 198, 240, 262, 292
Swedish Space Research Committee, 178
Sweeney, Dr. Stephen B., 160
Swihart, John M., 259–260
Swiss National Committee for Space Research, 68
Switzerland, 5
Swords Into Plowshares Award, 117
Sydney, Univ. of, 171
Symington, Sen. Stuart, 239–240, 245–246
Symposium on Remote Sensing of Environment, 85
Syracuse University Research Corp., 82, 308
Systems engineering, 66, 157, 238, 297

T

T-38 (jet aircraft), 128
Ta Khli AFB, Thailand, 69, 80
Table Mountain, Calif., 6
Taccomsat 1 (communications satellite), 189
Tactical photographic image transmission (TAPIT) system, 77
Tahiti, 180
Tallahassee, Fla., 128
Talkeetna, Alaska, 290
Tananarive, Malagasy Republic, 146
TAPIT. See Tactical photographic image transmission system.
Taschek, R. F., 91
Taurus (constellation), 273
Taylor, Hal, 197
Taylor, Dr. J. H., 161
Taylor, Gen. Maxwell D. (USA, Ret.) 45
TD-1 (ESRO solar astronomy satellite), 95
TD-2, 95
Teague, Rep. Olin E., 51, 99, 103, 222
Technical Information Services Co., 245

Technological Innovation in Civilian Public Areas (study), 27
Technology, 16, 63, 90–91, 103–104, 139–140, 178–179, 210, 235–236, 237, 246, 249–250, 271, 273, 299
benefits, 27, 98, 195–196, 274–275, 301
Federal support, 58–59, 126, 139, 237–238, 294, 314
gap, 105, 244
misuse of, 16, 119
U.S. policy, 135, 330–331
Technology utilization, space (see also Space results), 36, 48, 89, 91, 94
benefits, 2–3, 20–21, 62, 103–104, 141, 178–179, 232, 237–238, 246, 249, 274–275, 297
Teledyne Systems Co., 82, 144
Telemetry, 51–52, 83, 164–165
Teleoperators and Human Augmentation (NASA SP-5047), 82
Telescope (see also Apollo Telescope Mount), 12–13, 119, 128–129, 219, 284, 290, 294, 334
astrometric, 151
balloon-borne, 116
gamma-ray, 52
infrared, 184, 219
spectrographic, 240
vacuum, 296
Telespazio (Italian space communications company), 80
Television, 249, 276, 308
Apollo 7, 250–252, 312, 334–335
Apollo 8, 318–319, 323–324, 334–335
educational, 196–197
Soyuz 3, 264–265
space probe, use of, 3, 6, 24, 29, 84, 153
via satellite, 10, 80, 103, 111–112, 154, 155, 196–197, 204, 216, 227, 244, 251–252, 257, 315, 323
military use, 190–191
Teller, Dr. Edward, 165
Temperature, 17, 230, 243–244
Tennessee, Univ. of, Space Institute, 267
TERLS. See Thumba Equatorial Rocket Launching Station.
Terzian, Dr. Yervent, 171
Tether, inflatable, 145
Tetr I (*Tts I*) (Test and Training Satellite), 112, 274
Tetr II (TETR B), 274, 335
Texas, 249, 255
Texas Instruments, Inc., 87
Texas Tech, 115
Texas, Univ. of, 1, 82–83, 115, 188, 206, 294
TF-39 (jet engine), 15, 146
TFX. See F-111.
Thailand, 69, 72, 80, 84, 87, 102, 108, 109, 119, 137
Thant, U, U.N. Secretary General, 72, 189–190, 327
Thiokol Chemical Corp., 259
Elkton, Md. Div., 308

Thomas, David D., 175, 194
Thomas, Paul G., 174
Thompson, Dr. Floyd L., 15, 17, 51, 65, 101, 204
Thor (booster), 59
Thor-Agena (booster), 101, 138, 181, 216, 244, 271, 309
Thor-Burner II (booster), 120, 261
Thor-Delta (booster), 9, 95
Thorad-Agena D (booster), 116, 138, 216, 243
Thrust-Augmented Delta (booster), 7, 42–43, 153
Thrust-Augmented Improved Thor-Delta (DSV–3E) (booster), 273
Thrust-Augmented Long-Tank Delta (booster), 314–315
Thrust-Augmented Long-Tank Thor-Delta (booster), 191, 311
Thrust-Augmented Thor-Agena D, 9–10
Thrust-Augmented Thor-Delta (booster), 244, 300
Thule AFB, Greenland, 154
Thumba Equatorial Rocket Launching Station (TERLS), India, 30–31, 60, 75–76, 273
TIFS. See Total In-Flight Simulator.
Tillinghast, Charles C., Jr., 187–188
Tiros Operational Satellite (TOS) system, 1, 50, 149, 191–193, 311
Tiros I, 193, 311
Tiros III, 193, 311
Tiros, M, 49
Titan (booster), 148
Titan III, 162
Titan III–B, 127, 208, 272, 300
Titan III–B–Agena D, 10, 61, 86, 127, 179–180, 272
Titan III–C, 25, 132, 148, 189, 228
Titan III–D/Centaur, 302
Titan III–M, 20, 335
Todd Shipyards Corp., 310
Tokyo Astronomical Observatory, 155
Tokyo, Japan, 82–83, 200–201,
Tolansky, Samuel, 158
Tomahawk (sounding rocket), 31
Tompkins, Chargé d'Affaires Edward K. (U.K.), 295
Tornado observation, 62, 229
Torreon, Mexico, 295
TOS. See Tiros Operational Satellite.
TOS–E (Tiros Operational Satellite). See *Essa VII*.
Total In-Flight Simulator (TIFS), 81
Toulouse-Blagnac, France, 198
Townes, Dr. Charles H., 299, 312
Townsend, Dr. John W., Jr., 131
Tracking, 17, 30, 34–35, 50, 65, 112, 113, 144, 178, 256
 aircraft, 12
 deep space, 5, 6, 10, 12–13, 118, 203, 209, 237
 laser, 6, 259, 335
 MSFN, 5, 12, 97–98, 258, 274, 312, 335

radar, 7, 12–13, 14, 64–65, 133, 199, 214, 248, 291
 ship, 12, 258
 STADAN, 5, 7, 12, 143, 146
 station, 205–206
  Alaska, 35
  Europe, 137
  Pakistan, 18
  Spain, 312
  U.S., 114, 133, 171, 291, 312
Trade Expansion Act of 1968, 134
Trans World Airlines, 72, 172, 187–188
Transportation, 185, 187, 235, 323
Transportation, Dept. of (DOT)
 air traffic control, 189, 194, 196, 199–200, 206
 airports, 137–138
 budget, 19, 168
 noise abatement, 154–155, 271
 R&D, 187
Transradio, 293
Treasury, Dept., 144
Treaty
 nuclear nonproliferation, 11, 32, 62, 92, 128, 130, 132, 138, 149, 158, 159, 165, 210, 215–216, 228, 230, 249, 295
 nuclear test-ban, 228
 "ocean space" (proposed), 56
 space law, 12, 127, 258, 281
 space rescue, 89–90, 127, 161, 215, 225, 246, 258, 299
Triethylborane (TEB), 45
Trimethylaluminum (TMA), 45
TRIPLTEE. See True temperature tunnel.
Trippe, Juan T., 107–108
Tristan de Cunha, 37
Trowbridge, A. B., 40
True temperature tunnel (TRIPLTEE), 39–40
Truman, President Harry S., 250
Truszynski, Gerald M., 17, 34, 50, 65
TRW, Inc., 94
 Systems Group, 146, 217
Tts I. See *Tetr I*.
Tu–144 (U.S.S.R. supersonic aircraft), 126, 194, 259–260, 332–333, 335–336
Tucker, Dr. Gardiner L., 137
Tuke, John B., 111
Tulane Univ., 209
Tunisia, 118
Turcat, André, 198
Turin, Italy, 126
Turtle experiment, 283
Tycho (lunar crater), 6, 50

U

UCLA. See California, Univ. of, at Los Angeles.
Udall, Secretary of the Interior Stewart L., 7, 290–291
UFO. See Unidentified flying object.
Uganda, 69
U.K. See United Kingdom.
Ulybshev, Boris N., 324
U.N. See United Nations.

Underground nuclear test, 303–304, 317
Unidentified flying object (UFO), 99, 151, 169–170, 172, 227, 254
Unidentified satellite, 9–10, 16, 61, 101, 120, 127, 138, 179–180, 181, 208, 244, 261, 271, 272, 309
United Air Lines, 164
United Aircraft Corp., 2, 38, 316
    Pratt & Whitney Div., 123, 130, 197, 201, 248
United Kingdom (U.K.), 46, 91
    aircraft, 9, 32, 165, 256
        Concorde, 49, 115, 198, 217, 260, 294
    cooperation, defense, 9, 165
    cooperation space, 5, 95, 113, 137, 214, 240, 279
    House of Commons, 9
    launch
        missile, 55
        sounding rocket, 70, 123
    Ministry of Technology, 49, 118, 121, 256, 286
    nuclear nonproliferation treaty, 102, 132, 138, 295
    satellite, 1, 9, 60, 105, 121, 137
    science and technology, 46, 54, 83, 134, 139, 170–171, 200, 211–212, 253, 285, 290, 298
    space program, 60, 73
    space rescue treaty, 89, 299
United Nations (U.N.), 12, 117–118, 133, 172, 189, 199, 204, 281, 306, 323
    Committee on the Peaceful Uses of Outer Space, Legal Subcommittee, 127
    Conference on Exploration and Peaceful Uses of Outer Space, 189, 204, 211
    Disarmament Conference, 11, 32, 59, 62
    Economic and Social Council, 72
    General Assembly, 11, 62, 89, 102, 132, 138, 158, 161, 317
        Political Committee, 130
    Secretary General, 72, 189, 327
    Security Council, 59, 138, 159
United Nations Educational, Scientific and Cultural Organization (UNESCO), 27
United States (U.S.) (see also appropriate agencies)
    award, 38, 49, 198, 217, 221, 266, 271, 281, 290, 298, 309
    budget, 15–16, 18–19, 233
    communications, 27, 76, 307
    defense, 9, 18–19, 20, 29, 137, 242, 278, 308
    disarmament, 11, 59, 132, 149, 164, 166, 171, 188, 191, 210, 301
    education, 16, 331
    election results, 271–272
    health, 2, 301, 312–313
    international cooperation, 2–3, 9, 59, 70, 73, 99, 117–118, 127, 128, 138, 141, 156, 157, 159, 177, 189, 196–197, 247, 257, 306–307
    medical research, 2, 165, 186, 325
    nuclear nonproliferation treaty. See Nuclear nonproliferation treaty.
    nuclear tests, 108, 110, 303–304, 317
    oceanography, 27, 56, 105, 117, 177, 284, 300, 325
    pollution abatement, 10, 62, 301, 331
    research and development, 10, 16, 20, 101–102, 127, 131–132, 139–140, 141, 152, 156–157, 202, 208, 244, 271, 290, 291–292, 314, 333
    science and technology, 10, 16, 27, 38, 57–58, 60, 62, 70, 73, 93, 94, 99, 101–102, 115, 119, 154, 175, 178–179, 180, 194, 208, 244, 255, 271, 284, 291, 299, 301, 305, 314, 317, 325, 329, 331
    space rescue treaty. See Space rescue treaty.
    space program. See Space program, national
    transportation (see also Supersonic transport), 168, 187, 207, 235, 323
    Vietnam War. See Vietnam War.
United Technology Center, 3–4, 36–37
Universe, 87, 184, 197
Universities (see also individual universities), 11, 102, 271, 301, 329
    and space effort, 55, 68, 85, 235, 267, 269
    Federal support, 110, 291–292, 301, 314
    grants to, 44–45, 129, 151
    NASA program, 35, 39, 48, 115, 139–140, 151, 162, 249–250, 275, 280–281
Uppsala (Sweden) Ionospheric Observatory, 262
Upson, Ralph Hazlett, 188
Uranium, 166, 298
Uranus (planet), 79, 148
Urban Coalition, 117
USAF School of Aerospace Medicine, 11
USAF Space and Missile Systems Organization (SAMSO), 111, 121, 173, 194, 264
U.S. Air Force (USAF) (see also individual bases, centers, and commands, such as Air Force Systems Command, Arnold Engineering Development Center, Edwards AFB), 39, 40, 55–56, 110, 112, 135, 146, 164, 170–171, 207, 209, 211, 232–233, 283, 298–299
    Aero Propulsion Laboratory, 15, 135
    aircraft (see also individual aircraft, such as C–5A, C–130, F–111A, X–15, XB–70), 75, 112, 130, 146, 171–172, 239–240, 245–246, 259, 286, 313
        accident, 2, 8, 72, 80, 84, 87, 97, 102, 108, 109, 117, 119, 134, 195, 224
        sonic boom, 154–155, 227, 313
    anniversary, 11
    astronaut, 11–12
    award, 15, 104–105, 259, 299, 300
    booster, 9–10, 20, 61, 86, 111, 127, 132, 148, 162, 189, 228, 302, 335

budget, 201
communications satellite, 79, 111, 229
contract, 2, 9, 11, 45, 59, 61, 123, 173, 194, 201, 231, 232–233, 264, 272, 298, 334
cooperation, 42, 63, 70, 83, 118, 233, 237–238, 247, 276–277, 299, 317–318, 335
launch, 189, 208
  failure, 193
  missile, 3, 193
  reentry vehicle, 292
  satellite, 9–10, 16, 53, 61, 80, 86, 101, 120, 127, 132, 138, 159, 179–180, 181, 208, 216, 228–229, 244, 261, 271, 272, 300, 309
missile program, 3–4, 18, 61, 148, 159, 193, 195, 209, 262–263, 312–313
MOL, 20 69–70, 103, 122, 126, 166, 171, 208, 255, 293, 308–309, 335
Nimbus B Review Board report, 243
nuclear propulsion, 156
organization, 9, 150
parachute test, 112
personnel, 102, 152, 172, 207, 233
research, 70, 150, 163, 170, 181, 302, 312–313
test, 3–4, 77, 156, 181
training, 158
UFO, 99, 151, 169–170, 172
Vietnam War, 71, 72, 84, 104, 109, 155, 195, 224
U.S. Arms Control and Disarmament Agency, 27, 166, 295, 334
U.S. Army (USA), 293
  Advanced Ballistic Defense Agency, 43
  Cold Regions Research and Engineering Laboratory, 36
  missile, 43, 72
  nuclear clock, 170–171
  Sentinel System Command, 72
U.S. Atlantic Fleet Anti-Submarine Warfare Force, 29
U.S. Bureau of Mines, 95
U.S. Coast and Geodetic Survey (C&GS), 37, 89
U.S. Geological Survey, 118, 181–182, 191, 254–255, 325
U.S. Information Agency (USIA), 239
U.S. Navy (USN), 8, 29, 30, 41–42, 83–84, 146, 170–171
  aircraft, 187, 201, 245–246, 314
  F-111B, 14, 71, 75, 98–99, 123, 130, 167
  contract, 167, 187, 201, 227, 302–303, 310, 314
  cooperation, 50, 101, 300
  Deep Submergence Search Vehicle, 227
  missile, 191
  nuclear submarine, 160, 169, 261, 302–303
  oceanographic ship, 310
  Operation Tektite, 101
  Sealab III experiment, 41–42, 143, 295
  spacecraft recovery, 77–78, 251, 320

USNS *Huntsville*, 258
USNS *Point Barrow*, 30
USNS *Watertown*, 258
U.S. Public Health Service, 82–83
U.S.S. *Discoverer* (Coast and Geodetic Survey ship), 144
U.S.S. *Essex*, 251
U.S.S. *New Jersey*, 146
U.S.S. *Okinawa*, 77–78
U.S.S. *Pueblo*, 109
U.S.S. *Yorktown*, 320
U.S.S.R. (Union of Soviet Socialist Republics) (see also Soviet Academy of Sciences, etc.), 15–16, 42, 89–90, 92, 141, 145, 148, 210, 215–216, 233
  agreement, 127, 149–150, 157, 159, 161, 164, 166
  aircraft, 40, 70, 112, 126, 130, 156, 157, 163, 171–172, 194, 195, 231, 239–240, 259–260, 282, 297, 332–333
  accident, 70
  airlines, 127, 157, 163
  anniversary, 14, 45–46, 273, 336–337
  antimissile defense, 88, 185
  Aviation Day, 195
  booster, 34, 36, 186–187, 209, 289
  budget, 309
  communication satellite, 10, 88, 110, 155, 189–190, 244
  cooperation, 127, 128, 138, 154, 159, 202, 254, 281
  cooperation, space, 2, 10, 31, 52, 66, 99, 127, 202
  Cosmonaut. See Cosmonaut.
  disarmament, 32, 59, 127, 145, 149, 159, 161, 164, 166, 188, 191, 195, 210
  exhibit, 190
  launch, 23–24, 334, 335–336
  probe
    Luna XIV, 81
    Zond IV, 54
    Zond V, 211–212, 230
    Zond VI, 277–278
  satellite
    Cosmos, 9, 11, 32, 42, 57, 61, 63, 67, 76, 82, 84, 87, 88, 92, 93–94, 95, 106, 120, 123, 126, 128–129, 130, 132, 135, 139, 143, 154, 158–159, 163–164, 165, 172, 178, 180, 183, 201, 206, 211, 213, 219, 223, 239, 240, 245, 248, 259, 267, 269, 279, 289, 295, 296, 298, 308, 310, 313, 317, 335–336
    Molniya I-8, 88
    Molniya I-9, 155
    Molniya I-10, 244
    Proton IV, 284–285
    Soyuz II, 263
    Soyuz III, 264–265
  lunar exploration, 46, 54, 112–113, 116–117, 178, 186–187, 222, 223, 226–227, 254, 329–330
  meteorological satellite, 2
  missile and rocket program, 29, 42,

45–46, 55–56, 93–94, 102, 178, 188, 191, 210, 272, 273, 289
nuclear nonproliferation treaty, 11, 32, 59, 92, 102, 123–124, 128, 132, 133, 138, 149, 210, 215–216
science and technology, 46, 63, 93, 97, 103–104, 159, 163, 195–196, 244, 247, 254, 272, 275–276, 285, 292, 295, 298, 309, 316, 317, 324
space, biology, 200–201, 211, 283, 286–287, 295, 324
space rescue treaty, 89–90, 127, 215, 225, 299
space program, 15, 17, 22–24, 34, 36, 37, 40, 42, 46, 83, 85, 87, 93–94, 95, 102, 105, 112–113, 116–117, 119, 173–174, 178, 180, 190, 200–201, 202, 209, 211, 218, 222, 223, 226, 229–230, 248–249, 254, 262–263, 264–265, 266, 267–268, 269, 272, 275–276, 277–278, 283, 286–287, 289, 295, 298, 310, 327, 329–330, 332, 336–337
space station, 54, 119, 211–212, 222, 226, 230, 277–278, 283
spacecraft. See U.S.S.R., satellite; and individual spacecraft, such as *Luna IV*, *Molniya I–10*, *Soyuz III*.
submarine, 160, 224, 239–240
supersonic transport, 332–333
test, 116, 121–122
nuclear, 108
weapons, 29, 42, 45–46, 55–56, 59, 93–94, 102, 128, 178, 180, 185, 200, 205, 208, 210, 239–240, 255, 273, 309
U.S. Tactical Air Power Program (report), 305
UV: ultraviolet.

V

Vaccaro, Michael J., 283
Vaeth, J. Gordon, 233–234
Valparaiso, Chile, 119
Van Allen, Dr. James A., 22, 26–27, 51
Van Allen radiation belt, 26–27, 29–30, 131, 203, 228
Vandenberg AFB, Calif. (see also Western Test Range), 116
  contract, 11
  launch
    Advanced Ballistic Reentry System, 292
    failure, 193
    missile, 3
    satellite launch vehicle
      Atlas-Agena D, 179
      Atlas-Burner II, 193
      Atlas-F, 80, 159
      Long-Tank Thrust-Augmented Thor (LTTAT)-Agena D, 16, 61
      Scout, 53
      Thor-Agena, 181
      Thor-Agena D, 101, 138, 216, 244, 271, 309
      Thor-Burner II, 120, 261
      Thrust-Augmented Thor-Agena D, 9–10
      Titan III-B, 127, 208, 272, 300
      Titan III-B-Agena D, 10, 61, 86, 127, 179–180
    MOL launch complex, 335
Vanderbilt Univ., 326
Vanguard (program), 152
Vanguard I (satellite), 63–64
Vavilov, Nikolay I., 298
Vavilov, Sergei I., 298
Vega (star), 310
Vega Baja, Puerto Rico, 64–65, 329
Vela (constellation), 67
VELA (nuclear test detection satellite), 19, 25
Veli, India, 31
Venezuela, 160
Venus (planet), 49, 184
  atmosphere, 17, 199, 302, 329
  exploration of, 24, 30, 66, 188–189, 215, 248–249, 287–288, 327, 334
  gravity, 120–121
  landing, soft, 85
  life on, 17, 302
  magnetic field, 328–329
  map, 37, 171
  probe, 163, 188–189, 212
  rotation, 37
  surface, 17, 37, 171, 188–189, 199, 275–276
  temperature, 17, 163
Venus IV (U.S.S.R. probe), 17, 36, 85, 199, 275–276
Vereinigte Flugtechnische Werke, 151
Veronique (sounding rocket), 172
Verschuur, Dr. Gerritt L., 208–209
Vestine, Dr. Ernest Harry, 166–167
Veterans of Foreign Wars Space Award, 199
VFX (USN supersonic fighter), 187, 201
VFX–1. See F–14A.
VFX–2, 171–172
Victorialand, 89
Vienna, Austria, 190–191, 204, 211, 228
Vietnam, North, 71, 84, 146
Vietnam, South, 88, 132, 137, 146
Vietnam War, 62, 133, 149–150, 245
  aircraft, 69, 71
  budget for, 309
  effects on R&D, 296
  effects on space budget, 14–15, 56, 99, 146, 215, 245, 336
Viking, Project, 302
Vincent, Robert K., 250
Virgin Islands, 101
Virginia, Univ. of, 38, 151
Volcano, 191
Volpe, Gov. John A., 309
von Braun, Dr. Wernher, 28, 112, 303
  electric power system, space use, 275
  space program, 24, 57–58, 66, 94–95, 98
von Eshleman, R., 275–276
von Kármán, Dr. Theodore, 267

Voskhod I (U.S.S.R. spacecraft), 286
Voskhod II, 81
Vostok (U.S.S.R. spacecraft), 17
Vostok I, 70, 309
Vostok VI, 148
Voyager (program), 18–19, 34, 41
v/stol aircraft, 33, 166, 177
 landing tests for, 168
 research, for, 34, 47, 134, 238, 288–289
 wind tunnel for testing, 219, 224
vsx (antisubmarine aircraft), 187
vtol aircraft, 16, 33, 47, 116, 224, 231, 276–277
 foreign, 105, 231
 landing pads for, test, 135
 wind tunnel for testing, 39–40, 219
Vucinich, Alexander, 63
Vungtau, South Vietnam, 88

W

Wakstein, Dr. Charles, 117
Walker, Dr. J. C. G., 328
Wallace, Gov. George C., 271–272
Wallops Station (nasa), 30–31, 192
 award, 283
 hailstone model tests, 247
 launch
  Explorer XXXVII, 56–57
  ram c–ii, 198–199
  Reentry F experiment, 97
  sounding rocket, 1, 64–65, 160, 216, 288, 308, 328–329
   Aerobee 150 A, 141, 289–290, 291
   Arcas, 86, 214
   Astrobee 1500, 88
   Black Brandt IV, 106
   Javelin, 82
   Nike-Apache, 45, 64–65, 70, 168, 181, 182–183, 273, 287
   Nike-Cajun, 9, 29, 168, 213–214, 287
   Nike-Tomahawk, 13, 31, 60, 62–63, 64–65, 183, 198, 256
   Pacemaker, 144
 runway test, 118, 125, 286
Warnke, Paul C., 138
Warsaw, Poland, 161
Warsham, James E., 15
Washington Airlines, 224
Washington, D.C., 26, 82–83, 103, 122, 137, 145, 200, 217, 279, 306, 327
 awards presented at, 49, 235–236, 300
 meetings, 22–23, 38–39, 46, 55–56, 79, 86, 91, 94–95, 108–109, 111–112, 112–113, 187–188, 216–217, 239, 274–275, 280–281, 284, 293–294, 301–302, 315–316
 nuclear nonproliferation treaty signed at, 149, 295
 space rescue treaty signed at, 89–90
Washington National Airport, 179, 189, 203, 224
 air traffic, 194, 199–200, 206, 300
 lights, 122
 modernization, 231
 runway test, 324
Watson, Postmaster General W. Marvin, 113
Weakley, v/a Charles E. (usn, Ret), 29
Weapon systems, 55, 59, 191, 195, 239, 255
Weather modification, 101–102, 195
Webb, James E., 26, 101, 118–119, 215, 244–245,
 appointments by, 5–6, 17, 91–92, 160, 206, 209, 233
 award to, 306
 budget, 19–20, 33–34, 48, 93, 140, 161–162
 lectures by
  Diebold, John, 232
  McKinsey Foundation, 103–104, 109, 115
 resignation, 212–213, 218, 221–222, 229, 244–245, 336
 space cooperation, 201
 space program, 97, 209, 212–213
 tribute to, 66, 215, 218, 221–222, 229, 246–247, 306
 U.S.S.R. space program, 222
 visit to msc, 53
wefax (Weather Facsimile Experiment), 155
Weidner, Hermann K., 303
Weightlessness, effects of, 200–201
 animals, 46, 86, 151, 218, 286–287
 bacteria and viruses, 46, 325–326
 chromosomes, 46, 325–326
 human beings, 81, 190, 218, 268, 286–287
 plants, 46
Weinberg, Dr. Alvin M., 28, 331
Wells, Edward C., 5
Welsh, Dr. Edward C., 32, 105–106, 112–113, 141, 202, 254, 274, 307–308
Wenk, Edward, Jr., 284
West Virginia, 219, 230
Western Co., 95
Western Electric Co., 72
Western New York Nuclear Research Center, 31
Western Telegraph (U.K.), 293
Western Test Range (wtr) (see also Vandenberg afb, Calif.)
 launch, 1, 179
 contract, 59, 289
 failure, 116, 224, 242
 satellite
  launch vehicle
   Scout, 113, 182, 240
   Thorad-Agena D, 116, 244
   Thrust-Augmented Delta, 7, 42–43, 153, 210–211
   Thrust-Augmented Long-Tank Thor-Delta, 191–192, 311
 X–15 experiment, 164, 198, 211, 262
Western Union International, Inc., 154, 307–308

Westinghouse Defense and Space Center, 145
Westinghouse Electric Corp., 26, 272
Weston, Ill., 93, 154
Wetzel, Col. Albert J. (USAF, Ret.), 209
Wheeler, Gen. Earle G. (USA), 159
Wheeler, Dr. John Archibald, 221, 298
White, L/C Edward H., II (USAF), 306
White House, 7, 51, 57, 150, 212, 235–236, 298, 299, 306, 326, 332
White, Col. Maynard E. (USAF, Ret.), 9
White Sands Missile Range (WSMR), N. Mex. 296
  launch
    Aerobee 150
      radiation, 121
      solar astronomy, 40–41, 62, 128, 224
      stellar data, 30, 67, 104, 115
      ultraviolet astronomy, 115
      x–ray astronomy, 30, 40–41, 62, 87, 104
    Aerobee 150 MI
      atmospheric data, 238
      infrared data, 49–50, 316
      solar astronomy, 97, 118, 232, 290
      stellar data, 90, 282
      x-ray astronomy, 169, 170, 303
    balloon, 209–210
    Nike-Apache, 126–127, 300
      micrometeoroid sampling, 129, 185–186, 197, 310
    Nike-Cajun, 32–33
  test
    instrumentation, 17, 128, 238
    parachute, 32–33, 126–127, 227, 261, 309
    rail launch, 211, 214
    Solar Pointing Aerobee Rocket Control System, 65
    varibooster, 211, 214
White, Gen. Thomas D., Award, 104–105
White, Gen. Thomas D., Space Trophy, 122
White, William S., 149–150
Whittaker, Philip N., 172
Whittle, Sir Frank, 260–261
WHO. See World Health Organization.
Wible, M. Keith, 92
Wiesner, Dr. Jerome B., 139–140, 159, 303–304
Wilford, John N., 85, 186–187, 265
Williams AFB, Ariz., 92
Williams, Maj. Clifton C., Jr. (USMC), 128
Williams, Don, 77
Wilson, Andrew, 189
Wilson, George C., 7, 180, 261, 284, 316
Wilson, Prime Minister Harold (U.K.), 9
Wilson, T. A., 217
Wilson, Dr. William J., 200
Wind tunnel, 39–40, 42, 219, 281
Wing, aircraft
  delta, 332–333
  fixed, 19, 158, 206, 217, 220–221, 270, 285, 335
  fixed delta, 259, 270, 285
  swept, 14, 167, 259–260
  swing, 75, 207, 220–221, 259–260, 263, 270
  test, 75, 155
  tilt, 276
  variable, 206, 220–221
Wings Club, 66–67, 315
Wisconsin Regional Space Center, 68
Wisconsin, Univ. of, 38, 93, 115, 219, 229, 304
Withbroe, George L., 243–244
Withington, H. W., 259–260
Witkin, Richard, 259
Woeller, Fritz H., 16–17
Wolfe, Dr. John H., 138–139
Women's National Democratic Club, 94–95
Wonsan, North Korea, 109
Woods, George D., 16
Woods Hole, Mass., 275
Woomera Rocket Range, Australia, 296
World Affairs Council, 210
World Bank, 51, 53
World Health Organization (WHO), 291
World Meteorological Organization, 306–307
World Military Expenditures, 1966–67 (report), 334
World War II, 146
World Weather Watch, 19, 127, 155, 306–307
Wrench, Edwin H., 131
Wright Brothers Day, 313–314
Wrights Brothers Lecture, 38–39
Wright Brothers Memorial Dinner, 209, 286, 314
Wright Brothers Memorial Trophy, 209, 314
Wright, Orville, 313–314
Wright, Wilbur, 313–314
Wright-Patterson AFB, Ohio, 112, 181
WSMR. See White Sands Missile Range.
WTR. See Western Test Range.
Wykes, Raymond P., 151–152
Wyld, James H., Propulsion Award, 133

X

X–3 (supersonic aircraft), 293
X–15 (rocket research aircraft), 299, 300, 317–318, 335
  accident, 12, 164–165
  flight
    No. 1, 53, 78, 95, 132, 164, 198, 211, 212, 262
    No. 2, 259
  funding, 19
  Hypersonic Research Engine, 47
  pilots, 11–12, 12, 299, 300
  record, 11–12, 317–318
    altitude, 317–318
    speed, 317–318

test
  alpha cone, 198, 211
  electrical system, 53
  fluidic probe, 198, 211
  horizon measurement, 132
  horizon scanner, 198, 211
  insulation, 78, 95
  Saturn components, 95
X-24 (lifting-body vehicle), 42
XB-5A (v/STOL aircraft), 276-277
XB-70 (supersonic aircraft), 19, 158, 313
  contract, 149
  flight, 8, 38, 50, 67, 131, 269, 335
  instrumentation, pilot, 186
XB-70A, flight, 145, 167, 208, 258, 298-299
XC-142 (VTOL aircraft), 276-277
XE (nuclear rocket engine), 25-26
XLR-11 (rocket engine), 279
X-ray, 62, 104, 256-257, 280-281
  source, 30, 273, 303
XV-5A (v/STOL *aircraft*), 177
XV-5A (v/STOL aircraft), 166, 177, 276-277
XV-4B (Hummingbird II) (VTOL research aircraft), 230-231

Y

Yale Univ., 38, 328-329
Yardney Electric Corp., 121
Yavne., Aleksandr, 119
Yegorov, Dr. Boris B., 286
Yellowstone National Park, 154-155
Yelyan, Eudard V., 332-333
Yeshiva Univ., Graduate School of Science, 119
YF-12A (jet interceptor), 207
York, Dr. Herbert F., Jr., 40
Yosemite National Park, 154-155
Young, Cdr. John W. (USN), 173, 279-280
Young, Pearl I., 134-135

Z

Zeta (star), 67
Zeuschner, Robert B., 44
ZF-15A (fighter aircraft), 130, 232-233, 272, 334
Zohar, Dr. Shalhav, 171
Zond I (U.S.S.R. space probe), 212
Zond II, 54, 212
Zond III, 54, 212
Zond IV, 54, 151, 212
Zond V,
  launch, 211-212
  press comment, 222-224, 226, 230
  reentry, 230, 277
  results, 248-249, 283
Zond VI, 277-278, 287-288, 298, 332
Zwick, Charles J., 255, 286
Zwicky, Prof. Fritz, 287
Zworykin, Dr. Vladimir K., 43

NASA HISTORICAL PUBLICATIONS

HISTORIES
- Robert L. Rosholt, *An Administrative History of NASA, 1958–1963*, NASA SP–4101, 1966, $4.00.*
- Loyd S. Swenson, James M. Grimwood, and Charles C. Alexander, *This New Ocean: A History of Project Mercury*, NASA SP–4201, 1966, $5.50.
- Constance McL. Green and Milton Lomask, *Vanguard: A History*, NASA SP–4202 (1970).
- Alfred Rosenthal, *Venture Into Space: Early Years of Goddard Space Flight Center*, NASA SP–4301, 1968, $2.50.
- Edwin P. Hartman, *Adventures in Research: A History of the Ames Research Center, 1940–1965*, NASA SP–4302 (1970).

HISTORICAL STUDIES
- Eugene M. Emme (ed.), *History of Rocket Technology* (Detroit: Wayne State University, 1964).
- Mae Mills Link, *Space Medicine in Project Mercury*, NASA SP–4003, 1965, $1.00.
- *Historical Sketch of NASA*, NASA EP–29, 1965 and 1966.
- Katherine M. Dickson (Library of Congress), *History of Aeronautics and Astronautics: A Preliminary Bibliography*, NASA HHR–29, for sale by Clearinghouse for Federal Scientific and Technical Information, Springfield, Va. 22150, $3.00.

CHRONOLOGIES
- *Aeronautics and Astronautics: An American Chronology of Science and Technology in the Exploration of Space, 1915–1960*, compiled by E. M. Emme, Washington: NASA, 1961.
- *Aeronautical and Astronautical Events of 1961*, published by the House Committee on Science and Astronautics, 1962.
- *Aeronautical and Astronautical Events of 1962*, published by the House Committee on Science and Astronautics, 1963.
- *Astronautics and Aeronautics, 1963*, NASA SP–4004, 1964, $1.75.
- *Astronautics and Aeronautics, 1964*, NASA SP–4005, 1965, $1.75.
- *Astronautics and Aeronautics, 1965*, NASA SP–4006, 1966.
- *Astronautics and Aeronautics, 1966*, NASA SP–4007, 1967, $1.50.
- *Astronautics and Aeronautics, 1967*, NASA SP–4008, 1969, $2.25.
- *Project Mercury: A Chronology*, by James M. Grimwood, NASA SP–4001, 1963.
- *Project Gemini Technology and Operations: A Chronology*, by James M. Grimwood and Barton C. Hacker, with Peter J. Vorzimmer, NASA SP–4002, 1969, $2.75.
- *The Apollo Spacecraft: A Chronology*, Vol. I, *Through November 7, 1962*, by Ivan D. Ertel and Mary Lou Morse, NASA SP–4009, 1969, $2.50.

---

*All titles with prices can be ordered from the Superintendent of Documents, Government Printing Office, Washington, D.C. 20402.

www.ingramcontent.com/pod-product-compliance
Lightning Source LLC
Chambersburg PA
CBHW081715170526
45167CB00009B/3587